Process Modelling

Springer
Berlin
Heidelberg
New York
Barcelona
Hong Kong
London
Milan
Paris
Singapore
Tokyo

B. Scholz-Reiter · H.-D. Stahlmann
A. Nethe (Eds.)

Process Modelling

Editorial staff:
Andreas Noack and Thomas Bachmann

With 245 Figures
and 36 Tables

Springer

Prof. Dr. B. Scholz-Reiter
Prof. Dr. Hans-Dietrich Stahlmann
Dr. Arnim Nethe

Brandenburg Technical University of Cottbus
Chair of Industrial Information Systems
Universitätsplatz 3–4
D-03044 Cottbus
Germany

ISBN 3-540-65610-3 Springer-Verlag Berlin Heidelberg New York

Library of Congress Cataloging-in-Publication Data
Die Deutsche Bibliothek – CIP-Einheitsaufnahme
Process Modelling: with 36 tables / B. Scholz-Reiter ... (ed.). – Berlin; Heidelberg; New York; Barcelona; Hong Kong; London; Milan; Paris; Singapore; Tokyo: Springer, 1999
ISBN 3-540-65610-3

Printed in Germany

Hardcover-Design: Erich Kirchner, Heidelberg

SPIN 10715429 42/2202-5 4 3 2 1 0 – Printed on acid-free paper

Preface

Nowadays, the view on systems is more or less process orientated regardless of the kind of system that is under consideration.

This is also true for social and business systems, for logistics and production systems, for engineering systems, for medical, biological, ecological and other natural science systems, just to mention only some of the potential application areas.

To handle the development, improvement and management of complex systems it is often necessary to build a model of the system to reduce the complexity to the relevant parameters.

Moreover, the key for reducing and solving problems in the system and also for significant potentials for improvement lies in optimising the processes of the system.

For these reasons a process model is very often the main and kernel model for system analysis, design and management in various application areas.

Whilst general systems modelling requires acquisition and assessment of the knowledge, as complete as possible, about a process, with process modelling it is the limit of the process parameters which are of most importance. The process model has only to be as precise as necessary within the parameters of the individual field of application, whereas the precision externally is less important. This has considerable consequences for process modelling, because this makes it easier and open for structuring. Simulation can then be applied to consider the dynamic behaviour of the process.

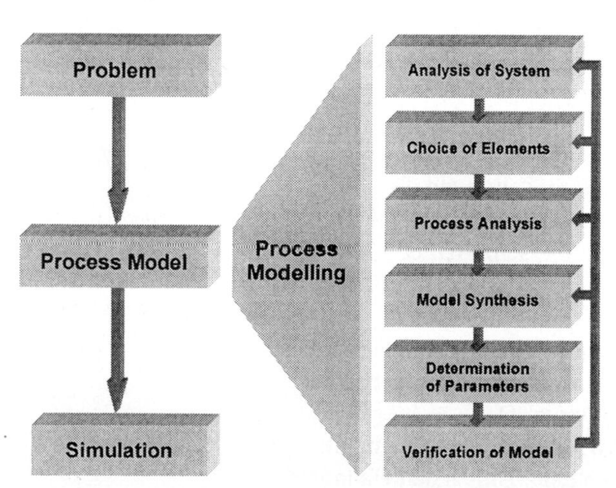

Despite the fact that process models are very often used in practise in various disciplines, it can be observed that the development and the construction of the process model itself is an individually different and mostly unstructured process and varies extremely between the different disciplines. Generally spoken, there is no common methodology existing and applied for process modelling.

This volume should be considered as a first step towards the establishment of an interdisciplinary common sense and methodology for process modelling. It intends to contribute to giving decision support regarding the construction of process models and to initiate an interdisciplinary structuring process.

For this volume 36 contributions have been selected, which cover the process modelling theme from different perspectives and in various application areas. The editors hope that on the basis of this book and on a related conference the discussion of process modelling in a common sense can be opened in the future.

The editors would like to thank, beside the contributors, all the members of the editorial committee and the numerous reviewers for their support in evaluating the high number of submissions for this volume.

Cottbus – Germany

February 1999

Bernd Scholz-Reiter, Arnim Nethe and Hans-Dietrich Stahlmann (Editors)

Editorial Committee

Process Modelling

Part 1 Common Approaches

Part 2 Business Processes Modelling

Part 3 Logistics Processes

Part 4 Engineering Processes

Part 5 Production Processes

Part 6 Modelling and Simulation of Water Systems

Part 1 Common Approaches

Survey of a General Theory of Process Modelling

Arnim Nethe and Hanns-Dietrich Stahlmann

Chair of Electromagnetic Theory, Brandenburg Technical University of Cottbus, Universitätsplatz 3/4, D-03044 Cottbus

Abstract. *Simulation is regarded as a strategic tool in the field of process development and process control. It is used to present certain aspects of an existing system, or a system to be developed, as models, or to reproduce them. It allows to examine systems which are too dangerous, too expensive or impossible to be tested. Precondition for simulation is the existence of a general or reduced process model which describes the important characteristics of the process with mathematical exactness. While in general process modelling it is in the foreground to collect and evaluate the knowledge on a certain process as completely as possible, in the reduced process model the boundaries of the process parameters are important. Such a process model has to be as exact as necessary within these parameters, while beyond the validity of these parameters exactness is of secondary interest. So as not to confine this introduction to process modelling to the technological, production-oriented area only, the introduced standardised term process model is now generalised as a brief outlook. A limitation in the area of nature is arbitrary and disadvantageous, hence one has to make an attempt to put process models in a systematic order in other areas, too.*

Keywords. *general process model, reduced process model, theory of process modelling*

1 Introduction

Simulation is regarded as a strategic tool in the field of process development and process control. It is used to present certain aspects of an existing system, or a system to be developed, as models, or to reproduce them. It allows to examine systems which are too dangerous, too expensive or impossible to be tested. Precondition for simulation is the existence of a general or reduced process model which describes the important characteristics of the process with mathematical exactness. Often such process models do not exist or are not yet in a stage of development which would allow a person who is not a specialist to apply it without any problems.

The following list shows that process modelling concerns all fields:

- Production: Processes in production, material flow, business, production development, logistics and production procedures,
- Natural sciences: information technology, mathematics, physics, chemistry, meteorology, earth science,
- Engineering science: electrical engineering, mechanical engineering, ecology, civil engineering,
- Medical science and psychology,
- Economics and social sciences.

Since the development of both general and reduced process models requires profound specialist knowledge on the one hand as well as creativity and intuition on the other hand and hence can be regarded as an art, the articles of this book are meant to contribute to defining and systematising the procedure of process modelling as such.

2 General and Reduced Process Models

First the difference between a general process model and a reduced process model has to be brought out in order to analyse the procedure of process modelling. The following definitions serve this purpose.

- The *general process model* of a system is an object which allows the observer to answer interesting questions on the system by means of this object. It is a reproduction of the succession of states and state transitions of a system. A general process model is a means to describe the experienced reality as completely as possible.
- A *reduced process model* is a model with a low degree of detailedness which reflects only the characteristics, which are important for a definite application within the confines (boundaries) of the process parameters, with sufficient exactness.

For applying general and reduced process models priority has to be given to gaining a profound understanding of the static and dynamic behaviour of a system on the one hand and the prediction of its behaviour under defined conditions. This allows a forecast of the system's behaviour and control possibilities under normal conditions. Moreover, it can be predicted how the system will behave under extreme or defective conditions (cf. (Profos, 1997)).

In this context a reduced process model in this sense can be regarded as a model which describes a process in its relevant section as simply as possible and as exactly as necessary in order to serve as a basis for planning work. This is called a general fundamental or theoretical process model which can be deducted from physical or natural scientific laws (deduction) (Schuler, 1995). Precondition for developing a general theoretical process model is a sufficiently exact qualitative knowledge of the process to be described as well as knowledge of the corresponding physical laws.

> The developed process model is faster in evaluating and easier to handle than the general process model, hence it can be applied in production preparation almost without any problems. Its results within the framework of process parameters are as reliable as the results of the general process model which is more difficult to develop.

3 Introduction to Process Modelling

Before the transition from the general process model to the reduced process model one has to find out what a general or reduced process model is. Even though in the interesting field of technological and non-technological processes models are described by abstract mathematical connections, we understand by a model in general (Minsky, 1965):

> "A model (M) for a system (S) and an experiment [E] is anything to which E can be applied in order to answer questions about S".

In derivation of this a reduced process model is a reproduction of nature with an emphasis on the characteristics which are considered relevant and with neglecting the aspects which are considered peripheral. It is a means to describe the experienced reality. Process models, however, do not exist as a matter of course but have to be made in a procedure of modelling.

On this basis one has to go a bit more into detail why process modelling is important or what the purpose of process models is in research, development and production, respectively. In the first place is the gain of knowledge (Profos, 1997) as well as a profound understanding of processes and procedures. This knowledge arouses the desire for a prediction of a system's behaviour, i.e. that output parameters can be determined after the input quantities are known. In the relevant process modelling two basically different procedures have to be considered which can be denoted deductive or inductive modelling, respectively.

- *Deductive modelling* is characterised by a deduction of a mathematical formulation of the process with the help of fundamental laws of natural sciences in order to find a quantitative description of a qualitative image.
- In *inductive modelling* the available input and output parameters of the process are observed with measurement first. By means of a statistical analysis these observations are to be correlated to a mathematical model in order to establish a connection between the relevant input and output parameters and to receive an empirical model.

That means that an inductive process model is made by identifying process parameters and quantities (process identification). The resulting process model can be used to predict the input and output behaviour, however, it does not allow to make statements on the function (the functional connection) of the process. It is also important in this context that the process models made in this way do not allow to draw physical conclusions.

3.1 Application of Process Models in Simulation

Purpose of simulations is the prediction of planned (production preparation) or running processes. Simulation can be presented as a mathematical experiment with the advantage that in general it is easier and more economical to make simulations instead of practical experiments.

In simulation often very exact mathematical descriptions of a process are used which lead to good simulation results also beyond the field of process parameters. This increased computation effort requires more computation time or more efficient hardware. Both force simulation costs up.

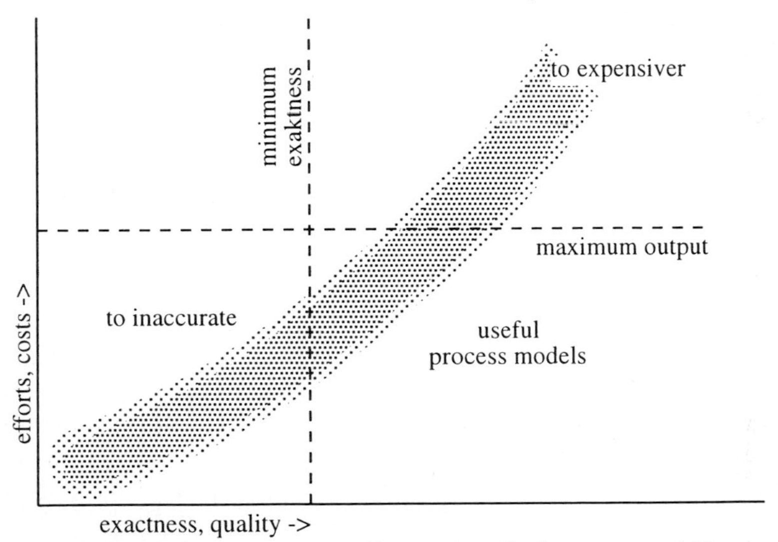

Figure 1: Correlation between expenditure and quality in process modelling

Figure 1 describes the problem how with increasing exactness costs are rising, too, so that only in a small section the correlation between expenditure and exactness is given for useful process models. If in this place a conception is applied in which the required exactness is given only in the field of the process parameters costs are declining and one arrives in the desired field of useful process models, its consequence are reduced process models.

3.2 Transition from the General to the Reduced Process Model

While in general process modelling it is in the foreground to collect and evaluate the knowledge on a certain process as completely as possible, in the reduced process model the boundaries of the process parameters are important. Such a process model has to be as exact as necessary within these parameters, while beyond the validity of these parameters exactness is of secondary interest. That means that the degree of detailedness should be kept as small as possible and that the process model deliberately reflects only the characteristics important for a definite application with sufficient exactness, so that the modelling purpose is just served. This has enormous consequences since it makes process modelling easier and suitable for structuring.

The structuring concerns the following aspects which are important for the splitting of the system and from which the selection of the relevant elements follows[1]:

- laws
- co-ordinates
- geometrical parameters
- space dependence
- boundary conditions
- assigned process parameters
- distributed parameters
- concentrated parameters
- structured decoupling.

Now the mentioned aspects will be described more in detail in order to allow to evaluate them in the course of process modelling.

- The *laws* imply natural laws in physics, chemistry, biology (deductive modelling), but also experimentally found connections of the process parameters (inductive modelling).
- *Co-ordinates* imply to select a co-ordinate system which is suited for the mathematical description, in most cases it is one of the three elementary (Cartesian, circular cylindrical, spherical) co-ordinate systems. It is, however, possible to widen this notion in a sense that it does not only comprise the position co-ordinates but also time and other physical quantities.
- *Geometrical parameters* are dimensions, areas and volumes in the ambient

[1]The order of sequence is not a criterion for importance. The mentioned aspects refer to the considerations which are made later on.

physical space (co-ordinate system). So the notion of geometrical parameters as well as co-ordinates can be widened.

- *Space dependence* is the dependence of the parameters on the space point, i.e. the considered problem in the area of process parameters can be one-, two- or multi-dimensional (see distributed parameters).
- To *boundary conditions* all conditions and values can be subsumed which are valid on the area boundaries and outside or which have to be met. Moreover, the initial conditions are important boundary conditions for time sequences.
- *Assigned process* parameters are input and output parameters such as manufacturing parameters.
- *Distributed parameters* are spatial parameters (one- or multidimensional, respectively).
- *Concentrated parameters* or point parameters are yielded by integration of the distributed parameters. In the simplest case they yield electronic components.
- The desired *structural decoupling* to reduce the complexity results from the assumption that the parameters in the area boundaries are allocated and present known input and output parameters. This process is often described with specific examples only and is generally denoted as simplification or reduction. (VDI/VDE, 1992)

The above items are the basis for postulating a system of equations in order to get a physically founded and structured general process model. All mentioned items within the model structure require intuition and experience in implementing the reduced process model. They are of decisive importance for the complexity and quality of such process models, however, nevertheless they are only based on a skilful application of basic physical or scientific laws. A potential applicant should not be deterred by that since in most cases there are measurement results for certain parameters and arrangements as checking device. Of course a reduced process model can also be checked by means of an existing general (more expensive) process model. In the final analysis the exactness of a process model is measured in the practical test. Hence the ultimate requirement of a process model is that the results are absolutely *foolproof*. Scientific perfection is left to research.

Finally, we explain the difference between a reduced process model and an approximation in the conventional sense. A mathematical approximation is based on more or less complicated functional analytical or numerical connections and represents them by means of simple, again functional connections, e.g. splines, i.e. purely mathematical knowledge and pure systematology are the background of such an approach. A reduced process model, however, considers the function of the parameters and is thus based on the physical background.

4 Development of a Process Model

Process modelling consists of the components presented in figure 2 which should be all of equal importance.

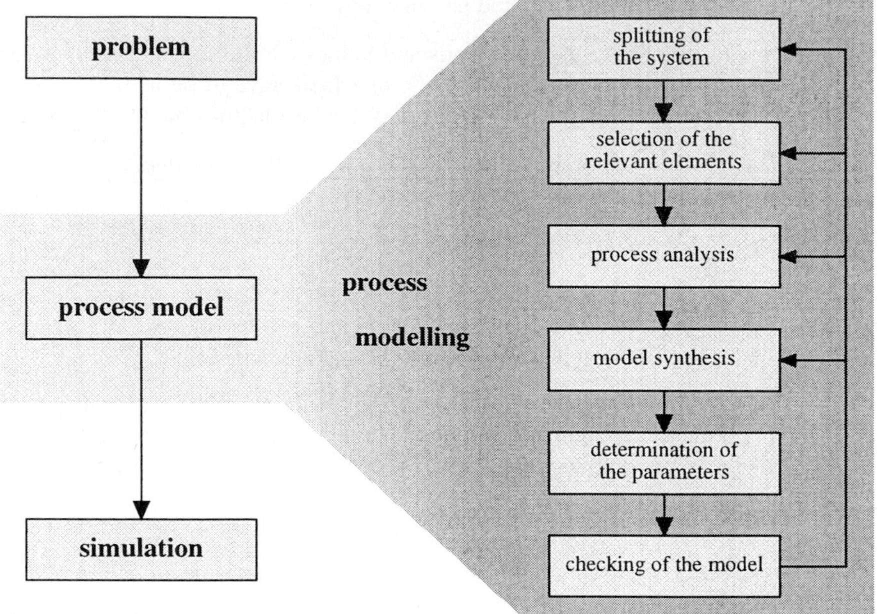

Figure 2: Order of sequence in process modelling as a necessary intermediate step for simulation

The first step, the splitting of the system in elements, is coupled with the precise formulation of the problem. At this stage it is possible to consider the selected problem as a whole or to split it into partial problems, for each of which a process model is to be made and which have to be put together later on. This, however, involves the danger of losing touch with one's goal because of diversification. If one decides in favour of the first possibility, the splitting has to be done very carefully. This should not just be a pure collection of information since an unstructured accumulation of information increases confusion and does not help to make a decision. In the second step one has to select the parts which are relevant for the mathematical description. Helpful to this end are the structuring points mentioned in the previous section. Possible occurring errors are to be corrected in a further iteration step.

In the now following process analysis the individual elements of the system are considered. Their impact on process parameters or vice versa the impact of process parameters on the individual elements has to be analysed. Goal of this section of process modelling is the evaluation and weighting of the influencing factors on the whole process model. In the synthesis the influencing factors are summarised with the areas of the parameters taken into account. After developing the process model to this point it is necessary to determine the parameters (parameter identification) in order to check it thereafter. This validation is the final act in process modelling. If the required criteria are not met it leads back to the previous stages. If they are met, then the process model is now complete and there are no more obstacles to its application in simulation.

The bio-cybernetician Ludwig van Beralanffy expresses this more generalised as follows (v. Bertalanffy, 1968):

> "Oversimplifications, progressively corrected in subsequent development are the most potent or indeed the only means toward conceptual mastery of nature:"

After process modelling was treated only in a general way up to now, the described systematology is now made clear with examples. Figure 3 shows a diagram with the help of which such an event can be formalised. Based on the problem a general process model and after subsequent reduction a process model are developed.

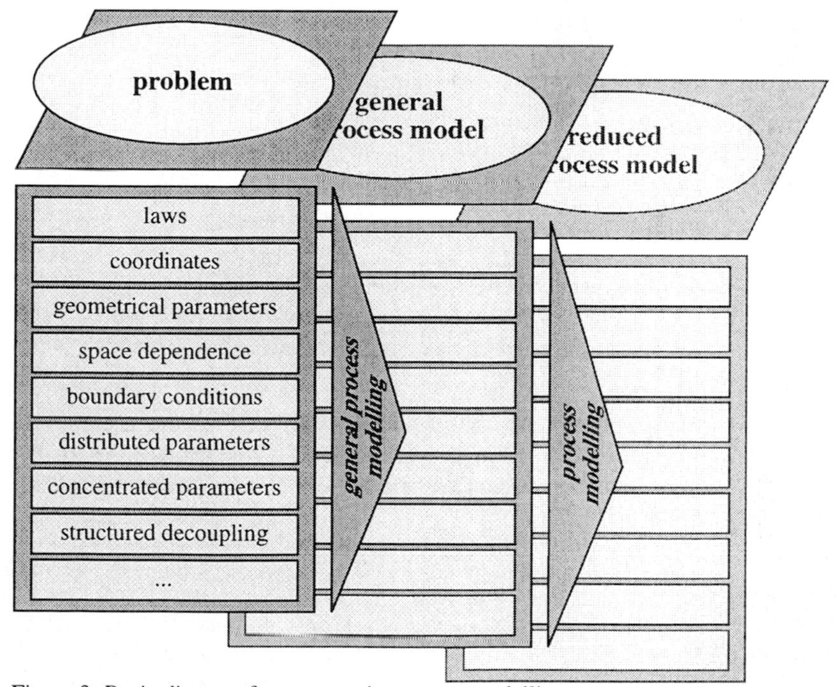

Figure 3: Basic diagram for systematic process modelling.

As an example we now consider an arrangement for induction surface hardening. In this case the reduced process model is confined to the field theoretical induction part of the hardening device for which there is still no satisfactory solution in the production process.

5 Case Example of Systematic Process Modelling

5.1 Structure of the Process Model and its Essential Elements

In order to get from the arrangement of the hardening device (figure 4) to the field theoretical process model the earlier mentioned points for the selection of the relevant elements have to be implemented accordingly.

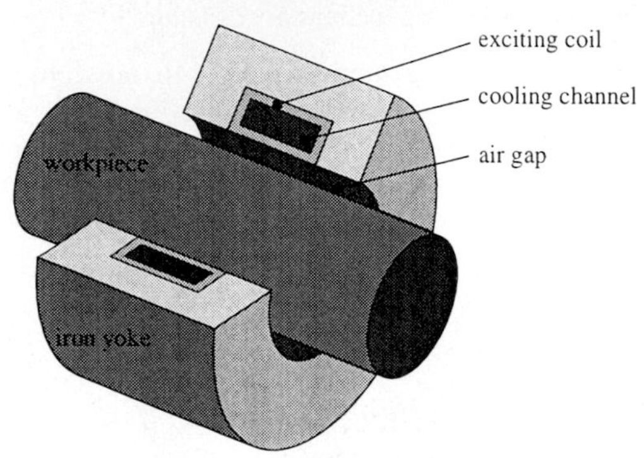

Figure 4: Technological arrangement for induction hardening of surfaces

For the electrodynamic part, which is considered here, Maxwell's equations are the laws which are transformed by means of field theoretical approaches into Laplace or skin equations, respectively. The resulting differential equations which are to be solved are Helmholtz equations. The co-ordinate system taken as a basis for the calculations is easily obtained because of the rotational symmetry; it is the circular cylindrical co-ordinate system.

The six existing spaces (see figure 5) are to be interpreted as the geometrical parameters of the problem which emerge from surfaces of constant co-ordinates from the given arrangement in the co-ordinate system. Each space has its own field equation to be satisfied.

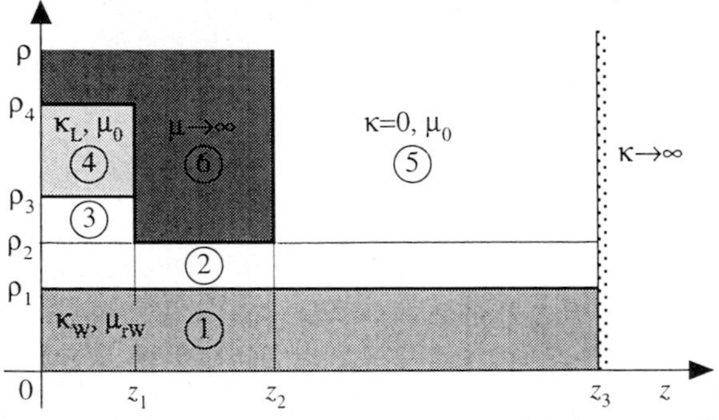

Figure 5: Field theoretical arrangement with the space distribution in circular cylindrical co-ordinates.

As boundary conditions of course the generally valid continuity conditions of the magnetic field strength or induction, respectively, are assumed. Only the highly permeable pole face is an exception from the rule, because here only a surface boundary condition can be assumed. However, we must also not neglect Ampère's law which combines the field parameters with the impressed exciting current.

The indication of distributed parameters requires a thorough knowledge of the physical events. To this end induction or current density in the conducting spaces, respectively, is needed for all six spaces. Since the process model is developed for induction hardening, the Poynting vector or the dissipated power density can be indicated additionally.

In contrast to that there are no concentrated parameters in such a form. If considering this arrangement together with the current / voltage supply, impedances would have to be indicated instead. With such a process model the electric behaviour could be considered from the point of view of control and regulation.

5.2 From Process Analysis to Model Synthesis

Goal of the process analysis is the evaluation of the individual components which the analytical field problem is composed of. One has to find out if one or several components can be simplified or neglected. As was mentioned in the beginning process modelling requires intuition and experience; for that matter it is not sensible in developing the process model to proceed schematically in the order of sequence of the above mentioned points, but it is absolutely legitimate to deal with the *simple* items first, i.e. with those which can be handled quickly, and then turn to the more difficult points.

First the question has to be answered whether it is possible to change from the rotational symmetric problem to a plane problem. This would simplify calculations considerably, since in the area of non-orthogonal functions this is comparable to a change from cylindrical functions to exponential or hyperbolic functions. In order to find a statement in this context one has to consider the arguments and the asymptotic developments of the Bessel functions. The arguments consist of the product of the eigenvalues and the co-ordinates. That means that they are large due to the process parameters, however, they do not change dramatically with the co-ordinate. This first statement can be checked easily by simple substitutions, while the second statement is not immediately comprehensible. Here it is taken into account that with the applied high frequencies a current distribution emerges in approximation. Consequently, the eddy currents flow directly under the surface and hence the dissipated power emerges there, too. Furthermore, the airgap is selected to be as small as possible in order to transmit much power into the workpiece. Logically, without a relevant impact on the result, i.e. with accepting a minor error, one can change from the rotational symmetric to the plane arrangement.

This is shown clearly in figure 5 for the rotational symmetric arrangement and in figure 6 for the plane field theoretical arrangement.

In the following we consider the groove depth, this is the area between the edge of the pole face and the exciting conductor. This space originates from the manufacturing of the pole face since the pipe of the exciting conductor is pressed into the pole face. The relevant magnitude is a few tenth millimetres; now the question occurs if this space is really relevant for the process model. Considering the field patterns in this area one will find out that the fields change considerably for large groove depths. This phenomenon can be explained with the highly permeable pole face which focuses the magnetic field and makes it come out mainly at the ends of the pole faces. In this context one can state further that on this basis the induction in the ambient space is nearly zero. From these observations one can deduce that the two mentioned spaces are not relevant for a process model in the given arrangement.

What is left up to now are the two conductors, the air gap as well as the pole face. A simplification of the air gap is out of the question since its width has a considerable influence on the field distributions and the transmitted power. The pole face, too, is of decisive importance for the distribution of the magnetic field. It prevents the emergence of stray flux, at its ends the field comes out vertically.

Considering the two conductors with the currents and eddy currents flowing in them one can see that they flow only in a very thin area below the surface. Consequently, the interior of the conductor is nearly field-free. Now one could apply a method which does not take the whole conductor into account but only its effect on the ambient spaces. This implies to find a boundary condition which takes into account the exponential decline of the current density toward the centre of the conductor on the one hand, but also the discontinuity in permeability existing in the workpiece on the other hand. With this boundary condition the field parameters of the ambient space only depend on one universal parameter. It contains all material parameters as well as the frequency of the exciting current. (Nethe, 1996; Nethe, 1997a; Nethe, 1997b).

Summarising the above conclusions one arrives at the process model presented in figure 5.

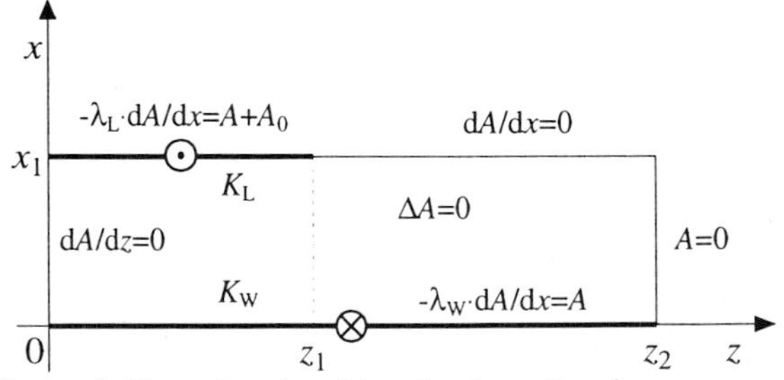

Figure 6: The reduced problem for the reduced process model.

It takes the effect of the exciting current and the repercussions of the eddy currents on the field in the air gap into account. Then the above mentioned boundary condition is applied which implies the exponential decline of the current density towards the centre of the conductor. Moreover, the other boundary conditions cause the magnetic field lines to come out of the pole face vertically and the ambient space remains field-free.

In this place we should like to point out again that changes of the material parameters or of the geometry can increase the error of the reduced process model or lead to completely incorrect values ; i.e. this process model is always valid for a certain parameter range of validity only (production parameters).

5.3 Diagram of Process Modelling

Summarising the explanations above and putting them in a systematic order leads to the diagram in figure 7. It is meant to illustrate the course of process modelling and at the same time serve as checking device. The left part of the figure shows the real arrangement for induction hardening of surfaces including the points from which the decisive elements are selected. The mentioned points include only those which are relevant for the field theoretical problem. On this basis the general process model was developed which is shown in the centre of the figure. It contains nearly all components of the arrangement, this allows to make calculations which may lie outside the range of validity of the process parameters. It can also be used to validate the process model. After an appropriate evaluation of the individual points and the structured decoupling one arrives at the process model (right part of the figure). Because of the tabular form of the diagram the reduction required for the process model is quite apparent.

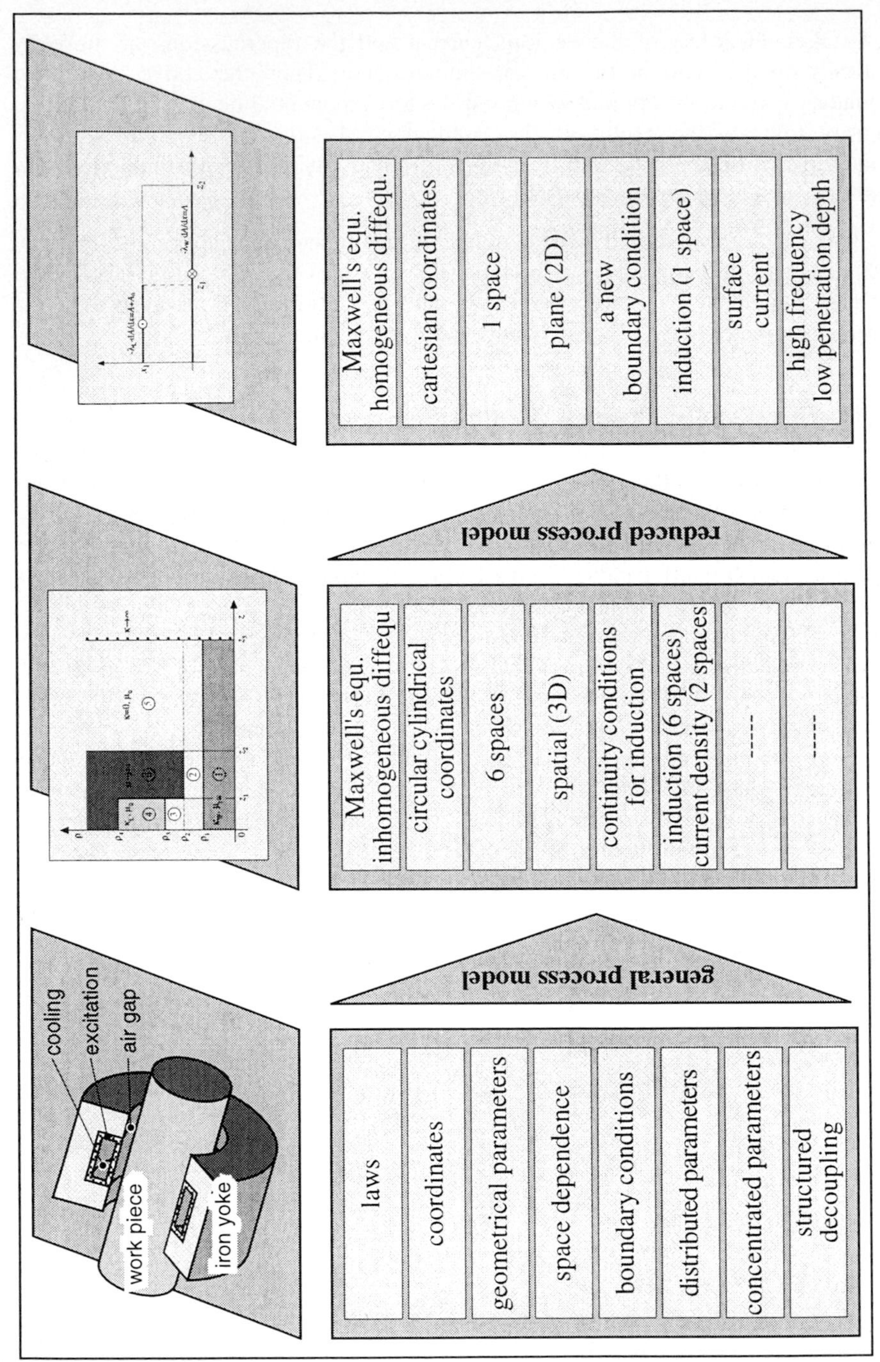

Figure 7: Diagram of process modelling: from the real arrangement via the complicated general process model to the process model

5.4 Outlook

So as not to confine this introduction to process modelling to the technological, production-oriented area only, the introduced standardised term process model is now generalised as a brief outlook. Initially it was mentioned that the considered process model and in particular the exemplary process model we are dealing with are physical process models. This limitation in the area of nature is arbitrary and disadvantageous, hence one has to make an attempt to put process models in a systematic order in other areas, too.

Such a project, of course, has to lead to an explication of the term process model which overcomes such a limitation.

An explication like this is rational if the explained object is indeed a better, efficient intellectual tool for this type of problem solutions, i.e. that with the help of the explained object natural laws can be postulated which would not be so obvious without it. In technology this does not require a long searching until one finds the term process structures which rids the term process model of its solid spatial structure, since process structures, as a principle, can not be lead back to components (e.g. holography). If one tries this, nevertheless, research can be led astray. The layer-theory in psychoanalysis serves as an example for this. In this theory the consciousness is divided into an "it", "super-ego" and "ego"; as a consequence it had been assumed until a few years ago that the terms of colloquial psychology in some way really reflect what is going on in our brain. According to modern science this is not true. Such processes are not a rule-governed handling of symbols, but only a sub-symbolic process of a process structure which is difficult to describe with rules (cf. also M. Spitzer (Spitzer, 1996)). From this point of view it is probably a promising approach to integrate chaos-theoretical affect-logical structures of psyche by Luc Ciompi.

Certainly, the interested specialist will find a lot more examples in his field. It is probably a worthwhile task to expand the above-mentioned rules of process modelling extending across the disciplines.

References

v. Bertalanffy, L. (1968): General System Theory. New York: Braziller, 1968

Minsky, M. (1965): Matter, mind and models. W. A. Kalenich (Hrsg.): Information Processing 1965. Proc. IFIP Congress, New York City, May 1965, 1 (1965) S. 45-49. Washington: Spartan Books, 1965

Nethe, A. (1996): Wirbelstromverluste bei der induktiven Erwärmung. Electrical Engineering, 79 (1996), S. 157-164

Nethe, A.; Stahlmann, H.-D. (1997a): Entwurf von Prozeßmodellen in der Theoretischen Elektrotechnik. Forum der Forschung, 5.1 (1997), S. 47-52

Nethe, A.; Stahlmann, H.-D. (1997b): Einsatz von Prozeßmodellen in der Produktionsvorbereitung. Industrie Management, 13 (1997) 2, S. 56-58

Profos, P. (1997): Modellbildung und ihre Bedeutung in der Regelungstechnik. VDI-Berichte Nr. 276. Düsseldorf: VDI, 1977

Schuler, H. (Hrsg.) (1995): Prozeßsimulation. Weinheim: VCH, 1995.

Spitzer, M. (1996): Geist im Netz – Modelle für Lernen, Denken und Handeln. Heidelberg: Spektrum Akademischer Verlag, 1996

VDI/VDE-Gesellschaft Mess- und Automatisierungstechnik (1992): Modellbildung für die Regelung und Simulation. Düsseldorf: VDI, 1992

Automatic Model Generation in Process Modelling

Johann-Adolf Müller

Hochschule für Technik und Wirtschaft Dresden, Postfach 120701, 01069 Dresden, Germany

Abstract. *Most important for a more sophisticated process modelling is to limit the involvement of users in the overall modelling process to the inclusion of existing a priori knowledge while making this process more automated and more objective. Automatic model generation is based on these demands and is a powerful way to generate models of ill-defined problems. "KnowledgeMiner" is a powerful and easy-to-use modelling tool designed to support the modelling process on a highly automated level and which presently implements three advanced self-organising modelling technologies: GMDH, self-organising fuzzy-modelling and Analog Complexing.*

Keywords. *Self-organising modelling, automatic model generation, GMDH-algorithm, neural networks, fuzzy modelling, Analog Complexing*

1 Introduction

Obviously, mathematical modelling forms the core of almost all decision support systems. However, in economy, ecology, biology, sociology etc. many objects are complex ill-defined systems that can be characterised by inadequate a priori information on the system, great number of unmeasurable variables, noisy and short data samples and fuzzy objects with linguistic variables. Related to modelling this means

- to realise a systematical holistic approach,
- to take in consideration the inadequate a priori information about the real systems and
- to describe the vagueness and fuzziness of linguistic variables (qualitative description).

Problems of modelling complex objects can be solved by deductive logical-mathematical or by inductive sorting-out methods. In the first case, models can be derived from existing theory (theory-driven approach or theoretical systems analysis) considering the fact that any theory itself is already a model of the world, and was seen from a special viewpoint. Deductive methods have advantages in cases of simple modelling problems. Here, the theory of the object being modelled is well known and valid, and it thus is possible to develop a model from physically-based principles employing the users knowledge of the problem.

To deal with ill-defined systems and, in particular, with insufficient a priori information, there is a need to find possibilities, with the help of emergent information engineering, to reduce the long time demanded by model formation before actual task-solving can begin. Computer-aided design of mathematical models may soon prove highly valuable in bridging the gap. The experience in using such theory driven tools has shown: the dialog-assisted model selection tool helps users who seek access to the process of model selection, who have time for dialogue and who want to have, besides good results concerning selected items, some information about the inherent logic of movement or the process investigated.

Using inductive sorting-out methods, models can be derived from data (data-driven approach or experimental systems analysis). Special data mining techniques and tools can assist humans in analysing the mountains of data. Knowledge discovery from the data and can assist humans in turning information located in the data into successful decision making. In many cases it is impossible to create models by theoretical systems analysis. Obviously, experimental systems analysis methods cannot solve an analysis of causes of events for such fuzzy objects but clever applications of these tools may reveal carefully guarded secrets from nature. A pragmatical solution to the modelling problem is a unification of both methodologies.

In general, if there is only a little a priori information for ill-defined, complex systems, the user may take some interest in the model results proper and may have little knowledge of mathematical, cybernetic and statistical techniques, and little time for the computer dialogue, yet have still to process problems with a large scope of time series or systems. Then, automatic model generation, that is based on data driven approach, is a powerful way to generate models.

2 Automatic Model Generation

Most important for a more sophisticated process modelling application is to limit the involvement of users in the overall modelling process to the inclusion of existing a priori knowledge while making this process more automated and more objective. Automatic model generation is based on these demands and is a powerful way to generate models of ill-defined problems. Soft computing, i.e.,

Fuzzy Modelling, Neural Networks, Genetic Algorithms and other methods of automatic model generation, is a way to mine data by generating mathematical models from empirical data more or less automatically (Müller, 1999). Associated mathematical models include the following sections.

2.1 Regression Based Models

Commonly, statistically based principles are used to select parametric models (Cheeseman, 1994). The goal is to find a model that predicts a response variable from predictors and that does well on new data. In addition to the epistemological problems of commonly used statistical principles for model formation, there are several methodological problems that may arise in conjunction with the insufficient a priori information on the system to be modelled. In conjunction with this indeterminacy of the starting position, which is marked by the subjectivity and incompleteness of the theoretical knowledge and by an insufficient data basis, there are several methodological problems such as those described in (Müller, 1999).

Elder and Pregibon (Elder, 1996) have provided an overview on statistical techniques applicable to knowledge discovery from data. They reviewed some major advances in statistics in the last few decades but also some influential classical and modern statistical methods for practical model induction. Recently, it can be observed that, after a period of model estimation, the statistic community is now focusing on model selection. For modern methods, the model search is processed on structure space and on parameter space as well. "It is not uncommon now for many thousands of candidate structures to be considered in a modelling run - which forces one to be even more conservative when judging whether improvements are significant, since any measure of model quality optimised by a search is likely to be over-optimistic" (Elder, 1996).

2.2 Nonparametric Models

Parametric methods replace sample data with a model representation, a global consensus of the model that the data represents. The parametric model is the best tool for a function's approximation and forecasting of deterministic objects where all inputs and outputs are measured. For ill-defined objects with very large noise, better results should be obtained by nonparametric models, such as patterns or clusters. Nonparametric models are selected from a given variable set by Analog Complexing (Müller, 1999) representing one or more analogous patterns of a trajectory of past behaviour, or by Objective Cluster Analysis representing one or more clusters. For optimal pattern recognition and clustering only partial compensation is necessary. More of what we are interested in is to minimise the degree of compensation to get more accurate results. Forecasts are not calculated in the classical sense but selected from the table of observed data. These methods are denoted as nonparametric because there is no need to estimate parameters. "Nonparametric 'model-free' methods instead keep the data around and refer to it when estimating the response or the class of a new point" (Elder, 1996).

2.3 Rule Based Models

For some applications, explanations of the reasoning process are important, sometimes even being legal requirements. Since Neural Networks have inherent difficulties to provide explanations, many users are increasingly examining rule induction type approaches that provide explicit explanations of reasoning (Goonatilake, 1994). Rule induction from data uses Genetic Algorithms to induce rules operating on data with linguistic categories. The representation of models is in the familiar disjunctive normal form. Local rules are generated, such as:

IF *some subset of the independent variable satisfies particular conditions*

THEN *a certain behaviour of the dependent variable is to be expected.*

This produces extremely easy, understandable, transparent decision models interpretable by decision makers.

More important for ill-defined applications such as economic, ecological and others will be fuzzy modelling. We can interpret the fuzzy modelling as a qualitative modelling scheme by which we qualitatively describe a system behaviour using a natural language. Zadeh has suggested the idea of fuzzy sets since he found difficulties in the identification of complex systems based on differential equations with numeric parameters.

2.4 Neural Networks

One development direction that cares about the practical demands on modelling is represented by automatic regression based modelling which depends on:

- the black-box method as a principle approach to analyse systems from input/output samples;
- the connectionism as a representation for complex functions by networks of elementary functions.

Separate lines of development of these scientific foundations are the theory of Neural Networks (NN) and the theory of Statistical Learning Networks (SLN). Whereas Neural Networks were developed mainly as real-time adaptive signal processors that must be trained to compensate for fluctuations in an "intelligent", that is, an adaptive manner during normal working conditions, Statistical Learning Networks are a kind of statistical method for data analysis and prediction. Here, all observations are given and forecasting can be made upon earlier analysis of similar data. Its "learning" is analogous to the process of estimation used in many statistical algorithms. Repetitive “learning” can lead to adaptive behaviour as well.

Statistical Learning Networks and in particular GMDH (section 3.1) can overcome some following most significant problems for successful NN-based modelling.

1. Models obtained by most commonly used NN’s are implicit models distributed over the network (internal connective representation of knowledge). However, analysts are usually interested in interpreting their model and in extracting some meaning from the model structure.

2. The required knowledge for designing the architecture of NN’s is not at the command of the users. The Multilayer Perceptron, for instance, requires to adjust many parameters. They represent the minimum number of design decisions, supposing not only knowledge about the theory of NN’s but also time and experience in this field. As a result, the qualitative model structure is, a priori, predefined as very complex.

3. NN’s do not use as much a priori information as other modelling methods, since ”black boxes” are theoretical abstractions only in reality. So, if a priori knowledge is available from systems theory, for example, it can hardly be used for NN model improvement. It is not applicable without a transformation into the world of NNs. The rules of this translation, however, are unknown.

4. NN learning techniques are a kind of statistical estimation, often using algorithms that are slower and less effective than algorithms used in statistical software [Sarle, 94]. Some of the known problems of NN learning algorithms like backpropagation are [Zell,98]: symmetry breaking, learning may stop in local minimum of criteria, stagnation on flat error plateau, oscillations.

5. If noise is considerable in a short data sample, it can be shown that these models are overfitted multivariate multiple non-linear (in special cases linear) regression functions. “Statisticians have been suspicious of artificial neural networks due to their overzealous promotion, but also because they appear so over-parameterised, and the weight adjustment procedure is a local gradient method (missing global optima), sequential (allowing early cases to have too much influence), and interminable (leading to a crude type of regularisation wherein moderating the runtime becomes the principle way to avoid overfit).” (Elder, 1996)

2.5 Complex Structured Process Models

Self-organising structured modelling uses a symbolic generation of an appropriate model structure (algebraic formula or complex process models) and optimisation or identification of a related set of parameters by means of genetic algorithms. This approach assumes that the elementary components are predefined (model based) and suitably genetically coded.

In an attempt to solve some of the above mentioned NN's parameterisation problems there has been considerable research effort into combining NN's and Genetic Algorithms. Genetic Algorithms have been used for weight training, for supervised learning, for selecting data, setting parameters and designing network architectures (Kingdon, 1997). Kingdon has developed an Automated Neural Net Time Series Analysis System (ANTAS) that combines feed-forward NN and Genetic Algorithms to automate NN time series modelling for an arbitrary time series (Kingdon, 1997). Mostly, however, application of Genetic Algorithms is very time-consuming (beyond NN's time requirements) and they claim to find an optimum in a highly multi-dimensional search space which is very tricky.

3 Self-Organising Modelling

In contrast to Neural Networks that use

- Genetic Algorithms as an external procedure to optimise the network architecture and
- several pruning techniques to counteract overtraining,

self-organising modelling (Müller, 1999) introduces principles of evolution - inheritance, mutation and selection - for generating a network structure systematically enabling automatic model structure synthesis and model validation. Models are generated adaptively from data in the form of networks of active neurons in an evolutionary fashion of repetitive generation of populations of competing models of growing complexity, their validation and selection until an optimal complex model - not too simple and not too complex - has been created. That is, growing a tree-like network out of seed information (input and output variables' data) in an evolutionary fashion of pairwise combination and survival-of-the-fittest selection from a simple single individual (neuron) to a desired final, not overspecialised behaviour (model). Neither, the number of neurons and the number of layers in the network, nor the actual behaviour of each created neuron is predefined. All this is adjusting during the process of self-organisation, and therefore, is called self-organising modelling (Müller, 1999).

Self-organising modelling algorithms realise, in an objective way, the following steps automatically:

1. generation of alternative models with different variables and growing complexity in each layer;

2. for parametric models: estimation of unknown parameters on a training set and performance validation on a testing set using at least two different criteria;
3. selection of a number of best models in each processed layer on the basis of external information;
4. as long as performance is growing, increase complexity and repeat steps 1-3, otherwise, selection of a final model of optimal complexity.

This methodology guarantees the objectivity of the selection procedure and its advantages are primarily seen in modelling of large, complex systems with many variables (>50). In the software tool "KnowledgeMiner" (section 4) two alternative ways are used to realise such a self-organising data mining:

- by applying GMDH-type Neural Networks and/or
- Analog Complexing.

3.1 GMDH-Type Neural Networks

The GMDH (group method of data handling) algorithm is based on adaptive networks. Self-organisation is considered as building the connections between the units by a learning mechanism to represent discrete items. For this approach the objective is to estimate networks of the right size with a structure evolving during the estimation process. A process is said to undergo self-organisation when identification emerges through the system's environment. We want to apply the principles of self-organisation for automatic creation or organisation of a mathematical model on the computer proceeding from samples of input and output. In this context, self-organisation sets have the least demands on information needed a priori: observations and the verbal formulation of the objective of the studies suffice in the extreme case.

To realise a self-organisation of models from a finite number of input-output data samples the following conditions must be satisfied:

First condition: there is a very simple initial organisation (neuron) that enables the description of a large class of systems through its evolution.

The recent version of the „KnowledgeMiner" software (see section 4) has included a GMDH algorithm that realises for each created neuron an optimisation of the structure of its transfer function (Active Neuron). As a result, the synthesised network is a composition of different, a priori unknown, neurons and their corresponding transfer functions have been selected from all possible linear or non-linear polynomials:

$$f(x_i, x_j) = a_0 + a_1 x_i + a_2 x_j + a_3 x_i x_j + a_4 x_i^2 + a_5 x_j^2 . \qquad (1)$$

In this way, neurons themselves are self-organised resulting in a significant increase in the flexibility of network function synthesis.

Second condition: there is an algorithm for the mutation of the initial or already evolved organisations of a population (network layer).

There are Genetic Algorithms working on more or less stochastic mutations of the model structure by means of crossover, stochastic changes of characteristic parameters and others. In the GMDH approach, a gradual increase of model complexity is used as a basic principle. The successive combination of different variants of mathematical models with increasing complexity has proven to be a universal solution in the theory of self-organisation. To apply this principle, a system of basic functions (simple elementary models or neurons) is needed. Their appropriate selection, the ways the elementary models are combined to produce more complicated model variants along with a regulation of their production, decides the success of self-organisation.

The core mutation process works as follows: All possible pairs of the m inputs are generated to create and validate the transfer functions of the m*(m-1)/2 neurons of the first network layer. In "KnowledgeMiner", for example, each transfer function is adaptively created by another self-organising process and they may differ one from another by their number of variables used and by their functional structure and complexity.

Third condition: there is a selection criterion for validation and measure of the usefulness of an organisation compared with its intended task.

According to this condition, several best models - each consisting of a single neuron only in the first layer - are ranked and selected by the external selection criterion The selected intermediate models survive and they are used as inputs for the next layer to create a new generation of models, while the nonselected models die.

The principle of selection is closely linked to the principle of self-organisation in biological evolution; it is very important for the evolution of the sorts. In our case, it is applied when the number of all possible intermediate models in one generation is going to become too large for a complete induction. One way for realising a selection of models is defining a threshold value subjectively, or tuning it heuristically. Using this threshold value, those intermediate models that are best in the sense of a given quality function are selected and are then considered as inputs for next generations' pairwise combination. The following theorem can be formulated in this context: the probability that the best solution was lost due to selection, and therefore the result is only a sub-optimal solution, is smaller, the more the respective power of this solution exceeds the threshold value.

The procedure of inheritance, mutation and selection stops automatically if a new generation of models provides no further model quality improvement. Then, a final, optimal complex model is obtained. In distinction to Neural Networks, the complete GMDH Network is, during modelling, a superposition of many alternate networks living simultaneously. Only the final, optimal complex model represents a single network while all others die after modelling (Fig. 1).

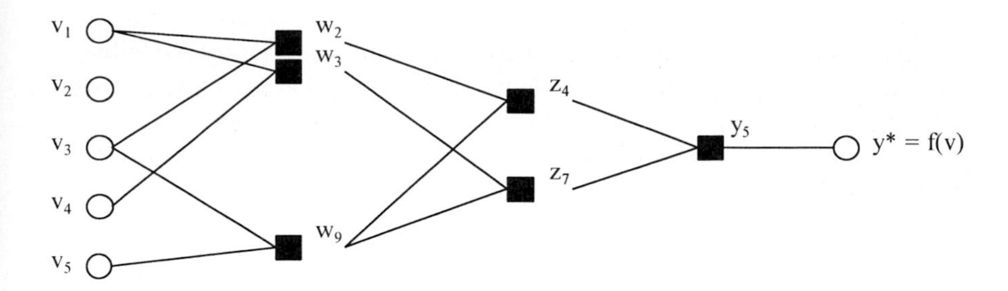

Fig. 1: Network after selection of an optimal model y*

GMDH objectively selects the model of optimal complexity using an inductive approach shown above. The important feature of such an inductive approach is the use of an external complement, which can be a selection criterion. Applied to the ill-posed task of selecting a model from the set of possible models, this principle is as follows: a "best" model can be selected from a given data sample only if additional external information is used. External information is information or data not yet used for the model creation and parameter estimation, usually done on a training data set. This means, an external criterion is necessary that evaluates the models' quality based on fresh information (a testing data set, for example).

KnowledgeMiner (section 4) has implemented a selection criterion that produces powerful predictive models. One very efficient solution here provides the Prediction Error Sum of Squares (PESS) criterion:

$$PESS= \frac{1}{N}\sum_{t=1}^{N}(y_t - f(x_t,\hat{\underline{a}}_t))^2 . \tag{2}$$

It is an external criterion but does not require users to subdivide data explicitly since it employs cross-validation techniques internally. Therefore, it is appropriate for under-determined modelling tasks or for modelling short data samples.

With increasing complexity, that is, with increasing number of variables included in the model or with increasing polynomeous degree of the model variants, the least-square fitness on the training data set continuously increases. By the application of an external selection criterion along with the principle of external completion, now models of different qualities occur at all stages; the external criterion - one example is the PESS fitness - has a minimum with increasing complexity. Several best models have to be selected at each stage to conform with the principle of selection. If a selection stage cannot improve the external criterion (model quality), a best model has been found and the modelling process stops.

3.2 Self-Organising Fuzzy Modelling

This GMDH approach can be used to generate fuzzy models. Fuzzy modelling is an approach to form a system model using a description language based on fuzzy logic with fuzzy predicates. Such a description is qualitatively able to describe a system behaviour using fuzzy quantities. In the following, we suggest considering a multi-input and single output system and the following type of fuzzy model for this system:

$$R^i : \text{IF } x_1 \text{ is } A_1^{j_1} \text{ and } x_2 \text{ is } A_2^{j_2} \text{ and .... and } x_n \text{ is } A_n^{j_n} \text{ THEN y is } B^i$$

where R^i is the i-th rule and A_j^i, B^i are fuzzy variables.

In the black box approach of automatic fuzzy model selection from data, we have to build a dynamic model using only empirical input-output data $x_1, x_2, ..., x_n$, y, where x and y are the input - output data respectively of a dynamic system. Commonly, the task of identification is divided in two tasks: structure identification and parameter identification. In self-organising fuzzy modelling, there are the following steps.

3.2.1 Fuzzification

Fuzzy quantities are expressed in terms of fuzzy numbers or fuzzy sets associated with linguistic labels. In this step the numerical observations of inputs $\underline{x} =(x_1, x_2, ..., x_n)$ and output y must be transformed into fuzzy vectors $(\underline{x}^1, \underline{x}^2, ...,\underline{x}^m)$ with $x^j = \mu_{A^j}(x)$ and $y=(y^1, y^2, ...,y^m)$ with $y^j = \mu_{B^j}(y)$. The fuzzy membership functions $\mu_{A^j}(x)$ and $\mu_{B^j}(y)$ used in the following have a triangular shape.

3.2.2 Structure Identification: Rule Generation

Given a class of models (description language) and the data type (fuzzy sets), the task of system identification is to find a model that may be regarded as equivalent to the objective system with respect to input-output data. Such a task of structure identification has to solve two problems: to find out input variables and to find input-output relations.

Self-organising fuzzy modelling solves both tasks, selecting a finite number from possible input candidates and adaptively creating a fuzzy model with an optimal number of fuzzy rules. The fuzzy rules have two parts: the premise part and the consequent part. The premise structure implies a fuzzy partitioning of the input space; therefore self-organising fuzzy modelling has to find out how the input space should be divided into fuzzy subspaces. To solve this task it is possible to use GMDH algorithm but also objective cluster analysis (OCC).

The rules are written in an IF/THEN-form. Commonly they are collections of expert know-how. In the black box approach we have to build dynamic models using only input-output data of the system. According to the number m of output fuzzy variables, m static or dynamic fuzzy models are to be generated (for m=7: y-NB, y-NM, y-NS, y-ZO, y-PS, y-PM, y-PB, where NB-negative big, NM-negative medium, NS-negative small, ZO-zero, PS-positive small, PM-positive medium, PB-positive big).

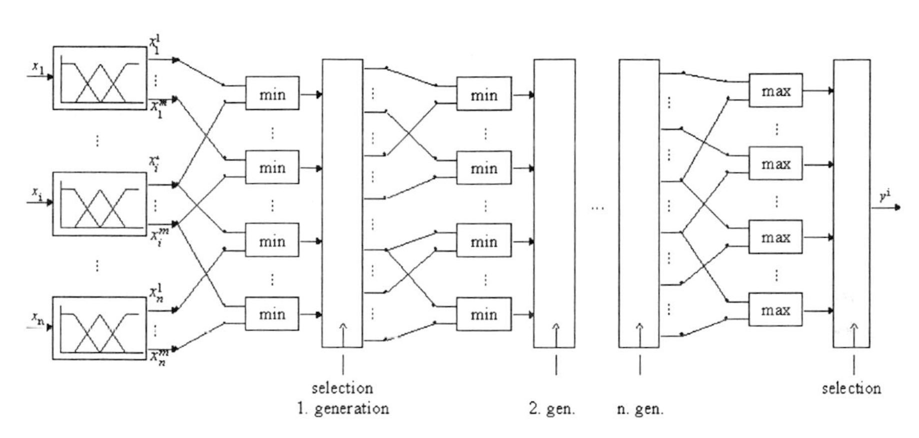

Fig.2: Multilayer self-organising fuzzy modelling

In the self-organising fuzzy modelling using GMDH algorithms (Fig. 2) in the first layer, every input represents an input fuzzy set. The number of inputs in the first layer is determined by the total number of fuzzy sets (m) for input variables (n). Therefore, in a static model there are $n^{\times}m$ neurons where n is the number of inputs and m the number of fuzzy variables. If the model is a dynamic one, there are $n^{\times}(L+1)^{\times}m$ input neurons, where L is the maximum time lag.

In the GMDH-theory two principles of generation of models with growing complexity are known: combinatorial and multilayer algorithms. Both can be used to generate fuzzy models with growing complexity. Fig. 2 shows a multilayer architecture, where every neuron has two inputs (x_i^j, x_k^l) and one output (y^r), which realises:

$$\text{IF } x_i^j \wedge x_k^l \text{ THEN } y^r(i,j,k,l). \qquad (3)$$

Using fuzzy logic, those links establish the antecedent relation that is an "AND" association for each fuzzy set combination. The method for fuzzy inference uses the most general max-min method: $y^r(i,j,k,l) = \min(x_i^j, x_k^l)$.

For the fuzzy output variable y^r (r=1(1)m) in the first layer and for all pairs of inputs (i=1(1)n, k≥i, j=1(1)m, l=1(1)m) the fuzzy model outputs y^r_t (i,j,k,l) are evaluated for all realisations t = 1(1) N. After generating all possible combinations, the F best rules of two fuzzy variables can be selected and used in the following second layer as inputs to generate fuzzy rules of 2,3 or 4 fuzzy variables. The following selection criterion can be used

$$Q_p(i,j,k,l) = \sum_{t=1}^{N} \left| y_t^r(i,j,k,l) - y_t^r \right|^p . \tag{4}$$

Such a procedure can be repeated up to an increasing sum of F values for the criterion of selected fuzzy models. After this in a second run of self-organisation, disjunctive combinations of F best models are generated, which can be evaluated by

$$y^r_m(i,j) = \max(y^i_n, y^j_n), \tag{5}$$

where y^i_n, y^j_n are outputs of the n-th generation (last layer), i=1(1) F, j≥i .

3.2.3 Defuzzification

For defuzzification, known methods, such as like the gravity method, can be used to transform the estimated vector of fuzzy outputs y^r , r=1(1)m back into a crisp value y*. Alternatively, the fuzzy output y^r can be transformed back into the original data space by a third run of self-organisation using GMDH. As a result, an optimised transformation $y^* = f(y^1, y^2, ..., y^r)$ will be obtained that excludes redundant or unnecessary fuzzy outputs.

Since only relevant fuzzy sets are considered in this way, information on the optimal number of fuzzy sets is also provided implicitly for the given membership function. Using this information for an optimised fuzzification, a complete new run of the rule induction process may result in an increased descriptive and predictive power of the models.

3.3 Self-Organisation of Logic Based Rules

The algorithm of self-organising fuzzy-rule induction described above can also be employed for generating logic based rules. In this special case, the variables x_i are of Boolean type, for instance, $x_i = 0$ or $x_i = 1$. Instead of the n(L+1)m input neurons used for self-organising fuzzy-rule induction (Fig. 2), there are now only

2 n(L+1) input neurons using both the Boolean variables x_i and the negation NOT x_i. In this way, logical IF-THEN rules can be induced from data, such as

IF B_bmw(t-1) & B_dj(t-6) OR B_dj(t-1) & NOT_B_dax(t-2) OR B_dj(t-1) & NOT_B_dax(t-10) OR B_bmw(t-1) & NOT_B_dj(t-3) OR B_bmw(t-1) & B_dj(t-6)
THEN buy_BMW(t),

where buy_BMW is a trading signal (buy) in a trading system, which was generated by means of „KnowledgeMiner" (section 4) in the period April 1, 1997 through to March 6, 1998 on base of DAX, Dow Jones and BMW stock indexes, and the DOLLAR/DM exchange rate. The Boolean variables are

$$B\_x(t) = \begin{cases} 1 & x(t) > 0 \\ 0 & \text{else} \end{cases}, \text{ where } x(t) = \frac{X(t+1) - X(t)}{X(t)}. \tag{6}$$

3.4 Analog Complexing

The basis for the Analog Complexing method is the assumption that typical situations exist, i.e., each actual period of state development for a given multi-dimensional time process may have one or more analogues in history. If so, it will be likely that a prediction could be obtained by transforming the known continuations of the historical analogues. It is essential that searching for an analogous pattern is not only processed on a single state variable (time series) but on a set of representative variables simultaneously and objectively. This method was developed by Lorence (Lorence, 1969) and was successfully applied to meteorological forecasts. Recently it has been enhanced by an inductive, self-organising approach and an advanced selection procedure to make it applicable to evolutionary processes, too (Müller, 1999). In Analog Complexing, the observed process itself is used for forecasting. Forecasts are not calculated in the classical sense but selected from the table of observational data. The main assumptions are:

- the system to be modelled is described by a multidimensional process;
- many observations of data sample are available (long time series);
- the multi-dimensional process is sufficiently representative, i.e., the essential system variables are included in the observations;
- it is possible that a past behaviour might repeat in future.

If we succeed in finding one or more parts of the past (analogous pattern) which are analogous to the most recent part of behaviour trajectory (reference pattern), the prediction can be achieved by applying the known continuation of this analogous pattern to the reference pattern.

4 KnowledgeMiner

„KnowledgeMiner“ (http://www.scriptsoftware.com/km/index.html) is a powerful and easy-to-use modelling tool which was designed to support the knowledge extraction process on a highly automated level and which has implemented three advanced self-organising modelling technologies at present: GMDH, Analog Complexing and Fuzzy/binary rule induction using GMDH (Müller, 1999).

The GMDH implementation employs active neurons and provides in this way networks of active neurons at the lowest possible level already. It can be used to create linear/non-linear, static/dynamic time series models, multi-input/single-output models and multi-input/multi-output models as systems of equations and, also, systems of fuzzy rules or logic based rules from short and noisy data samples. All obtained models are described analytically. Systems of equations are necessary to model a set of interdependent variables objectively and free of conflicts. They are available both analytically and graphically by a system graph reflecting the interdependent structure of the system.

For modelling and prediction of fuzzy objects „KnowledgeMiner“ provides an Analog Complexing algorithm. It is a multi-dimensional search engine to select the most similar, past system states relative to a chosen (actual) reference state. This means, searching for analogous patterns in the data set is usually not only processed on a single time series (column) but on a specified, representative set of time series simultaneously to extract significant hidden knowledge. Additionally, it is possible to let the algorithm search for different pattern lengths (number of rows a pattern consists of) within one modelling process. All selected patterns, either of the same or different pattern length, are then combined to synthesise a most likely prediction. „KnowledgeMiner“ performs this in an objective way using a GMDH algorithm to find out the optimal number of patterns and their composition to obtain a best result.

Here, a model consists of a composition of similar patterns. Since several most similar patterns are always used to form a model and a prediction by a synthesis, an interval of uncertainty of prediction will be produced simultaneously. This is of special importance when using predictions for decision support.

All models created in a document are stored in a virtually unlimited model base. This means, every variable in KnowledgeMiner’s database can have five different model representations in a single document - a time series model, a multi-input/single-output model, a system model (multi-output model), a nonparametric model obtained by Analog Complexing and a rule based model. To provide alternate models when using GMDH, up to three best models are created and stored in the model base separately. They are equally accessible and applicable. Every model can be used immediately after modelling for status-quo or what-if predictions within the program creating or using new data. Comments can be added to models by either writing text or recording spoken remarks.

All data are stored in a spreadsheet with core functionality including simple formulas and absolute or relative cell references. The data can be imported in two ways: as a standard ASCII-text file or via the clipboard by copying/pasting data. Several mathematical functions are available for synthesising new data to extend the data basis optionally.

5 Application

5.1 Application Fields

The application field is decision support in economics (analysis and prediction of economical systems as well as for market, sales, financial predictions, balance analysis) (Müller, 1999; Lemke, 1997) and in ecology (analysis and prediction of ecological processes such as temperature of air and soil, air and water pollution, water quality, growth of wheat etc., drainage flow, Cl- and NO_3-settlement, influence of natural position factors on harvest) (Müller, 1996) but also in other fields such as medicine/biology, sociology, engineering, meteorology with small a priori knowledge about the systems.

A very important example gives the prediction of effects of experiments. The GMDH algorithm solves two problems: calculation of effects for a given experiment, calculation of parameters which are necessary to reach optimal effects. Similar problems can be solved for experiments in many research areas. It means that the realisation of experiments can often be replaced by computer experiments. But it is necessary to have in disposition the data sample of similar experimental observations.

5.2 Model of the National Economy

One important task given to economic sciences is to improve the quality of planning and management on all national economic levels. In economic research and elaboration of plans and strategies, the analysis of economic systems are gaining importance. The analysis of economic systems is the making of studies on the level of developments achieved so far and on existing deviations from a plan decided before, and the prediction of economic indices, that is, the determination of potential development possibilities of the economy under study. The purpose of such studies is to create preconditions appropriate for expected future developments, and to find the laws and factors of influence causing these developments. This example shows the status quo prediction of 19 important characteristics of the German national economy (Lemke, 1995).

Twenty six yearly observations (1960 - 1985) are given of 19 variables (Table 1). This data set was used as is without additional data pre-processing. The variables are not separated into endogenous or exogenous a priori, although some variables such as exports/imports clearly expresses the influence of other national economies.

Variable		a	b	c	D
gross domestic product	x_1	1.84	1.57	1.98	1.80
gross national product	x_2	0.03	1.13	0.67	0.61
national income	x_3	2.65	2.46	2.92	2.68
gross income dependent work	x_5	0.79	1.25	1.69	1.24
gross wages and salaries	x_6	0.1	0.88	0.73	0.57
weekly gross wages (workers)	x_7	0.76	1.22	10.1	1.00
unemployed persons	x_8	4.11	2.04	5.18	3.78
employed persons	x_{10}	0.2	0.19	0.2	0.2
savings	x_{11}	0.27	1.18	2.46	1.30
cash circulation	x_{12}	2.11	7.81	11.55	7.16
personal consumption	x_{13}	0.35	0.99	2.25	1.20
state consumption	x_{14}	0.32	0.77	2.8	1.30
sum of investments	x_{15}	0.97	3.18	2.57	2.24
sum of credits	x_{17}	3.53	1.44	3.82	2.93
mean		*1.29*	*1.87*	*2.85*	*2.00*

Table 1: Long-time prediction errors (MAD [%]) (a: 1986; b: 1986-88; c: 1986-1990; d: mean)

Since the noise dispersion of the data is small, a description by a linear system of equations using GMDH was chosen as the appropriate modelling method. For setting up the modelling, the only parameters required here were the data length (26 observations), the maximum dynamics (4 years) and the model type (linear system of equations). Other parameters such as when the process has to stop, penalty terms, learning rates or topology settings are not necessary.

Using this data set (19 columns, 26 rows) and the chosen system dynamic of up to 4 years, the information matrix for modelling is constructed automatically by the software in the background. It consists of 94 columns (18 non-lagged variables and 76 lagged variables) and 22 rows and contains already normalised values. Here, a linear, dynamic system of equations with 94 input variables and 19 output variables was created autonomously by GMDH:

$$\underline{x}_t = A\,\underline{x}_t + \sum_{j=1}^{4} B_j \underline{x}_{t-j} \,. \qquad (7)$$

The obtained system model was used then for predicting all output variables ex ante 4 years ahead in a single step. After a single modelling run that took about three hours for all 19 variables, the system of equations was generated. For each output variable an analytical model equation is available (for the structure look at (Lemke, 1995)) that was transformed back by the algorithm into the original data space. For example, the equations generated for the gross domestic product and for the unemployed people are:

$$x_{2,t} = 6.170 - 0.0210\, x_{8,t} - 0.0110\, x_{10,t} + 0.4081 x_{11,t} + 1.2765\, x_{15,t} + 0.1175 x_{2,t-1} + 0.1435 x_{11,t-1} + 0.2056 x_{4,t-3}$$

$$x_{8,t} = -0.2950 - 0.2668 x_{9,t} - 2.5033 x_{15,t} + 0.1536 x_{8,t-1} + 6.1348 x_{14,t-1} + 0.9117 x_{1,t-2} - 2.6524 x_{11,t-2} + 1.1648 x_{17,t-3} - 0.1916 x_{8,t-4}.$$

It is evident that only a subset of the most relevant variables are included in the models making them parsimonious and robust. Table 1 shows the prediction errors when predicting the system.

Although the dimension of the modelling task was much larger here, the duration for creating the system of fuzzy rules was about three hours too. This is because there are no parameters to estimate and the structure of the transfer functions are faster to optimise. For the gross domestic product and employed people the rules shown in Table 2 were synthesised.

IF PB_EmpPers$_{t-3}$ & NS_Invest$_{t-4}$ **THEN** NS_GrossDom$_t$
IF PB_EmpPers$_{t-3}$ & PS_Wag/Emp$_{t-4}$ OR PS_GrossDom$_{t-2}$ & PB_EmpPers$_{t-3}$ OR PB_Inhab$_{t-1}$ & ZO_Invest$_{t-1}$ & ZO_StCons$_{t-3}$ & ZO_Invest$_{t-4}$ **THEN** ZO_GrossDom$_t$
IF NB_Unemp$_{t-3}$ OR ZO_Savings$_{t-2}$ & ZO_Savings$_{t-1}$ & ZO_Ex/Im$_{t-2}$ OR ZO_Unemp$_{t-3}$ & ZO_Credits$_{t-1}$ & PB_Inhab$_{t-2}$ & PS_Invest$_{t-4}$ **THEN** PS_GrossDom$_t$
IF NS_Savings$_{t-1}$ OR PS_Invest$_{t-3}$ & ZO_Savings$_{t-4}$ OR ZO_Savings$_{t-1}$ & PB_Inhab$_{t-2}$ OR PB_Inhab$_{t-2}$ & ZO_EmpVac$_{t-4}$ **THEN** NB_EmpPers$_t$
IF PB_Credits$_{t-4}$ OR PS_Ex/Im$_{t-3}$ & PS_Savings$_{t-4}$ OR PS_Credits$_{t-2}$ & PB_GrossDom$_{t-1}$ & PS_Invest$_{t-4}$ **THEN** ZO_EmpPers$_t$
IF NS_EmpPers$_{t-4}$ & PS_Invest$_{t-2}$ & NS_Inhab$_{t-3}$ OR ZO_EmpVac$_{t-1}$ & PB_StCons$_{t-2}$ & NS_Inhab$_{t-2}$ & PB_CashCirc$_{t-3}$ & PS_Invest$_{t-2}$ **THEN** PB_EmpPers$_t$

Table 2: Generated fuzzy rules for the gross domestic product and employed persons

5.3 Water Quality of the River Elbe

From a measuring station in the river Elbe, primary data of some water characteristics were measured automatically every three hours: O_2-oxygen concentration, TW- water temperature, Q-rush through and TR-dimness. Figure 3 plots these four characteristics from October 1, 1974 to November 13, 1974. In this example, models for two time periods were generated: a: October 29 to November 11, 1974 (100 observations) and b: October 1 to November 11, 1974 (330 observations). The models were then used for long-term prediction of 11 and 22 steps ahead.

Using "KnowledgeMiner", linear and non-linear systems of difference equations were generated for the four characteristics. For the period b), for example, the following linear system model was generated (max. time lag was 40; 163 input variables; 4 output variables):

$$O_2(t) = 0.82 + 0.96\, O_2(t-1) - 0.084\, O_2(t-3) - 0.033\, O_2(t-10) + 0.031\, O_2(t-22) - 0.048\, TW(t-6) + 4.10^{-4}\, Q(t-5) - 0.076\, TR(t-4)$$

$$TW(t) = 0.06 + 0.96\, TW(t-1) + 0.109\, TW(t-7) - 0.174\, TW(t-10) + 0.034\, TW(t-23) - 0.039 TR(t)$$

$$Q(t) = 14.16 + 1.18\, Q(t-1) - 0.188\, Q(t-6) - 1.179\, TW(t-14) + 3.626\, TR(t-35)$$

$$TR(t) = 0.0214 + 0.978\, TR(t-1).$$

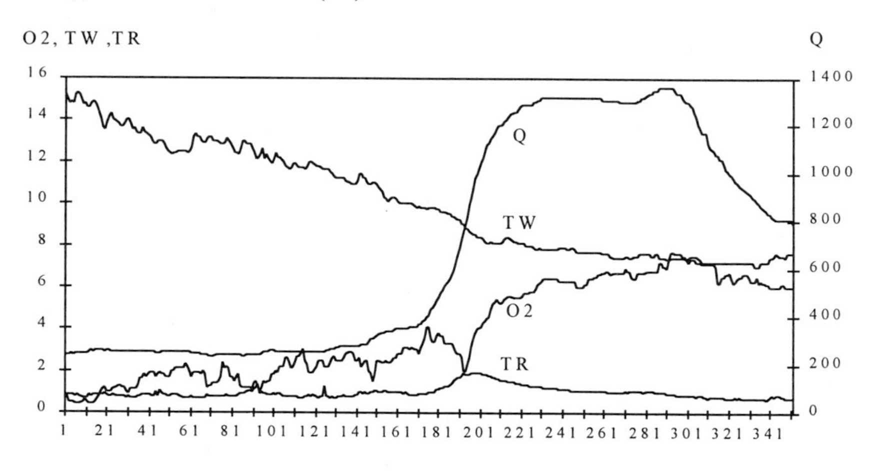

Fig. 3: Water characteristics: O_2-oxygen concentration, TW- water temperature, Q-rush through, and TR-dimness measured for the Elbe river.

Table 3 contains the prediction error (MAD[%]) for two periods and a forecast horizon of 11 and 22 observations (a/11, for example, is the result of period a) and a forecast horizon of 11). Using the self-organising fuzzy rule induction for the given periods, several fuzzy rule-based models were generated (max. time lag 5) such as shown in Table 4. Here, differences of the input variables $\Delta x(t)=x(t)-x(t-1)$ and output variables were used for fuzzification.

	GMDH				Fuzzy			
	a/11	A/22	b/11	b/22	a/11	a/22	b/11	b/22
O_2	4,47	6,4	1,50	1,14	4,42	6,22	5,41	8,13
TW	1,13	2,66	1,53	3,98	1,34	3,52	2,85	7,55
Q	3,00	8,01	2,72	6,26	4,13	13,11	0,50	2,83
TR	6,04	7,39	5,49	7,52	4,94	11,86	3,62	5,94

Table 3: Long-time prediction error (MAD [%]) (GMDH and Fuzzy models)

IF NOT_N_tw(t-2) & ZO_tw(t-4) & P_o2(t-3) & N_trueb(t-4) & NOT_ZO_q(t-3) & ZO_tw(t-2) OR P_q(t-1) & P_o2(t-3) & P_o2(t-5) OR NOT_N_q(t-2) & NOT_P_o2(t-2) & NOT_N_o2(t-1) & N_o2(t-1) & P_o2(t-3) & ZO_o2(t-5)) & NOT_P_tw(t-1) & ZO_o2(t-1) & ZO_o2(t-3) & NOT_N_tw(t-2) & NOT_N_tw(t-4) & NOT_P_o2(t-3)
THEN N_o2(t)
IF ZO_q(t-4) OR N_o2(t-4) OR P_tw(t-2) OR N_tw(t-5) OR P_tw(t-3) OR ZO_q(t-3) OR ZO_q(t-1) OR NOT_P_trueb(t-1) & NOT_N_trueb(t-4) & NOT_P_trueb(t-5) OR ZO_q(t-1) OR NOT_ZO_trueb(t-1) OR NOT_P_trueb(t-1) & NOT_P_trueb(t-5) & NOT_P_trueb(t-1) & NOT_N_trueb(t-5)
THEN ZO_o2(t)
IF P_trueb(t-1) & P_q(t-3) OR N_trueb(t-2) & P_tw(t-3) OR N_o2(t-3) & N_trueb(t-4) OR P_tw(t-1) & P_q(t-5) OR NOT_P_q(t-2) & NOT_P_tw(t-2) & NOT_P_o2(t-5) & P_tw(t-3) & P_trueb(t-3) & NOT_P_o2(t-3) & NOT_P_o2(t-1) & ZO_o2(t-2)
THEN P_o2

Table 4 : Generated fuzzy rules for O_2

6 Conclusions

According to the above described approaches to automatic model generation, there exists a range of self-organising modelling. All kinds of parametric, nonparametric, algebraic, binary/fuzzy logic models are only simplified reflections of reality. Some models exist with a sufficient degree of adequacy for a given sample of data, but every model is an abstraction, a one-sided reflection of some important features of reality. A synthesis of alternative model results gives a more complete reflection.

If models are obtained in a short time, it is possible to generate several alternative models, which can be used to estimate the vagueness of model results. Also, it is possible to select the best models or to generate their combination (synthesis) with GMDH-algorithms. Based on forecasts obtained by these models, a continuous parameter adaptation of the GMDH - type Neural Networks as well as a frequent adaptation of nonparametric models is possible. Using the estimated vagueness, frequent updates of models are advisable. Another direction of further research for improving the described approach is the implementation of fuzzy, combining, or synthesis methods.

The methodological foundation for all approaches gives the soft systems methodology. The inductive approaches of automatic model generation described in this paper increase the a priori information by including qualitative model components and computer-based modelling (self-organisation of mathematical models). Such a soft-modelling approach using soft computing improves the modelling technology and objectifies the results obtained.

References

Cheeseman, P.; Oldford, R.W.(eds) (1994): Selecting models from data. Springer-Verlag, New York 1994

Elder IV, J.F., D. Pregibon (1996): A Statistical Perspective on Knowledge Discovery in Databases. In „Fayyad, U.M. et al: Advances in Knowledge Discovery and Data Mining. AAAI Press/The MIT Press. Menlo Park, California 1996", pp. 83-116

Goonatilake, S., Feldman, R. (1994): Genetic Rule Induction for Financial Decision Making. In „Stender, J. et al. (Eds): Genetic Algorithms in Optimization, Simulation and Modelling." IOS Press 1994

Kingdon, J. (1997): Intelligent Systems and Financial Forecasting. Springer. London, Berlin, .. 1997

Lemke, F. (1995): SelfOrganize! - software tool for modelling and prediction of complex systems. SAMS 20 (1995), 1-2, pp.17-28

Lemke, F., Müller, J.-A. (1997): Self-Organising Data Mining for a Portfolio Trading System. Journal for Computational Intelligence in Finance vol.5 (1997) No.3 pp.12-26.

Lorence, E.N. (1969): Atmospheric predictability is revealed by naturally occurring analogues. J. Atmos.Sci. 4 (1969), S.636-646

Müller, J.A. (1994): Automatic Model Generation. SAMS vol.31 (1998) No. 1-2, pp. 1-32.

Müller, J.A. (1996): Analysis and prediction of ecological systems. SAMS vol.25 (1996), pp.209-243

Müller, J.-A. (1999): Lemke, F.: Self-Organising Data Mining. Numerical Insights into Complex Systems. Gordon & Breach 1999.

Sarle, W.S.: Neural Networks and Statistical Models. In: Proceedings of 19th Annual SAS User Group International Conference. Dallas. (1994) pp. 1538-1549

Zell, A. (1998): Einführung in Künstliche Neuronale Netze. In: „Biethahn et al (ed.) Betriebswirtschaftliche Anwendungen des Soft Computing. Vieweg. Braunschweig 1998". pp. 3-34.

The Modelling Process and Petri Nets: Reasoning on Different Approaches[1]

Nicola Mazzocca, Stefano Russo and ValeriaVittorini

Dipartimento di Informatica e Sistemistica, Università di Napoli "Federico II", Via Claudio 21, 80125 Napoli, Italy

Abstract. *Well defined procedures for the construction of models are necessary in order to support the real applicability of modelling techniques in the industrial settings where practical engineering means are required to model and evaluate complex systems and productive processes. This paper describes the application of two different approaches to the construction of Petri Nets models and provides some general hints on the modelling process. In particular, starting from a simple working example we suggest that a* behavioural approach, modularity *in the model construction phases, and* integration *of different modelling techniques could contribute to improve the efficiency in many fields of research, development and its controlling.*

Keywords. *Behaviours, complex systems, modelling process, Petri Nets, traces*

1 Introduction

Modelling a real life complex system -whether it is a specific application or a generic development process- is hard task since many different requirements have to be addressed which are difficult to coexist. The construction of formal and executable models of such systems often assumes a fundamental importance as it allows formal analysis and verification. In spite of this, there is a great need of defining rigorous approaches to modelling orientated to the real applicability of such means in the industrial settings, where practical engineering tools are required to model and evaluate complex systems (Bowen, 1995; Hall,1990).

---

[1] This work has been supported in part by grants from MURST: "Metodologie e strumenti di progetto di sistemi ad alte prestazioni per applicazioni distribuite. Sottoprogetto Sistemi Reattivi".

Petri Nets (PN) are a well known and established formalism in specification and design, and they are widely used to model different classes of systems. A major drawback is the *construction process* of a PN model, since it is often built in a "naïve" way rather than resulting from rigorously defined steps. The aim of this work is twofold:

a) to describe the application of our event-driven approach to modelling and show how it is possible to automate the building of PN models;

b) to derive some general hints about modelling. We do this by introducing a different approach developed by Meyer and Sanders (Meyer, 1993) and reasoning about the features of both.

In particular we refer here to Stochastic Petri Nets (SPN) as we often deal with time-dependent,probabilistic systems. Table 1.1 shows some SPNs classes that are supported by available tools for the editing and the analysis of the models[2] Among them, Generalised Stochastic Petri Nets (GSPNs) and Stochastic Activity Networks (SANs) allow both numerical and simulation analysis.

In this paper, our approach is applied to the construction of GSPN models, whereas the second approach is based on the construction of SAN-based models. A simple industrial electronic system is used as a working example. It is not a "process modelling" case study, but it belongs to a very specific class of applications; nevertheless it is interesting to observe that the resulting remarks are universally valid and apply to every kind of process modelling. In the following, we suppose that the reader is acquainted with the Communicating Sequential Processes (CSP) theory (Hoare, 1985) and the GSPN (Ajmone-Marsan, 1987) and SAN (Meyer, 1985) formalisms; however some hints about SANs are given in Section 2 where the two approaches are briefly described. In Section 3, we introduce the working example and build the SAN and the GSPN models. Finally in Section 4, we briefly discuss the two approaches and suggest that *a behavioural approach*, *modularity* in the model construction phases and *integration* of different modelling techniques could contribute to improve the efficiency in many fields of research, development and its controlling.

SPN class	**Tool**
Deterministic/Stochastic Petri Nets	DSPNExpress (Lindemann, 1992)
Generalised Stochastic Petri Nets (GSPN)	GreatSPN (Chiola, 1991), SPNP (Ciardo, 1989
Stochastic Activity Networks (SAN)	METASAN (Sanders, 1986; ltraSAN (Couvillo 1991)

Table 1.1:. Some SPN classes and related CASE tools

2 A good source of information about the Petri Nets world is at the URL: http://www.daimi.aau.dk/PetriNets/index.html where a wide PN bibliography is also available

2 Modelling Approaches

A complex system is not necessarily complex in size, but maybe it has a complex behaviour and/or it asks to cope simultaneously with different requirements (real time issues, safety, dependability, performance, fault tolerance). Moreover, it could require hardware and software co-design (e.g. embedded systems). In these cases, it is not possible to model the system without providing some means which help us to define what we want to model and build the model. Thus a modelling process should match some general criteria:

1. it must be effectively supported by available tools or must be proved to be automatable to a large extent;
2. it should have a "vocation" for integration according to two different aspects: a) building of a unified framework for specifying different requirements, i.e. techniques and formalisms integration; b) building a unified structure for the definition and evaluation of different indices and performance figures, i.e. measures integration;
3. it should support the automatic construction of the model of the system. Thus, the model should be not solely dependent on the fantasy and intuition of the designer: once the level of abstraction is fixed, there do not exist as many different models as designers;
4. proper techniques must be available to manage the complexity in size of the resulting models.

Of course the above list of criteria is not exhaustive nor definitive. These are hints suggested by our experience with specifying and modelling systems. In Section 4 we will discuss these remarks on the basis of the application of the event-driven and SAN-based modelling approaches.

2.1 The GSPN Event-driven Modelling Approach

The event driven GSPN-based approach integrates the trace logic of the CSP theory (Hoare, 1985) and PNs in order to define a practical engineering methodology for the constructions of complex systems models. It works according to the following steps:

1. The components (processes) of the system are identified by starting from an informal description of the system;
2. a list of events and a list of states are derived for each process;
3. a list of constraints on the events is specified. They can involve one (internal constraints) or more (external constraints) processes;

4. a trace-based specification is derived for each process, where a trace describes a behaviour of the process. In this step the internal constraints are formally expressed as logical predicates that we call internal assumptions and used to determine the feasible traces of the processes;
5. by means of proper translation rules the set of the feasible traces of each process is translated into a PN subnet whose behaviour is equivalent to the behaviours described by its trace set;
6. the integration of the PN subnets is performed by means of linking rules and taking into account the external constraints. They can be formally expressed as interactions between processes or by means of linear temporal logic formula. We call them *external assumptions* and finally we call PN *skeleton* the net obtained by linking the PN subnets;
7. the PN *skeleton* of the system is transformed into a stochastic net by supplementing it with quantitative parameters.

This approach works under the following constraints:

a) dynamic creation of processes is not allowed, with the exception made for recursion; b) sequential composition of more processes in conjunction with recursion is not allowed; c) the set of the different events in which each component can be involved is finite; d) exactly one initial event is allowed for each component, but more distinct final events are possible; e) the behaviour of each component is sequential: if a process performs parallel actions, this means that the related component must be further decomposed in more elementary sub-systems.

Non-deterministic behaviours are allowed and concurrency is introduced in the integration step, as well as causality relations among events, which are introduced through the external assumptions.

This means that the system itself could not behave as a CSP process, but its single components modelled in isolation must do. Indeed, the above restrictions ensure that the behaviours of the system components we deal with can be described by the trace sets of CSP processes and that they have a finite state space. In particular, the second constraint is necessary since Hoare (Hoare, 1985) has shown that sequential composition used in conjunction with recursion can define a machine with an infinite number of states. Nevertheless, to ensure that the PN skeleton also has a finite state space, we will also require that it is a K-bounded net. This implies that further constraints will be added in modelling data and in iteration of events as well in introducing external assumptions.

A detailed description of the event driven GSPN-based approach is in (Mazzocca,1997a; Mazzocca, 1997b).

2.2 The SAN-based Modelling Approach

SANs are a probabilistic extensions of Activity Networks (ANs) which, in turn, are generalised Petri Nets with timed and instantaneous activities (i.e. transitions), extended with the use of *gates* and *cases*.

Fig. 2.1 illustrates the SAN graphical components by means of a sample net. Places hold tokens and represent the state of the modelled system, activities represent actions. In Fig. 2.1 T1 is a timed activity and T2 is an instantaneous activity. *Cases* are associated with activities and model uncertainty related to the event following the activity completion. They are drawn as small circles on one side of the activity, T1 has two cases.

The *cases* of an activity have a discrete probability distribution that can be marked dependent. If only one *case* is associated to an activity, it has probability of one and no circles are drawn, just a standard output arc. A SAN changes its marking when an activity is completed. Input and output *gates* are introduced to enhance flexibility in the defining of enabling and completion rules. Input gates are drawn as triangles with their point connected to the activity they control and have enabling predicates and functions. They respectively control the enabling of the related activity and the change of the marking that will occur upon the completion. of the activity. In Fig. 2.1 G1 is an *input gate*. If a place is directly connected to an activity, like P1, the related arc behaves as an input arc of standard Petri Nets. Output gates are drawn as triangles with their flat side connected to an activity or a case and only have functions. They define the marking changes that occur when the related activity completes. In Fig. 2.1 G2 is an output gate.

SANs allow one to build performability models. Performability copes with the integration between performance and dependability analysis of systems (Meyer, 1995). A performability model is defined through a probabilistic class of measures which allows the evaluation of the systems capability to perform in presence of faults when safety and/or dependability requirements have to be satisfied.

Often, the dynamic behaviour of the system can be described by a stochastic process, called *base model* of the system. The *base model* could be a non-Markovian process or it could not be analytically solved. In this case it could be used as a simulation model. Performability variables can be defined on a SAN model leading to an extension of the "reward models" (Ciardo, 1993; (Howard, 1971; Qureshi, 1994), which consist of a stochastic process, a reward structure and a performance variable defined on them.

There are two types of rewards: *rate* rewards are associated with the marking of a SAN and they are related to the time spent in a state; *impulse* rewards are associated with each state change and they are related to activity completions.

Enabling rules and an exhaustive definition of SANs structure and behaviour can be founded in (Meyer, 1985).

The SAN-based approach we apply here builds a composed model from SAN submodels: the model of the system consists of a performance submodel and a structure submodel which are joined over a set of common places.

This approach is defined and described in (Meyer, 1985; Sanders, 1991) The structure submodel must contain all the activities which represent modification in the system due to a change in its structure.

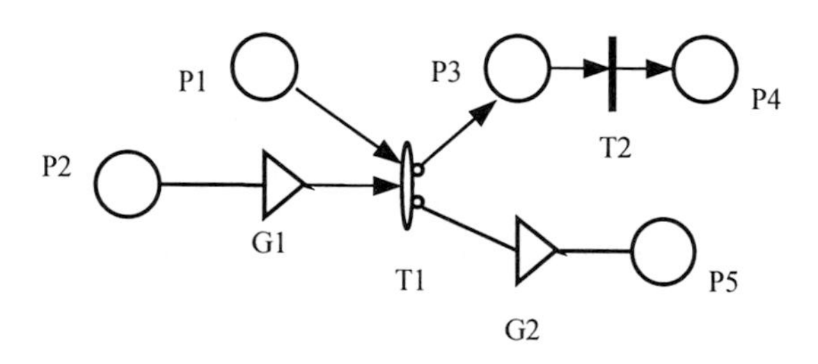

Fig. 2.1: SAN components

The performance submodel must contain all the activities which represent modification in the internal state and environment of the system, excluding the above mentioned structure related activities. The base model is correctly defined in term of the state behaviour of the structure submodel and the reward rates associated to the performance submodel iff the SAN satisfies four conditions (Meyer, 1985) informally stated below:

- the state transition set of both the structure and performance submodels contains a finite number of distinct states;
- the marking of a common place is not modified by the completion of any activity of the performance submodel;
- no activity of the structural submodel depends, either directly or indirectly, upon a marking of a place belonging to the performance submodel;
- the performance submodel reaches the steady-state conditions between the occurrence of structure related timed activities.

3 Building Models of a Voting System

The system we deal with is a classic industrial electronic control system which consists of one module replicated three times, (instances are named A, B and C), a recovery-from-repair unit, and a mechanism implementing a voting algorithm over the A,B,C modules. The replicated modules perform the same data processing and produce an output. A voting algorithm checks for the coherence among the results of the processing and states if the output is correct. It assumes that the result is correct if at least two modules have produced the same output, and it resets the system accordingly.

These systems are known as *M/N* voting systems, where *M* is the minimum number of outputs nedeed to be equal and *N* is the number of the replicated units. We consider a *2/3* voting system. The final state (of an *M/N* system) can be *safe* or *unsafe*: the safe state includes both the case of correct output and the case of safe resetting. Moreover, the system has a self-exclusion and recovery mechanism, i.e. if one of the replicated modules fails, a given percentage of faults can be detected and a repair action can be taken. Meanwhile, the system behaves like a *2/2* voting system. It is possible which a undetected anomaly occurs that causes coherent yet wrong results, for example the three outputs are equal, but wrong. In this case the system enters into an unsafe state.

3.1 The Event-driven GSPN Model

From the previous description five components are identified: the voting unit, the recovery unit and the three replicated modules; in the following they will be respectively referred to as VT, RP, AM, BM, and CM. Each component is informally described by means of a list of the events in which it can be involved and a list of its feasible states. Events are not listed according to any precedence order.

They are merely all the events in which each component of the system can be involved. The events set of each component formally corresponds to the alphabet α(P) of its related CSP process *P*. Here we name the CSP processes by means of the component identifiers introduced above, thus the CSP specification of the *2/3* voting system consists of the five processes VT, RP, AM, BM and CM.

For brevity's sake in the following we show the application of the event driven GSPN-based approach described in Section 2.1 only through the replicated module, whose list of events is in Table 3.1

The elements of the alphabet of a process are named following the event identifiers introduced in the first column. For instance, the alphabets of AM are:

$\alpha(AM)= \{E1S_A, E2S_A, E3S_A, ..., E11S_A, E12S_A\}$

Note that the events have been subscripted to emphatise that they describe the same behaviour over the three replicated modules but they belong to different alphabets. Each component of our system is a non-terminating process: it is activated once and immediately returns to its initial state after a finite sequence of state transitions.

Table 3.2 summarises both the informal description and formal definition of the states of the replicated module. A feasible state *S* of a process *P* is defined by means of the subset of the events which cause the transition of the related component in the state *S*.

Event name	**Description**
E1S	Process data
E2S	Self-test
E3S	Failure occurrence
E4S	No failure occurrence
E5S	Failure detection
E6S	No failure detection
E7S	Send the auto-exclusion signal to VT
E8S	Receive a repair-done signal from RP
E9S	Send data message to VT
E10S	Reset to restart computation (normal)
E11S	Reset to restart computation (after system failure)
E12S	Reset to restart computation (after module failure)

Table 3.1. Events of the replicated module.

State name	**Description**	**Definition**
compute	Processing data (initial state)	$<> \vee s' \in \{E10S,E11S,E12S\}$
check	Performing the self-test	$s' \in \{E1S\}$
failstate	Detecting the failure	$s' \in \{E3S\}$
endcheck	Evaluating the check results	$s' \in \{E2S\}$
next1	Resetting after a normal execution	$s' \in \{E9S\}$
next2	Resetting after a repair action	$s' \in \{E8S\}$
send.failsig	Sending the failure signal	$s' \in \{E5S\}$
send.msg	Sending data	$s' \in \{E4S,E6S\}$
send.repsig	Sending the repair request	$s' \in \{E7S\}$

Table 3.2. States of the replicated module

For example, the definition of the state **compute** on the first row of Table 3.2, says that either **compute** is the initial state at the activation of the process (<> is the empty trace) or it is determined by the occurrence of one of the events s' belonging to the subset {E10S,E11S,E12S}. Internal and external assumptions are the last information of the system informal description needed to derive the formal trace based specification of the processes.

Internal assumptions define the precedence constraints between events that belong to the same alphabet and they are formally expressed by specifying the precedence relations between states. Table 3.3 summarises the informal description of the internal constraints over the replicated module and their related formal definitions (assumptions). For example, the internal assumption *intS3* says that the event E3S occurs (i.e. the replicated component enters into the state **failstate** if the previous state was **endcheck**, i.e. if the previous event s' was E2S (see Table 3.2). This describes a precedence constraint between E2S and E3S.

External assumptions impose constraints related to events which belong to different alphabets (i.e. involving different components) and this means that they describe the dependences of the state of a component from the state of other components. Besides precedence constraints, external assumptions also describe constraints that can not be formally expressed in terms of traces, such as timing constraints involving the concept of simultaneity or overlapping in time of states and events.

External assumptions are introduced directly on the equivalent Petri subnets during their integration. The problem we deal with in this paper requires the consideration of five external assumptions that we do not describe for brevity's sake. The reader can refer to (Mazzeo,) for a detailed discussion about the external assumptions.

Once the internal assumptions have been defined on the states of the process P, the set *traces(P)* of the feasible traces of P are determined. The set of the feasible traces of a process is built by means of an algorithm which uses the formal definitions of the states and the internal assumptions to build a state matrix and an event matrix. Table 3.4 and Table 3.5 show the resulting matrices of the AM component.

The columns of the events matrix of AM describe the possible behaviours of AM from the initial state **compute** to the states **next1**, **next2** and **send.repsig** from which it returns to its initial state. We call *basic sequences* such traces. The set *traces(AM)* is not finite because AM is a non-terminating process, but it can be expressed as a recursive process (in Hoare's sense) starting from its basic sequences. Let *bs1* and *bs2* be two basic sequences, we denote by:

traces(P) = $\{\text{bs1},\text{bs2}\}^*$

the trace set of a non-terminating process P which only behaves according to the sequence *bs1* or *bs2*. We have:

$traces(AM)=\{<E1S_A,E2S_A,E3S_A,E5S_A,E7S_A,E8S_A,E11S_A>,<E1S_A,E2S_A,E3S_A,E5S_A,E7S_A,E12S_A>,$
$<E1S_A,E2S_A,E3S_A,E6S_A,E9S_A,E10S_A>,<E1S_A,E2S_A,E4S_A,E9S_A,E10S_A>\}^*$

Assumptio	Description	Definition
intS1	The unit is in the state **compute** iff has just been actived or the previous state was next1 or next2 or send.repsig	$\{<>\backslash\} \vee (s'=E10S \Leftrightarrow next1(s'')) \vee (s'=E11S \Leftrightarrow next2(s'')) \vee (s'=E12S \Leftrightarrow send.repsig(s''))$
intS2	The unit is in the state **check** iff the previous state was compute	$s'=E1S \Leftrightarrow compute(s'')$
intS3	The unit is in the state **failstate** iff the previous state was endcheck	$s'=E3S \Leftrightarrow endcheck(s'')$
intS4	The unit is in the state **endcheck** iff the previous state was check	$s'=E2S \Leftrightarrow check(s'')$
intS5	The unit is in the state **next1** iff the previous state was send.msg	$s'=E9S \Leftrightarrow send.msg(s'')$
intS6	The unit is in the state **next2** iff the previous state was send.repsig	$s'=E8S \Leftrightarrow send.repsig(s'')$
intS7	The unit is in the state send.**failsig** iff the previous state was failstate	$s'=E5S \Leftrightarrow failstate(s'')$
intS8	The unit is in the state **send.msg** iff the previous state was **endcheck** or failstate	$(s'=E4S \Leftrightarrow endcheck(s'')) \vee (s'=E6S \Leftrightarrow failstate(s''))$
intS9	The unit is in the state **send.repsig** iff the previous state was send.failsig	$s'=E7S \Leftrightarrow send.failsig(s'')$

Table 3.3. Internal assumptions of the replicated module

compute	-	-	-
next2	-	compute	compute
send.repsig	compute	next1	send.repsig
send.failsig	next1	send.msg	send.failsig
failstate	send.msg	failstate	failstate
endcheck	endcheck	endcheck	endcheck
check	check	check	check
compute	compute	compute	compute

Table 3.4. State-Matrix

E11S	-	-	-
E8S	-	E10S	E12S
E7S	E10S	E9S	E7S
E5S	E9S	E6S	E5S
E3S	E4S	E3S	E3S
E2S	E2S	E2S	E2S
E1S	E1S	E1S	E1S

Table 3.5. Event-Matrix

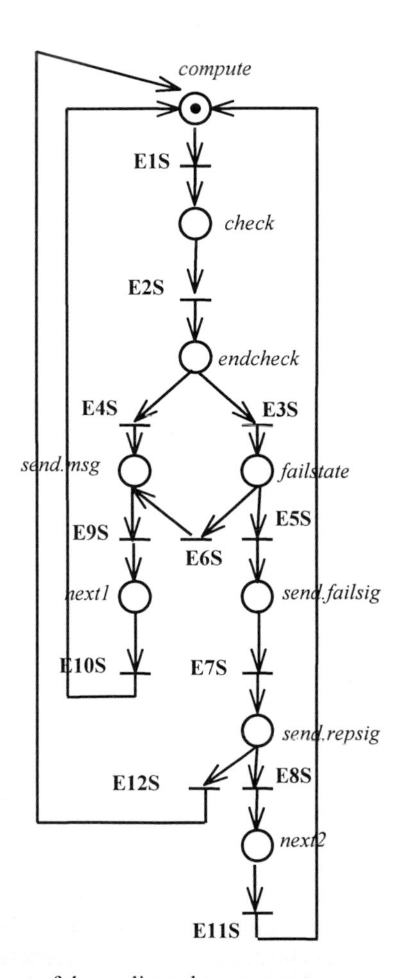

Fig. 3.1. PN equivalent subnet of the replicated component

The trace set is the basis on which the PN subnet of the process is built by means of the automatic translation rules introduced in (Mazzeo, 1997). It is called PN *equivalent* subnet since it describes the same feasible behaviours of the process which are described by its trace set.

In the PN equivalent subnet of a process, the events correspond to transitions and states correspond to places. Fig. 3.1 shows the resulting PN specification of the AM module from its trace set where the four basic sequences are easily recognisable by following the paths of the corresponding transitions from the place **compute** to the places **next1**, **next2** and **send.repsig**.

The PN equivalent subnets of RP and VT are also obtained by applying the steps described above; for brevity's sake they are not shown but all the subnets must be integrated to build the PN system specification. Since there are external assumptions to be introduced in the system specification, the integration will be performed in two steps:

- introduction of the proper PN subnets between processes to model the constraints expressed by the external assumptions;
- integration of the PN equivalent subnets to model the concurrent behaviour of the system;

These two steps are also accomplished by applying automatic rules, respectively external assumptions translators and composition rules. They are described in (Mazzeo,). For completeness, Fig. 3.2 shows the PN *skeleton* resulting from the integration phase after the temporisation. It is a GSPN net derived from the PN *skeleton* by changing the transitions which represent time consuming events into timed exponentially distributed transitions. Information about these events must be supplied by the informal description of the system together with their firing rates. The same holds for the probabilities associated with the immediate transitions which represent the alternatives of a choice.

3.2 The SAN-based Model

Fig. 3.3 and Fig. 3.4 respectively show the performance and structure submodels of the voting system, built according to the SAN-based approach described in Section 2.2. We refer to these nets as "system" and "structure" respectively. In the following activity names are uppercase and place names are bold. The set of common places consists of the places named **A, B, C, sync** and **ok**.

The timed activities TA, TB, TC in Fig. 3.3 represent the processing, whereas I5 and I6 represent idle times at the start and end of the processing phase. The timed activity TK of the structure net of Fig. 3.4 represents the integrity control by the voter component, TRp represents the repair action in case of a transient fault and TRs models the reset process in case of total failure; TV1 and TV2 model the comparison of the outputs in case of normal behaviour of the system and in case of a self-exclusion, respectively.

The common places **A, B**, and **C** model the availability of the related module if they are marked by a token, otherwise it means that the related module is faulty (it has excluded itself). A token in the common place **ok** means that the system is working, whereas the common place **sync** is used to synchronise the behaviour of the two subnets: a module can process a message only after its availability has been checked, i.e. after the completion of TK. Thus the *input gate* G1 of the system net (Fig. 3.3) enables I6 if there is a token in **ok** and no token in **sync**, whereas the function of the *output gate* G2 depends on the marking of **A, B, C** since the completion of I6 must enable only the processing activities related to available modules. The structure net (Fig. 3.4) is always enabled. Upon the completion of TK one of the following *cases* can be verified:

1. no faults have occurred, a token goes into the place **duesutre** which represents the *2/3* voting;
2. one module (there is one *case* for each module) is faulty, the function of the corresponding output gate (G8, G7, or G6) removes the token from the corresponding common place (**A, B** or **C**), puts a token into the place **repair** (it enables TRr) and a token into the place **duesudue** which represents the *2/2* voting. Upon the completion of TRr the repaired module is made available (i.e. the related common place is marked again);
3. more self-exclusions have occurred, a token goes into the place **stop** which models the total blocking of the system. The reset activity TRs is enabled, its completion takes back the system to the initial configuration.

Activities TV1 and TV2 are in mutual exclusion. If TV1 is enabled the system performs a *2/3* voting, so that five *cases* are possible:

1. the three outputs differ, a token goes into the place **stop** and causes the blocking and resetting of the system;
2. the three outputs are equal and correct, a token goes into the place **SAFE** which models the safe state of the system. The instantaneous activity I3 puts a token into the place **ready** so that a new execution of the structure net may start;
3. the three outputs are equal but wrong, in this case the voting fails since it is not possible to detect the error, a token goes into the place **UNSAFE** which models the unsafe state of the system. The instantaneous activity I4 puts a token into the place ready so that a new execution of the structure net may start;
4. one output differs fron the others two outputs which are correct. The net behaves as in case 2;
5. one output differs fron the others two outputs which are wrong. The net behaves as in case 3.

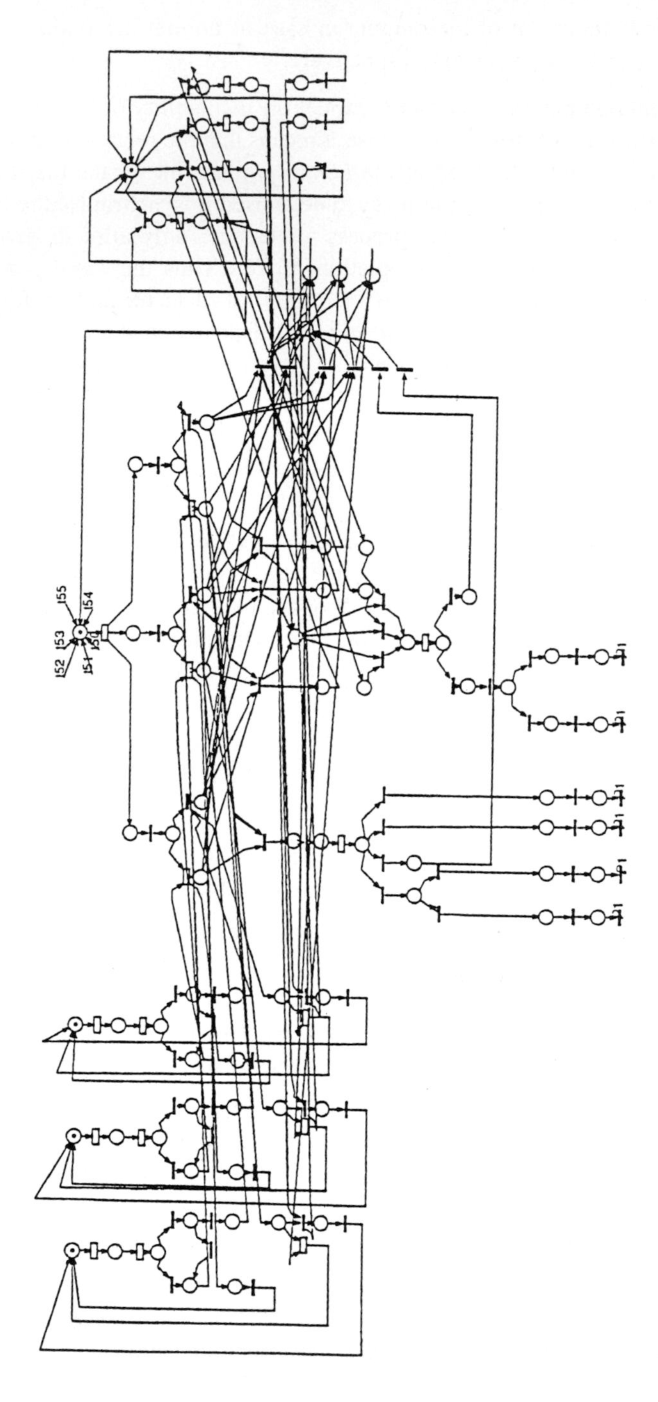

Fig. 3.2. GSPN model of the system

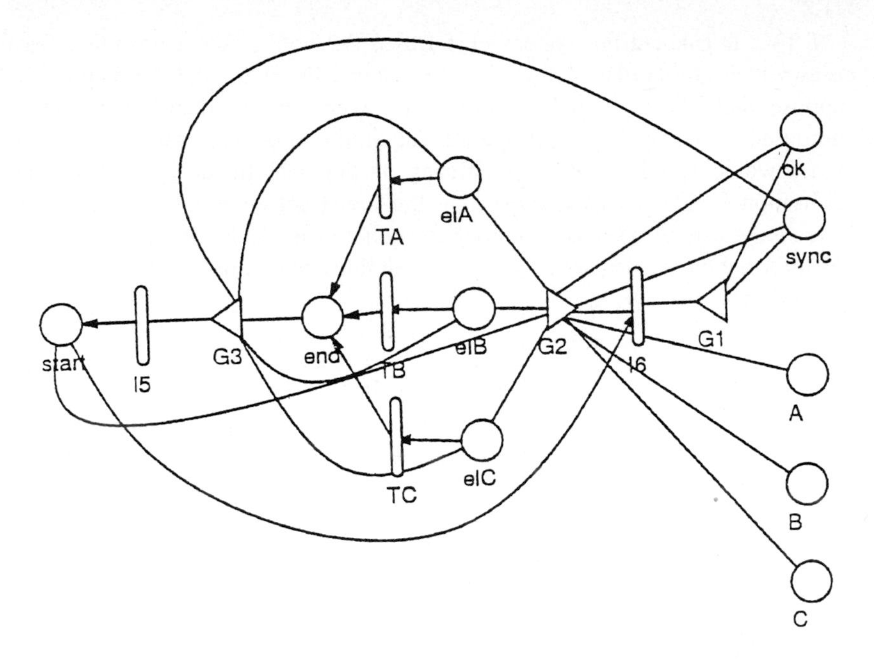

Fig. 3.3. System net

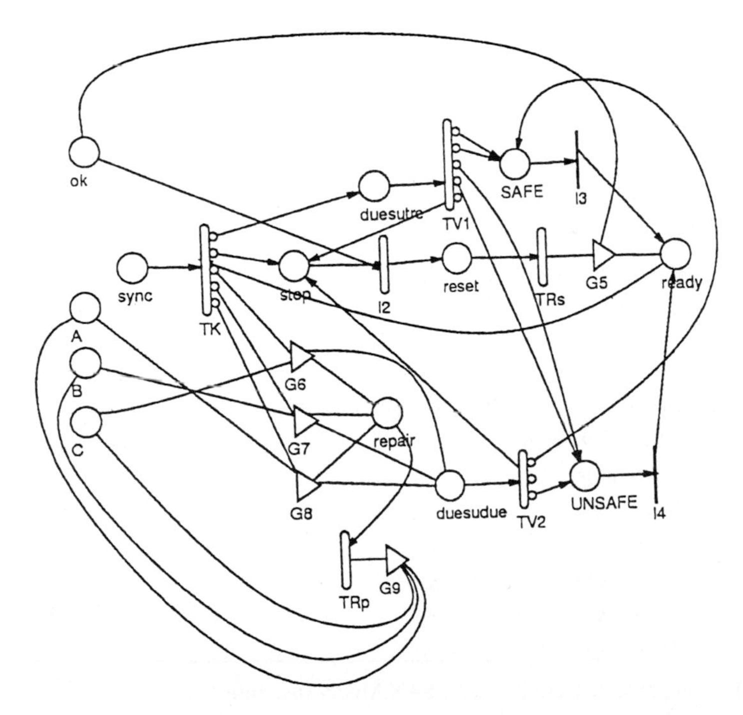

Fig. 3.4. Structure net

If TV2 is enabled the system performs a *2/2* voting: there are three feasible *cases* which are similar to the first three items listed for the *2/3* voting. If we suppose that all the timed activities are exponentially distributed, the SAN composed model verifies the properties required in order to obtain a Markov *base model* which can be solved by means of the analytic solvers of UltraSAN (Couvillon, 1991). If we suppose that the timed activities are deterministic the *base model* can be used as a simulation model. The Tables from 3.6 to 3.9 give the information needed to complete the definition of the structure and system submodels.

Place	*Marking*
A	1
B	1
C	1
ok	1
ready	1
start	1

Table 3.6. Initial Markings for the composed SAN Model

Gate	*Definition*
G5	*/*reset the structure model after termination or stop */* *MARK(ready)=1;* *MARK(ok)=1;*
G6	*/* switch on 2/2, enable the performance model and repair C */* *MARK(duesudue)=1;* *MARK(repair)=1;* *MARK(C)=0;*
G7	*/* switch on 2/2, enable the performance model and repair B */* *MARK(duesudue)=1;* *MARK(repair)=1;* *MARK(B)=0;*
G8	*/* switch on 2/2, enable the performance model and repair A */* *MARK(duesudue)=1;* *MARK(repair)=1;* *MARK(A)=0;*
G9	*/* reset a module after repair */* *if (MARK(A)= =0) MARK(A)=1;* *if (MARK(B)= =0) MARK(B)=1;* *if (MARK(C)= =0) MARK(C)=1;*

Table 3.7. Output Gate Definitions for SAN Model *structure*

Gate	*Definition*
G1	<u>*Predicate*</u> */* The system performance model is enabled */* *((MARK(ok)= =1) && (MARK(sync)= =0))* <u>*Function*</u> */* do nothing */*
G3	<u>*Predicate*</u> */* The messages have been elaborated */* *((MARK(end)>0) && (MARK(sync)= =0) && (MARK(elA)= =0) && (MARK(elB)= =0) && (MARK(elC)= = 0))* <u>*Function*</u> */* Reset the system after messages elaboration */* *MARK(end)= 0;* *MARK(sync) = 1;*

Table 3.8. Input Gate Definitions for SAN Model *system*

Gate	*Definition*
G2	*/*the active modules only */* *if ((MARK(A)= =1) && (MARK(B)= =1) && (MARK(C)= =1))* *{* *MARK(elA)++;* *MARK(elB)++;* *MARK(elC)++;}* *else if ((MARK(A)= =0) && (MARK(B)= = 1) && (MARK(C)= = 1))* *{* *MARK(elA)=0;* *MARK(elB)++;* *MARK(elC)++;}* *else if ((MARK(B)= = 0) && (MARK(A)== 1) && (MARK(C)= = 1))* *{* *MARK(elA)++;* *MARK(elB)=0;* *MARK (elC)++;}* *else if ((MARK(C)= = 0) && (MARK(A)= =1) && (MARK(B) == 1))* *{* *MARK(elA)++;* *MARK(elB)++;* *MARK(elC)=0;}* *else {* *MARK(start)=1;}*

Table 3.9. Output Gate Definitions for SAN Model *system*

4 Reasoning about Modelling

An immediate remark on the modelling exercise which we have described is that by applying two PN formalisms we have obtained two very different models of the same system. Indeed, they are so different that the results of their analysis cannot be compared! The fact matter is that we implicitily started from two different sights of reality. The first one is "behavioural" and "compositional": it describes the system as a set of distinct, interacting objects (components, processes etc.) whose behaviours can be defined through states, events, constraints, durations of actions. The second one looks at the system as a whole and focuses on its properties (fault tolerance and performance in our case).

Thus the **first lesson learnt** through this exercise in "modelling comprehension" is that it is necessary to become aware of the viewpoint we are consciously or unconsciously adopting, and decide if it suits the goals we want to reach by modelling. This step comes before any decision about the abstraction level of the models we build, because it determines the means and the techniques to be used.

A second remark is that the behavioural event-driven approach is very general since it can be used as a ground for a wide class of process modelling: every where a process can be described by means of states and events. The resulting model of the system depends on which set of translation rules is defined. It could be a transformation from traces to PN subnets as well as traces to timed or untimed automata, or traces to some event-driven metalanguage...

In this sense the event-driven approach fits the first three criteria of Section 2 quite well. Thus the **second lesson learnt** is that we need modularity in process modelling: it should consist of distinct phases, each of them should be implemented by a "black box" that can be replaced according to needs.

On the other hand, the SAN-based approach copes better with the measures integration (see point 2b in Section 2) and the complexity of the resulting models (point 4). Indeed it allows to define classes of variables which suit the specification of different measures of performance, dependability, performability and models complex behaviours in a very concise way. Moreover, it may of benefit from the reduced base model construction methods to generate the state space of the stochastic process.Thus the **third lesson learnt** is that we would be able to integrate the two modelling approaches in order to combining the advantages of both.

This is a very important issue. To be able to integrate different formalisms and techniques means to be able to cope with complexity and provide the designer with a systematic (i.e. non "hand-made") and automatic construction of the system models.

There is a long distance to cover in this way. This is one of the challanges in modelling and analysing complex systems today.

References

Ajmone-Marsan, M., Chiola, G. (1987): "On Petri Nets with Deterministic and Exponential Transition Firing Times," Lecture Notes in Computer Science, Vol. 266 [24], 1987, pp. 132--145.

Bowen and Hinchey, M.G. (1995): "Seven More Myths of Formal Methods", IEEE Software, Vol.12, n.4, July 1995.

Chiola (1991): "GreatSPN 1.5 Software Architecture"; Proc. of the 5th Int. Conf. on Modelling Techniques and Tools for Computer Performance Evaluation, Torino,13-15 Feb. 1991, pp. 117-132.

Ciardo, J.K. Muppala, and K.S. Trivedi (1989): "SPNP: Stochastic Petri Net Package", in Proc. Int. Workshop on Petri Net and Performance Model, pp.142-150, Kyoto, Japan, IEEE CS Press, Dec. 1989.

Ciardo, A. Blakemore, P.F. Chimento, J.K. Muppala, and K.S. Trivedi (1993): "Automated generation and analysis of Markov reward models using Stochastic Reward Nets", in: Linear Algebra, Markov Chains, and Queueing Models, IMA Volumes in Mathematics and its Applications, Vol 48, pp. 145-191, Heidelberg, Germany, Spring-Verlag 1993.

Couvillon et al. (1991): "Performability Modeling with UltraSAN", IEEE Software, Vol. 8, No. 5, pp. 69-80, Sept. 1991.

Hall (1990):"Seven Myths of Formal Methods", IEEE Software, Vol.7, n.5, September 1990.

Hoare (1985): "Communicating Sequential Processes", Prentice-Hall, Englewood Cliffs, NJ, 1985

Howard (1971): "Dynamic Probabilistic Systems", in: Semi-Markov and Decision Processes, vol.II, New York, Wiley, 1971.

Lindemann (1992): "DSPNexpress: A Software Package for the Efficient Solution of Deterministic and Stochastic Petri Nets", in Proc. 6th Int. Conf. on Modeling Techniques and Tools for Computer Performance Evaluation, pp. 15-29, Edinburgh, Great Britain, 1992.

Mazzocca, S. Russo, V. Vittorini (1997a): "Integrating Trace Logic and Petri Nets Specifications", in: Proc. HICSS-30, Hawaii Int. Conf. on System Sciences (Software Technology Track), IEEE CS Press, vol.1, pp.443-451, 1997.

Mazzocca, S. Russo, V. Vittorini (1997b): "Formal Methods Integration for the Specification of Dependable Distributed Systems", Journal of System Architecture, vol.43, n.10, pp.671-685, September 1997.

Mazzeo, N. Mazzocca, S. Russo, V. Vittorini (1997): "A Systematic Approach to the Petri Net Based Specifications of Concurrent Systems", Real Time Systems, vol.13, pp.219-236, November 1997.

Mazzeo, N. Mazzocca, S. Russo, C. Savy, V. Vittorini; "Formal specification of concurrent systems: a structured approach", accepted for publication in The Computer Journal.

Meyer, A. Movaghar, W.H. Sanders (1985): "Stochastic Activity Networks: Structure, Behavior and Application", in Proc. of the Int. Conf. on Timed Petri Nets, Torino, Italy, pp. 106-115, 1985.

Meyer and W.H. Sanders (1993): "Specification and Construction of Performability Models", in: Proc. of the 2nd. Int. Work. on Performability Modeling of Computer and Communication Systems, Mont. Saint-Michel, France, June 28-30, 1993.

Meyer (1995): "Performability Evaluation: Where It Is and What Lies Ahead", in: Proc. of the IEEE Int. Computer Performance and Dependability Symposium, pp.334-343, Erlangen, Germany, 1995.

Qureshi and W.H. Sanders (1994): "Reward Model Solution Methods with Impulse and Rate Rewards: An Algorithm and Numerical Results", Performance Evaluation vol.20, pp. 413-436, 1994.

Sanders and J.F. Meyer (1986): "METASAN: A Performability Evaluation Tool Based on Stochastic Activity Networks", in: ACM-IEEE Fall Joint Computer Conference, pp. 807-816, Dallas, TX, 1986.

Sanders and J.F. Meyer (1991): "A Unified Approach for Specifying measures of Performance, Dependability and Performability", in: Dependable Computing and Fault Tolerant Systems, vol.4, Springer-Verlag, pp. 215-237, 1991.

High-Level Modelling of Development Processes

Ansgar Schleicher

Department of Computer Science III
Aachen University of Technology, Ahornstraße 55, 52074 Aachen, Germany
e-mail: schleich@i3.informatik.rwth-aachen.de

Abstract. *In this paper a high-level, user-friendly, yet executable process modelling language is introduced. The modelling of a process' structure is supported through ER-like diagrams, whilst its behaviour can be specified in terms of state transition diagrams with guarded transitions, refined operations, and event handlers. It is enforced that process models consist of well separated process fragments which are subject to reuse. The enviroment supports the implantation of knowledge gathered during process execution into the process model to enhance assistance for the project management.*

Keywords. *process modelling, process reuse, dynamic task nets*

1 Introduction

The development of process modelling tools and languages has received increased attention in recent years. As a result there exist high-level modelling languages with restricted simulation and process enactment capabilities (cf. Rombach, 1995). On the other hand, formal process programming languages with well-defined execution semantics were developed (cf. Cen, 1994; Sutton, 1995) which lack the expressiveness of their counterparts.

To reach acceptance for a process-centered environment, I believe that both levels are required:

- The level of *formalised, executable languages* with sufficient complexity to express all situations that may occur during process enactment.
- The level of *user-friendly, high-level modelling languages* that abstract from the complexity of the afore mentioned formalised languages in order to increase usability, to offer semantic analyses, to shorten the process development cycles and to make reuse of models possible.

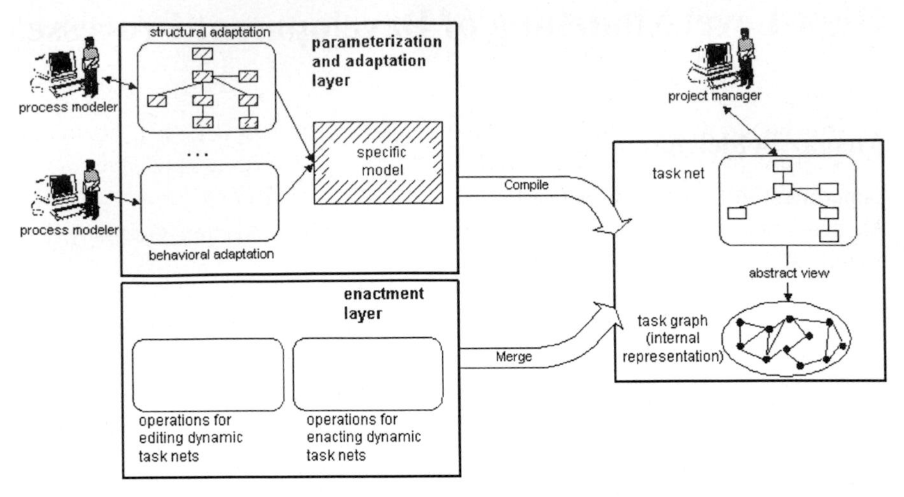

Fig. 1: Overview of the administration environment MADAM.

The process modelling environment MADAM (**M**anagement and **Ad**aptation of **A**dministrational **M**odels) which is presented in this paper offers both: High-level graphical modelling languages enriched with formalised textual languages on an abstract level to maintain user-friendliness. The MADAM-environment's structure is presented in Fig. 1. It consists of the following parts:

- The top left corner represents the *parameterisation and adaptation environment* that offers graphical and textual modelling languages to support the process modeller in defining the structure and the behaviour of a specific development process on the type level.
- The bottom left corner represents a *semantically rich generic process engine.* It provides all necessary mechanisms to edit and enact dynamic task nets which allow for the instance-level modelling of processes (cf. section 2).
- The execution semantics of processes modelled in the adaptation environment are defined through a transformation into the internal languages of the process engine. Linking the transformed model and the process engine together results in a *domain specific administration environment.* It is constrained by the process model and offers uniform editing and execution operations to the project manager, who is responsible for guiding and monitoring a process at runtime.

This paper will focus on the adaptation environment which supports process modellers with varying levels of expertise. Highly customised models can be defined by an experienced user with a set of special purpose textual and graphical languages. These allow for the specification of both general and context specific reusable process model fragments. Unexperienced modellers shall later be enabled to simply combine and refine these fragments to model their development processes on a high level of abstraction.

The remainder of this paper is structured as follows: Section 2 contains a short introduction of dynamic task nets, while sections 3 and 4 deal with structural and behavioural process modelling, respectively. Section 5 provides an overview of related work. A conclusion and my visions for future work are given in section 6.

2 Dynamic Task Nets

As a means to describe and enact processes on the instance level, the DYNAMITE (**Dynamic** **T**ask Nets) model has been developed (cf. Heimann, 1996; Heimann, 1997). It provides mechanisms to model creative development processes and their inherent dynamics which can be subsumed as follows:

- *Forward development*: The process structure may be dependent on the product structure which evolves gradually.
- *Concurrent Engineering*: To shorten development cycles, concurrent engineering methods may be applied to the domain. This results in highly dynamic work contexts for the participating developers since they are often working on preliminary versions of input documents.
- *Feedback*: Errors may be detected in later phases of the development process, which leads to changes to results of earlier phases. The consequences of feedback cannot always be predicted as even the process model itself might be affected.

Dynamic task nets consider these dynamics as they may evolve at runtime and can be executed and edited in an intertwined fashion without interruption of the process or the risk to introduce inconsistent states.

Modelling a process as a dynamic task net means splitting the overall process into subprocesses and clarify their respective relationships and behaviour. A task consists of the description of what is to be done, the task interface, and of the description of how it is to be done, the task realisation. Dynamic task nets may contain complex and atomic realisations, the first of which consists of a net of subtasks whose results lead to the fulfilment of the superior task.

Three kinds of task relations are provided by dynamic task nets:

- *Control flow* relations introduce a temporal ordering on tasks and are similar to ordering relations in PERT-charts.

- *Cooperation* relations mark task sets which cooperate with regard to time and contents to fulfil a common goal. They allow to model peer-to-peer cooperation.
- *Feedback* relations are always directed oppositely to control flow relations and are introduced to mark iteration or exception steps of a process.

Outputs to be produced and inputs needed by a task are modelled as parameters of the task's interface. Between parameters data flow relations may be introduced which indicate the flow of documents through a task net.

Fig. 2 gives a tiny example of a dynamic task net's structure and its execution. At the beginning of a software project little is known about the development process. A *design* and an *installation* task are introduced into the net, while the intermediate structure remains unspecified as it is dependent on the design document's internal structure (part i). As soon as this is produced, the complete structure of the task net can be specified (part ii).

Output documents of tasks are versioned and can be released on a task-by-task basis. Part iii) shows how a new version of a design document is produced after feedback occurred from the task implementing module B to the design task. This new version is at first only selectively released.

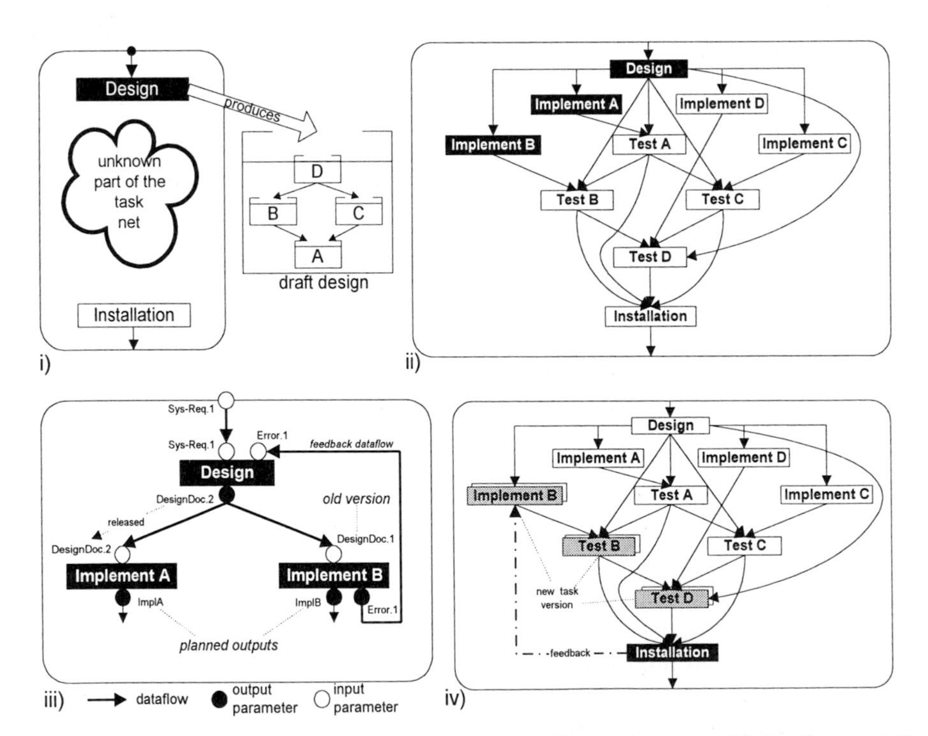

Fig. 2: Examples of dynamic task nets: **i)** initial net, **ii)** completed net, **iii)** detail view with feedback to active task, **iv)** new task versions after feedback to terminated task

If feedback occurs to terminated tasks, these are not reactivated. Rather, a new task version is derived from the old one and the old work context is reestablished (part iv). Using derivation rather than reactivation of tasks results in the creation of a process execution's trace, which can later be examined by the project manager to find indications for misplannings.

The example contains information about interface and parameter types, different ways of reacting to feedback or releasing document revisions to succeeding tasks. All this structural and behavioural information is domain specific and cannot inherently be part of the DYNAMITE model. Rather it has to be specified using an adaptation environment.

3 Structural Adaptation

To support the project manager in handling a process, a process modeller can define a process' structure and behaviour as a process schema on the adaptation layer. This section introduces MADAM's capabilities for modelling process structures.

The following requirements must be met when designing a modelling language to offer a process modeller adequate support:

- A model must be *intuitively comprehensible* and should therefore be visualised graphically.
- The *effort* to define a task schema should be reduced to a *minimum*.
- The creation of *reusable process fragments* should be possible so that evolutionary modelling of process schemas through the reuse and refinement of stored fragments is enabled.
- The modelling language should be designed independently from a fixed generic model to reach a certain degree of *robustness* against changes of the latter.

To define the syntax of MADAM's language for structural process models, a meta scheme has been developed in UML (cf. Fowler, 1997) which is illustrated in Fig. 3. According to this meta-scheme, process schemes are separated into a type and a composition layer. In the type layer the context free properties of task types are specified, while contextdependent properties are fixed in the composition layer.

The type layer thus contains declarations of new task interface types together with their possible parameter profile and the fitting available task realisation types. Parameter types may be marked with optional, obligate and set cardinalities.

Context free behaviour of interface types is equally defined on the type layer but will be described in the next section.

The composition layer allows for the definition of task net structures such that control and data flow between interfaces and their respective parameters are constrained.

The static semantics of this structural language have been formally defined in the specification language PROGRES (cf. Schürr, 1989) and are described in Krapp 1998. They shall not be explained in this paper. Rather, I would like to give an informal description of the associations and aggregations used in the meta-schema.

Type layer:

- *may-have* aggregations link parameter declarations to their interface type.
- *may-realise* aggregations form the set of possible realisation types for an interface type.

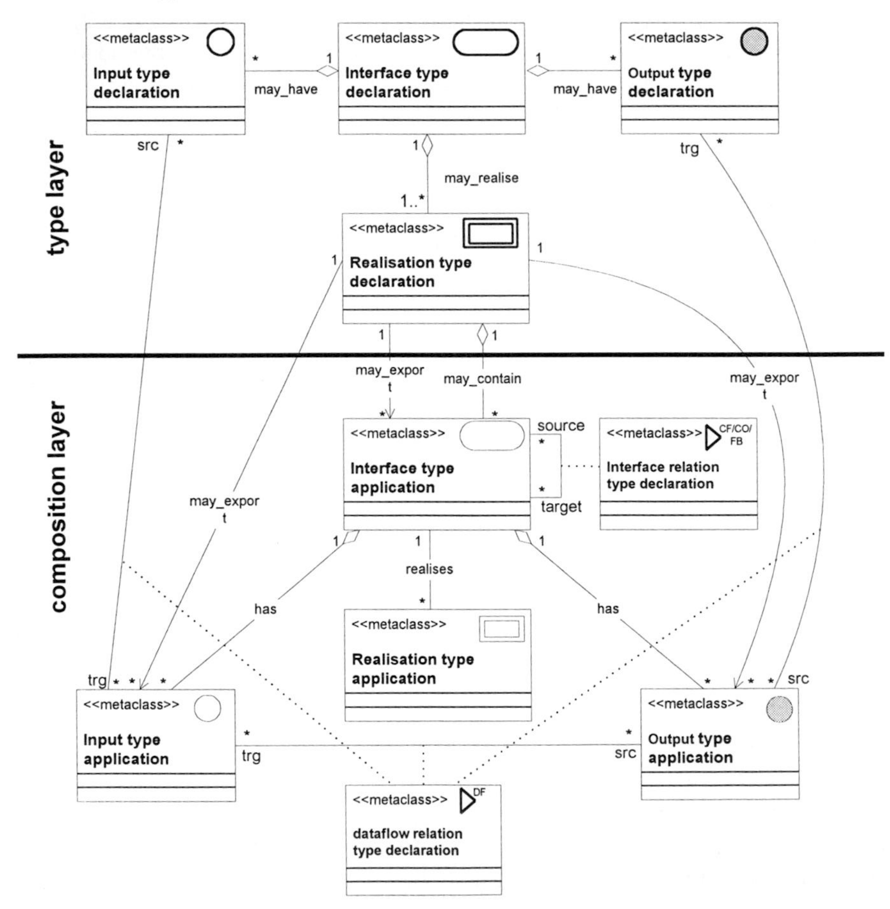

Fig. 3: Process meta-scheme

Intersection of type and composition layer:

- *may-contain* associations form the set of interface types that may be used to build a realising task net for a realisation type.
- *may-export* associations mark all interface- and parameter types that are visible outside of a realisation at enactment time.

Composition layer:

- *has* aggregations allow for the context sensitive limitation of an interface type's parameter profile.
- *realises* associations allow for the context sensitive limitation of an interface type's possible realisation types.
- *control flow*, *cooperation*, *feedback* and *data flow* association classes are the schematic counterpart of the instance level relations.

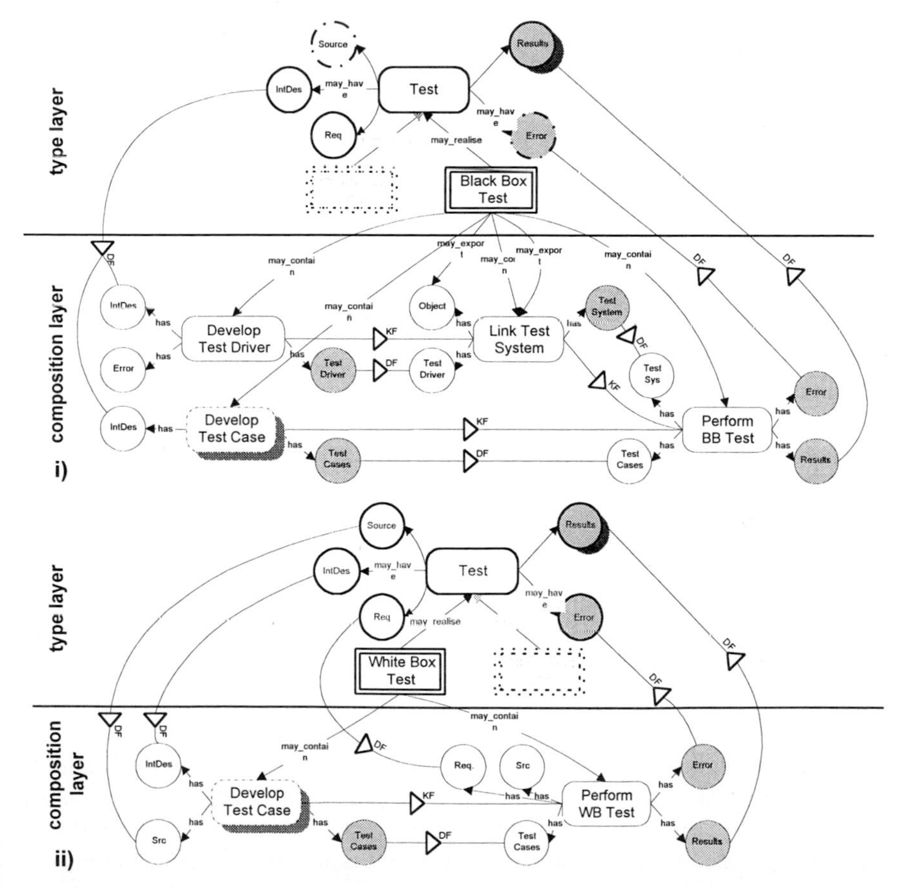

Fig. 4: Test interface type with **i)** black-box and **ii)** white-box realisation types. The type layer is defined only once. It is shown twice for better illustration purposes.

The proposed meta schema for process models has many inherent advantages. Interface types abstract from a multitude of possible realisation types. A project manager may decide at process enactment time which realisation type is appropiate. Through realises-associations the use of certain realising subnets can be prohibited in a specific context by the process modeller. This is useful if certain realisation types have proven less successful in this very context. In the field of archiving and reuse, the separation of type- and composition layer proves to be very efficient as it leads to distinct process model fragments. Besides reuse and modularization, this allows the distributed development of process models.

An example for a process model fragment defined with MADAM is given in Fig. 4. A test may alternatively be realised as a black- or a white-box test. The composition of task types, which are declared elsewhere, forms schematic task nets for the execution of a black-box or a white-box test.

To perform a black-box test, a *test driver* and various *test cases* have to be developed. Test cases may be developed parallel by a group of test engineers. The test driver is linked together with a module's object code to build an executable *test system* which is then fed with the mentioned test cases. The test results and possible execution errors form the result of the superior test task.

The shown white-box test is performed without computer support. Using the source code, a set of test cases is developed that covers all its conditional branches. These test cases are then applied to the code during a code review.

4 Behavioural Adaptation

MADAM offers sublanguages that allow to specify a process' behaviour during enactment through conditions, refined operations and event handlers.

4.1 Conditions

The execution semantics of dynamic task nets are based on communicating state machines. One state transition diagram is prescribed for all tasks (cf. Fig. 5). This may seem a little restrictive, but in allowing each task type to define its own state transition diagram would imply the need to model the interrelations between state transition diagrams for every pair of task types. Since this increases the complexity of modelling a process a lot, it would counteract my aim to simplify the modelling task. Tasks can be *defined, activated, suspended and replanned.* Every task can terminate in one of two final states, one of which marks its successful completion (*Done*), while the other one marks a task's failure (*Failed*).

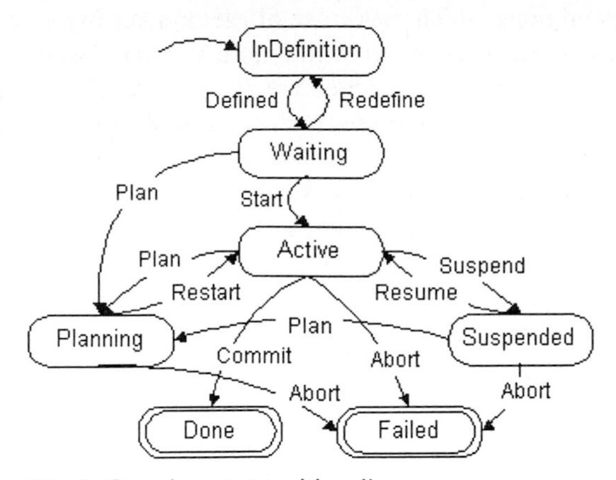

Fig. 5: Generic state transition diagram

The following standard behaviour is established with respect to the shown transition diagram:

- Transitions consider the states of preceeding and succeeding tasks as well as of their subordinate and superior tasks.
- Suspension or abortion of a task leads to suspension or abortion of all subordinate tasks.
- The enactment layer's generic operations on tasks are dependent on certain states of all participating tasks.

To model a task's specific behaviour, a condition can be attached to every transition of the state transition diagram. These conditions provide the means to influence the way a task net is executed. Different development policies like *simultaneous* and *sequential engineering* can be realised or certain states can be eliminated from the generic state transition diagram.

In order to simplify the definition of conditions, a specialised order of sorted logical language was developed which reflects the structure of a set of analysed conditions in example process definitions. Its basic building blocks are shown in Fig. 7. Through the use of selectors, qualified variables can be bound to a set of process objects. The resulting set can optionally be reduced to all objects of a specified type. The use of predefined, parameterised predicates and standard boolean operators enables the user to formulate restrictions on the bound variables. The generic model provides a standard set of predicates which can be enhanced by the process modeller.

Specialised conditions which make use of the optional type check are usually attached to one interface type only, while more generally formulated conditions may be kept in condition libraries which are available to all other projects. Fig. 6 shows an example condition for the start transition of the *perform test* task. It may be started as soon as the test system is available and some predecessing *develop test cases* task have been carried out so that some test cases can already be checked.

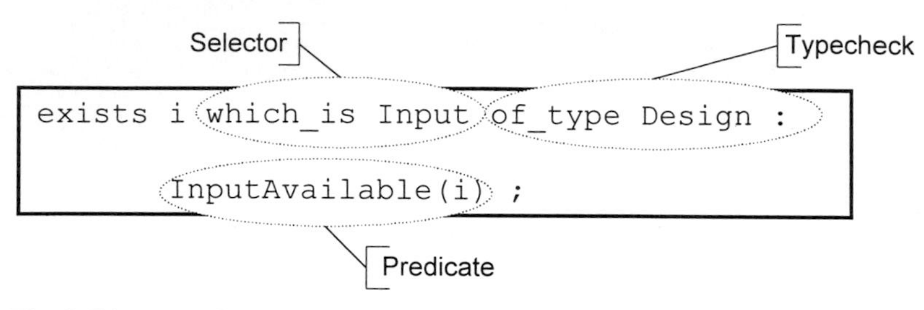

Fig. 7: Elements of a conditional expression

```
exists i which_is Input of_type Test_System :
  InputAvailable(i);
and
exists t which_is Predecessor of_type Develop_test_cases:
  t.state = Done;
```

Fig. 6: Example condition of the *perform test* task

4.2 Refined Operations

The enactment layer offers a uniform set of operations for editing and enacting dynamic task nets for all development domains. Each of these generic operations and every transition may be refined by the process modeller and enriched with additional functionality. MADAM offers a language to specify type specific execution patterns which may include:

- *Calls to other operations*, i.e. the inherent consumption of an input when the start transition is performed,
- language constructions to *operate on sets* of values,
- *conditional branching* to react to differing contexts,
- *nondeterminism* to allow the specification of alternatives which are executed if an action failed.

```
for all inp which is Input do
  try Consume(inp)
end;
If AllInputsConsumed(task) then
  generic_start(task)
else
  suspend(task)
endif;
```

Fig. 8: Refined operation for the *link test system* task

Fig. 8 shows the refined start transition of the *link test system* task. It is supposed to consume all inputs at start time because it is executed automatically. The *try*-statement avoids failure of the transition if an input could not be consumed. Instead, the *suspend* transition is activated immediately afterwards.

4.3 Context Free Event Handling

Every generic operation and transition sends out an event as soon as it is completed. The enactment layer offers an event/trigger mechanism that triggers the execution of event handlers. These event handlers enable process objects to react to actions performed in their individual context. This leads to a clean separation of proactive (refined operations, transitions) and reactive (event handling) behaviour. I identified two ways for handling events: context free and context sensitive event handling.

Context free event handlers are specified for an interface type and can only react to events raised by specific interface type operations. Thus all editing operations are excluded since they embed a process object into a certain context. A context free event handler inherits the event's parameter profile. Actions performed by the event handler may then operate only on these parameters and the interface type that the event handler has been attached to.

Even if the context is not known, at least the roles that a typed interface may play during process enactment are well known. Interfaces can play the role of parent interface to a realising subnet. They can be children to such a parent interface, thus part of a realising subnet. Inside of a subnet they can play the role of predecessor and successor to other interfaces with respect to the control flow relation. It is important to notice that instances of an interface type do not have to occur in all of these roles.

Yet, this knowledge enables the user to include an event's source relative position, with respect to the event handling interface, in a context free event handler. This is very important indeed, because interfaces should react quite differently to events raised by their parent, their children or their neighbours.

An example event handler for the *link test system* interface as shown in Fig. 9 illustrates these issues. A *link test system* task can automatically be started as soon as its preceeding task has been terminated successfully. It is important to react only to the *commit* event if it was sent out by a predecessor as a successor's commit should not trigger the activation of the *link* task.

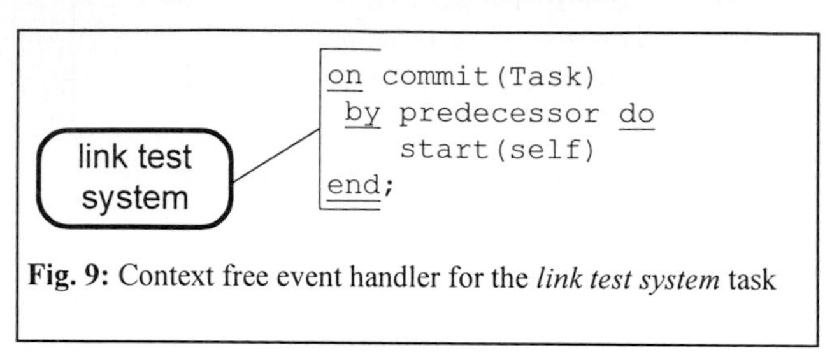

Fig. 9: Context free event handler for the *link test system* task

4.4 Context Sensitive Event Handling

Context sensitive event handling is a more powerful mechanism to specify reactive behaviour as it is expressive enough to manipulate a whole realising task net. Conceptually a new layer is needed below the composition layer. It is called the *rule layer* as it offers constructions to define reactive task net transformations rules. This layer is constrained by the type and composition layers, thus dependent on the defined process scheme. Its main task is to keep task nets consistent with respect to the wanted policies.

Rules are always uniquely associated with a realisation type and allow to specify integrity constraints and complex structural tansformations on the realising subnet. A net transformation rule consists of the following parts:

- *Input parameters* identify the location for a rule's application.
- A *net pattern* defines the process objects (tasks, parameters, relationships etc.) that have to be existent in order to further execute a rule and their respective states (precondition of the rule).
- An optional *net template* consists of all elements which will be introduced into the task net when the rule is fired.
- An optional *operation sequence* is applied to objects of the net pattern and template when the rule is fired.
- A *boolean return state* indicates whether the specified net transformation took place.

A rule may contain a net template, an operation sequence or both. MADAM includes a sublanguage to define net transformation rules in a declarative, graphical and therefore user-friendly way. The rule shown in Fig. 10 gives an example for the language's expressiveness. The rule automatically handles feedback between an *implement* and *design* task of a software process which might occur quite frequently.

The pattern elements establish the necessary context for handling a *CreateFeedback* event. The event's parameters *source* and *target* are part of the net pattern as well as the intermediate *module design* tasks, their succeeding *implementation* tasks and the parent task *develop software system.* Elements of a net pattern may be marked with the usual cardinalities to allow for optional parts and the retrieval of object sets.

Various conditions regarding states and types must hold on to the net pattern's process objects. The previously presented condition language is therefore embedded here.

The feedback is handled through a structural manipulation, namely the creation of two error parameters and a connecting dataflow, and some operation sequences which perform the following actions:

- The terminated *design* task is derived and reactivated.
- The intermediate *module design* tasks revoke all previously released revisions from their succeeding *implementation* tasks. They are also suspended.
- The feedback's source and all *implementation* tasks working with unsecure *module designs* are suspended.

The process modeler is not concerned with specifying an order of execution for the operation sequences as the separate sequences are executed nondeterministically. This reduces the chance for errors and enables a more declarative specification of rules.

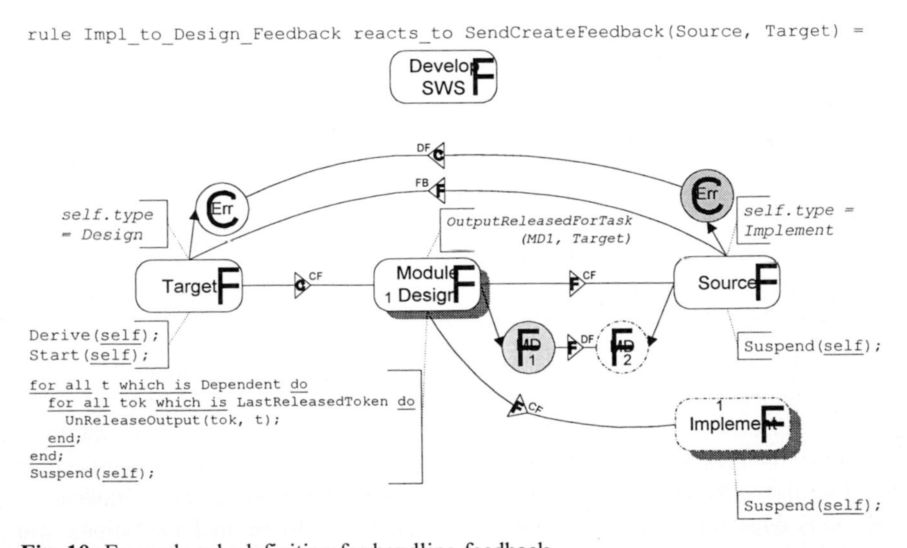

Fig. 10: Example rule definition for handling feedback

Rule executions are triggered either by the specified event or through an explicit call by the project manager. Currently, two different execution semantics are offered for event handlers. Synchronised semantics force the handling of the event as soon as it has been raised, while asynchroneous semantics queue event handlers and execute them when the system is unbusy.

5 Related Work

MADAM provides a high-level modelling language with well-defined execution semantics. There is a certain trend towards the enhancement of semantically rich process modelling languages with user-friendly interfaces. This section will compare the DYNAMITE/MADAM-approach to related approaches including EPOS, JIL and MERLIN/ESCAPE. All of these enable the execution of process models and offer modelling languages which abstract from the underlying execution machinery.

EPOS (cf. Conradi, 1994) is a plan based approach with an underlying object orientated database. The object orientated language SPELL (cf. Conradi, 1992) is used for the definition of process models on the type level. EPOS separates *entities* and *relations*. Processes are modelled as textual representations of *ER-diagrams*. On the type level static and dynamic pre- and postconditions of a task type, its executable code (i.e. starting of a tool), its parameter profile and the types of possible subtasks can be specified. The underlying *execution manager* is very flexible and able to handle well scheme evolution and the intertwined editing and execution of task nets. However, EPOS' capabilities to model reactive behaviour are somewhat limited in comparison to MADAM's. The offered event/trigger-mechanism corresponds to MADAM's context free event handling. Automated complex net transformations can not be specified.

JIL (cf. Sutton, 1997) is based on the process programming approach. As a successor to APPL/A, it provides more flexible control flow alternatives than the latter and enables proactive and reactive control. JIL is a formal, rich textual language which makes programming a process rather complicated and time consuming. It definitely requires a certain degree of expertise in the field. To reduce the complexity different visualisation mechanisms for JIL are currently being investigated but cannot be subject to evaluation yet.

The rule based, product centred approach MERLIN (cf. Junkermann, 1994) has been enhanced by the adaptation environment ESCAPE, which combines *EER-nets* for specifying the product structure and *state machines* to define their respective behaviour (cf. Junkermann, 1995). While the technical approach is quite similar to the MADAM environment, the system has some major drawbacks. ESCAPE does not offer corresponding mechanisms to refined operations and context sensitive event handling. The use of ESCAPE as a modelling environment will not significantly shorten the process development cycles since the reuse of fragments is not supported.

6 Conclusion and Visions for Future Work

This paper presented an adminstration environment that offers high-level, graphical and abstract but formalised, textual modelling sublanguages. Structural process schemes and complex net transformations can be specified graphically and in a very declarative way. The graphical languages are enriched with formal, textual languages for the definition of operations, conditions and context free event handlers, which are designed to be executable.

All languages have been formally specified in the graph transformation based on the specification language PROGRES or compilers have been built. A framework for graph based software engineering tools has been used to build a prototypical adaptation and a management environment. These environments and their underlying formalisms have been validated in the field of software engineering, plastics processing and machine engineering. They will now be developed further within the collaborative research centre IMPROVE and applied to the field of chemical engineering.

Feedback from engineers using the environment has shown that the offered languages remain too complex. My future work will therefore concentrate on further support for the process modeller. The MADAM environment shall be enhanced to provide different levels of modelling which allows every user to use the environment with regard to his/her degree of expertise in the field. Non-expert users can rely on "out-of-the-box" process model fragments that can be refined and adapted to a specific context. Libraries of predefined conditions, refined operations and event handlers can be used to adapt a process' behaviour.

With growing expertise and anticipation to define more adequate process models, a user may use the shipped languages to enhance the libraries or define very specialised process characteristics. To increase the general comprehension of modelled processes, I will evaluate possibilities to replace MADAM's special purpose languages with a standard modelling language, namely the UML. The mentioned framework for graph-based applications is currently redeveloped using the user interface tool kit ILOG JViews. Using the tool kit simplifies the task of implementing specific application graphical symbols, which allows us to implement a UML editor using this framework.

Future work will thus concentrate on further simplification of the offered languages and mechanisms to support archiving, reuse and refining of graphical process model fragments and textual expressions. Mechanisms to support intercompany developments must equally be adressed by future research as those for scheme evolution.

My long term vision is to offer an integrated process modelling environment, whose modelling support is much above a modelling language editor and a compiler that transforms models in to internal executable representations.

References

Cen, J.Y.; Tu, C.M. (1994):CSPL: a process centered environment. Information and Software Technology 36(1), (1994)

Conradi, R.; Hagaseth, M.; Larsen, J.; Nguyen, M.N.; Munch, B.P.; Westby, P.H.; Zhu, W.; Jaccheri, M.L.; Liu, C. (1994): EPOS: Object-oriented Cooperative Process Modelling. In: Finkelstein, A., Kramer, J., Nuseibeh, B. (ed.): Software Process Modeling Technology. Research Studies Press, Taunton, Somerset, England (1994)

Conradi, R.; Jaccheri, M.L.; Mazzi, C.; Aarsten, A.; Nguyen, M.N. (1992): Desing, use and implementation of SPELL, a language for software process modeling and evolution. In: Derniame, J.C. (ed.): Proceedings from EWSPT'92, Trondheim, Norwegen. Lecture Notes in Computer Science, Vol. 635, Berlin Heidelberg New York: Springer 1992, pp 167-177

Heimann, P.; Joeris, G.; Krapp, C.A.; Westfechtel, B. (1996): DYNAMITE: Dynamic Task Nets for Software Process Management. In: Proceedings of the 18th international conference on Software Engineering, (1996)

Heimann, P.; Joeris, G.; Krapp, C.A.; Westfechtel, B. (1997): Graph-based Software Process Management. International Journal of Software Engineering and Knowledge Engineering (7), World Scientific Publishing Company (1997)

Junkermann, G.; Peuschel, B.; Schäfer, W.; Wolf, S. (1994): Merlin: Supporting Cooperation in Software Process Development through a Knowledge-based Environment. In: Software Process Modeling and Technology, Research Studies Press, Somerset, UK (1994)

Junkermann, G. (1995): ESCAPE Eine graphische Sprache zur Spezifikation von Sofware-Prozessen. Ph.D. Thesis, Department of Computer Science, University of Dortmund, (1995)

Krapp, C.A. (1998): An Adaptable Environment for the Management of Development Processes. Ph.D. Thesis, Department of Computer Science, Aachen University of Technology, (1998)

Nagl, M. (1996): Building Tightly Integrated Software Development Environments: The IPSEN Approach“, Lecture Notes in Computer Science, Vol. 1170, Chapter 5.1. Berlin Heidelberg New York: Springer 1996

Rombach, H.D.; Bröckers, A.; Lott, C.M.; Verlage, M. (1995): MVP-L Language Report Version 2. Internal Report. Department of Computer Science, University Kaiserslautern. (1995)

Schürr, A. (1989): Introduction to PROGRES, an Attribute Graph Grammar Based Specification Language. In: Nagl, M. (ed.): Proceedings of the 15th International Workshop on Graphtheoretic Concepts in Computer Science. 1989, pp 151-165

Sutton, S.M.; Heimbigner, D.; Osterweil, L.J. (1995): APPL/A: A language for Software Process Programming. ACM Transactions on Software Engineering and Methodology 4(3), pp. 221-286, (1995)

Sutton, S.M.; Osterweil, L.J. (1997): The Design of a Next-Generation Process Language. CMPSCI Technical Report 96-30, Department of Computer Science, University of Massachusetts, Amherst (1997)

Fowler, M.; Scott, K. (1997): UML Distilled - Applying the standard object modeling language. Reading, MA, Addison Wesley (1997)

Aspects and Potentiality of Unconventional Modelling of Processes in Sporting Events

Jürgen Perl [1]

Johannes Gutenberg-University of Mainz, Institute for Computer Science, FB 17, 55099 Mainz

Abstract. *This paper describes how inexact processes as presented in sporting events can be recorded, analysed, and evaluated by means of neural networks and fuzzy modelling.*

Keywords. *process modelling, sporting event, neural network, fuzzy modelling*

1 Introduction

A technical process as presented in production and administration ("work flow") is normally characterised by sequential and parallel structures of separated components. These process components can normally be activated only under specific conditions. Their effects are widely determined.

Behavioural processes (e.g. processes in sporting events) on the other hand are characterised by an absence of precision: Decisions are made spontaneously and without deep analysis. Activity contexts are fuzzy and variable. Positions and techniques of actions are vague. The inherent absence of precision in connection with an extremely high variability of each possible process step results in two main consequences: On the one hand, analysis and synthesis of processes in sporting events cannot be done in the way technical and natural sciences are. On the other hand, the extremely large number of possible process realisations makes it very difficult to record, analyse, and evaluate any of them.

Under these circumstances, it has not yet been possible to find a feasible way for a scientific game analysis using computers and media in sport sciences – although recording, analysis and evaluation of sporting events have been dealt with for more than 25 years.

---

[1] Chair for Applied Computer Science; Speaker of the Section Computer Science in Sport of the German Society of Sport Science

However, increasing performance of personal computers in connection with the development of unconventional algorithms and modelling paradigms now present new ways of solving such problems. Our working group "Computer Science in Sport" has been discussing these new possibilities with sport science and practice for a couple of years now.

In this paper, first of all the paradigms *Fuzzy Logic* and *Neural Networks* are introduced – as far as they are necessary in describing and handling sporting events. This introduction is followed by a demonstration of how neural networks (in particular: Kohonen Feature Maps) can be used to record, analyse and evaluate the processes of such events. Finally, the connection with aspects of fuzzy logic is discussed: Here not only fuzziness of positions and of selection but realisation of activities is of interest. In addition we discuss how fuzzy modelling can help use the vagueness of phase and technique oriented game processes for classification and thereby prepare processes for being identified by neural networks.

2 Unconventional Modelling Paradigms

Briefly said, modelling paradigms are called "unconventional" or "new" if they try to solve complex problems in a "biological" way, i.e. done without algorithmic precision and determination. The idea is to recognise and simulate strategies of individuals and populations of how to transfer information and to solve problems.

Meanwhile, these approaches have been documented in various ways. Therefore, this paper only outlines some examples. In particular, we have to do without discussing technical details of developing and designing practical procedures. Moreover, there are generators, which support building such procedures by selecting feasible design parameters. In WILKE (WILKE, 1997) a large list of references is given.

Additional references with regard to computer science and sport science are given in the bibliography. Further explanations, approaches, and references of modelling and unconventional paradigms in *Computer Science in Sport* can be found in PERL, LAMES & MIETHLING. (PERL, 1997)

2.1 Fuzzy Logic

The apparent precision of calculated results often leads to an only apparent understanding of the real facts. This in turn can badly influence the understanding of a problem and thus makes it even more difficult to find ways to solve the problem.

For example, it is no longer a serious technical problem to record each player's exact position on the playing field continuously over time. But - with regard to the relevance of such precise information - there is no sense in recording it: There is no way to control a player by means of centimetres and milliseconds. On the contrary, trying to do it this way would confuse the player and make him loose his inspiration.

Accordingly, it would be inadequate to call a team "successful" if its score only reaches 75 points on a 100-point-scale (see fig. 1). Such characterisations by means of yes-no-conditions or by 0-1-sets (compare the bold horizontal lines in fig. 1) may make sense from a mathematical point of view. For characterising terms that contain an inherent vagueness, however, *fuzzy*-sets are much more qualified. In fig. 1 an example of a fuzzy-set representing the property "successful" is shown: Each score between 0 and 100 has the property "successful" – with a degree of membership between 0 and 1, depending on the particular model that is used. This way, *each* team can have the property *successful* – depending however, on its current score with different degrees of memberships. Moreover, each team has not only one such attribute, but can be *unsuccessful*, *rather successful*, *successful* and *very successful* with characteristic different degrees of memberships at the same time.

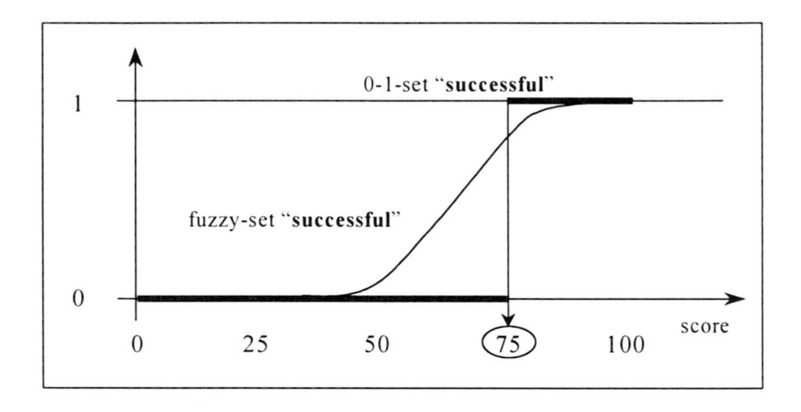

Fig. 1 : Fuzzy-set of the score-attribute "successful"

The advantage of using fuzzy-sets shall be demonstrated by an example which will play an important role later on in section 3.2: Observing and recording playing processes in a sporting event due to analysing and improving the tactical playing structure can, amongst others, result in events (e.g. pass, flank, dribbling, ...) and positions (e.g. left, centre-left, centre, ...). As mentioned above, a precise way of recording and analysing was to accurately measure the positions using technical tools and to compute precise values to control the playing process. As one result, the coach could recommend running a particular process with a certain frequency, concerning particular positions. Without any problems, frequencies as well as positions could be calculated very precisely by the computer. The players, however, can neither act as precisely as a computer calculates nor can they anticipate the computer calculated results. So, the process of recording, analysing, anticipating, and running shows a remarkable inconsistency: The observable concrete actions are fuzzy. The inexact information about actions are "made precise" by recording. Precise input is processed by the computer, resulting in precise output. The coach then has to make the computer output inexact to make the players able to anticipate it. The players finally run the process using fuzzy and inexact actions (also see section 3.2, fig. 8).

As pointed out in section 3.2, it seems to be better to avoid bypasses like this. Instead of using apparent precision it is necessary to use a homogenous level of vagueness that is oriented on the fuzzy view and understanding of coaches and players.

2.2 Neural Networks

It is the basic idea of neural networks to identify and specify objects. For this purpose, the nodes or *neurons* of the network are *trained* in a way that the objects are mapped onto the neurons. In addition, certain important relations between the objects are mapped onto the neurons and their connections. A number of network types can be classified where mainly the internal structures and the organisational ideas of training are different. In thc following section and in section 3.2 so called *Kohonen Feature Maps* are dealt with. The characteristic property of these type of neural networks is that they are *self-organising*. This means that the process of training runs without any controlling activities.

In fig. 2, a mapping from objects to neurons under the aspect of "similarity" is demonstrated: Similar patterns are assigned to neighbouring neurons in a way that clusters of neighbouring neurons represent assemblies of similar patterns. The model generating conception behind this approach is the idea of how the brain works – i.e. that associated experiences, impressions, pictures or patterns are associated with connected areas of the brain.

Having been trained, the *recognised* objects are associated with the neurons of the network. Now the trained network can be used to recognise or identify new objects, which have not yet been learnt, and which will be associated by the network with its respective best fitting neurons. Moreover, the best fitting neuron itself is member of a cluster, which in turn classifies a type of "similar" objects. This way, the new objects are not only identified but classified to a type as well.

One example of how to use neural networks for sporting events is to represent game processes by neurons (compare the example in section 3.2). The aim then could be to identify and analyse the opponent's activities with regard to their strategic meaning in order to determine optimal counter-strategies.

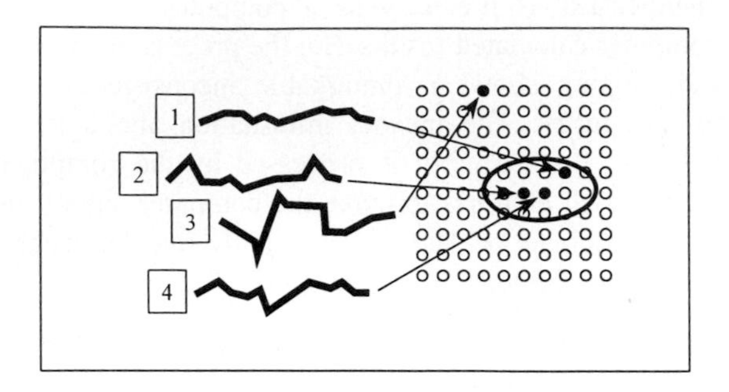

Fig. 2 : Stimulating a Kohonen Feature Map in the training phase

The advantages of using neural networks instead of manual work are given under at least two aspects: On the one hand, even extremely complex assemblies of data can be reduced to handy sizes. For example some 10 thousand process patterns can be reduced to about 15 significant process types within seconds, using a 200 MHz pentium II personal computer. On the other hand, the data resulting from the identification process can immediately be transferred into data bases or an expert system. This means that after having recorded the game, all the observed processes or activities are identified and classified into significant types, usable e.g. for statistical analysis or qualitative evaluation using expert systems.

3 Process Model

3.1 Foundations

The following model for describing, recording, and analysing processes shall demonstrate how systems with complex activity structures can be mapped, and which role neural networks and fuzzy logic can play in reducing their complexity.

Observation and analysis of complex sporting events often lead to the following central problem: Restricting the recording to simple facts, techniques, positions, or results can reduce the necessary expenses extremely – the profit, however, is reduced as well. If events and states of a game are not recorded in their particular process connections and contexts, important aspects of the actions (e.g. assumptions and conditions or time-dependent, logical, and structural relations) remain unrecognised and cannot be analysed in order to understand what has happened. If on the other hand all these aspects are recorded, then the number of observed processes increases enormously. One of the consequences then is that the frequency of any of the items is insignificantly small and cannot be used for statistical analysis.

A very simple example may make the problem clear: Assume that an attacking process in football is characterised by a sequence of just 5 positions, namely *left, centre-left, centre, centre-right,* and *right.* Assume further that the normal length of a process is 10, i.e. the process is described by a sequence of 10 positions (compare fig. 3). Then about 10 million different attacking processes would have to be classified. Even if one of these processes happened each minute, one could not record more than 100 processes from one game. Therefore, it follows that the expectation of recording one particular process during a game is about 10^{-5}. This is really not a good platform for a statistical analysis of successful attacking processes.

To avoid this conflict, processes have to be grouped into classes or clusters of similar processes. Of course, in this paper it is not possible to reflect all of the different deterministic or stochastic clustering concepts and methods. One well-tested approach from the area of unconventional paradigms, the above mentioned neural networks, will be presented instead.

3.2 Position Processes

The concrete example we investigated uses a positional process description where *similarity* is deduced from the natural similarity of positions. The net type we used is a Kohonen Feature Map, which is trained with a sufficient number of processes, as has been mentioned in section 1.3.

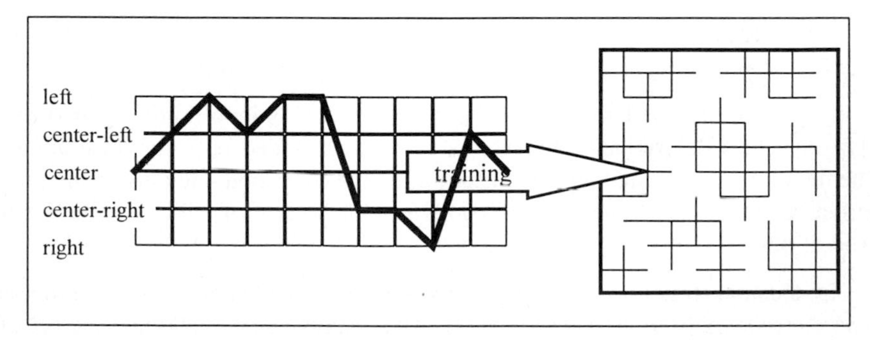

Fig. 3 : Neuron-clusters effected by training

What "sufficient" exactly means depends on the necessary resolution and precision of the results. In our tests, we used nets with a resolution of 20x20 neurons, and we trained them with 5000 up to 10000 process items.

As a result of the training the net develops clusters of neurons which are associated with assemblies of similar processes (see fig. 3). If a neuron of a trained net is activated it presents the process pattern it is "responsible" for. As fig. 4 shows, such a *pattern* is not exactly what was defined to be a position process (i.e. the steps of the process pattern are not exactly the positions left, centre-left and so on but are in a fuzzy way distributed over the range of defined positions). This means that the net develops a two-level-clustering: On the first level, similar processes are fused into "fuzzy" process patterns and assigned to one identifying neuron. On the second level neurons that identify similar patterns are organised into connected clusters. These clusters then characterise what we call process *types*.

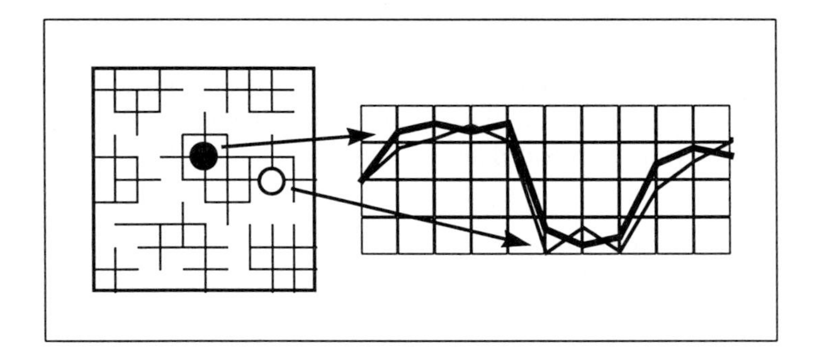

Fig. 4 : Cluster, neurons and identified process patterns

The following two figures show how a trained net typically works:

In fig. 5 an observed and recorded position process is input into a trained net. (The attributes describing techniques or actions are added only for illustration.) The net reacts on the input by first determining the corresponding cluster and then activating the particular neuron, the corresponding pattern of which fits the input process best.

Fig. 6 demonstrates how the input process might be embedded in to the process type corresponding to the neuron cluster. The resulting reduction from the set of processes to the set of process types has a lot of obvious advantages: First of all, the number of types is much smaller than the number of processes; at least, the number of types is limited by the number of neurons. In reverse, the frequency of types is much larger and, therefore, more suitable for statistical analysis than that of process items. Furthermore, the identification of a process, i.e. its assigning to a type, is automatically done by the net. The recognised type can, therefore, be processed automatically, – e.g. as an input into a data base with a subsequent analysis using expert systems.

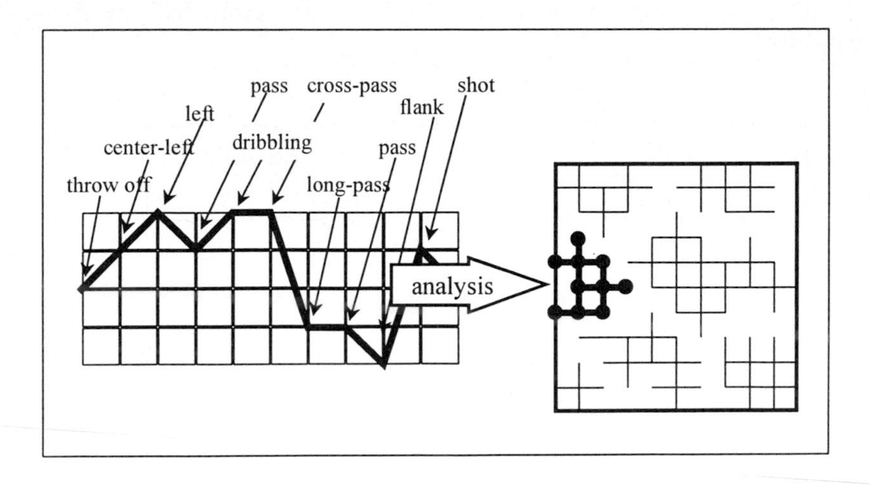

Fig. 5 : Process analysis (process with additional attributes)

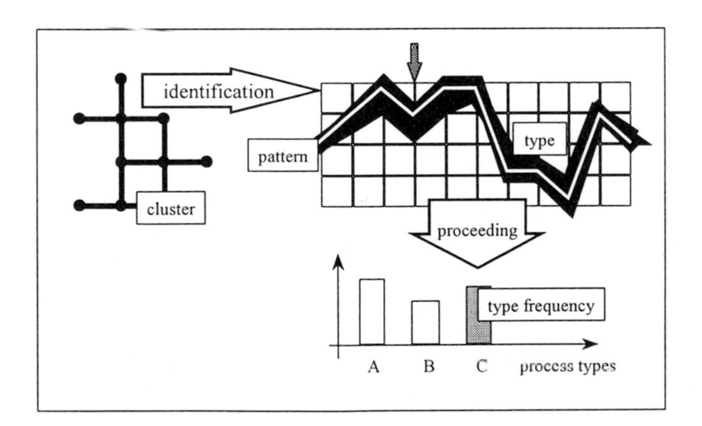

Fig. 6 : Process identification (step 4 marked by an arrow)

But there is still another aspect, which is interesting and can be used advantageously for recording and analysing: As mentioned above, the process patterns determined by the net do not show the defined position in an exact way. In fig. 6 for example, the position in the fourth step is stretched over a range from *centre* over *centre-left* to almost *left*. For this type of vague description the possibility of fuzzy modelling has been introduced in section 2.1. Fuzzy modelling enables the quantification of vagueness by measuring the degrees of memberships and so allows mathematical, numerical, or statistical analysis. In fig. 7 both the types of vague description with regard to the example "centre-left" are shown in comparison.

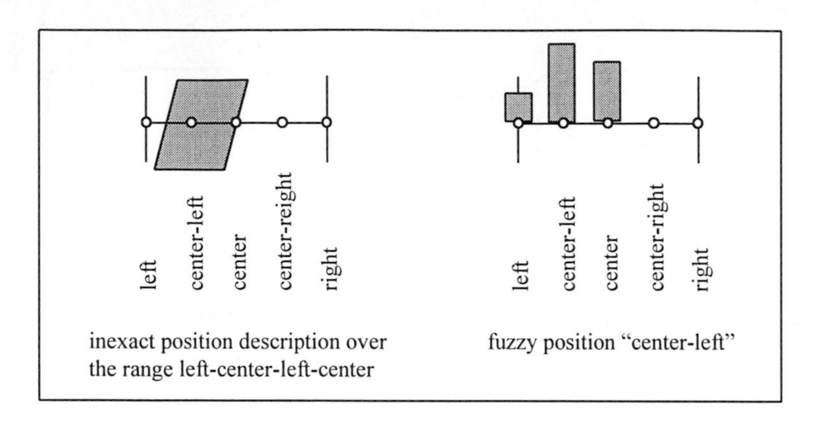

Fig. 7 : Qualitative vagueness and fuzzy position description

The other way round, defining positions to be inexact fuzzy positions results in a number of advantages:

First of all inconsistencies in the precision of information transfer can be avoided which often appear in computer aided game analysis (see fig. 8).

	action *player*	observation *coach*	input *computer*	output *computer*	controlling *coach*	action *player*
Conventional						
Object	situations	facts	numbers	numbers	facts	situations
Precision	low	medium	very high	very high	medium	low
Fuzzy	homogeneous level of vagueness					

Fig. 8 : Precision of information transfer in the process of observing and anticipation

In this way, apparent precision, which is coined by the use of computers or by a particular type of technical thinking, can be also avoided. If - in the context of an action - a position on the ground can only be taken with a tolerance of ±2 meters then there is no sense to calculate or to think in terms of higher precision.

The consequence is that even in the phase of data recording unrealistic high precision can and should be avoided: If a reasonable clustering reduces the number of processes from 10 million to 20 process types, which can be handled easily and better anticipated, then a more precise and costly recording is senseless.

Together with simplified observation and recording anticipation and controlling can also be simplified: The results of computer aided analysis use the same vague terms, which the observer uses during data recording, and which are common to the players. So the cost for the information transfer from calculated results to its anticipation and controlling can be significantly reduced.

However, the purpose is not to give the impression that precision recording in general is senseless or useless. If the focus is set on single events or actions instead of long and complex processes then of course a high precision data recording can be useful and necessary. Two examples, amongst others, are given by biomechanical models for motion analysis and by dot cluster models for the analysis of activity profiles.

3.3 Phase Oriented Processes

Finally we will give some ideas of how the use of the above mentioned approach can be extended. The use of neural networks in the standard case is limited to attributes that define a "natural" similarity of objects. If, however, the process of a game like tennis is modelled using phases – e.g. 1st service, 2nd service, baseline, net rally, point – this assumption of similarity is not valid: Obviously, a baseline stroke is not at all similar to a 2nd service. Therefore it would not make any sense to identify them by means of a common cluster. As senseless as a common cluster is a process type, the fourth step of which stretches fuzzily from 2nd service over baseline to net rally.

However, similar to fig. 7, it makes sense to define a fuzzy distribution with respect to the degrees with which the action of the fourth process step is a member of the phases 2nd service, baseline, and net rally, respectively (see fig. 9). This way the fuzzy distribution in fig. 9 means that the current action of the process type being looked at belongs with high degrees to the phases baseline or 2nd service; a membership of net rally is possible but with a very low degree. Note, however, that – as is mentioned in section 2.1 – the degree of membership does not mean any probability: Fig. 9 does not mean "*In his second action the player will most likely play a baseline stroke.*" but "*The process type belongs with his second step to the base line phase with a high degree of membership.*"

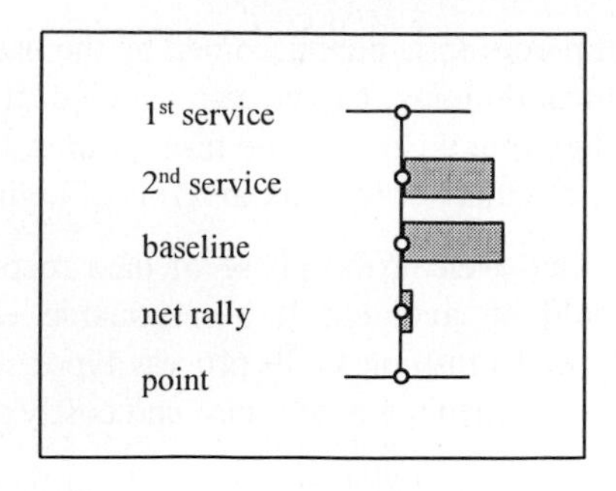

Fig. 9 : Fuzzy characterisation of a phase-type-action

Our working group succeeded in getting this new approach to work using fuzzy neural networks. At first results seem to be very promising. Currently, we are running a project together with the University of Rostock applying fuzzy neural networks for recording and analysing the processes of beach volleyball. It comes out that the network with a surprisingly high precision is able to cluster the processes and to determine the significant standard types.

4 Conclusion and Outlook

Central problems of recording and analysing game processes in sport, of inherent vagueness and high variability, can be solved using neural networks for clustering processes into types. The network-generated, neuron-associated patterns show a characteristic type of vagueness which is the reason for certain fuzzy approaches: On the one hand, observing process attributes can be simplified to observing fuzzy terms like "centre-left" and so reduce the recording costs by far. On the other hand, recording, proceeding, and anticipation can use the same level of precision, which reduces the problems of information transfer. Finally, fuzzy modelling can help in the case of processes where similarity cannot be a reason for a clustering method. Here, fuzzy distribution of attributes can help to cluster processes by means of neural networks.

Currently, the interface between neural network and data base is being worked out. Using this interface it is possible to identify, register, and analyse recorded processes automatically. Typical queries then could be: "Which of the attack process patterns that Miller took part in were successful?" In this connection, the focus will be set on the use of expert systems to support comfortable query forms. Finally, the network itself will be an object of user interaction: The training-defined network has to be modifiable by the user, in order to adapt the model optimally to the concrete needs with respect e.g. to the selection of types or the degree of resolution.

References

DEMANT, B. (1993): Eine Einführung in die Fuzzy-Theorie und Gedanken zu ihrer Anwendung im Sport. In: PERL, J. (Hrsg.): Sport und Informatik III. Köln 1993, 49–54.

KÜNZELL, S. (1995): Motorik und Konnektionismus. Neuronale Netzwerke als Modell interner Bewegungsrepräsentationen. Köln 1995.

LAMES, M. (1994): Systematische Spielbeobachtung. Münster 1994.

LIESEGANG, W. (1996): Ein Fuzzy-Modell für Angriff-Abwehr-Interaktionen im Handball. In: QUADE, K. (Red.): Anwendungen der Fuzzy-Logik und Neuronaler Netze. Köln 1996, 33–40.

PERL, J. (1996): Grundlagen der Modellbildung und Konzepte der Umsetzung. In: QUADE, K. (Red.): Anwendungen der Fuzzy-Logik und Neuronaler Netze. Köln 1996, 5–17.

PERL, J. (1997a): Möglichkeiten und Problem der computerunterstützten Interaktionsanalyse am Beispiel Handball. In: PERL, J. (Hrsg.): Sport und Informatik V. Köln 1997, 74–89.

PERL, J. (1997b): Modelling in Sports: Unconventional Aspects and Approaches. In: Proceedings of the 2nd Annual Congress of ECSS. Kopenhagen 1997.

PERL, J.; LAMES, M.; MIETHLING, W.-D. (Hrsg.) (1997): Informatik im Sport: Ein Handbuch. Schorndorf 1997.

QUADE, K. (Red.) (1996): Anwendungen der Fuzzy-Logik und Neuronaler Netze. Köln 1996.

RITTER, H.; MARTINETZ, T.; SCHULTEN, K. (1994): Neuronale Netze. Bonn 1994.

SCHÖLLHORN, W.; BAUER, H.-U. (1997): Linear vs Nonlinear Classification of Complex Time Course Patterns. In: Proceedings of the 2nd Annual Congress of ECSS. Kopenhagen 1997.

WILKE, P. (1997): Soft Computing - Prinzip, Simulation und Anwendung. Habilitationsschrift. Erlangen-Nürnberg 1997.

ZIMMERMANN, H.-J. (1991): Fuzzy Set Theory – and Its Applications (2nd ed.). Boston 1991.

Studying of the Influence of Phytoviruses on Plants' Organism in Conditions of Ecological Instability and its Estimation by Mathematical Modelling in CEIS System

Yuri V. Zagorodni [1], **Volodimir V. Voytenko** [2] **and Anatolij L. Boyko**[3]

[1] State Academy of Agriculture and Ecology of Ukraine
[2] Zhitomir Institute of Engineering and Technologies
[3] Kiev National University of Taras Shevchenko, Faculty of Virology.

Abstracts. *The computer ecological intellectual system (CEIS) is developed in the investigated regions on the basis of the object-orientated approach with the purpose to be used in systems of processing ecological information. (Voytenko, Zagorodni, 1997). The system has developed dialogue means, which allow setting base data maps for their processing through internal and external models of natural and agricultural processes. The result of such processing is new maps of the economic, ecological or social characteristics. The maps are created on the basis of researched region division into geographical elements for the supervision of necessary characteristics changes of dynamics. Thus, there is an opportunity to realise the analysis and forecasting. CEIS is logically and structurally subdivided into the subsystems. The subsystem is an abstraction. Its subclasses are versions of objects that represent the global characteristics of any mental or real territorial subsystem. According to such project and hierarchy of objects, it is obvious, that objects of derivative classes from the abstract generalised class and can submit the specialisation of the subsystems. Hence, at the centre of consideration –we have a subsystem, that displays a real current state of a pattern, which answers one of the characteristics, say, of current radiating loading in view of accommodation of pollution sources, natural-climatic conditions, sociologic-economic structure and so on. The minimal specification of the project is determined as follows: - Biological subsystem (ground, microorganisms, viruses, plants, animals, birds); - Subsystem of environment (climatic conditions, pollution); - Subsystem of social development; - Subsystem of managing and economy. One of the external models, which is represented in CEIS, is a model of a haricot bean plant development in ontogenesis, nodule nitrogen fixing during a virus infection in different ecological conditions (Zagorodni, Boyko, Beiko 1995). As it is known, virus is a very sensitive system at change of the natural factors, therefore it can be used as the indicator of the environment state. The model uses the initial information as maps of the environment subsystem of CEIS* *(temperature, humidity, PH of soil, radiating, pestisidal and industrial soiling) and is a basis for the construction of new maps - maps of infection, productivity and its quality, pollution of a crop and so on. Also, through such a model, it is possible to find optimum conditions for the cultivation of a healthy vegetative material. The processes, which pass in vegetative organism, are separately given in the model. For each process, it is constructed of a function of potential force and a function of real force, which is defined under conditions of action of the laws of conservation of substances and energy. Thus, CEIS can be applied in the research of ecological processes in different conditions of investigated regions. Then the region is broken into geographical elements with the given set of abstract subsystems. As an example, a model of development of the plants in different conditions of Ukraine is shown.*

Keywords. *CEIS system, data maps, real force of process, external and internal models.*

1 Methodology, Construction and Structure of Computer Ecological Intellectual System (CEIS 1.0)

CEIS 1.0 is the computer system in the branch of mathematical modelling. It is developed on the basis of the object-orientated. approach for the description of certain natural patterns, their fillings, functioning and usage, with the purpose of the approached comparison with the real world. It is intended for discrete -event modelling of systems, which are characterised by conceptual and physical parallelism of processes of the information processing. The system has the advanced dialogue means, which allow to set the initial data for the model, to reconsider the model in a step-by-step mode with the stops on specific event, value of the model time. It gives an opportunity to supervise its development in dynamics, and realise the data analysis as well as forecasting for the future period.

CEIS is developed on the basis of the object-orientated. approach for the operational system Microsoft Windows 95 with use of the programming language Borland C ++ 4.5. It is realised as a set of C++ modules. The modules are documentary with the English and Ukrainian user-interface.

The CEIS system includes:

- The accessible graphic interface of an operational environment MS Windows 95;
- Contextual - sensitive system of support and help;
- Realisation by necessity of partial transition from point methods to vector methods, for the analysis of the display of the graphic information;

- Complete supervising process of the model dynamics;
- Creation of external files of the model data for further usage by some models.

The first version of CEIS rather successfully functions in the environment of Windows 95, at the laboratory of modelling of ecological systems of State Academy of Agriculture and Ecology of Ukraine. It is used for simulation of ecological pollution development in territory of northern areas of Zhitomir region.

The nucleus of CEIS functioning is subsystem.

The subsystem is a pure abstraction. Its subclasses are versions of objects, that represent the global characteristics of any mental or real territorial subsystem. According to such project of objects and hierarchy, it is obvious, that objects of derivative classes from the abstract generalised class can submit the specialisation of subsystems (Boyko, 1994; Zagorodni, 1997; Voytenko).

Hence, at the centre of consideration, lies the subsystem, that displays a real current pattern state, which answers one of the characteristics, say, current radiating pollution in view of accommodation of sources, natural-climatic conditions, social-economic structure and so on. Each subsystem is completely independent from another, and is realised in parallel way.

The minimal specification of the project includes:

- Biological subsystem (ground, microorganisms, viruses, plants, animals, birds and others);
- Subsystem of environment (climatic conditions, pollution);
- Subsystem of social development;
- Subsystem of managing and economy.

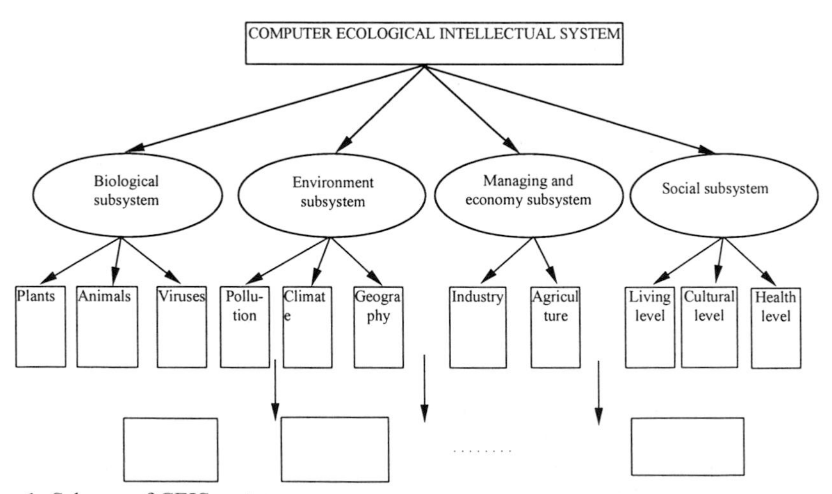

Fig. 1: Scheme of CEIS system

System Databank

The generated data received in a research way, makes a system databank. Considering their state in dynamics, an opportunity of their research, use and analysis occurs, which searches some important influential factors for models in ecology, revealing interesting dependencies between different parameters. The creations of the data external files, that are accumulated by target results of a specific pattern, gives an opportunity of their further analysis both in CEIS system and separately.

Openness of the system and its application for various models.

CEIS is in general the structure for the formation (generating) of complex computer ecological systems. The generalised object-orientated. modelling of a skeleton for formatting complex models provides for the future using of the uniform interface for simulation of development of different systems, abstracting from their difference in those places, that is not dominant at the given consideration, and allows to support any number of hierarchical levels of considered subtasks. Using the object-orientated approach, it allows to use the same formalism for the description of real models, except the intermediate consideration of model, that simplifies its description and interpretation of results. The skeleton provides a uniform set of rules for designing systems of any behaviour - only initial data defines the life cycle of system.

If we consider the system for using various models, its functioning can be considered in two conditional modes - forecast and information. The mode of the forecast provides opportunities to adapt existing models for the certain objects of subsystems, to adjust parameters of models and to connect new (external) models. The mode of the information provides support of hierarchical structure of carriers of the information, and their units are in the data files. The openness of system follows from the above mentioned process.

Mode of functioning (visualisation).

One of the main questions is the modelling of the basic ecological processes of the entered objects (models of transformation of pollution and production models). It includes development of information - graphic system for supporting ecological modelling, which includes an electronic map, and database adhered to it, and contains the basic characteristics of ecological objects and processes. On a given basis, it offers the finding of solutions of tasks of restoration of the missed data in an ecological database, tasks of various statistical processing, destruction of ecological objects and mathematical modelling of ecological processes with the visualisation of results of modelling on an electronic map.

The system has an opportunity to create and to write down on a disk the certain maps as files, which includes the specific data of a regional attributes state; to reconsider the diagrams of dynamics of pollution development in the certain area, chosen by the user, of a card through the given number of steps of recalculation with current time; to create multi-map interface, which enables simultaneously to feign the development of events in different investigated regions and so on. With the help of Windows and a choice of attributes value it is possible to create necessary conditions to check for various hypotheses about the distribution of the pollution of an environment.

The use of imitating modelling (modes of visualisation) helps to imagine precisely and investigate complex mathematical models and their characteristics, what the system gives. The display of a degree of pollution is reached by two ways - appropriate colour of a palette or numerical value, fixed for each cell (unit of measurements of the area). The calculation of a current level of pollution is defined in view of dynamics in time, taking into account the usual diffusion effect - interaction of the next crates a distribution behind a wind (direction and force of air weights).

There is also an opportunity to reliably supervise the parameters which are peculiar to subsystems. Their number is fixed, yet absolutely unessential to each consideration. The Main consideration of the problems is the selecting of subsystem attributes for opportunity of their adequate representation in CEIS system.

2 Viruses - Biological Indicators of a Condition of Environment

It is known, that the virus begins its own reproduction by getting into a cell of the owner. Significant changes in the metabolite of vegetative organism can be observed at this time. The organism is a set of various cells and their functions. It is known, that the attributes of any one organism depends both on a genotype, and on external conditions of its existence. The modern ecological researches show that the change of ecological conditions of biosphere leads to the changes of properties of viruses (Boyko, 1985; Boyko, 1996; Boyko, 1994). Viruses are strong mobile genetic systems, therefore the problem of ecological instability of environment today gives also the problem of sharp changes of pathogenic viruses properties. From what is mentioned above, it follows that the researchers should be close to probable flares of viruses in ecologically unstable zones of our planet.

In natural conditions the distribution of the viruses epiphitoties depends on the many factors of external environment, but nowadays, the scientists have not complete representation about the mechanism of action of the conditions on reproduction process of viruses. Phytoviruses put the considerable loss to agricultural manufacture in the certain ecological conditions. It is proved that it is possible to allocate from plants different shtams of the certain virus, which are characterised by heterogeneity in natural conditions. Thus some shtams amaze one plants latently, and cause strong pathological changes on others. The measure of infectious influence depends not only on a kind of shtam of a virus, the plant resistance, but also on the certain ecological factors.

The distribution of a virus is strongly influenced by the various temperatures, humidity, and condition of plants' cultivation, measure of radiating, pesticide and industrial pollution. So, for example, the layered structures, that are localised in the tomato plant cells infected by the virus of a tobacco mosaic, in conditions of Chornobyl accident are revealed in the zone. Also, it is investigated, that in those regions of Ukraine, where the plants of a string bean have large intensity of nodule nitrogen fixing process, a virus of a yellow mosaic reproduction is a lot weaker (Boyko, 1996). This dependence can be described by the formula

$$I (Bh) = 19.128-0.174Bh,$$

where Bh - average number of the nodules at a healthy plant in region, I (Bh) - biological titre of a virus of the patient of infected plant (number of necroses in the plants leaves).

Thus, it is possible to consider, that the ecological situation of the environment influences the virus reproduction processes, and the structural changes of virions. It can be useful for the solving of theoretical and practical tasks both in area of virology, and in the field of general ecology, when the virus can be used as the indicator of an ecological state of agrithenoses and natural environment.

3 Research of the Phytoviruses Influence on Plants Organism in Conditions of Ecological Instability with Help of CEIS System

Applications of CEIS system for research of the constructed model of development of nodule nitrogen fixing process on roots of infected plant in different ecological conditions is represented in the following scheme:

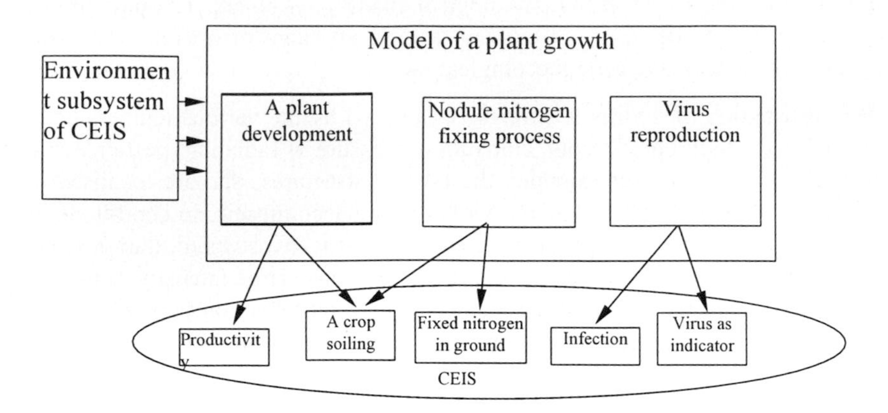

Fig. 2: Scheme of a plant growth model using CEIS System

The model of plant development as a virus influence can be presented in the following way:

M = (X, Q, E, P),

where X - set of variables of the model state, Q - set of the model processes, that take place in researched system, E - set of variables of environment condition, which are initialised by CEIS system, P - set of the model coefficients.

The set of variable of the model state X is divided into a subset of concentration value of important chemical connections, and on a subset of values of the development of separate organism's bodies. The first subset includes the concentration in a vegetative cell of the following substances: water, carbonic acid, oxygen, ATP, carbohydrates, amines, virus, hydrogen in nodules, carbohydrates in nodules, protons in nodules, ATP in nodules, fixed nitrogen, nodule amines, electrons in nodules, nodule oxygen. The other subset includes parameters of development of chloroplasts, mithohondries in a cell, parameters of development of plant stalk and root, general integrated value of nodules development on roots of plant.

The set of processes Q includes the following processes: photosynthesis, breath of a plant, construction of chloroplasts, mithohondries, general parts of plant, water exchange, virus reproduction, disintegration of the cell structures, inactivation of the virus, construction and development of the nodules, nitrogen fixing, synthesis of the nodule ATP, exchange between plant and nodules, having dug molecules of hydrogen. Each process model has two variables, which describe forces of intensity - potential and real. The potential force of a process is defined by the certain function, proceeding from the current state of the plant organism (value of elements of the set X) and from the current state of the environment. The influence of the environment is described by elements of a vector E, such as temperature, humidity, light exposure, PH, productivity, and radiating and industrial pollution. This vector has seven elements E = (e1, e2,..., e7). The real force of the process is constructed with the account of both potential force and laws of preservation of substances and energy.

Let us assume that the function of potential force for a process $q \in Q$ is $F^q(X,E)$. Such a function is positive - a certain one for own arguments. Then for each substance $x \in X_1$, that is spent in process q, the function of its expense is defined as $R_q(x) = R_x(F^q(X,E))$. Then the general value of expenses of **substance x for all processes of the set Q is defined by the formula:**

$$R(x) = \sum_{q \in Q} R_x(F^q(X,E)).$$

In the model the laws of conservation of substances and energy should be carried out, therefore the real forces of all processes are constructed so that the following condition was carried out:

$$R(x) \le x.$$

For example we shall consider process of photosynthesis. Schematically it may be written so:

$$[H_2O] + [CO_2] \xrightarrow{F^1(X,E)} [CH_2O] + \frac{1}{2}[O_2]$$

Its potential force is defined by function $F^1(X,E)$. Two substances - water (concentration - [H2O]) and carbonic acid (concentration - [CO2]) are spent during this process. Then the real force of the process can be defined by the formula:

$$m_1 = \min(\frac{a_{11}F^1(X,E)[H_2O]}{R([H_2O])}, \frac{a_{21}F^q(X,E)[CO_2]}{R([CO_2])},$$

where $a_{11} = a_{21} = 1$.

If any substance is not spent in a process q, then $m_q = F^q(X,E)$.

Thus the real forces of all the processes of the model can be constructed in such a way. Then it is possible to prove, that the following conditions will be carried out:

1) The sum of all expenses of a substance $x \in X_1$ on all processes of the model is smaller or is equal to the value of its current concentration (law of preservation);

2) If the substance $x \in X_1$ is spent in the process q and its concentration is equal to zero, then the real force of the process is also equal to zero;

3) If we add one more l process into the model, $v \in Q_1 = Q + \{v\}$, and the same substances are spent both on this process and on a process $q \in Q$, then the value of real force of process q will decrease;

4) Increasing the value of the potential force of process $q \in Q$ at recalculation of the model conducts the increasing of the value of its real force.

Thus, the process of the virus reproduction can be considered as an additional process in the system of the plant, for which the condition 3 is right. Therefore other processes are halted, and the plant growth is halted. Thus the model is represented in the shape of system of the dynamical discrete equations:

$$x_i^{k+1} = x_i^k + \sum_{q \in Q} h_{iq} m_q \text{ , i = 1, N k = 0, T,}$$

where x_i^k is the value of i-s components of the set X on a step of recalculation k, T - number of days of the vegetation period.; $h_{iq} = 1$, if the value of concentration of the substance increase during the process, $h_{iq} = -1$, if the value of concentration of the substance decreases during the process, and $h_{iq} = 0$, if the process q does not influence the value of concentration of the substance.

The integrated parameter of nodules development [S] is divided into a vector N = (n1, n2, … …, nk ….) of each nodule development:

$$n_i^{k+1} = n_i^k + \frac{n_i^k}{\sum_{j=1}^{[N]_k} n_j^k + e} ([S]^{k+1} - [S]^k),$$

where - $[N]^k$ is the number of the nodules on step k.

At such distribution it is ease to see, that

$$\frac{n_i^{k+1}}{n_j^{k+1}} = \frac{n_i^k}{n_j^k}, \forall i, j \leq [N]^k$$

This dependence shows, that the speed of a nodule development depends both on the value of the development of the integrated parameter [S], and on the value of this nodule development on the previous step. The presence in the formula of parameter e guarantees the distribution of the certain part of integrated parameter to the occurrence of new nodules.

With the result parameters of the model, it is possible to consider the infection of the virus reproduction process under the certain ecological conditions influences, which would be set by a vector E; productivity of the culture, pollution of a crop, state of development of cultural plants. Therefore, with help of scheme 2, it is possible to say, that the functioning of such a model in CEIS gives an opportunity to build new maps of the regions, on the basis of initial maps, which are presented by scheme 1. It makes CEIS system a real intellectual system.

Thus, the model will simulate the process of virus reproduction and nodule nitrogen fixing for new conditions at the dynamics of changes of values of parameters E in a separate cell of a maps.

4 The Results and Conclusions

CEIS can be used for the research of different regions of Ukraine or other countries. So, the territory is divided into the given number of elements, where for each one, all characteristics of subsystems of CEIS are considered. For example, it is possible to set the characteristics of a subsystem of environment for each element proceeding from the current real data. Then the system will already change such characteristics as pollution of territories with the help of the known laws of transformation and diffusion. Thus in each element there is a database of a subsystem of environment.

Fragment of CEIS subsystem functioning is submitted below. The line marks a way of researches of the plants development:

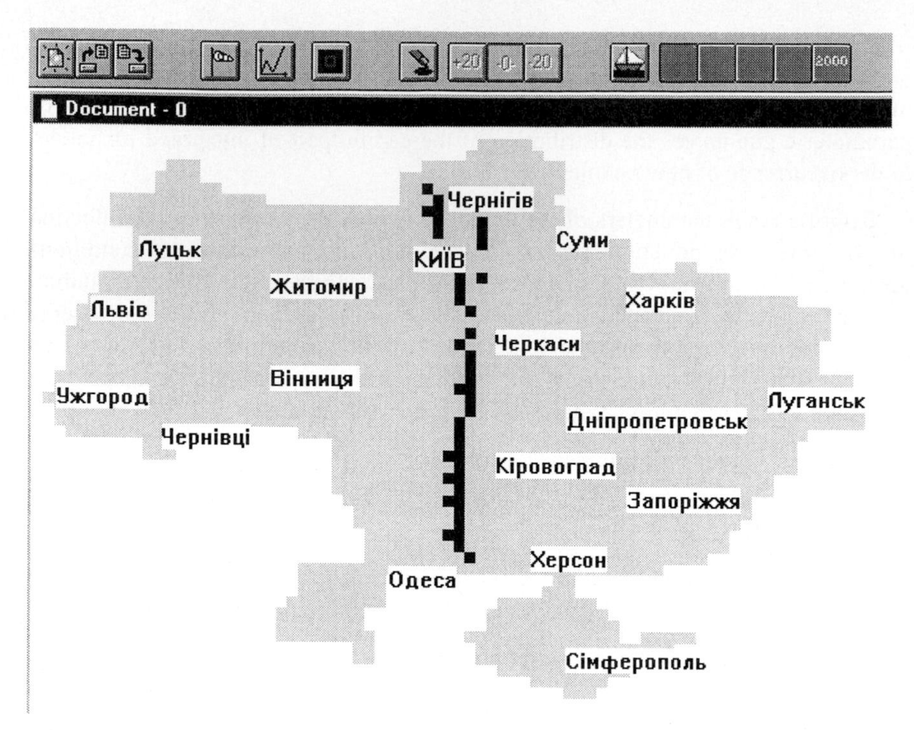

The model of the plant development was identified for the data observation of the nodule nitrogen fixing process on roots of haricot bean plants at a defeat by a virus of a yellow mosaic in seven regions of Ukraine behind a route Chornobyl - Odesa. The observations were carried out over three years. After identification it became possible to find optimum conditions of the development of a healthy vegetative material (features 3.4).

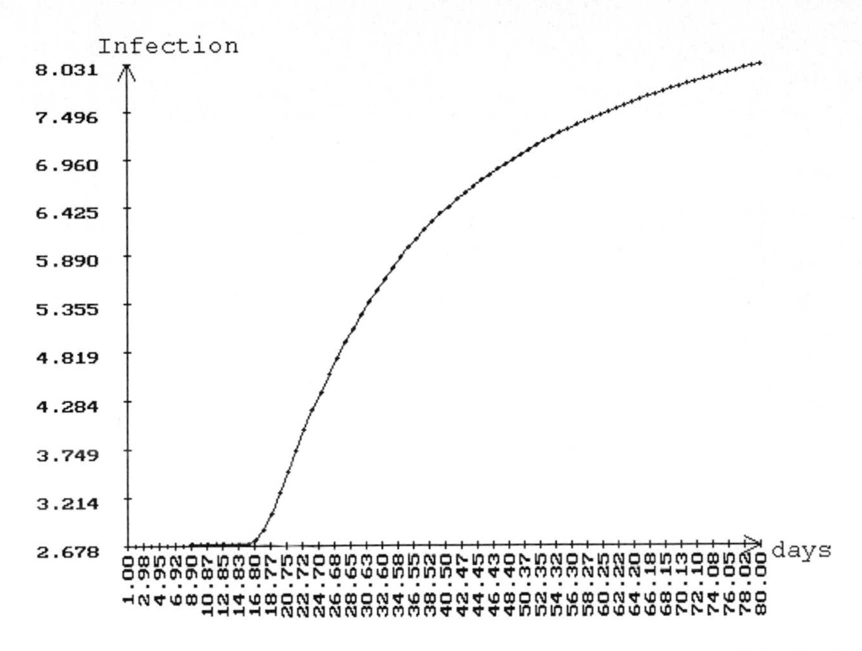

Fig. 3: Development of the infection process during vegetation period in the found optimum conditions

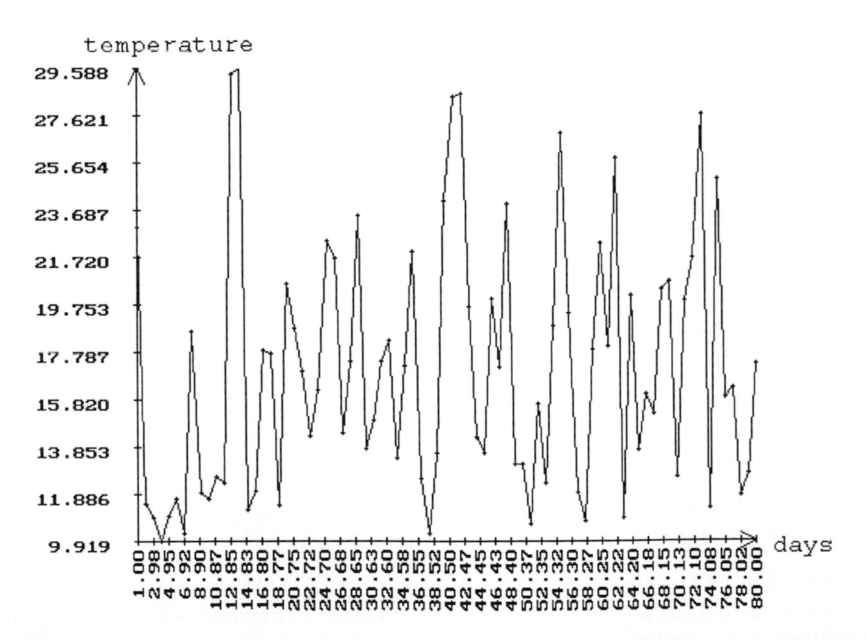

Fig. 4: Change of temperature during vegetation period from a subsystem of optimum conditions.

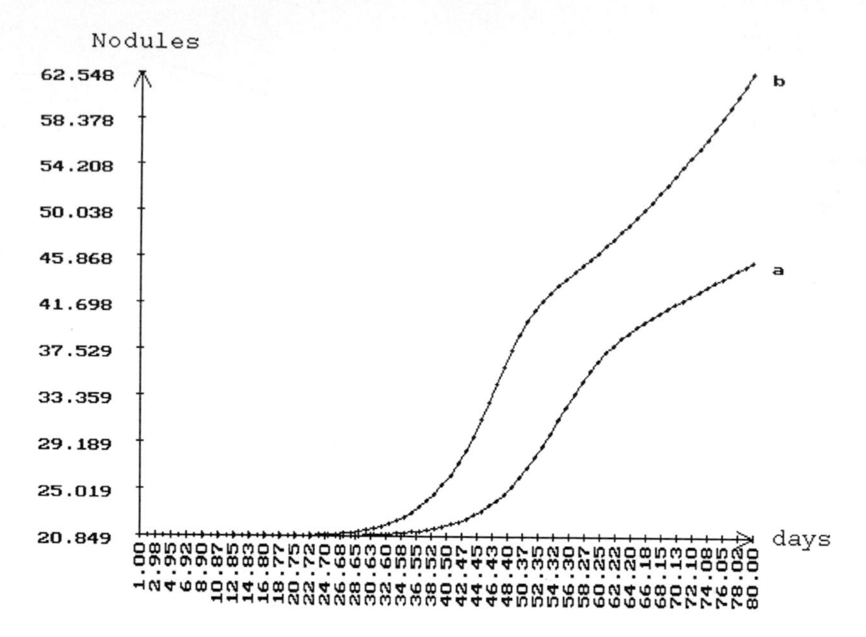

Fig. 5: Dynamics of increasing of nodules number during the vegetation period for healthy (a) and for infected (b) plants.

Thus, CEIS system can be applied in research of ecological processes in different conditions of researched regions, when the region is divided into geographical elements within the given set of abstract subsystems.

Reference

Boyko, A. L. (1985): Ecology of Viruses of Plants; Kyiv: Higher school; 1985 (in Russian)

Boyko, A. L.; Zagorodni, Yu. V.; Beiko, I. V. (1996): Critical State of Plants Under the Influence of Viral Infection and Ecological Disbalance. // Arch. Phytopath. Pflantz, 1996, Vol. 30, pp. 367-370.

Boyko, A. L.; Prister, B.S.; Zagorodni, Yu. V.; Polichuk, V.P. (1994): The virus infection on the cultivated plants at the different ecological niches. // Fundamental and Applied Problems in Phytovirology. Abstracts, Ukraine, Yalta, 1994.

Voytenko, V. V.; Zagorodni, Yu. V. : Creation of computer ecological intellectual system (CEIS). // Zhytomir. The ZIET Visnyk, vol 5, pp. 120-124.

Zagorodni, Yu. ; Boyko, A.; Beiko, I.; Skrygun, S. (1997): Construction of Computer Simulation of Critical Plants State Under Influence of Phytoviruses Infection and Ecological Unstability. // Papers of 15 IMACS World Congress on Scientific Computation, Modelling and Applied Mathematics, Berlin/Germany, August 1997.

Part 2 Business Process Modelling

Guidelines of Modelling (GoM) for Business Process Simulation

Christoph v. Uthmann and Jörg Becker

Westfälische Wilhelms-Universität Münster, Institut für Wirtschaftsinformatik
Steinfurter Str. 107, D-48149 Münster, Germany
Tel.: (0251) 83 38-064/-100, Fax: (0251) 83 38-109
{ischut | isjobe}@wi.uni-muenster.de

Abstract. *Due to necessary specifications of manifold co-ordination mechanisms for managing resource interdependencies and corresponding control data flows, the design of simulation models describing business processes often tend to generate considerably high complexity itself. With regard to economic efficiency of process modelling and simulation studies it is essential to cope with this complexity on a systematic basis. The paper will take up this problem, which has so far rarely been regarded. Coming from the Guidelines of Modelling (GoM), a project of the University of Münster, three major mutually dependent topics are focused upon, and are applied in a case study: View-orientated phases of the model construction, the use of model components and the application of modelling conventions.*

Keywords. *Business Processes, Business Process Modelling, Business Process Simulation, Guidelines of Modelling, Petri Nets, Simulation*

1 Guidelines of Modelling (GoM) and their Contribution to Business Process Simulation

Business process models should be able to serve as a communication basis for all involved persons in all phases of process design and control. The considerations of modelling concentrate on syntactical questions and nave only paid some attention to the semantics of models. Against this background the Guidelines of Modelling (GoM)[1] give design recommendations, whose application increases the quality of models beyond the fulfillment of syntactic rules (Becker, 1995; Schütte, 1998).

---

[1] The GoM are developed within a research project. Numerous publications of BECKER ET AL. can be found under http://www-wi.uni-muenster.de/is/projekte/gom/ .

The Guidelines of Modelling contain six principles to ameliorate the quality of information modelling (see fig 1). These are the principles of correctness, relevance, economic efficiency, clarity, comparability, and systematic design. The term GoM has been chosen as an analogy to the Generally Accepted Accounting Principles (GAAP). On the one hand, the GoM result from the selection of the relevant aspects for information modelling from the GAAP. On the other hand, they adapt elements of existing approaches for the evaluation of information models. While the observance of the principles of correctness, relevance and economic efficiency are a necessary precondition for the designers and users of models, the principles of clarity, comparability and systematic design have a mere additional character. The architecture of the GoM consists of three levels. Every general guideline (level 1) contains recommendations for different modeling views (level 2, e.g., process models) and for different modeling techniques (level 3, e.g., Event Driven Process Chains (EPC) or Petri Nets, see below).

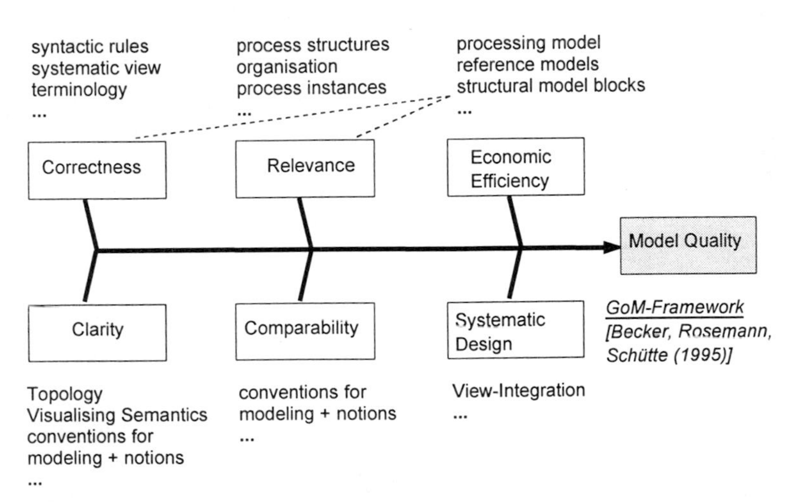

Fig 1: Goal Framework of the Guidelines of Modelling (GoM) including some examples

The design of co-ordination mechanisms in business processes (in the following just called 'processes'), e.g., logistical and manufacturing processes, is often very complex, because the occurring material and information flows show strong interlocks in the involved resources. Therefore, process models in this area often require detailed specifications and visualisations of control flows (Rohrer, 1997). Moreover, complexity increasing details like interactions with the process environment, deterministic or stochastic time variances, queue disciplines and resource choice rules have to be taken into account.

Simulation can serve as an adequate descriptive and experimental instrument for the design of such complex processes. A *process* - for clear differentiation sometimes called process *instance* - will here be understood as a succession of discrete states and functions (activities) respectively, with which an (physical and/or informational) object flow corresponds (here it is not meant the object term in Software Engineering) (Becker, 1996). These object flows result from the mechanism, that output objects (e.g., a drilled workpiece) of functions (e.g., drill workpiece) enter as input objects (e.g., workpiece to be milled) in subsequent functions (e.g., mill workpiece). *Model construction* is understood here as the transformation of existing (system oriented) imaginations of processes into graphical models described in simulation languages. A *process structure model* (process scheme, process type) determinates all possible process instances on type level through topologies of directed and (if necessary) cycling graphs. Going beyond pure static description, a characteristic feature of simulation models, is their *executability*. For this, beginning from an initial state, process constituting model states have to be incrementally derived (feed-forward), according to rules underlying the respective simulation model. Therefore, process structures *and instances* have to be described explicitly.

Process instances of the complexity outlined above can often not be “simply“ designed merely by modelling process structures, which implicate “intuitively“ a description of process instances. The relation between process structures and instances is in complex processes often beyond the modellers intuitive understandings (v. Uthmann, 1997). Consequently, the construction of simulation models tend to generate considerably high complexity itself. With regard to economic efficiency of process modelling and simulation studies it is essential to cope with this complexity on a systematic basis. It is true that there are various phase models for conducting process modelling and simulation studies, but there is only a little attention paid to the complexity management of model construction in the narrow sense mentioned above.

This paper will take up this problem applying the Guidelines of Modelling (GoM). The previous GoMs for process models refer mainly to the construction of process structure models. As mentioned above, dynamic simulation models make higher demands on such a framework of principles. This paper discusses some specific aspects of constructing simulation models which have to be taken into account in addition to the previous GoMs. Three major mutually dependent topics are focused: View-oriented construction phases, the use of model components and the application of modelling conventions. Paragraph 2 contains a general explanation of these concepts. How to them into practice is shown on the basis of a case study in Paragraph 3.

2 Concepts of GoM for Business Process Simulation

2.1 View-orientated Construction Phases

The complex construction problem of simulation models can be decomposed in less complex partial problems by focussing on different aspects of the process description in *views*. Traditionally, from *function view* a process is understood as a succession of object-using, modifying and/or deleting functions (activities), and from *data view* as a succession of states corresponding to the existence of process-related objects (v. Uthmann, 1997). Within this context it will be distinguished between process and resource objects[2]. Whereas *process objects* (more exactly their treatment) are the reason for carrying out processes (e.g., parts to be manufactured or assembled), *resource objects* are only used for carrying out functions (e.g., tools, transport devices or employees). Therefore, a process object is the input and output object of each function in a corresponding process. The logical relations between functions (sequence, AND, XOR, etc.) can consequently be determined within the exclusive view of process objects, abstracting from resource objects.

Because such a *separated view of process object flows* is directed to the purposes of processes it simplifies their identification. Object flows in processes, which do not have connections to other process chains or resource scarcity, run linearly or parallelly, but not complexly with interlocks (see above). Consequently, they can be designed relatively simply. Moreover, the construction of such processes can be done in a first step function-orientated without explicitly looking at the data view. That means, in terms of data, only informal descriptions of split conditions are sufficient. In view of the widely used structure- and function-orientated process descriptions (e.g., work plans and network plans) and process modelling tools (Hess, 1996) (for example PROPLAN, which was created by the WZL (Aachen, Germany) in particular for process management in industrial companies (Eversheim, 1995)) it can be assumed that such a view of processes is intuitively easier to understand, especially by non scientists, and therefore, better accepted than a function-data-integrated thinking in process instances applied in simulation models. A further advantage of a separated view of process object flows is that corresponding process models can be hierarchically generalised or refined without difficulties in the description of indirect resource assignments.

For the model construction a *combined top-down-/bottom-up-approach* is recommended. At first the whole process is refined to independent partial process

[2] Compare to (v. Uthmann, 1997). A similar differentiation can be found in classifications of business rules (Taylor, 1995).

models (modules), which are then separately developed through further refinement and embedding. Finally, these developed modules will be rejoined. Resource object flows only have to be included in the model level, at which model execution is supposed to be run. The links between process and resource object flows can not be constructed with pure structure-orientated model approaches because for this a description of process instances is necessary (see below).

The view-orientated construction phases presented below offer a framework for breaking down the problem of constructing simulation models (table 1). Each of the phases is characterised by certain facts to be modelled. These are always described in the first line of each phase description in table 1.

2.2 Model Components

The aim of using (best practice) model components, which will be composed, is a more efficient and secure model construction. There are reference simulation models in the form of *context-related* model components (Carrie, 1992; Mertins, 1996). Complementary to these ones here are presented model components, which are not related to concrete organisation or engineering problems, but systematise *context independent structure analogies*. Their higher abstraction level allows the use of these components for a simplified individual construction of *unique* simulation models *from scratch*. This is often important, in particular for the detailed conception of business processes. The model components are related to single views, and therefore, are assigned to certain construction phases. Here only a few components can be presented exemplary (table 1).

2.3 Model Conventions

Methods of process modelling contain generally just a set of syntactic rules, which give modellers many liberties to construct models. Therefore, one fact can be depicted in different models. Model conventions are supposed to restrict this liberty. One important intention of this is to ensure a uniform, clear and unequivocal understanding of models by all users involved. Besides this *semantic* aspect *technical* aspects like the performance of simulation models have to be taken into account. Hence, the application of model conventions can increase the efficiency of using simulation models. Table 1 contains some exemplary conventions for the construction of simulation models.

Phase 1	**Process Object related Structures in the Function View**
(Partial) Processes	Finding (partial) processes results from identification of process objects (see above). In industrial processes, these are often parts to be manufactured, assembled or transported and corresponding information objects like manufacturing plans or NC-programs. In the administrative domain pro-

	cess objects are often documents - like invoices, offers or complains - which initiate workflows (see Workflow Management).
Functiona l Structures	Functions are object transforming elements, which refer to single tasks within processes regarding these declaritively as black-boxes. The time required for the execution of functions can be declared by fix values, stochastic distributions or dynamic values respectively (see below). Functions can refer to complex activities, which are subdivided by hierarchical refinements. Logical relationships between functions are derived from process object flows (see above).
Control Data Flows	The internal process interfaces, i.e. the connection between functions, are constructed by describing control conditions which trigger functions. Within this phase it is sufficient just to give informal descriptions about starting, splitting and ending conditions.
Description	Process structures described this way have to be defined in the used simulation language. Functions can be either represented by active (e.g., ''mill workpiece'') or passive, state related (e. g. ''workpiece is milled'') passive model elements.
Conponents	function, state/condition, sequence, XOR, OR, AND, iteration (loop), deterministic/stochastic split, horizontal/hierarchical connection, fixed/stochastic/dynamic time requirement for a function/state, number of process objects
Conventions	It has been shown to use the natural languages for process descriptions (phase 1.1) from the beginning, which is implemented (if necessary simultaneously) in a simulation language (phase 1.2). For the clarity of description (diction, inscriptions) general and internal terminological conventions are necessary (technical terms model). Within a representation of functions by passive model elements the active model components describe only timeless and resource free transitions of states. Such a data-orientated view is especially useful for the purposes of process automation, because control data related states have to be controlled and therefore be explicated. The first representation alternative (see section 'Description') is suitable to focus on active process elements with consumption of time. Apart from conventions for possible, self-explanatory terms of passive and active model elements (see above), the logic structures (AND, XOR, OR) should be explicated by appropriate comments on the elements. Regarding the validation of simulation models, stochastic distributions of time should be neglected because stochastic time consume in test simulations might be difficult to evaluate.

Phase 2	**Resource Object Flows**
Resource-objects	Resource Objects can be human, material, operational, informational, financial, energy and time resources. Relevant operational resources in industrial processes are often machines, transport devices, tools and other technical devices. The most important resources of administrative processes are human resources.
Coordination Mechanisms	For reflections on process instances resource interdependencies of parallel process chains resulting from supply shortfalls are particularly interesting. For the solution of conflicts between competing functions appropriate coordination mechanisms (e.g., priority rules) have to be specified on an instance level. Apart from supply shortfalls the effects of resource quality on functions (e.g., execution times depending on certain attributes of resources) might be relevant.
Components	store/stock capacity, resource occupation/release, resource amount, resource capacity, competing function, priority rule, queue discipline (FIFO, LIFO etc.)
Conventions	If the scarcity or the quality of resources does not have any impact on process instances the use of special resource related passive model elements is not necessary. Comments attached to the active model elements might be more accurate. The same is valid for state related passive model elements within linear function sequences if there are not any changes of process or resource objects. In these cases summarising elementary functions is sufficient. For getting an additional visualisation of the semantic context of passive model elements it is recommended to indicate them differently (e.g., different colors). Possible here is a differentiation in 'related to process objects', 'human resource objects', 'technical resource objects which are assigned to corporate divisions' and 'other technical resource objects'. In keeping the model transparent and avoiding overlapping vertices, the passive model elements referring to resources should be constructed by so called fusions (see below).
Phase 3	**Formal specification of the data model**
Data Model	The intention of data modelling is the description of process constituting of states related to the existence of certain objects. Within process structures data is described on type level, (e.g., as ERM or record), within process instances on attribute value level. From the process structure model there has to be a reference to the data model.
Montor-ing	For runtime monitoring and reporting the model must be added by passive and active model elements dedicated to the extraction of certain data.

Compo-nents	Control/application data /-object -types/-objects, attribute, attribute value, data extraction
Conven-tions	Although a distinction between control and application data is essential for process automation, their separated construction in different nodes is not necessarily advantageous, because they do not always belong to disjunctive sets. Nevertheless, data should be indicated as control or application data, for example within comments on active model elements. Often it might be sufficient to use unstructured, non distinguishable objects (e.g., black tokens in lower Petri Nets). Corresponding (trivial) data models have not to be explicated. Resulting simulation models work in general, quicker. On the other hand, the use of distinguishable objects (e.g., in PR/T-Nets) can reduce the complexity of models topology significantly.
Phase 4	**Procedural Transition Description**
Procedure	The input/output behaviour of transition of functions has to be specified through any script language. This also includes the assignment of stochastic and dynamic execution times (see above).
Compo-nents	programming language constructs, connection of procedures with adjacent states
Conven-tions	Besides minimising the number of elementary functions (compare the conventions of Phase 2) results from Software Engineering here are of special interest.
Phase 5	**Interdependencies with the process environment**
Input of Process Objects	Interdependencies with the process environment are constituted by flows of process and resource objects, which go in and come out the model via external interfaces. A core aspect within this context is the construction of input sources, which generate process objects (e.g., orders) for instantiating processes or reduce resource objects (e.g., night shift with reduced staff). Appropriate stochastic input source models are generally defined by stochastic distribution functions. The most important ones are the Poisson and the Exponential Distribution respectively. Within the context of simulation the inclusion of input sources requires particular simulation models, which often are provided as standard components. An interesting possibility of constructing deterministic input source models is the extraction of data tables or lists which may refer to real data.
Compo-nents	stochastic/deterministic input source, connection of input source models to the core model
Conven-tions	Regarding the transparency and variability of models particular input source models implemented as separated modules should be indicated appropriately.

Phase 6	Disturbances and Changes
Distur-bances / Changes	Besides planned process input and output disturbances have to be described. Important disturbance source models initiate the withdrawal of resource object describing tokens (e.g., the failure of a machine), incorrectly executed functions or faults in data transmissions. For the construction, the same principles are applicable as for input source models (see phase 5). Changes are in the same way modelled as disturbances.
Phase 7	**Initial Model State**
Initial State	Initial model states refer to starting conditions, from which process instances are supposed to be executed (see above). They have to be defined on instance level.
Compo-nents	initial state
Conven-tions	As Phase 5 (initialising models).

Tab. 1: Procedure of constructing a simulation model

3 Petri Nets as a Reference Method

The GoM for Business Process Simulation should be applicable to a variety of simulation tools. While the construction phases are independent of tools, the model components and model conventions have to be refined in terms of the underlying simulation modelling techniques intended for use in each case. That means they have to be put into concrete forms using the modelling technique specific construction elements and terminology.

For the starting considerations, Petri Nets (Reisig, 1985) were chosen as a reference Method. The reasons for this were, in particular, on to the following advantageous characteristics of Petri Nets:

1. *worldwide known*: The basic concepts of Petri Nets for describing explicitly and unambiguously process structures and instances are worldwide taught and applied within various fields.
2. *universally applicapable*: Petri Nets as a method can be applied independently of tools and application domains. Moreover, Petri Nets offer simultanously a data- and a function-orientated process view (see above).
3. *appropriate for the business administration domain*: The constructive concept of Petri Nets transforms the applied object based process understanding

perfectly[3] and is compatible to many Business Process Reengineering (BPR) model techniques. Within this context they are perfectly compatible to the Event Driven Process Chains (EPC) (Scheer, 1994) which are used in ARIS from IDS-Prof. Scheer GmbH[4] (the world market leader in BPR tools) and SAP R/3 (v. Uthmann, 1997). Moreover, there are several Petri Net based BPR[5] and Workflow management tools[6] (Van der Aalst, 1996a; Van der Aalst, 1996b).

4. *compatible to many simulation model techniques*: The constructive concept of Petri Nets are compatible to most major simulation model techniques.
5. *formal and automatic analysability*: The formal semantics of Petri Nets allows - complementary to simulative experiments - several formal and automatic analyses on structural and behavioural properties of models.
6. *extensively explored*: There are extensive world wide research activities within diverse aspects of Petri Nets, so that there is already a great fund of findings with respect to the construction of models.

Petri Net specific GoM are (with restrictions) transferable to other simulation modelling techniques due to the mentioned compatibility (point 4). Exemplary, Fig. 2 shows a comparison between Petri Nets and AweSim from Pritsker.[7] AweSim is known worldwide and is used in research as well as in industrial projects. AweSim originates from queuing systems analysis and uses function-arc-orientated graphs. AweSim itself is domain neutral. However, there is the very successful system Factor/AIM[8] dedicated to simulate manufacturing systems which is an adaptation of AweSim. Also, the world market leader in manufacturing simulation systems ARENA/ISSOP[9] has some similarities to AweSim which is based on common historical origins.

Petri Nets are a more general technique then AweSim: While the objective of Petrinets is based on all Kinds of systems (which they construct only with passive and active elements and vertices relating to these) AweSim focuses on queuing processes.

---

[3] In the business administrative domain an object based process understanding has been established for a long time (Nordsieck, 1934)

[4] http://www.ids-scheer.de/

[5] E.g.: Protos from Pallas Athena BV, the market leader in the Netherlands. http://www.pallas.nl/

[6] E.g. COSA Workflow from COSA-Solutions, one of the leading Workflow Companies in the World. http://www.cosa.de/

[7] http://www.pritsker.com/awesim.htm

[8] http://www.pritsker.com/aim.htm

[9] http://www.dual-zentrum.de/deutsch/simulat.htm

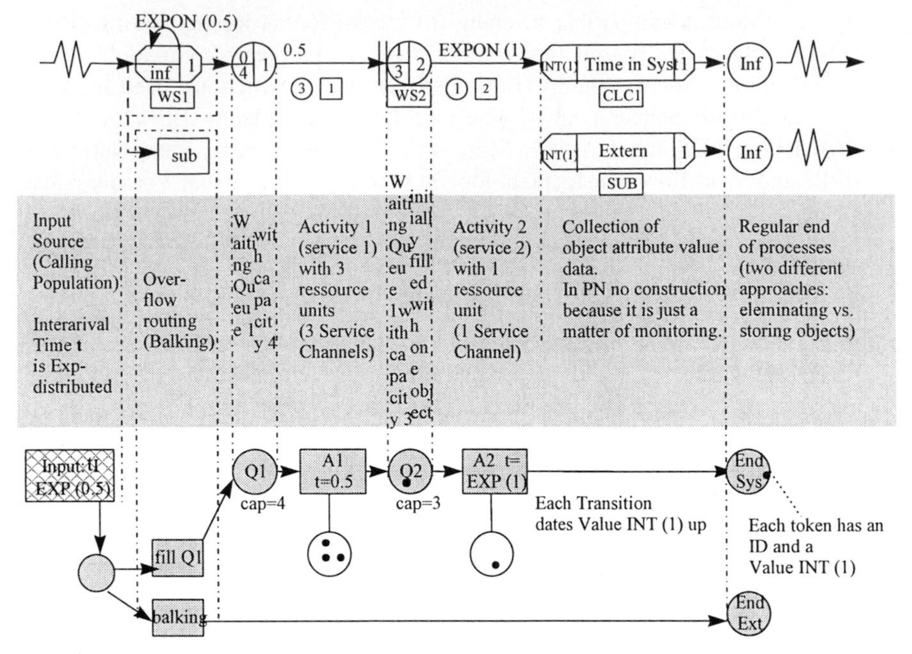

Fig 2: Comparison between AweSim and Petri Nets

AweSim has a higher variety and more specialised construction elements than Petri Nets. Therefore, some elements of AweSim are analogous to non elementary Petri Net components (see 2.2). This variety and specialisation effect increases with the domain orientation simulation tools (see Factor/AIM and ARENA/ISSOP. The additional (partly more complex) construction elements require additional modelling conventions. While the abstract components required in each phase remain the same, their realisation varies with the used model technique.

Specialised model techniques might be able to realise certain components in less complicated ways, sometimes only with one single standard element (on the other hand, users have to handle more construction elements). Nevertheless, to be flexible in terms of constructing different problems (e.g. office organisation, automated processes in manufacturing or workflows) these tools also provide constructive elements of a similar simplicity as the Petri Net elements. For example, AweSim Queue nodes correspond to PetriNet places with capacity; AweSim arcs correspond to Petri Net Transitions. In this respect the specific GoM for Petri Nets offer some kind of maximal set which can be flexibly adapted to other simulation techniques.

Regarding this and the reasons above, Petri Nets are certainly a reasonable choice to create a first base of GoM for Business Process Simulation. The case study which is presented in the following paragraph shows some GoM applied to Petri Nets.

4 Case Study

The presented approach was partly developed and applied during a simulation study about a flexible manufacturing system (FMS) (v. Uthmann, 1995). The intention was to find out the demand for specific resources (workers, machines, transport devices, etc.) as well as the development, specification and evaluation of alternative control concepts for manual and automatic production. The simulation study was conducted with the wellknown Petri Net based BPR- and simulation tool INCOME (Oberweis, 1996) from Promatis.[10] Today, the FMS is the central unit within the turbine production for manufacturing different screws and shafts. It consists of two identical manufacturing lines which contain various machines and one transport robot each. The processes which had been unknown at the beginning of the simulation study turned out to be highly complex. This was because the work pieces had to run through strongly different iterative work sequences, and because more than one work pieces had to be produced parallely on each line within various resource-interdependencies. Furthermore, a connection between the lines through commonly used machines had to be regarded. Starting point of the model construction was a feasibility study which contained the problem definition as well as design and project plans.

During **phase 1** (see fig 3) all important system components and the manufacturing process were described with natural language in a function orientated way. As Process objects were identified as work pieces which could be described in specific states (from raw material to finished work piece).

The whole manufacturing process was divided top-down in partial processes. These partial processes were presented by Predicate/Transition (P/N) Nets (a higher Petri Net type) on the lowest model level. Due to the process control focus it was important to describe all control data exactly. The interfaces between two different work operations were formed respectively by one transport process and vice versa. In correspondence to this all partial processes were horizontally combined within a (transport-) matrix-structure. The resulting model was enriched by embedding other partial processes, and the relevant organisational constraints were specified (i.a. splitting the process in two manufacturing lines, integration of preassembly processes, switch between manual/automatic production).

---

[10] http://www.promatis.de/

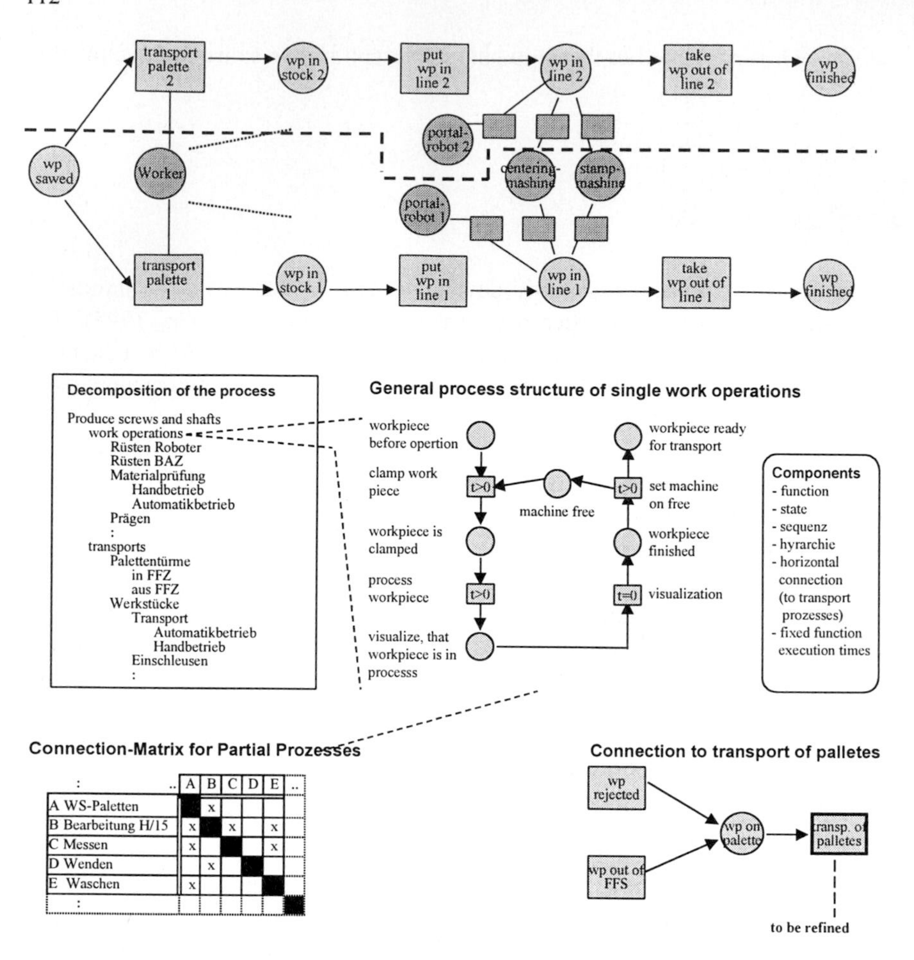

Fig 3: Partial Models in Phase 1 of the Simulation Study

Starting point of **phase 2** (see fig 4) was the static registration of all resource assignments to the partial processes described in phase 1. To realise the requirement that the portal robots or workers carry out parallelly other transports and operations respectively, it was necessary to specify appropriate coordination mechanisms. Moreover, coordination mechanisms to change a work piece through a turnable robot grab had to be modelled. In keeping the model transparent and for avoiding overlapping vertices the passive model elements referring to resources were constructed by so called fusions. Fusions represent a passive P/N Net element (channel) logically existing only once in multiple physical graphic expressions.

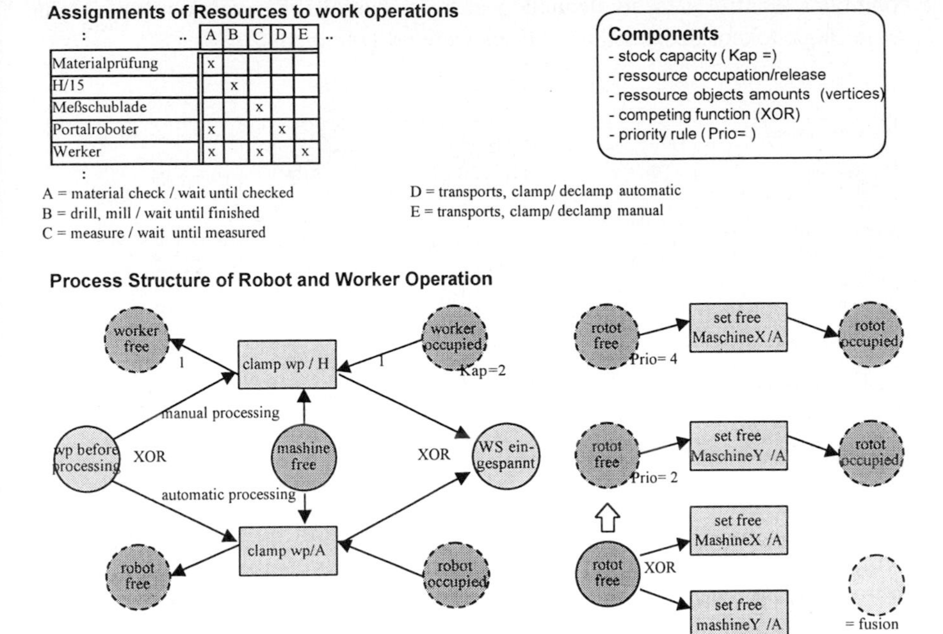

Fig 4: Partial Models in Phase 2 of the Simulation Study

Phase 3 (see fig 4): Numerous binary control data like 'machine free/occupied' or 'manual/automatic mode' were represented by structureless channels which were marked with black tokens on instance level. The impact of resource quality on the execution of functions (e.g., measure through a worker or a machine) could only be considered dichotomously. Therefore, it was sufficient to use simple splits of the functions, so that all resource object flows could be represented by black tokens. In contrast to this, the process object flows were represented through individual tokens whose structure was described through records on type level. These represented on the one hand all physical work pieces (predicative data), on the other hand they each contained performing operation sequence as control data (in accordance with work plans) and application data (here just execution time) for all individual operations (in accordance to NC-programs). The practical nonconformity of such an "encapsulating" of the physical and informational object flow was highly compensated by the minimised complexity of the model structure. For runtime-monitoring, reporting and model validation status attributes of each performing operation sequence ("not executed/executed") were added. After executing functions these were set appropriately through postconditions.

The procedural functions description in **phase 4** (see fig 5) was - in accordance to the tool - realised through PROLOG-programmes. To this belongs the setting of determined delays and performing status as well as the processing of split-conditions (control data) implemented in the transitions. For validating the status set in single tokens dedicated transitions were programmed.

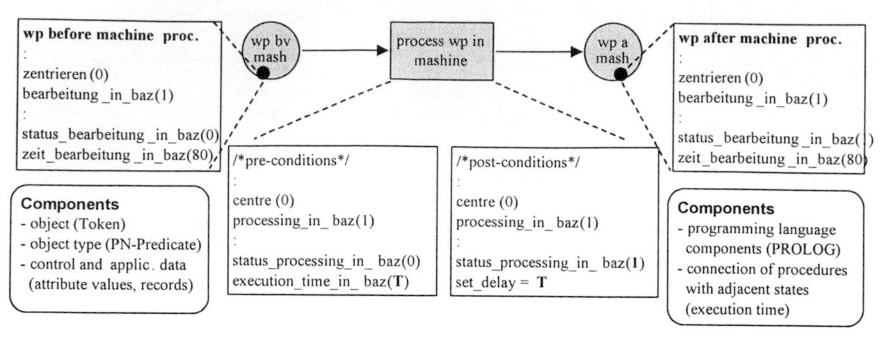

Fig 5: Data Model and procedural Function Descriptions

The interdependencies of processes with the environment specified in **phase 5** were constituted by the input of certain work pieces to be processed. An appropriate input source was modelled through reading dedicated files whose data records contained the attribute values of individual tokens (see fig 5). Different token mixes were generated with the help of a PASCAL programme which provided an appropriate dialogue.

Disturbances and changes (**Phase 6**) were at first not considered. In **phase 7** a particular initialising model (see fig 5) was constructed. It generated a start marking which represented two empty and completely functioning manufacturing lines.

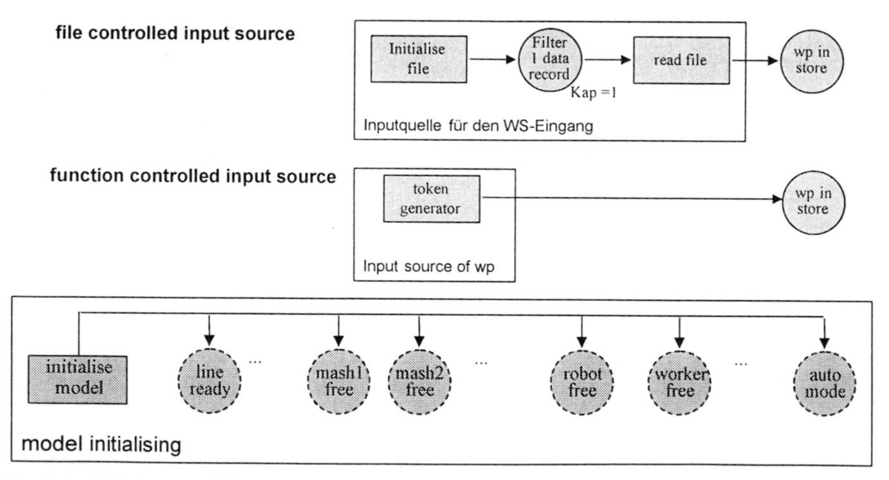

Fig 6: Initialising models

5 Conclusion

The concepts presented here offer, in particular for less experienced modellers some hints for a systematic and adequate construction of complex simulation models. They make for more sensitive than complex details which is particularly important to complex business processes. In this way, a participative construction of simulation models is made possible. By describing first function sequences in natural languages and/or simple graphical models specialists without modelling experience are also able to take part in the construction process.

Thus, complexity management of the construction of simulation models for business processes can make a significant contribution to the efficiency, clarity and security of process modelling and simulation studies. Moreover, it discloses the utilisation of complex simulations of companies with more limited human and financial resources.

Petri Nets are in many respects a proper reference method for developing GoMs for Business Process Simulation. The extensive findings in the Petri Net field can give some contributions to further research. Other interesting inputs can come from analysing, consolidating and integrating the numerous reference models for logistical and administrative processes.

References

Alan, A.; Pritsker, B.; O'Reilly, J.J.; LaVal D.K. (1996): Simulation with Visual SLAM and AweSim. Indianapolis, USA 1996.

Becker, J.; Rosemann, M.; Schuette, R. (1995): Grundsätze ordnungsmäßiger Modellierung, Wirtschaftsinformatik, 37, 4 (1995), pp. 435-445.

Becker, J.; Vossen, G. (1996): Geschäftsprozeßmodellierung und Workflow-Management: Eine Einführung. In: Geschäftsprozeßmodellierung und Workflow-Management. Hrsg.: G. Vossen, J. Becker. Bonn et al. 1996, pp. 17-26.

Carrie, A. (1992): Simulation of manufacturing systems. 2nd edition. Kilbride, Great Britain (1992).

Eversheim, W.; Heuser, T. (1995): Prozeßorientierte Auftragsabwicklung in der kundenorientierten Fabrik. ZWF, 90 (1995) 1-2, pp. 28-31.

Hess, T.; Brecht, L. (1996): State of the art des Business process redesign: Darstellung und Vergleich bestehender Methoden. 2. Aufl. Wiesbaden 1996.

Mertins, K; Rabe, M.; Friedland, R. (1996): Simulations-Referenzmodelle erschließen neue Potentiale. ZWF, 91(1996) 10, pp. 479-481.

Nordsieck, F. (1934): Grundlagen der Organisationslehre. Stuttgart 1934.

Oberweis, A.; Stucky, W.; Zimmermann, G. (1996): INCOME/STAR: Facing the challenges for cooperative information system development environments, in: W. König, K. Kurbel, P. Mertens, D. Pressmar (ed.): Distributed Information Systems in Business. Berlin et al. 1996, S. 17-34

Reisig, W. (1985): Petri Nets - An Introduction. Berlin 1985.

Rohrer, M. (1997): Seeing is believing: The importance of Visualization in Manufacturing Simulation. IIE Solutions, (1997) 5, pp. 24-28.

Scheer, A.-W. (1994): Business Process Engineering - Reference Models for Industrial Enterprices. 2nd ed. Berlin et al 1994.

Schütte, R.; Rotthowe, T. (1998): The Guidelines of Modeling - an approach to enhance the quality in information models. Proceedings of the ER'98. Singapoore 1998.

Taylor, D. A. (1995): Business Engineering with Object Technology. USA (1995).

v. Uthmann, C. (1997): Nutzenpotentiale der Petrinetztheorie für die Erweiterung der Anwendbarkeit Ereignisgesteuerter Prozeßketten (EPK). In: Oberweis, A.; Rump, F.: Proceedings zum Workshop "Formalisierung und Analyse Ereignisgesteuerter Prozeßketten (EPK)". Hrsg.: Oldenburg 1997. http://www-is.informatik.uni-oldenburg.de/~epk/.

v. Uthmann, C. (1995): Systemanalyse und -entwickung einer flexiblen Fertigungszelle mit Hilfe von Petrinetz-Simulation. Diplomarbeit, Universität-GH Paderborn. Paderborn, Germany 1995.

Van der Aalst, W.M.P.; van Heh, K.M. (1996a): Business Process Redesign: A Petri-net-based approach. Computers in Industry, 29 (1996) 1-2, pp 15-26.

Van der Aalst, W.M.P. (1996b): Petri-net-based Workflow Management Software . In Sheth, A. (ed.): Proceedings of the NFS Workshop on Workflow and Process Automation in Information Systems. Athens, Georgia 1996, pp 114-118.

Modelling and Evaluation of Processes Based on Enterprise Goals

Frauke Weichhardt

UBIS GmbH, Alt-Moabit 98, D-10559 Berlin

Abstract. *In order to support a more structured process design, many businesses today are designing their business processes using methods and tools for process modelling. Structured process design and planning increases the likelihood that newly designed processes can be realised in a timely manner, because process participants understand the need for the changes and understand how to integrate them with existing processes and resources.*

Keywords. *business process reengineering, process modeling, modeling tool, model, quality, process simulation, process visualisation, process analysis*

1 The Role of Process Modelling

Typical organisational changes are often spontaneously decided. Those impacted by these decisions seldom have the opportunity to give input or to understand the rationale behind the decision. Line employees simply have to learn to live with the new situation.

For example, the department head of Sales tells the Executive Management Committee that during the last month, 10% of all customers have changed to its competitors. Management decides that the company needs to strengthen customer relations. After a short brainstorming session, the committee decides that customer relationships will improve if service is delivered faster. The proposed solution recommends that all employees in contact with customers must deliver service in half of the current time. This decision is to be communicated immediately and employees are expected to immediately work faster. The Head of Sales Department is given full responsibility to see that correct actions are taken.

What will he do?

As he is not entirely informed about what is going on in his department, he might write a motivating letter to all employees that he thinks might be in contact with customers. In the letter he informs them about the dramatic situation and tells them the decision of the committee. He announces strict performance checks using a new performance measuring system. Four weeks later sales have decreased again as half of the remaining customers have switched to competitors. It is only a matter of time before there are no customers left.

On the other hand, if the department head had documented his processes, he could quickly identify the process models containing customer contacts, along with the names of the responsible employees. He could then discuss the current problems with those employees to develop an appropriate response. By effectively using process models, all those involved are able to identify potential problems and determine appropriate corrective actions. During the process focused discussion participants might find out that a new employee in accounting was not properly trained and it is the delay in bid calculation that is slowing down the process. As we all know, life is not that easy.

Here is another possible scenario: Two years ago the department head contracted two consultants to model processes in the sales department in order to identify potential cost reducing measures. After the project, the resulting models were stored in the desk of a department employee. This employee has since left the company. After hours of searching, the models are found again. After reviewing the models everybody reaches the same conclusion - they do not understand the models nor can figure out why the models were created in the first place.

Why is it so difficult to interpret old models?

Without benchmarks to show modeller's assumptions it is difficult to put a model into its proper context. The pictorial representations also look chaotic. Lack of context and models with too much information in a single model means that it will take an exorbitantly long amount of time to understand the models. In the end the models are put aside and ignored and the department head simply sits down with his team leaders and asks them what they would do.

Without the support of models a possible recommendation from the sales staff might be to hire new sales support people. The proposal is agreed upon by most of the participants and management decides to implement this suggestion. Sales no longer deteriorate, but sales costs rise to the point where the profit margin is unacceptably low. The executive committee, in turn, decides to implement a cost reducing program that recommends reducing the number of employees.

Could anybody have done anything differently in order to make use of the existing models?

2 Possible Problems Concerning Modelling

The scenario mentioned above is often found in companies that have already tried process modelling. Models are made with great effort and disappear into somebody's desk. The resulting attitude is that modelling takes too much time, costs too much money and does not deliver long-term benefits. At the same time there continues to be a diffuse sentiment that it is necessary to know about existing processes and to be able to talk about them. There is also the general conviction that graphical representation of processes is helpful. Up to now, existing methods have not been able to fulfill the promises process modellers tend to make. What are the major disappointments in process modelling?

- Most common process modelling symbols are hard to understand unless the viewer has an engineering or software development background because symbols usually have specific meanings that need to be learnt. They evolved from a technical view of the world. What cannot be clearly defined and measured is simply ignored. If we look around us in the companies where we work, it is difficult to find complete clarity or to represent all of the alternative ways of doing things. People communicate with words and pictures that often have multiple interpretations. All of these factors make it difficult to create useful, easy to understand, process models. Modellers need a tool that allows average employees to work with the models so that they are willing to participate and are able to easily find and correct the representation of the tasks that they perform.

- Models contain too much information and become chaotic, even for small processes. To be able to get an overview of the whole process and obtain enough detailed information requires the investment of a large amount of time. This additional time does not improve the willingness to work with the models. Modellers need a tool that has an additional navigational dimension to allow viewers to drive down deeper into detail or to be able to jump from one process to another.

- Models often do not conform to reality. It is difficult to maintain consistency when modelling the complexity of the real world. The same facts may be represented differently in one part of the model than in another. Determining which part is correct requires additional investigative work that takes time. If that time is not available the models cannot be used as a basis for decisions and they are therefore useless. Modellers need a tool that either helps them maintain consistent models or to help them properly document the actual inconsistencies that exist in the real world.

- Because there are so many different types of modelling goals, the purpose of the model is often not immediately clear. This makes it hard to understand the model in its appropriate context. The goal of the model needs to be clearly documented. Project success and planning depend upon the ability to bundle activities in a way that focuses on achieving the defined project objectives. The easiest way is developing a list of detailed questions that have to be answered by the modelling project. Only objects that are related directly to those questions have to be modelled. (Krallmann, 1998).

- Systematic maintenance of models requires appropriate resource allocation. Due to these additional expenses, it is either done poorly or not done at all. Modellers often try to maintain their own models but as soon as they leave or take other assignments, their successor seldom continues the work. The lack of continued support from others is due to unfamiliarity with the models and lack of general support from central management. This is especially true for individuals who do not have a modelling background.

How can we extract the use of process models, so that their value is clearly seen by everybody?

3 Modelling Based on Strategic Goals

Process modelling is done for various reasons. Models have goals and they have content. Goals need to be clearly conveyed to everyone that will be asked to interpret the models. For example, a goal might be generally stated as "strengthen customer relations" or it might be operationalised to "increase customer return rate by 50%". Operationalisation is necessary to set attainable targets. The "as is" model is used to find out the current company situation (e.g. it might be discovered that nobody checks to see if a customer has already been there before). A good model needs to include operationalised goals in order to determine the rate of change.

What types of goals are relevant in this context? The goals that are determined by the strategic plan of the company are the goals that should be used as an input for process modelling. These modelling goals have to be concrete strategic goals. They are determined by numbers like sales volume, profit, rentability, costs, market share etc. Strategic goals are the frame of reference for every economic concern in the company. What needs to be accomplished in order to meet these operationalised goals? The answer lies in the procedures used to develop the models.

Strategic goals should represent those process attributes that can be numerically measured. The first step is the determination of process requirements. These can be obtained by asking the following question: "Under which circumstances will we achieve these goals?". The requirements concerning process design defined in this way can be of varying importance by classifying them as critical success factors. In this manner priorities can be set while designing the process and conflicting requirements can be identified. The last step is the determination of the target numbers for the process to determine if strategic goals have been set. This check should be done by simulating the process models before the implementation of the process in order to avoid making unnecessary mistakes.

In order to obtain meaningful process models it is therefore necessary to define the questions that must be answered by the model. Focused process design requires detailed operational goals. One possible method is that developed by Hoffrichter (Hoffrichter, 1998):

Strategic goal	First concrete steps: Determination of circumstances	Second steps: Requirements concerning process design	Third steps: Critical success factor	Fourth steps: Characteristic number for the process
Contract finalisation via Internet	➪ Which conditions have to be fulfilled to make contracting via internet successful?	➪ security ➪ contract must be easy to fill out ➪ process has to be transparent	➪ graphical view of the process	➪ satisfaction concerning contract (accuracy): at least 95 % positive answers
Strengthening of customer relations	➪ Which one of our services and products makes the customer come back?	➪ Customer is served within 5 minutes ➪ Contact must offer valuable information	➪ Enough information needs to be provided to make a decision possible	➪ Customer satisfaction (percentage of customers that come again > 50%)

Table 1: Creating concrete strategic goals

It does not make sense to evaluate a model that has not been modelled with clear goals and clearly defined target values. Conforming to goals insures that models contain appropriate context information so that they can be interpreted in other contexts.

If questions regarding the content of the model and its strategic goals can be adequately answered, the purpose of the model is clear. For example, are the models designed to achieve process clarity, to perform quantitative analysis or do they describe IT system requirements? These criteria determine the necessary level of detail, the scope of the modelling project and which modelling option to use (e.g. loops can be shown as loops or just as one activity). The purpose of the model determines the model type.

Certain factors need to be taken into consideration from the beginning of a modelling project to be able to answers specific questions in regarding how close we are to attaining target values. Changing model direction later is much more complicated than setting and attaining clear objectives at the beginning. This is especially true when evaluating process alternatives.

Focused modelling projects allow models to be created that can be understood with or without the modeller being present. This is because the relevant context information is delivered with the model. Goals and objectives are implicit in the model design. This is also true of the graphic presentation of the data. How the models look is determined by the goals of the model.

4 Drawing Methods for Models

Model data can be represented in very different ways. Symbols might be used or a specific model structure may allow easy navigation through the model.

Symbols should be as simple to understand as possible in order to give casual users a chance to understand the pictures without learning the underlying method. Communication is the central purpose of models and only models that are easy to read are able to attain their optimal value to the organisation.

Established legacy notations such as EPC or IDEF and upcoming new standards such as the Unified Modelling Language (UML) may be well suited to describe software systems but they are not intuitive to most non-engineers. If these sorts of process diagrams are to be used by non-engineers, people will need additional time to learn about the notation and to alter their thinking to that of software engineers. A graphical language that can be quickly learnt by non-technical people is a better alternative. Graphic displays need to be kept simple, just a small number of symbols with each symbol mapping directly to a real world object.

Successful models also require structure. The use of clearly structured sub models allows casual users to navigate through the model content without the risk of getting lost in huge networks of nodes and links. Reducing complexity requires hierarchichal structures and multiple views. Specific aspects of models can be emphasised by showing how business objects are used in different scenarios. Using these technologies users are able to focus on the information they need and they can easily ignore information they do not need.

5 Modelling Methods

Hierarchical structures and multiple views split the model into different parts. Modelling time is reduced and consistency is maintained because the same objects are modelled in all views (e.g. both the data model and in the process model). Process refinements contain processes that are linked with all the other refined processes belonging to the same hierarchy. A modeller is only able to model one view at a time. Tool support is necessary to maintain consistency across all views.

Some existing tools provide views of the organisational structure, process flows and data models. Very few tools allow users to get from one view to another. This is because views are modelled independently and are often inconsistent. This is also true of refinements within hierarchical structures. The logical link between the model and its refinements is generally missing. Modellers are responsible for maintaining consistency between partial models and the model overview using numbering systems or other types of pointers. This often results in hard to identify inconsistencies that can only be eliminated by expensive and timely investigation.

6 Organising Models

Process models are often used to support activities outside of their original purpose or long after the modelling project was completed. How do model viewers determine if the models are valid in this new context? Optimally, modelling tasks should be clearly distributed along with the responsibilities for model maintenance and the preparation of the documentation of modelling standards. A data dictionary is also helpful as it explains all object names. If these conventions are followed, even modellers unfamiliar with the existing models can be quickly integrated into new projects so that they can best take advantage of the existing model library.

7 Example of Focused Model Design

This example is to clarify the focused modelling method.

A company funded a large project designed to improve its IT support. One part of the project is tasked with evaluating IT support in customer care. Given the changing circumstances (e.g. new technologies, new products, new market requirements), existing process definitions need to be reviewed. The strategic goals of the project are to increase sales by strengthening customer relation and to lower costs by standardising business processes across all customer segments and business units. Furthermore, the project is to define IT requirements based on these new process definitions. Care also needs to be taken to ensure that defined business processes can cope with the upcoming amount of data.

First concrete steps: What can be done in order to strengthen customer relation? Initial observations determine that the complaint resolution time is too long. This situation needs to be changed as soon as possible.

7.1 Project Goals

The first concrete step is to define general project goals. In this case it is to provide appropriate IT support and to reduce the number of loops in the process. The second concrete step is to make changes that will improve response time. The third concrete step is to prioritise the critical success factor while working on process design. The fourth concrete step is to describe the exact handling of incoming complaints.

7.1.1 Other Goals

- Integrate all existing customer services
- Combine process competency with associated responsibilities
- Create processes that are able to quickly and flexibly react to environmental changes
- Reducing waiting and handling times
- Improve the competency of consulting services
- Individualise the treatment of customers

The following process diagram is the "as is" customer service process. The following diagram is the "complaint handling" refinement. The purpose of the model is to evaluate duration times. The necessary simulation data is already entered into the model.

The inputs for this simulation are 200 daily telephone complaints and 15 daily written complaints distributed evenly over the working day. Resource capacities were modelled as follows:

- Back Office (9 employees = 72 h): 8:00- 16:00 (4), 16:00 - 24:00 (5)
- Call Centre (7 employees = 56 h): 7:00- 16:00 (3), 16:00 - 23:00 (4)
- Process manager (1 employee = 8 h): 9:00 – 17:00 (1)

Number of complaints	200 daily telephone complaints 15 daily written complaints
Existing resources	17 employees, 8 hours per day (= 136 h per day)
Requested put through time per complaint	Same day service

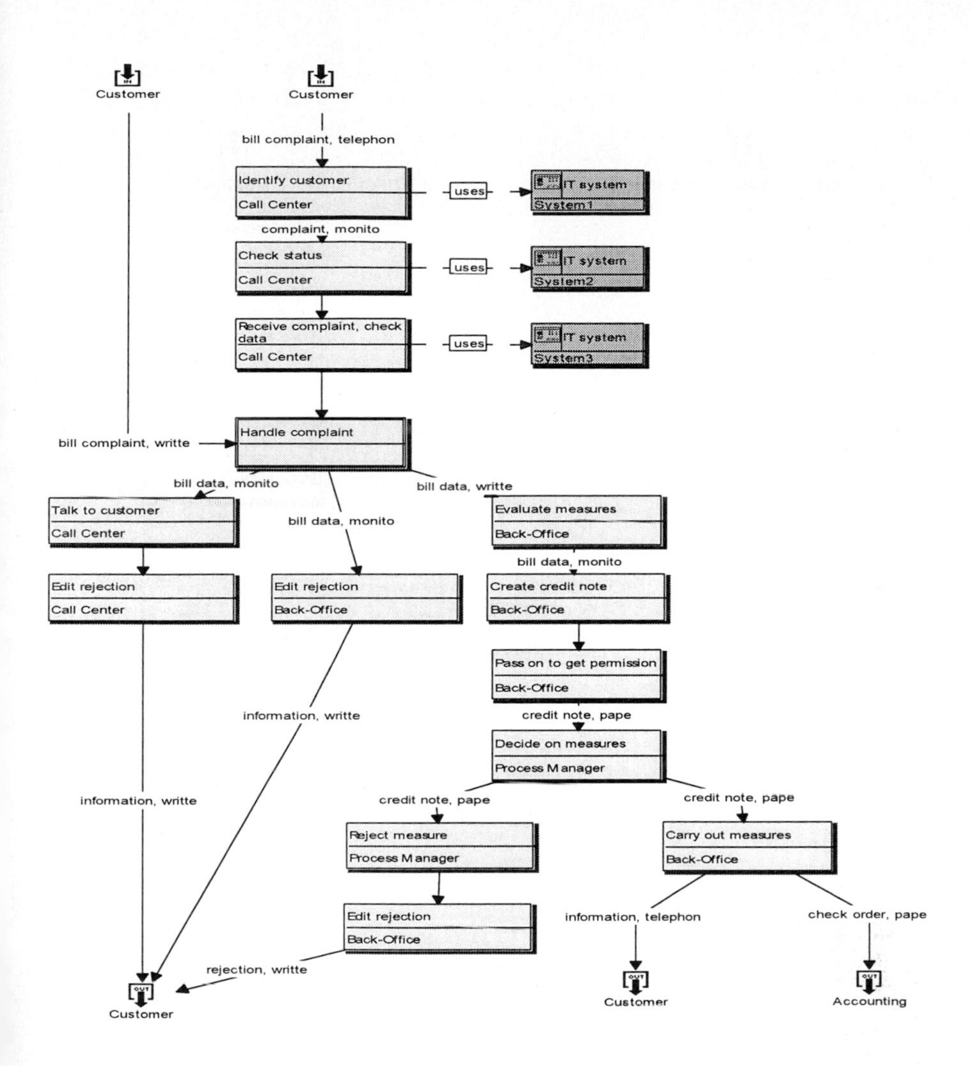

Fig. 1: Process diagram "customer billing services" – as is

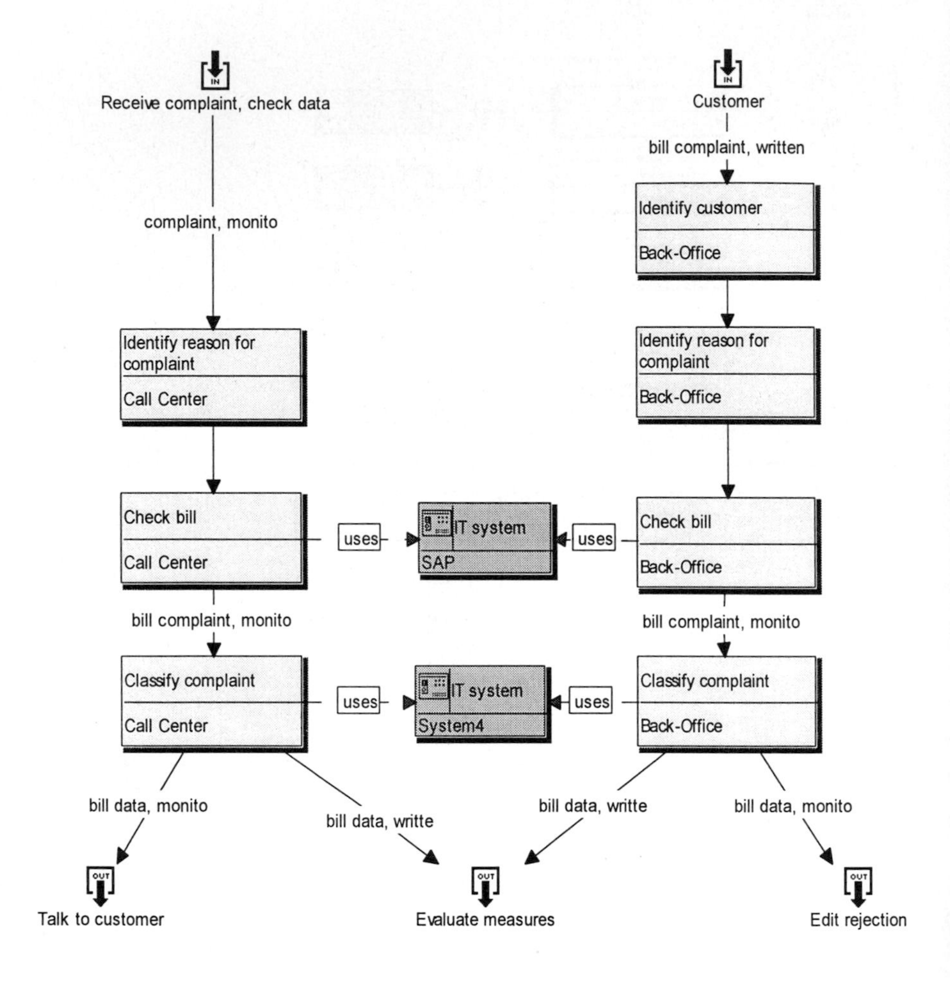

Fig. 2: Process refinement "Complaint handling" – as is

These two diagrams represent the process in a clear and understandable way. The information is already prepared to execute the necessary evaluations within the simulation.

Simulating the existing "as is" process, it takes 13 days to process 200 telephone and 15 written complaints.

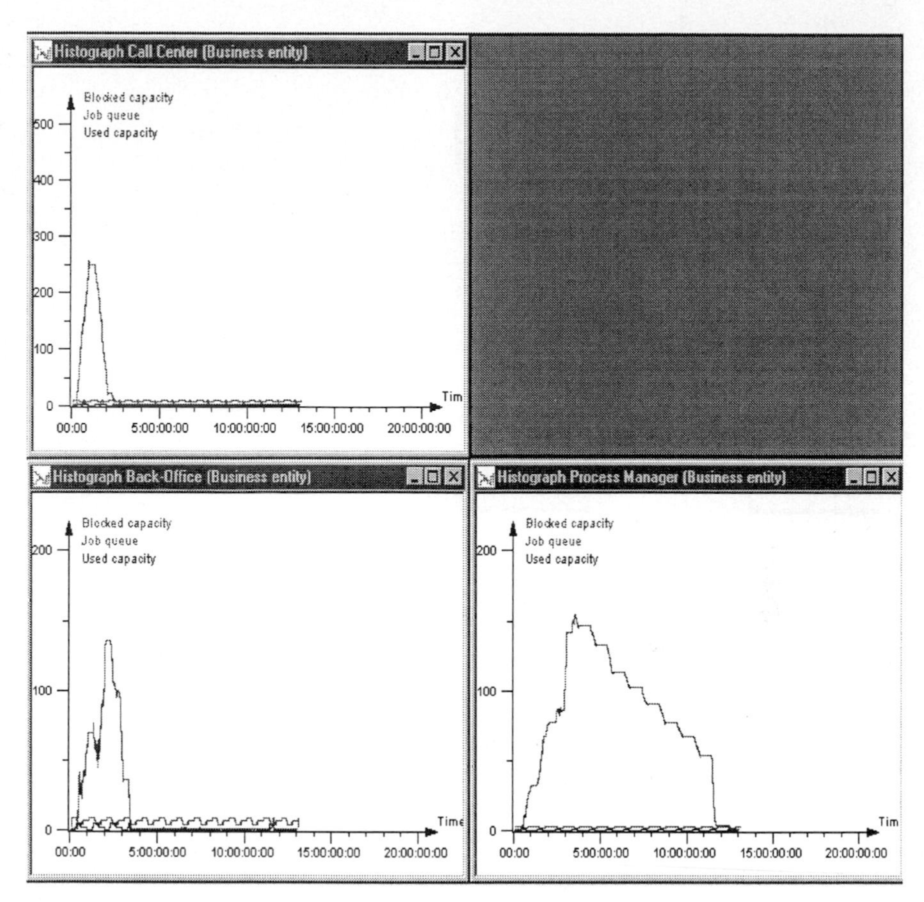

Fig 3: Simulation results – "as is" (Process duration in days)

The Same Day Service goal is far from being met. Using the information obtained from the simulation, the process flow needs to be redefined. The following "to be" process flows are the result:

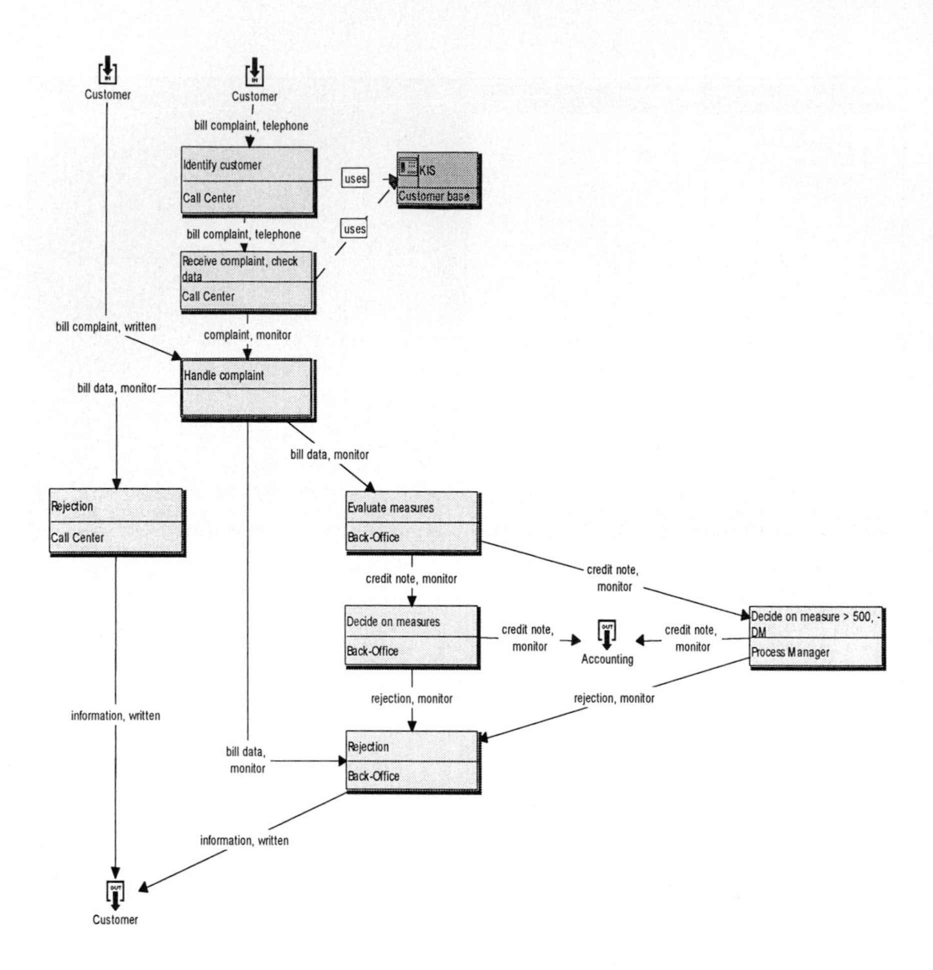

Fig. 4: Process model "Customer services concerning billing" - to be -

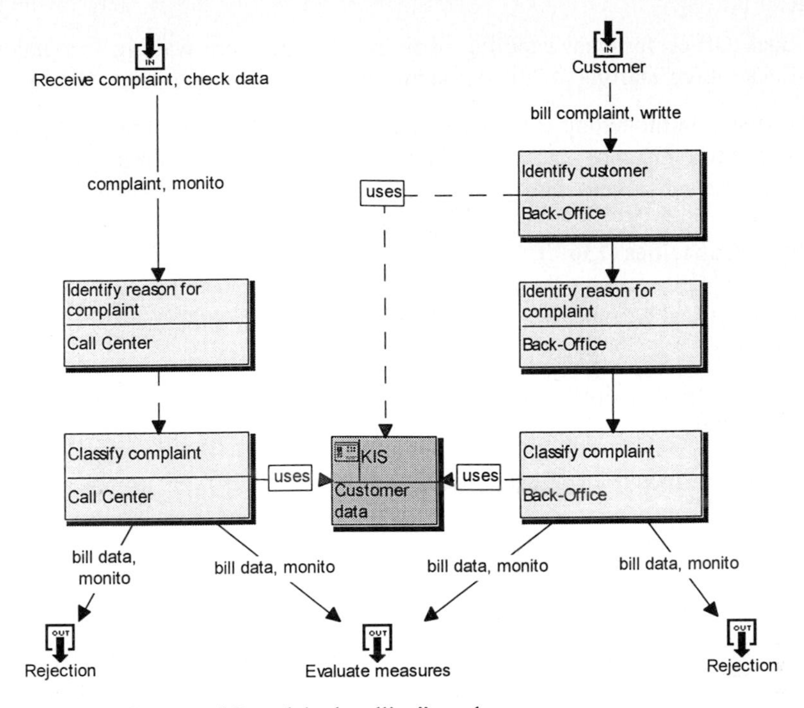

Fig. 5: Process refinement "Complaint handling" - to be -

In order to check the target times for "Handling time for complaints", times are taken from the "as is" model. Comparisons between the simulation results of the "as is" to the "to be" model are used to determine if the desired results have been reached.

7.1.2 Capacities (136 h)

- Back Office (9 employees = 72 h): 8:00 - 16:00 (4), 16:00 - 24:00 (5)
- Call Centre (7 employees = 56 h): 7:00 - 16:00 (3), 16:00 - 23:00 (4)
- Process manager (1 employee = 8 h): 9:00 – 17:00 (1)

7.1.3 Simulation Results:

- 1 employee as process manager – clear bottleneck
- Back Office and Call Centre are not able to achieve a day exact complaint handling. The reason lies in the complaints that come in late in the first day or between the first and second day (see histograph Call Centre). Complaints are handled in the Call Centre on the second day and they are then passed on to the

Back Office for final handling. This is why there are waiting complaints in Back Office, starting at 7:00 in the morning (see histograph Back Office).

Next step in modelling is the elimination of the bottleneck "Process manager". As capacities (136 h) cannot be changed, one employee from Back Office becomes second process manager.

7.1.4 Capacities (136 h)

- Back Office (8 employees = 64 h): 7:00- 15:00 (2), 9:00 - 17:00 (3), 16:00 - 24:00 (3)
- Call Centre (7 employees = 56 h): 7:00- 15:00 (2), 9:00 - 17:00 (2), 16:00 - 24:00 (3)
- Process manager (2 employees = 16 h): 9:00 - 17:00 (1), 16:00 - 24:00 (1)

7.1.5 New Simulation Results:

- Day exact handling of complaints

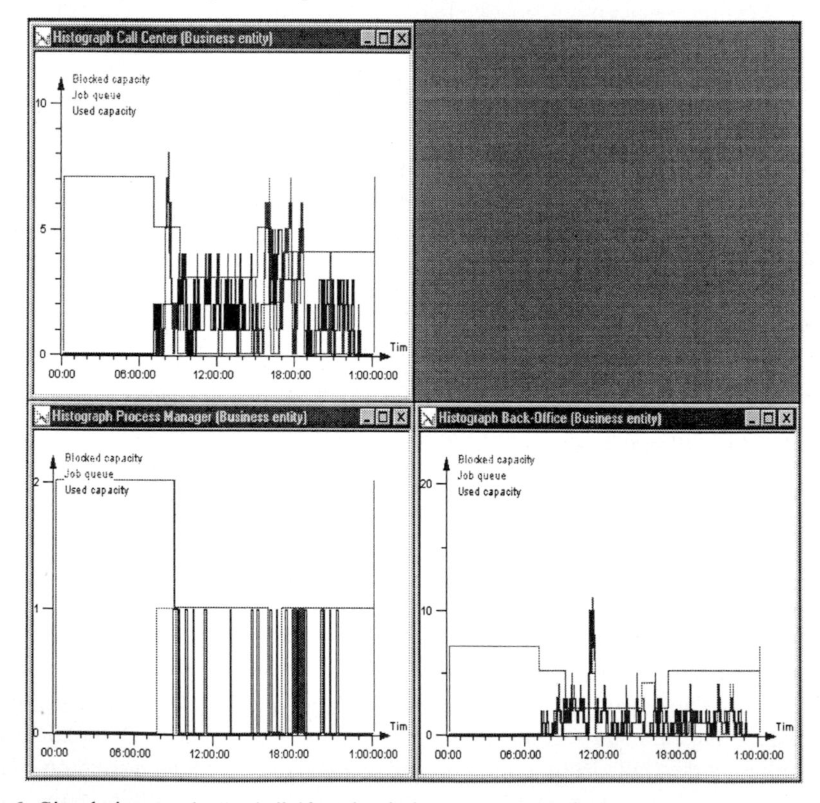

Fig 6: Simulation results "to be" (duration in hours up to one day)

Process modelling as shown above, has identified a project solution in a very easy, timely and understandable way. The implementation is supported effectively by the diagrams as they can easily be used to communicate upcoming changes to all employees. With the help of simulation, the solution found here was proven to achieve the original project goals.

8 Conclusion

With the modelling method shown above, models can be created in a focused manner, so that the model can answer exactly those questions that were asked at the beginning. By modelling with a clear focus, models were created that can easily be understood and evaluated. If model maintenance is established to keep the model up to date, created models can be effectively and economically used for other purposes than those defined in the original project.

References

Hoffrichter, M. (1998): Service Engineering – Prozesse gezielt gestalten; in: DIN Mitteilungen 12/98, Berlin

Krallmann, H.; Wood, G. (1998): Bonapart; in: P. Bernus, K. Mertins, G. Schmidt: Handbook on Architectures of Information Systems, Berlin/Heidelberg 1998

Performance Measurement Based Process Modelling

Christian Aichele

Abstract. *Performance Measures and performance measurement systems are the basis of quality and quantity-related generation. They control business objects such as organisational units and business processes. The enhancement of current process modelling methods with possibilities to model performance measures allows companies to create a prospective company controlling system. Therefore, it is necessary to visualise the strategic objectives linked with the key strategic measures and integrated with the performance measures of the business objects. This paper points out the use of process modelling extended with strategic and operational performance measurement for all types of organisations.*

Keywords. *Performance Measurement, Balanced Scorecard, Strategic Objectives, Key Performance Indicators, Business Object Analysis, Business Process Analysis, Strategic Enterprise Management, Prospective Enterprise Controlling, Indicator System, Management Information System*

1 Demands Placed on Process Modelling and Performance Measurement

The increased opening up of markets has, in almost all branches of industry, led to increased competition. Cost reduction, shorter order processing times and faster product development are, with quality remaining the same, important factors for success in achieving the survival of a company.

The overall objective of business object based indicator systems (indicators are performance measures related to business processes and/or organisational units and/or data objects) is to analyse and control the optimum situation of a company in a standardised, consistent and, above all, comprehensible way. Only organisational models including performance measurement indication and integration which fulfil these requirements can be applied to the optimisation of business processes for enterprises like producing companies and public services etc.

An optimal performance measurement considers the enterprise strategy with the derived strategic objectives. Therefore, the strategic measures which monitor the success in achieving the strategic objectives have to be linked with the key performance measures of the enterprise business objects such as the different organisational units and the core business processes.

2 Process Modelling with Performance Measurement Integration

Information, criteria, indicators and points of orientation will play an important role in economic life in the years to come.

In an economic system based on competition, it is vital to make out the future economic situation as early and as clearly as possible in order to take one's chance as soon as it presents itself. It is therefore a matter of great concern for vital companies to have methods and systems at their disposal, with the help of which statements about future combinations of important core competencies can be made.

The management of a corporation needs indicators or performance measures:

- To be able to distinguish, with the number and complexity of business processes being high in each corporation, what is important and relevant from what is unimportant and irrelevant and to be able to obtain aggregated information from a mass of disordered data;
- In order to think in a relational way rather than according to the monetary quantities of profit and costs and to pay more attention to productivity than to production;
- To see causal connections, cause and effect, and the interplay of positive and negative factors;
- To know the company's own position as a competitor in the market as well as the strengths and weaknesses of the company in comparison with the competitors, so that it is in a position to react timely to changes and new developments in the market;
- To be equipped with an instrument of management for targeted thinking and action;
- As an instrument to monitor and optimise the operationalised strategy of the company.

The indicator systems which have been developed so far are either incomplete, i.e. quantitatively important task areas of the management are left out of consideration, or their structure is not sufficiently logical. In the past, the management normally used financial metrics such as economic profit and variable production costs in order to measure the performance of their company. Whereas non-financial measures such as e.g. customer service level, market shares, order processing times, production lead times etc. were not considered in a strategic context or were hardly considered at all.

On the basis of a business object oriented performance measurement system, processes can be quickly analysed and evaluated through process models taking into consideration the enterprise vision, business strategies and corporate objectives as well as the factors of performance linked with them. The correlation of the individual indicator analysis with the business objects such as organisational units, data objects and business processes makes it possible to proceed to a subsequent targeted optimisation of the organisation structure and business process architecture of the company. Figure 2.1 shows an extract of modelling methods or modelling types which are required for a Performance Measurement based Process Modelling integrating the strategic and operational performance measures.

The relevant indicators of the strategic objectives linked with the key performance measures of the core business objects provides a performance measurement system which allows proactive and prospective enterprise controlling and management. Every change in business object related performance measures are now related to the strategic objectives of the enterprise and therefore supports the key management task in prioritising activities, initiatives and projects for the achievement of the business strategy.

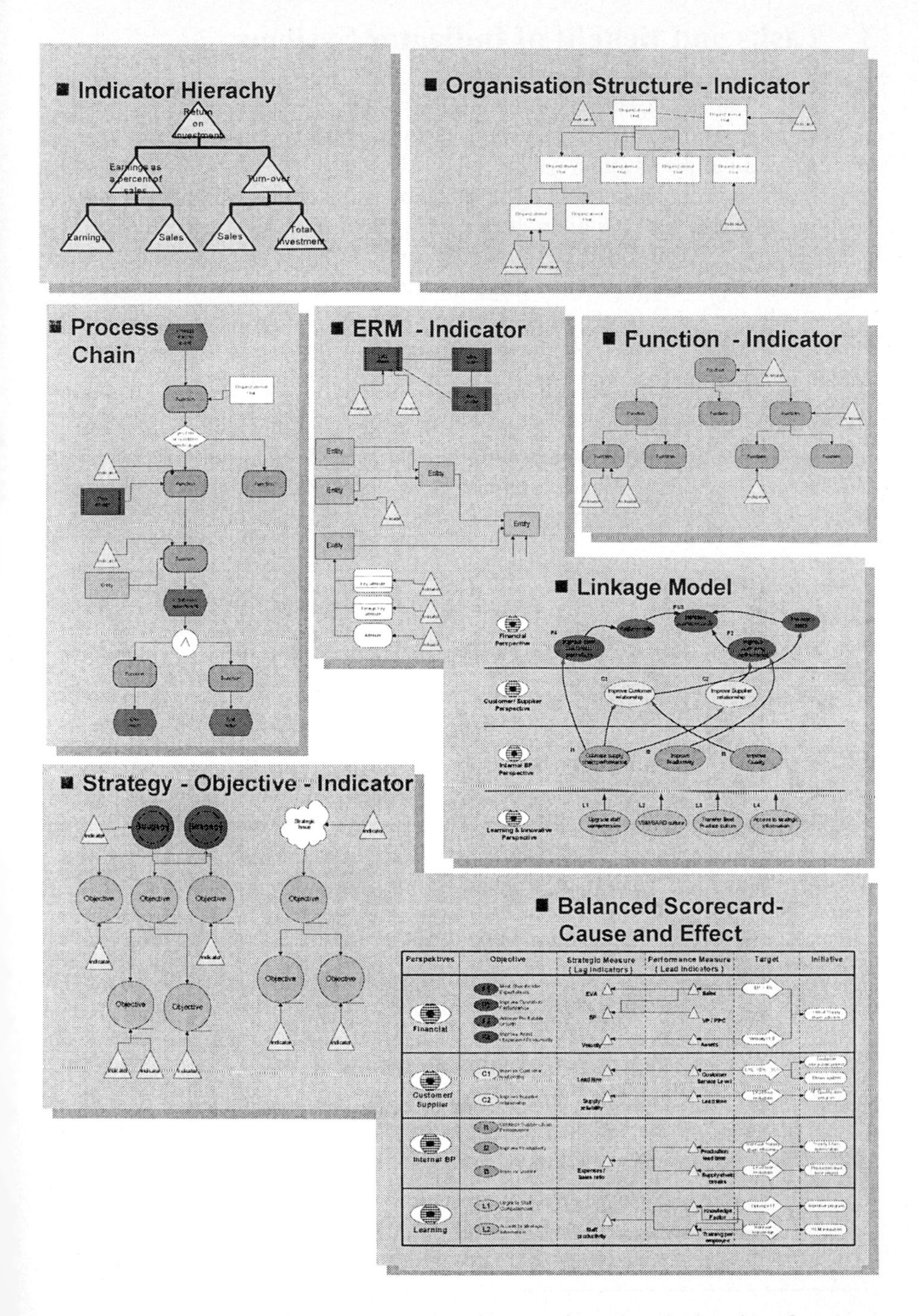

Fig. 2.1: Process Modelling integrated with strategic and operational performance measurement

3 Tasks and Benefit of Indicator Systems

Companies and organisations, and above all their management, must be in a position to obtain rapidly, from a few but instructive pieces of information, a precise picture of the situation within and outside the company. Information about cost structures, capacity utilisation, production programmes, volume of orders, operating position and liquidity is especially essential for the survival of a company or organisation. The strengths and weaknesses of a company or organisation in comparison with competitors in the same branch of industry have to be synthesised. Indicators which are well thought-out and carefully evaluated supply the tools for rapid answers to these complex questions. The formation of indicators is the most important way of condensing information.

It is therefore necessary to provide information systems in a company or organisation which supply relevant information for decision-making in a concentrated, aggregated and concrete form. Indicator systems or reference models with indicator systems are such information systems.

4 Performance Measurement Based Process Modelling

Well-established graphic forms of description are used for the representation of business objects (business processes, business functions, data objects, organisational units) related to performance measures. Functions are hierarchically grouped in function trees, organisation charts are used for the organisation view. To describe the data view, the Structured Entity Relationship Model (SERM) is utilised. The SERM is regarded as the description method which is, at present, best suited for describing data structures as its graphic form of presentation and its clear definitions make it especially user-friendly. The representation of the control and process views is mainly carried out through a derived modelling method of the extended Event-Driven Process Chain (eEPC) called Performance Measurement Process Chain (PMPC). For describing the business strategy and the strategic objectives with the linked strategic measures adapted modelling methods of the Balanced Scorecard approach like the Strategy-Objective diagram, the linkage model and the cause-effect relationship are used[1].

---

[1] Compare Kaplan, R.S. / Norton, D.P. (KN01)

The required modelling methods are:

– Performance Measurement Process Chain
– Function hierarchy
– Organisational chart
– Structured Entity Relationship model
– Indicator hierarchy
– Balanced Scorecard – Linkage Model
– Balanced Scorecard – Cause and Effect Relationship

All modelling methods are containing the object indicator. Figure 2.1. shows the modelling methods. Figure 4.1. emphasises on the integration of the enterprise strategy and the linked strategic measures with the enterprise business objects with the linked operational performance measures.

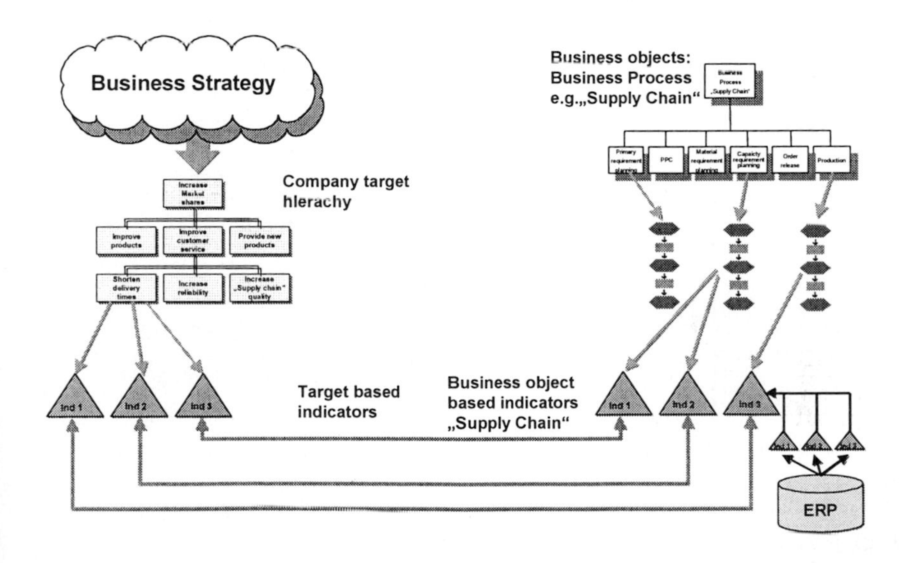

Fig. 4.1. Structure of a Business Object and Balanced Scorecard Performance Measurement System

5 Strategy Related Performance Measurement System for the Speciality Chemical Industry

A Swiss based Speciality Chemical Industry company started a major initiative in order to increase production quality, market share, customer satisfaction and above all to increase the economic profit. More than 35 parallel projects should deliver a very large increase in the enterprise performance.

For that, a prerequisite was to develop a performance measurement system which aligns the core business strategy, "Sustainable profitable growth", with the key performance indicators of the corporate projects, see figure 5.1..

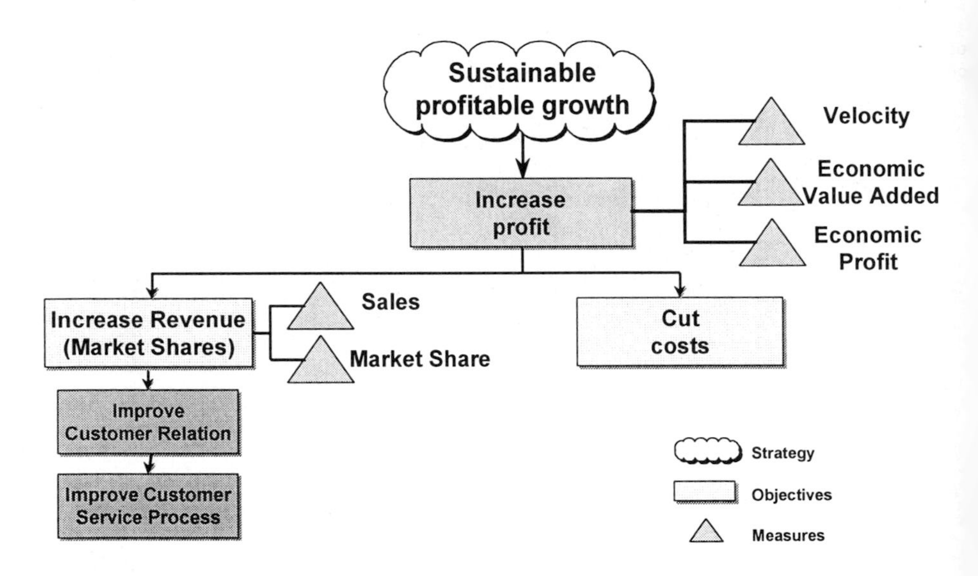

Fig.5.1: Strategy - Objective hierarchy

Figure 5.2. points out the integration of projects like "Production Lead Time reduction", "Supplier Management", "Global Demand Flow Management", "Customer Integration" and "AR initiatives" with the corporate strategy described by applying the balanced scorecard approach.

With that instrument the management is able to identify and prioritise core projects according to the extent of added value to the business objects.

Furthermore the cause-effect relationship of the business object performance measures to the strategic targets can also be evaluated.

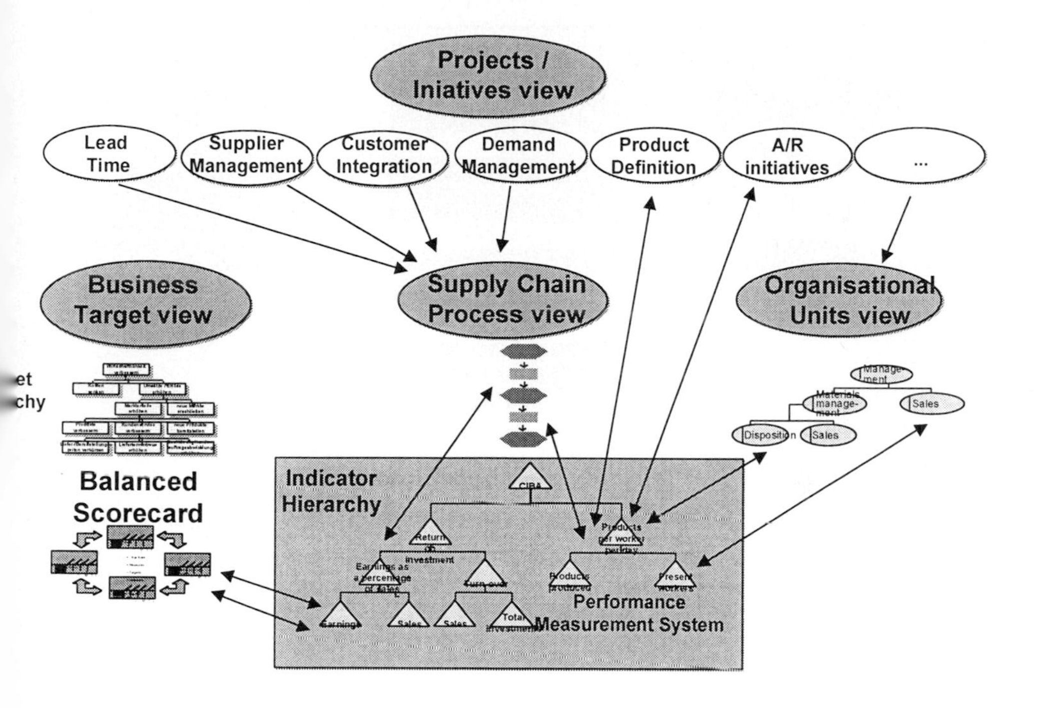

Fig 5.2: Integrated Performance Measurement System

The objective is to identify and generate transferable best practice to reduce the overall effort and development time as well as to increase efficiency and implementation benefits. In this respect single projects are carried out only, or mainly in one unit. The experience and know-how is then transferred to other organisational units. The management is monitoring and supporting the projects and ensuring that the results are in line and in balance with the global supply chain vision and divisional velocity targets.

The described modelling methods have been used for the development of the performance measurement system and also in order to reduce the manual effort of data collection and data aggregation as a specification for the generation of an IT based performance measurement system, see figure 5.3..

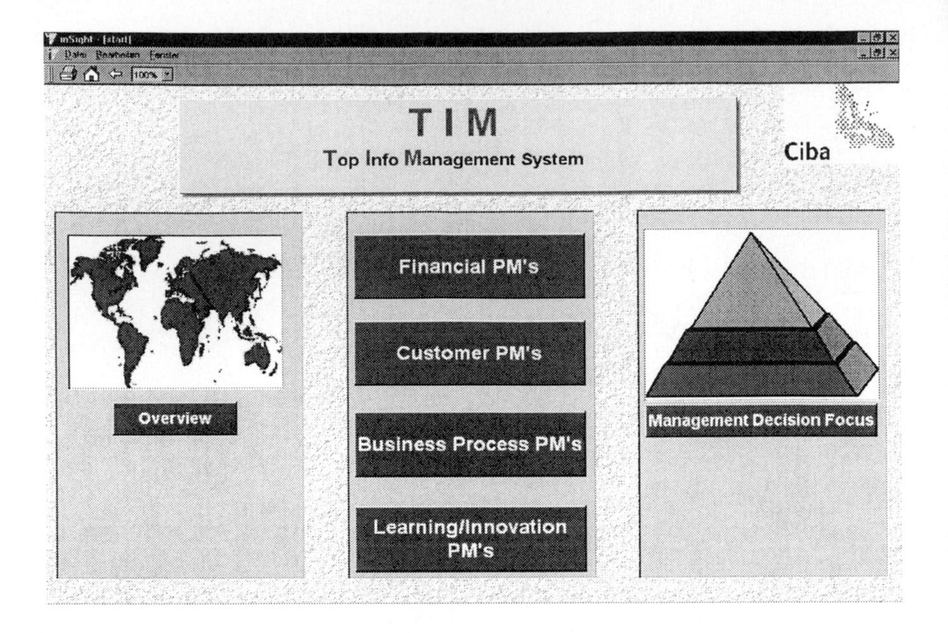

Fig. 5.3: Management Information System with organisational analysis, balanced scorecard related analysis and a management decision support system

References

Aichele, Christian / Kirsch, Jürgen (1995): Geschäftsprozeßanalyse auf Basis von Kennzahlensystemen; Management and Computer; 3. Jg.; Heft 2 1995

Aichele, Christian (1997): Kennzahlenbasierte Geschäftsprozeßanalyse; Gabler Verlag

Aichele, Christian / Böhning, Sebastian (1997): BPR and Benchmarking with ARIS-based Industry-specific Reference models; Proceedings of the 4th International Conference , Computer Integrated Manufacturing; page 119 –131; Springer; Singapore 1997

Aichele, C. / Elsner, T. / Thewes, K.-J. (1994): Optimierung von Logistikprozessen auf Basis von Referenzmodellen; Management and Computer; 2. Jahrgang; Heft 4 1994; S. 253-258; (Optimierung von Logistikprozessen)

Kaplan, R.S. / Norton, D.P.(1996): The Balanced Scorecard; 1996

Scheer, August-Wilhelm (1994): Wirtschaftsinformatik; Referenzmodelle für industrielle Geschäftsprozesse; Springer Verlag 1994

Tiemeyer, Ernst / Zsifkovitis, Helmut E. (1995): Information als Führungsmittel; Computerwoche Verlag GmbH; Edition 1995

Transparency and Flexibility: Towards an Adaptable Environment for Cooperative Modelling

Dörte Bastian and Bernd Scholz-Reiter

Chair of Industrial Information Systems, Brandenburg Technical University of Cottbus, Universitätsplatz 3/4, D-03044 Cottbus, Germany,
{bastian, bsr}@iit.tu-cottbus.de

Abstract. *Design and improvement of manufacturing systems is a very complex and creative planning task. In Small and Medium Enterprises (SMEs), it is carried out in the course of planning projects by a group of enterprise planners who belong to different departments but work in close cooperation. Each of the team members has already been supported by sophisticated modelling and simulation tools to work on a specialist part of the model of the manufacturing system. Influences and effects between the modelling activities of different planners arise dynamically during the planning process due to the creative nature of planning tasks. They are caused by logical interdependencies and overlapping data that exist between the partial planning models and specialist working fields of the planning experts. Transparency of all those modelling activities of planners that effect other collaboration partners is important for successful collaboration. An IT-enabled support of the group awareness and an intelligent reaction to mutually interrelated modelling activities is presented here. It is an essential part of a groupware-based toolbox for cooperative modelling, besides a concept for cooperative problem solving already developed and implemented in a prototype. Handling, controlling and coordinating the interdependent modelling activities between planning experts mainly depends on the planning issue, the characteristics of the planning project and the course of the whole modelling process. As planning issues are unique, requirements on a computer-based planning environment will change from one project to another. For example, the cooperation partners in a project group and their specific skills, the modelling tools used and the interdependencies existing between the specialist working fields may change. To realise cooperative modelling in SMEs, it is necessary to enable an adaptation of the groupware-based toolbox to changing organisational, information and technical demands.*

Keywords. *Business Process Modelling, Manufacturing Simulation, Distributed Cooperation, Tool Integration, Groupware*

1 Problem Field

Comprehensive and creative planning issues regarding the development of manufacturing systems are carried out by a group of planning experts belonging to different specialist fields such as factory planners, manufacturing engineers, organisers, systems analysts and software developers. The project members have a joint planning goal, they are responsible for solving the planning issue and deciding about the course of the planning and modelling process. Various interdependencies exist between the collaborating partners during a planning project. These interdependencies mainly depend on the particular planning issue and the specialist areas necessary to deal with it. Therefore, the interrelationships have to be identified for every particular planning project. They have continuously to be observed and taken into account during the project development.

1.1 Transparency during the Planning Process

Multiple disciplines are necessary to provide the knowledge needed to understand the diverse impacts and implications within a manufacturing system (O'Sullivan, 1994). Therefore, a planning issue is divided into partial modelling and simulation activities, which are assigned to the project members depending on their special working field. A number of sophisticated modelling and simulation tools are available to support planning experts to fulfil their project tasks. Manufacturing simulators support modelling and simulation of material-related processes. Business process modelling tools can be used to represent, analyse and simulate the processes through informational connection of humans, tasks and techniques. A large number of logical and methodical interdependencies exist between the two main directions of modelling manufacturing systems, i.e. the material-related and the information-related view.

Many influences and effects between the partial activities of planning experts occur due to the methodical interdependencies and overlapping data. Any modification on joint modelling data causes an effect on those planning experts who can model the same or interrelated data by means of their tool. Most data modifications arise flexibly because of the creative and rather poorly-structured nature of the development task and in addition to the dynamics of the planning process. It is important that modelling activities that exert effects on the working steps of collaboration partners are immediately considered during the planning process in order to achieve a high-quality result. However, the awareness of changes on interrelated or overlapping modelling data is poor. In addition, notifying affected project partners is mostly done manually by the planner who performed the modifications. This means that effected planners do not automatically and quickly know about any change. It leads to the consistency between partial models having not always been considered sufficiently and never automatically.

The need for an awareness of data manipulation actions also results in the necessity of passing on modelling data and information about modelling activities, or planning decisions, to another or all co-workers in the project team. A problem exists here in that planners sometimes forget to send all of their changes on logically shared data or they do not inform all the planners concerned, which often leads to missing information. Consequently, planners who modify an old state of the planning model often have to update their planning model with a lot of effort or even remove their changes done in the meantime.

It is necessary to improve the transparency of distributed modelling activities and the awareness of new working results in order to enable a quick reaction to changes on logically joint data within a project group. Thus, it is necessary to identify this logically shared data, to automatically discover modelling activities that perform modifications on shared data and to provide the planning experts with information about it. An IT-enabled support of these tasks should improve the accomplishment of planning issues. Therefore, a computer-based toolbox for cooperative modelling was developed.

1.2 Flexibility and Changing Project Requirements

The identified planning issue within a manufacturing system and the specialist fields concerned have a large influence on the project organisation, the composition and the structure of the project group and the disciplines of the project members. This influences the different working steps necessary to solve the problem and their handling and coordination during the planning process. The consequence of unique projects is that the project planning and realisation change considerably from one project to another. Therefore, the development of the computer-based toolbox for cooperative modelling has to deal with the architectural concepts of openness, flexibility and adaptation to organisational, information and technical demands.

Supporting the cooperative modelling of manufacturing systems by an IT-enabled environment calls for flexible and adaptable components, which work under consideration of the characteristics of a particular planning project. Hence, project-specific knowledge must be gathered at the beginning of every project and changing project requirements taken into account during the whole planning process. Establishing a meaningful connection and flexible coordination between the members of a project group mainly depends on the different planning experts involved in tackling a planning issue. Dependencies between the particular working fields of planners are based on the modelling tools used, the modelling data each tool can represent and the interrelated and overlapping data between the partial planning models created. Thus, making interdependent modelling activities and partial working results transparent to other collaboration partners is founded on these interrelationships between the planners and logically joint data between their partial planning models. Information modules had to be developed and implemented within the groupware-based toolbox to collect information about the planning group, the modelling tools used and the methodical dependencies between the group members.

Due to missing standardised interfaces of current modelling and simulation tools, it was necessary to design a way of flexibly integrating the application systems into the toolbox. Therefore, every modelling system that is used for cooperative planning has to go through a tool integration process. Here, tool-specific information is collected which is later used for transforming the modelling data into the object model and establishing a meaningful connection between the modelling tools. Furthermore, tool-specific knowledge enables a synthesis of, so far, isolated application tools with each other. Planning tools that already exist in a SME can flexibly be integrated into a tool-spanning planning environment where they are meaningfully linked with each other. This meets the user demand to select and work with tools that best suit the purpose of the enterprise planners. Thus, the modelling environment is provided with the feature of openness.

If project requirements change then the structure and organisation of a planning project must be adapted to the changes. These new project features must be considered by the toolbox when realising transparency of modelling activities, awareness of the working results and automatic reaction to modifications of joint data. The coordination of the distributed, creative modelling tasks must also be adapted to new requirements. Therefore, the components of the toolbox must be designed flexibly and universally to enable their initial configuration and adaptation to new project demands.

2 Aims of Research

The ability to respond to modifications on joint or interdependent modelling data, as well as to react to changing project requirements, should be greatly improved during the distributed cooperative business process planning. This takes the growing importance of the competition factor time into account (Ott, 1997]. In addition, the quality of the project result may be increased and the number of failed projects reduced. Therefore, a seamless, continuous planning process even through using separated, partial application systems must be realised and the cooperative development of planning projects improved. However, only a few up-to-date concepts, for example, new modelling methods or object-oriented methodologies have been established in the field of planning manufacturing systems and are implemented in computer-based modelling and simulation tools. They mainly focus on single model elements or on a special field for the planning and are currently used in isolation.

In contrast to this, foundations of seamless and continuous cooperative modelling are:

- awareness of the working results of collaboration partners,
- transparency of the modelling activities and
- reaction to modelling actions which exert an influence on project members.

Intelligent combining of planning experts and flexible coordinating of the modelling activities is based on continuously overseeing the state of interrelated and joint parts of the partial planning models. Logically shared parts of the planning models are these modelling data that can be processed and modified by more than one planner. Joint and interrelated modelling data represents data-oriented logical interdependencies between the project members and their partial planning models. Considering the changes on joint data is especially important during a creative modelling and simulation process in order to keep all the partial planning models consistent.

Therefore, it is necessary to overcome the departmental boundaries during the whole planning process, which exist only artificially between the enterprise planners of different specialist fields. The important development tasks were identification of joint data, realisation of routines for steady analysing joint data about changes followed by the design and implementation of a meaningful and automatic reaction. The result of the automatic consideration of any modification on joint data lays in improving the planning quality and accelerating the planning process.

Cooperative planning of manufacturing systems was improved by developing a groupware-based toolbox based on integration of existing modelling systems as well as tool-spanning coordination (Scholz-Reiter, 1995). In addition, the flexibility of the cooperative planning process must be improved which is especially important with regard to the growing international competition in producing goods.

Due to changing requirements on projects and the ever-changing project environment, it is also necessary to equip the computer-based planning environment with means of:

- configuration,
- adaptation and
- flexible integration of modelling tools.

The benefit of further using specific and highly sophisticated tools is that they do not have to be re-invented, but can be used in the same form in which they are available on the market. In addition, planners in various fields can continue to work with tools they are familiar with and which were developed for their individual tasks. Planning experts are not stretched by the mega-functionality of a comprehensive tool. Design and development of a complex modelling and simulation tool for integrated enterprise modelling are presented in (Mertins, 1997).

3 An Adaptable Toolbox for Cooperative Modelling

As a result of the above-mentioned reasons, the design of an adaptable groupware-based toolbox for distributed, cooperative modelling was necessary (Fig. 1).

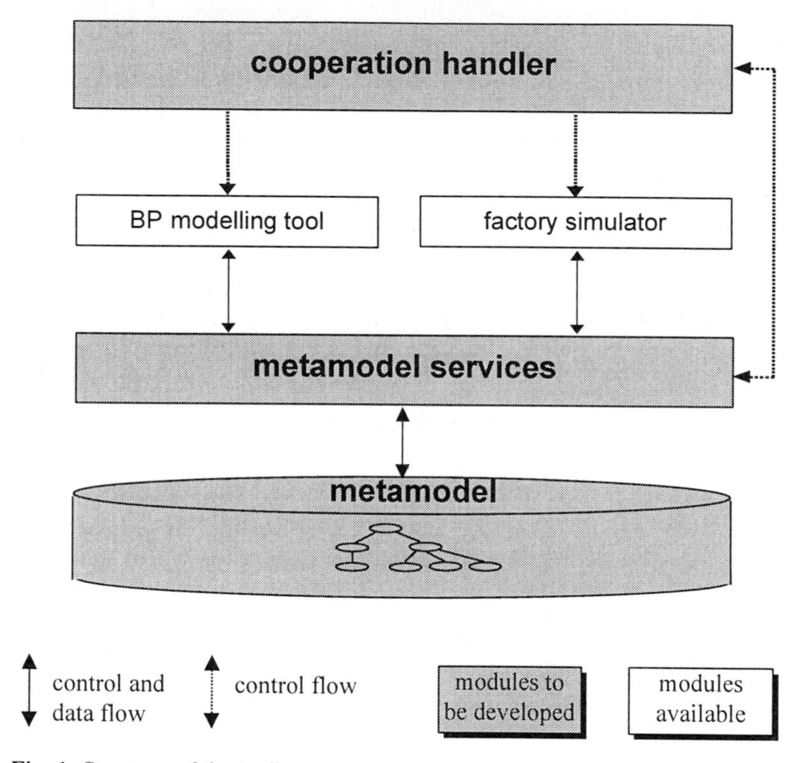

Fig. 1: Structure of the toolbox

The basis of the integration within the toolbox is a comprehensive, object-oriented metamodel of manufacturing systems. It allows all modelling elements and their interdependencies to be dynamically stored and given back from very different points of view. Metamodel services allow a modelling tool to present its data on the meta model and get back data from the metamodel prepared for its particular need. The metamodel services are also equipped with some "intelligence" for processing the multi-directional transformation and a sensible combination of modelling tools. The main functionality of the intelligent metamodel services lays in automatically identifying modifications on joint parts of the integrated model after a data conversion has been executed and intelligent reactions to any change. The cooperation handler manages the distributed groupwork with the help of a tool-spanning system for cooperative problem solving. The toolbox differs essentially from multi-user systems with a shared, but no cooperative usage of resources such as distributed databases or tools with client-server technology (Dohmen, 1994). The aim of cooperative, distributed problem-solving lies in supporting the coordination and management of the actions of the group members.

The process of continuous, seamless modelling can be improved by eliminating coordination problems caused by an inefficient consideration of the partial working results between the planning experts. Team-wide awareness of modifications on logically shared data and transparency of the modelling activities are important to realise a goal-directed collaboration and ensure working with updated partial planning models. Therefore, the cooperation environment must offer cooperation-accompanying functionalities in addition to the computer-based support of several types of collaboration. This includes functionalities to make the creative activities of project members visible to each other. Other functionalities offer access to knowledge about the planning project and realise a project-wide and quick information sharing within the project group.

If the project requirements during a planning project change then the structure and organisation of a planning project, as well as the project realisation by using the toolbox for cooperative modelling, must be easily adapted to the changes. Therefore, the components of the toolbox must be designed to work flexibly and in consideration of the particular project requirements, i.e. project-specifics and tool-specifics (Azuma, 1995).

A module for the integration of modelling systems into the toolbox must be developed which enables the combination of existing, but so far very isolated, application tools with each other (Janocha, 1996). Due to integrating existing systems, the planning experts can continue to work with tools they are familiar with. Additionally, they get better support and guidance in matching their interdependent partial models and coordinating their creative working steps in the planning process. The integration module must be designed to dynamically insert modelling and simulation tools in order to enable an adaptation of the toolbox to organisational, information and technical demands that may differ from one project to another.

3.1 Transparency and Information Sharing

Transparency of the activities within a group and awareness of the working results accomplished by the group members involved are necessary features of a group supporting working environment (Fuchs, 1996). This is necessary in order to enable the planning experts to be aware of other project member’s activities, their intentions and working results. The following aspects of handling events and information within a planning group are essential:

1. Awareness of working results
2. Transparency of modelling activities.

3.1.1 Awareness of Working Results

Every single planning expert in the project group creates an isolated partial planning model of the manufacturing system and a specific view of the modelling data by using a specialist sophisticated modelling and simulation tool. The different representations of the same aspect within the discourse field of manufacturing systems as well as interdependencies between the partial planning models must first be identified in order to integrate and meaningfully combine the modelling tools. After every modification of modelling data and its transformation into the integrated data base, the state of joint modelling data must be checked and the effected planners notified.

- **Dynamic determination of interdependencies between partial models**

The identification of logically joint modelling data in all partial planning models, which is the foundation of the meaningful connection between modelling and simulation tools, was not directly and explicitly realised. Instead, the integrated metamodel is used as an intermediate layer for defining interrelationships and overlapping data. The integrated metamodel represents the methodical dependencies between different directions of modelling manufacturing systems and integrates the different user views. Relationships between the conceptual data model of a modelling tool and the integrated metamodel of manufacturing systems must be defined. Therefore, the entities of a modelling tool and their corresponding objects of the integrated metamodel must be examined. The object classes of the integrated metamodel of manufacturing systems were extended in order to store these logical associations. New data members, namely ToolRepresentatives, were designed for every metamodel object class. These new data members were defined as collections that store a reference to each associated data entity of every integrated modelling tool. Consequently, a meaningful connection and the mutual interdependencies between different modelling and simulation tools can be established based on this semantic information.

The synthesis of partial models is carried out when integrating a modelling tool. It was realised by designing system-controlled user-dialogues within Lotus Notes. This is an implicit way of setting up a meaningful combination of planners, where the integrated metamodel and the metamodel services are used as middleware to connect the modelling and simulation tools.

- **Universal transformation of modelling data into the object model**

Means for transforming data from a modelling tool into the integrated database were developed by defining object classes for data conversion as part of the meta model services. Converter object types were created for each of the object classes of the metamodel of manufacturing systems (Fig. 2).

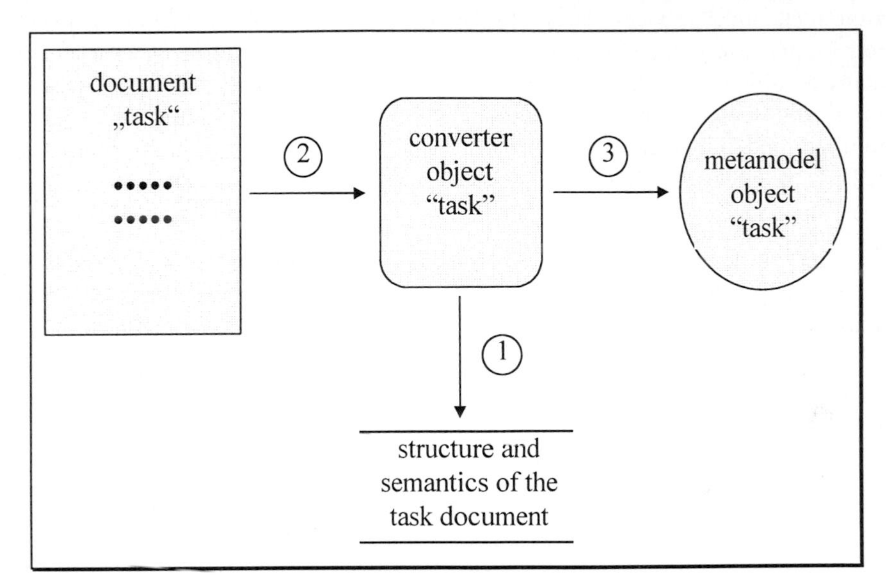

Fig. 2: Method of data transferring using a converter object

The data converters were realised universally, regarding classes of modelling systems in order to enable a flexible integration and further use of already existing modelling and simulation tools. A limitation of the universality of the data converters lays in the storage mechanisms and the access modes to the modelling data where it is distinguished between a data extraction from text files and an access to databases. Converter object classes for modelling tools providing data access through text files were implemented by using the predefined C++ file class CStdioFile. Extracting data from a database was realised by applying ODBC classes.

The data converter for a metamodel object, here a task object, requests information about the corresponding modelling entities of a modelling tool from a tool information module of the cooperation handler ①. Data, such as access type to the corresponding document or planning model, path, file type and structure of the document, is asked for. As a result of the request, a reference that stores the associating tool-specific information is handed over to the metamodel services and then used as a parameter of the converter object ②. Thus, the converter object reads, in conformance with structural and semantic information, the modelling data from the document. The actual transformation transformation of the data into the metamodel is afterwards accomplished by the converter object ③.

- **Flexible identification of changes on joint data**

Modifications on joint or interrelated modelling data are discovered with the help of the metamodel services. They observe the joint parts of the integrated object model after every transformation of modelling data into the object-oriented database, which is explained in more detail in (Scholz-Reiter, 1996). Possible modifications on the joint data stock could concern newly defined, deleted or modified modelling elements. These changes are identified by running particular methods of the metamodel object classes to search object collections that store the modelling data.

In order to decrease the effort of identifying a coordination demand, only such model components those modifications could effect other team members are examined. Therefore, the collaboration partners in a particular planning project, as well as joint and interrelated modelling data between their partial planning models, must be identified. To do this, the object classes of the metamodel of manufacturing systems are analysed in terms of a semantic correspondence that may exist to any of the modelling tools used in a particular planning project by one of the project members. Therefore, project-specific information and tool-specific data must be available within the computer-based toolbox.

The planning expert who carried out some modelling work and the planning project he worked on is determined. This information is used to identify all the planning experts who collaborate in the same project. The modelling systems, which the involved project members use, are identified. This data is used to identify those metamodel objects that contain a semantic correspondence to the modelling tools of the project members. Only these identified metamodel objects represent joint or interrelated data, and changes on them exert an influence on collaboration partners.

- **Preparing changed data corresponding to a particular modelling context**

The development and design of manufacturing systems is carried out by co-workers who belong to different special fields and several departments of the company. They work on a special part of the joint, integrated model of the manufacturing system in accordance with their working field and the modelling system used. Every modelling tool is characterised by these model components of a manufacturing system it can present and process (Heimann, 1997). Consequently, it is only this changed modelling data prepared for sending, which a modelling tool can represent and process. When the metamodel services identify a change on joint data, this modified modelling data, accessible by each of the involved planners, is identified.

The particular modelling data is prepared by the metamodel services for every planning expert in conformance with his view of the data (Fig. 3). This is the consequence of keeping the effort for using changed modelling data low. Therefore, it is necessary to know the semantic information and the structure of the modelling data of every planning tool.

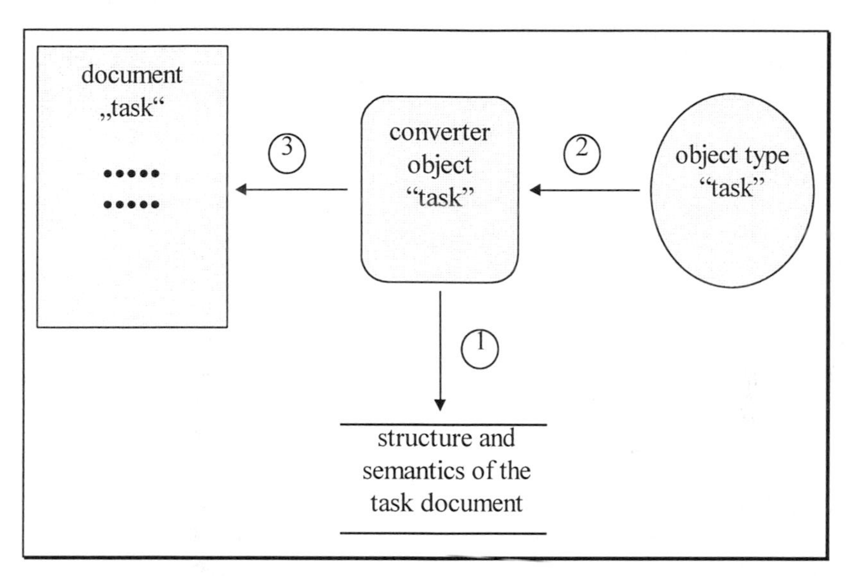

Fig. 3: Method of data preparation using a converter object

The metamodel services and their converter objects requests information about a particular tool, such as type of data storage, path, file type and structure of the file. In the next step, they write the changed modelling data in conformance with the structural information into a file. Modelling data are differentiated here in modified, deleted and newly created information.

Due to their implementation, the data converting functions of the metamodel services are universal for classes of modelling tools where the type of data storage, as well as the data access is the distinguishing criteria and the limitation for a general universality. Now, the feedback that enables the effects of changed model components on other planning experts and their modelling tools to be considered is realised.

- **Automatic notification of effected planners**

The consequence of not being able to predefine the workflow strictly and precisely during the whole modelling process is a communication-based, flexible coordination. In contrast to a well-defined workflow, dynamic notifications about relevant activities of collaboration partners are necessary for the seamless, coordinated modelling. Notifications are created and distributed dependent on the course of the modelling process. The presentation formats of notifications differ by the probability of recognising them as a direct consequence of how striking their display is. Forms of notification range from avoiding notification messages to a modal dialog box that the user must close over some intermediate presentation forms.

If changes on joint modelling data have been detected by the metamodel services, their next task consists of informing the cooperation handler by creating an event to trigger its notification service. To meet the dynamics of the planning process, the project members effected by changes on modelling data are automatically informed by the notification service. They also receive the changed data in an appropriate format. Determining the effected planners is again based on project-specific information. Structural tool-information is necessary to consider the specific user views and types of data access when preparing the data.

As a result, a flexible coordination of planning experts dependent on the state of the integrated model of the manufacturing system could be realised. The communication means necessary for sending notification messages to the cooperation partners who have been effected, as well as transmitting the changed data to them, are comprehensively explained in (Scholz-Reiter, 1997).

3.1.2 Transparency of Modelling Activities

If enterprise planners can mutually trace the activities of the collaboration partners during a planning project, it leads to a better understanding of the working steps of the project members and the creation of a context for one's own activities. Therefore, information about the modification activities on joint modelling data is created, stored and shared within the toolbox besides the actual changes on the planning model. Data analysing procedures and a notification service were established within the cooperation environment in order to handle changed modelling data. The management of information about these modelling activities required means to record and present the development of a partial planning model. This information about the processing of a planning document is called the history of the planning model.

An information module based on a database for handling the history of processing activities on the planning documents, a history function within the toolbox, was created.

This was implemented by using Lotus Notes. The history function causes an automatic and persistent storage of information regarding modelling activities such as the date and time of modifying a planning document, the editor's name and the number of the revision (Fig. 4). Additionally, comments about a processing step can be entered such as information about the method of problem solving or the decisions that were reached.

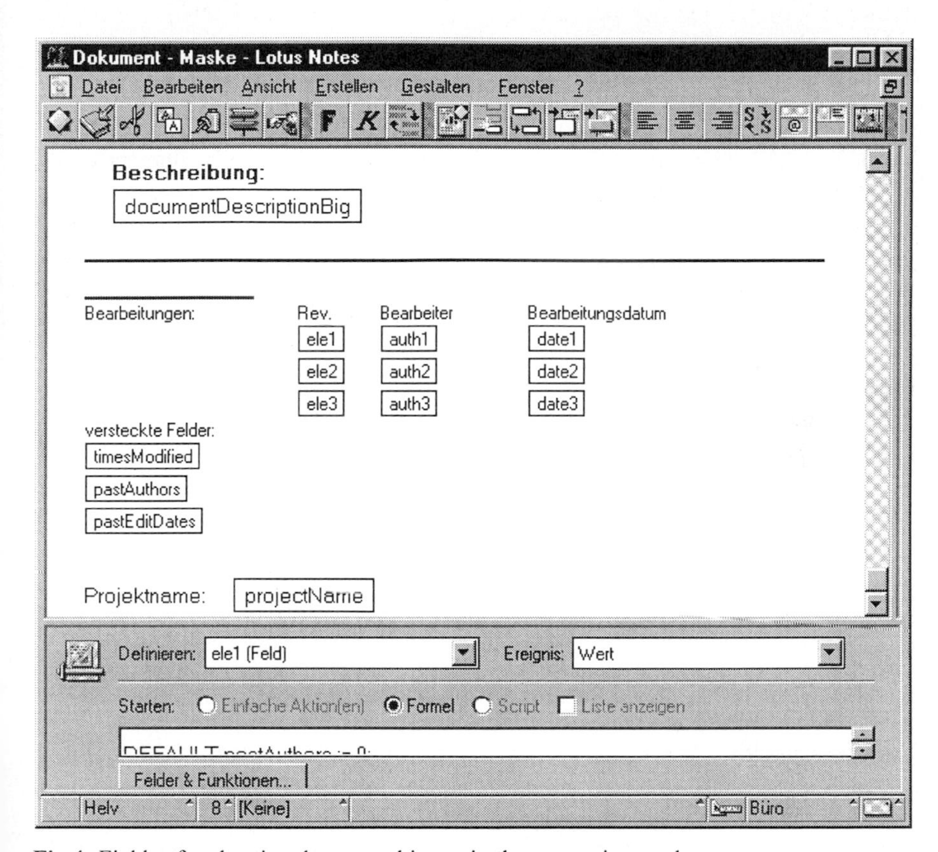

Fig 4: Fields of a planning document history in the processing mode

The information, which is stored when opening a planning model, guarantees that no modification of modelling data can occur without being noticed. To work on a planning model, the particular planning project must first be chosen from the Lotus Notes database to describe the project-specifics. The planning model must then be selected from the view „documents of the project". Whenever a planning expert works on a planning model to modify its content, history information is stored within Notes. However, only the last three entries are always shown in the Notes document because it is more user-friendly than a complete list.

There is a function available to obtain a comprehensive overview about all the information stored about a planning model, if it is required. The planning experts can quickly obtain an overview about past events and activities by means of the planning model and editor history. The history module can reconstruct sequences of modelling events in order to visualise special aspects within the planning process. Therefore, some masks were developed in Lotus Notes that enable the presentation of all activities that have been done on one planning model, or all the planning documents that have been edited by a particular planning expert.

3.2 Configuration and Adaptability of the Toolbox

Any modification of project requirements, for example, caused by changes in the manufacturing environment, or by the ever-changing customer demands can influence the project development. A refinement of the planning aim, or project goals in the course of a re-organisation project also occurs due to the ill-structured and creative nature of the designing task. A computer-based planning environment in such a dynamic environment has to guarantee its adaptability to different task and group structures in order to ensure a universal use.

3.2.1 Definition and Adaptation of Project-Specific Information

Project-specific data such as the structure of the project group, the tools used and the partial planning models created by the modelling tools are necessary information during the course of a planning project. Project-specific information is needed for:

- transferring modelling data into the integrated object model,
- coordinating the interdependent subtasks of different co-workers,
- keeping the consistency between partial planning models and
- preparing modelling data in accordance to a user's view.

Additionally, electronic availability of information about the project's structure and organisation is useful for the project leader and the team members because the requirements of planning projects are rather unique. Thus, it is important to get easy access to project-specific information that differs from one project to another.

Therefore, an information module was developed which gathers data for planning projects. It is based on a user dialogue and is realised by designing and implementing masks and views in Lotus Notes. Information about the structure of the project group and the modelling tools has to be articulated and stored at the beginning of a planning project. Groups can be established, tools allocated to group members, planning documents described and project information looked at by using the project module. Information about the project group, the team leader and the group members, as well as the modelling tools used and the planning documents are stored in project-related objects of the integrated metamodel. If project-specific information changes in the course of a planning project, it can be changed by a responsible person and sent to the collaboration partners, if required. As a result, the components of the toolbox, which realise the cooperative modelling, always consider the specific project requirements.

Information about the partial planning models or documents handled by a particular modelling tool, along with tool-specific knowledge, is requested by the metamodel services to transfer the modelling data into the object-oriented database. In addition, the metamodel services need project-specific information, i.e. what tool a planner uses, along with tool-specific data to prepare modified modelling data in accordance to each particular user view before data can be sent. Information about the project group and the modelling tools is also necessary to identify which planning expert is affected by changes on the integrated model of the manufacturing system. The IT-enabled consideration of relevant modelling activities of cooperation partners, which partly bases on project-specific information, relieves the project leader and the team members of necessary organisational and information tasks. The automatic execution results in benefits regarding an efficient coordination of subtasks during the planning process.

3.2.2 Flexible Integration of Modelling Tools

An integration component was realised as a part of the cooperative modelling environment which provides the means to integrate modelling and simulation tools that store their data either in textfiles with a table-oriented data structure or in a database.

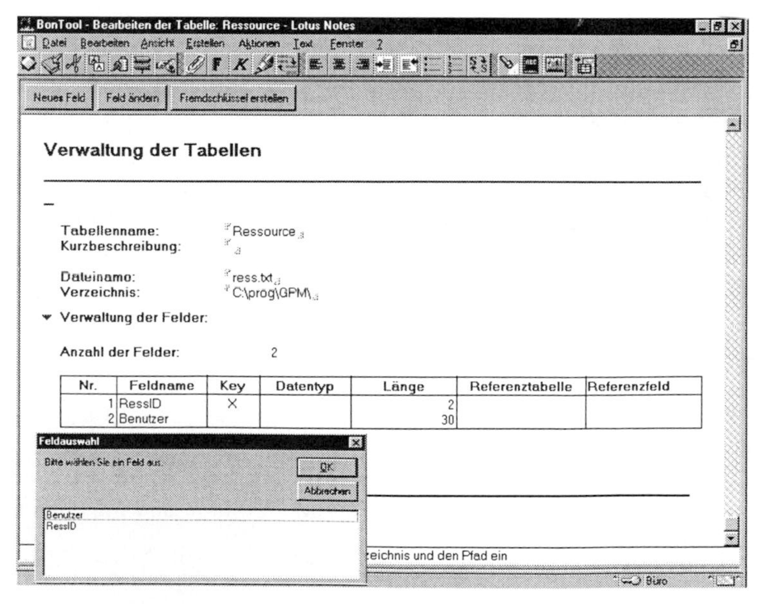

Fig. 5: Mask for managing tables

A dialogue-based module was developed in order to acquire the necessary tool-related information and store it afterwards within the groupware-based modelling environment (Fig. 5). Masks for data input, as well as views for presenting the information, were designed and implemented by using Lotus Notes. Thus, tool-specific knowledge regarding the structure of the modelling data and the organisation of the data storage is gathered. Information about tables, fields and relationships between the tables, as well as about files, directories and paths is asked for.

- **Storing structural tool information**

ToolIntegration object classes were designed to store tool-specific knowledge, i.e. the structure of the modelling data and the way of data handling within the modelling system. These classes were defined by means of Visual C++ and laid up in the object-oriented database. The development of tool-specific object classes was focused on application systems with a table-oriented data storage method (Fig. 6). Therefore, the object classes CTable and CColumn and relationships occurring between them were defined. Information about tables and the single table fields is held in respective objects, which are stored in collections.

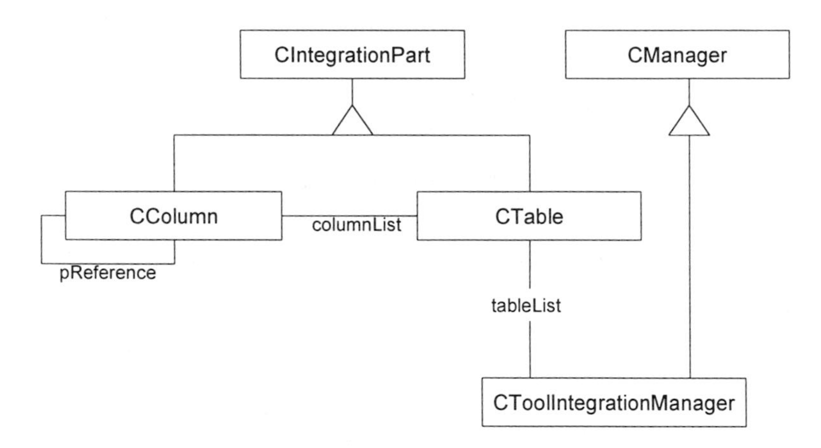

Fig. 6: Objects of the tool integration module

For managing all the objects regarding tables or table-oriented files belonging to a modelling system, the object class CToolIntegrationManager was created. This object class contains methods for creating, storing, changing and deleting object instances of the object classes CTable and CColumn and the relationships between them.

After describing the data entities of a modelling tool in Lotus Notes, an object of the respective object class, i.e. CTable, CColumn or a relationship, is built up for every modelling entity and then stored in ObjectStore. Therefore, the respective Create-method of the CToolIntegrationManager to dynamically generate a new object is called. It creates an instance of the respective object class for every data entity that was defined during the tool integration dialogue in Lotus Notes. The newly established objects are then laid up in collections.

The objects containing the information about the fields of a table are stored in a collection of the corresponding table object, namely the variable columnList. The table field objects declared by the type CColumn store data that is required to define the table fields. This is, for example, the name of the table field, the data type, the column number within the table and the type of the separator between columns if it is a table-oriented text file.

- **Gathering semantic connections**

After the physical view is gathered, semantic knowledge about the data entities of a modelling tool must be collected. Therefore, the semantic connections between the modelling entities of the tool and the objects of the object-oriented metamodel have to be identified. The tool integrator fulfils a system-controlled dialog in Lotus Notes to enter the necessary information about the interdependencies between the tool elements and the integrated metamodel.

The integration module offers prevalues for an association, where possible, in order to relieve the acquisition of information about the semantics of all the entities of a modelling tool. Therefore, the semantic information regarding each of the objects of the metamodel was analysed. As a result of examining the naming of the components of a manufacturing system, numerous technical terms used as synonyms for the same object were recognised. They were then stored in the respective metamodel objects. These synonyms are used within the tool integration user dialog to offer a prevalue for the semantic association between a tool component and its corresponding metamodel object. If a prevalue for an association does not meet the semantic correspondence between the conceptual schema of a tool and the object-oriented metamodel, then the tool integrator has to interfere with a correction into the dialogue.

The semantic associations between the data entities of a tool and the objects of the metamodel are laid up within the object-oriented database, after defining them in Lotus Notes. Therefore, each object class of the metamodel was extended by defining ToolRepresentative data members to store the semantic information. These ToolRepresentative data members hold references to the associated modelling elements of all the modelling tools integrated. This is the basis for the logical connection between the modelling tools where the metamodel objects are used as a linking layer.

After executing the dialogue-based tool integration process, the toolbox is capable of exchanging modelling data between the tools that have been integrated. The data transformation is done by using the semantic information about the assignment of the tool data and the metamodel objects as an intermediate layer. The objects containing the syntactical tool-specific data about a modelling system are passed to the transformation routines as a parameter. They are an important part, together with the semantic information about a tool, for realising the metamodel services, which can commonly be used, as well as for a user-suited preparation of data after changes on the joint data stock.

4 Conclusion

Organisations and also manufacturing companies find themselves in a most complex and turbulent environment, and must continuously adapt to this environment in order to be competitive on an ever-changing, global market. Therefore, it is necessary for a manufacturing company to use an adaptable and flexible planning procedure for the redesign of its manufacturing system. The groupware-based toolbox for cooperative modelling was extended to enable a quick and flexible response to any kind of change.

The integration of modelling and simulation tools, only used in isolation so far, as well as the meaningful connection of the collaboration partners based on awareness of working results, transparency of modelling actions and flexible coordination, shows the potential of groupware-enabled cooperative modelling of manufacturing systems. The use of information technology to realise group awareness, transparency of modelling activities and distribution of information within the toolbox results in essential advantages in contrast to conventional means. It leads to automatic initiating of the notification processes, speeding up of the data exchange and improving information sharing within a project group. In addition, it enables a relief from organisational, coordination and information tasks in the course of a planning project.

Processing a planning model must be started from the cooperation environment of the toolbox to avoid any unnoticed modification of modelling data. Information about the modelling activity is then created and stored and can be accessed later. Therefore, an information database was designed and implemented by using Lotus Notes. It manages the information about modification activities of modelling data and planning documents, namely the history of a planning model. It is helpful to trace and understand the modelling actions, which have been carried out on a planning model.

In addition, it is now possible that a planner can select a modelling tool from the toolbox that best suits his purpose. To meet this demand, a tool integration module was developed that helps to describe the physical view of an applications system as well as its semantic assignment to the corresponding metamodel objects. The tool-specific information gathered is also necessary to meaningfully combine the planning experts. Message services along with the semantic information and the metamodel are used as middleware to connect the modelling systems.

Another requirement of a computer-based toolbox for cooperative modelling existed in providing the means to adapt the planning environment to a particular planning issue and to changing project characteristics. Therefore, information modules were developed to enter and access project-specific data. This information is used within the toolbox to handle and control the modelling activities of a project group flexible and dependent on the state of the planning model.

References

Azuma, M. (1995): Towards the 21st Century's Software; State of the art and International Standard in JTC1/SC7; In: Lee, M.; Barta, B.-Z.; Juliff, P. (eds.): Software Quality and Productivity: Theory, practice, education and training; Chapman & Hall; London et al.; 1995

Dohmen, W. (1994): Kooperative Systeme: Techniken und Chancen; Hanser Verlag, München, Wien; 1994

Fuchs, L.; Sohlenkamp, M.; Genau, A.; Kahler, H.; Pfeifer, A.; Wulf, V. (1996): Transparenz in kooperativen Prozessen; Der Ereignisdienst in POLITeam. in: Krcmar, H.; Lewe, H.; Schwabe, G. (eds.): Herausforderung Telekooperation. Fachtagung Deutsche Computer Supported Cooperative Work 1996 (DCSCW '96); Berlin u.a., Springer; 1996

Heimann, P.; Westfechtel, B. (1997): A Generalized Workflow System for Mechanical Engineering; In: Proceedings Workshop "Arbeitsplatzrechner-Integration zur Prozeßverbesserung" der 27. Jahrestagung der GI; Aachen; 1997

Janocha, A. T.; Arnold, F. (1996): Architekturkonzepte integrierter CAx-Systeme; in: Proceedings of the International Conference of Computer Integrated Manufacturing (CIM '96); Zakopane; Poland; 1996; Volume III; pp. 183-190; 1996

Mertins, K.; Jochem, R.; Jäkel, F.-W. (1997): Integrated Business Process Modelling, Analysis and Supporting Tools; in: Doumeingts, G.; Browne, J. (Eds.): Modelling Techniques for Business Process Re-engineering and Benchmarking; Chapman & Hall; London et al.; 1997

Nastansky, L.; Hilpert, W.; Ott, M.; Riempp, G. (1995): Die Produktivität Groupware-basierter Anwendungen im Workflow Management; In: Krcmar, H. (Hrsg.): Tagungsband des 14. Baden-Württemberg-Kolloquiums "Computer Supported

Cooperative Work und Computer Aided Team - Verbesserung der Gruppenarbeit durch Computer"; Universität Hohenheim; Hohenheim; 1995

O'Sullivan, D. (1994): Manufacturing systems redesign: creating the integrated manufacturing environment; PTR Prentice Hall; Englewood Cliffs; 1994

Ott, M.; Nastansky, L. (1997): Groupware Based Organization Design for dynamic Workflow Management and Office Systems, Future development of organization design as a (tele-)cooperative, distributed teamtask; in: Abramowicz, W. (Hrsg.): Proceedings Business Information Systems '97 (BIS'97); Poznañ; Polen; 1997

Scholz-Reiter, B.; Bastian, D. (1995): Groupware based simultaneous organizational and manufacturing modelling; in: Bubnicki, Z. (Ed.): Proceedings of the 12th International Conference on Systems Science; Vol. 2; Wroclaw; 1995; pp. 225 – 232

Scholz-Reiter, B.; Bastian, D. (1996): Distributed, co-operative modelling of manufacturing systems; In: Scholz-Reiter, B.; Stickel, E. (Hrsg.): Business Process Modelling; Springer Verlag; Berlin u.a.; 1996

Scholz-Reiter, B.; Bastian, D. (1997): Groupwarebasierte Zusammenarbeit bei der verteilten Modellierung industrieller Prozesse; in: Tagungsband des Workshop „Rechnergestützte Kooperation in Verwaltungen und großen Unternehmen" im Rahmen der 27. Jahrestagung der Gesellschaft für Informatik (Informatik'97); Aachen; 23.09.1997

Stickel, E.; Hunstock, J.; Ortmann, A.; Ortmann, J. (1995): Verfahren zur werkzeuggestützten Integration von Datenbankschemata; In: König, W. (Hrsg.): Wirtschaftsinformatik ´95; Physica-Verlag; Heidelberg; 1995

Acknowledgement

The work described was funded by the German Research Association (DFG) - Scho 540/1-3.

Transformation of Static Process Models into Dynamic Simulations: Issues and Considerations

George M. Giaglis

Department of Information Systems and Computing, Brunel University,

Uxbridge, Middlesex UB8 3PH, UK

Abstract. *Static business process models (for example, flowcharts), as well as dynamic simulations, have been amongst the most widely used and effective methods for studying and analysing business processes. However, the transition from time-independent static descriptions of business processes to time-dependent dynamic simulation models may present modellers and decision-makers with a variety of problems, mainly related to data collection and experimentation. In this paper, we present a real-life case study of business process modelling. We illustrate the modelling process followed and the various challenges faced by modellers when enriching the static process models with the numerical data needed to drive the simulation runs. We present a number of specific issues that needed the attention of modellers, as well as some potential measures that can be taken to alleviate such problems. Drawing on the findings of our study, we discuss a novel approach for integrating models at different levels of abstraction to support the transition from static to dynamic process models.*

Keywords. *Static Process Modelling, Dynamic Process Modelling, Simulation, and Flowcharting.*

1 Business Process Modelling (BPM) and Business Process Simulation (BPS)

An organisation can be viewed as a collection of business processes that change or evolve with time in response to signals from the business environment. To allow organisations to study their processes and identify change opportunities more effectively, Business Process Modelling (BPM) has emerged as a set of techniques and tools for understanding and redesigning processes with greater ease and increased efficiency.

A process model can be viewed as an abstract representation of reality whose purpose is to reduce the complexity of the real world to allow users to study processes more effectively. A process model can be used for various application purposes, for example to facilitate human understanding and communication, support process improvement, support process management, automate process guidance, or automate execution support (Curtis, 1992). Of course different application objectives will focus on different process elements and therefore impose different requirements on the capabilities of the chosen BPM method. To facilitate such diverse requirements, various BPM techniques have been developed to provide modellers with access to a wide range of options to suit particular modelling needs.

BPM techniques can generally be classified into two categories, namely *static* and *dynamic* modelling techniques. Static techniques can provide only business process 'snapshots' at a specific point in time and they do not generally provide capabilities for analysing the process behaviour over a period of time. As a consequence, although they allow for understanding the nature and essence of a process, they cannot be used to predict the impact of proposed changes on process performance (Gladwin, 1994). On the other hand, dynamic BPM techniques allow for 'running' process models over a time period and therefore allow for studying time-related process phenomena, such as bottlenecks or throughput problems. However, dynamic models are generally much more difficult to develop, as they often require modellers to have an advanced level of expertise in the particular modelling technique chosen, and additional data are needed to calibrate the models.

Table 1 provides an overview of some widely used static and dynamic BPM techniques. It is not intended to provide a comprehensive account of such techniques, but rather an illustration of the wide range of alternatives provided to the prospective business modeller.

Amongst the dynamic BPM techniques, discrete-event Business Process Simulation (BPS) offers a theoretically attractive mechanism for modelling and studying complex phenomena in quantitative terms. BPS has already been identified as a suitable technique for process modelling and has been successfully employed in individual applications (Giaglis, 1998; Lee, 1996; Giaglis 1996; Mylonopoulos, 1995; Ninios; 1995). However, despite the significant advantages that BPS offers in theory, the development of valid and credible simulation models may present a number of practical difficulties.

In practice, BPS models are usually being developed as extensions to static process representations, for example flowcharts. This is reflected by the existence of many commercial off-the-shelf (COTS) software that allow users to develop static BPM models and then populate them with dynamic data that will drive process simulation runs. However, this transformation of static process models (flowcharts) into dynamic simulation models is not always trouble-free. Even if no modifications are needed on the flowchart itself as a result of the transition process, modellers should almost always expect problems when collecting and validating the numerical data needed for simulation. Even more importantly, problems may arise during the experimentation phase because of the inability to generate the quantitative data needed to drive the simulation runs of models that correspond to non-existing process situations.

	Functions	Typical Methods
STATIC BPM TECHNIQUES	• Show processes as sets of interconnected activities • Show the flow of entities (control, information, materials, etc.) throughout a set of activities	• Process Flowcharting • Data Flow Diagramming • IDEF0 (Function Modelling) • IDEF1x (Data/Information Modelling) • IDEF3 (Process Modelling) • Basic Petri Nets • RAD (Role Activity Diagramming)
DYNAMIC BPM TECHNIQUES	• Support analysis of and experimentation with alternative process structures • Support communication through visualisation and animation • Support user interaction with the process model	• Discrete-event Business Process Simulation (BPS) • System Dynamics • Timed Petri Nets • Workflow Modelling

Table 1: Overview of static and dynamic modelling techniques

In this paper we present a real-life BPM application to illustrate some of the problems that modellers may encounter during the transition from static flowcharts to dynamic process models. A number of specific issues will be presented, together with some ideas for potential remedy measures. Drawing on the findings of the study, we will present a novel approach for integrating simulation models at different levels of abstraction to support the transition from static to dynamic process models.

2 The Case Study

The study refers to a business process improvement effort jointly undertaken by two collaborating organisations in Greece: a major pharmaceuticals company (henceforth referred to as ABC S.A) and one of the regional distributors of its products (a small company we will call XYZ Ltd.). The project aimed at assessing the potential of redesigning the trading communications scheme between the two companies and evaluating the possibility of introducing IT applications for supporting the redesigned processes.

Company ABC is a subsidiary of a well-known multinational family of companies. The company employs more than 300 people in three sites. The headquarters are in Athens, while the plant and the warehouse are located a few miles north of the city, in Mandra. Furthermore, the company operates a smaller office in Thessaloniki (the second largest city in Greece). This office is responsible for managing ABC sales for Northern Greece.

ABC employs a network of collaborating distributors across Greece to deliver its products to customers. One of these distributors is XYZ Ltd. XYZ is a small company based in Thessaloniki and it is ABC's exclusive distributor for the whole of Northern Greece.

2.1 Scope and Objectives of the Study

Due to the special nature of the health care business and the subsequent urgency of most customer demands, XYZ has to operate within strict deadlines regarding deliveries. Each order has to be fulfilled with 24 hours if the products are to be delivered within Thessaloniki or within 48 hours for the rest of Northern Greece.

However, it has been noted that the aforementioned targets are virtually never met in practice. Since this represented a major source of customer dissatisfaction it was decided that a more in-depth study of the problem should be sought facilitated by the application of BPM and BPS. In this context, the study presented here was conducted. The objectives of the study were:

a) To examine in detail the existing processes of customer order fulfilment, invoicing, and warehouse management, where co-operation between ABC and XYZ is taking place.

b) To propose redesigned process structures to alleviate existing problems.

c) To evaluate the potential of introducing an appropriate IT infrastructure to facilitate the communication between the two companies.

2.2 The Business Processes

The overall process to be considered in this case study is the *Order Fulfilment Process (OFP)*. Fig. 1 shows the parties involved in the OFP as well as the existing communication between the participants (both physical and informational exchanges).

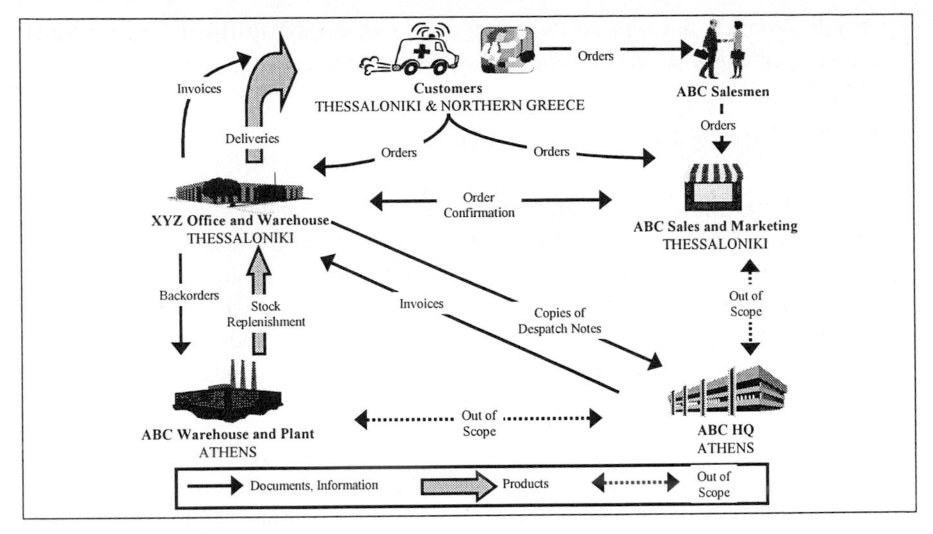

Fig. 1: Information and Material Exchanged in the OFP

The OFP consists of three inter-related sub-processes:

a) The **Order Taking Process (OTP)**. Customers can placc their orders *either* to XYZ *or* to ABC Thessaloniki office *or* directly to a salesman. Orders are given by phone or fax (or directly to a salesman who then delivers the order to ABC Thessaloniki). All orders have to be authorised by ABC Thessaloniki before they can be further processed.

b) The **Invoicing Process (IP)**. This process is triggered for every customer order and ends when the customer receives an invoice corresponding to that order. At regular intervals, XYZ collects and sends to ABC headquarters in Athens copies of all the Despatch Notes issued during the previous days (each Despatch Note contains information about products shipped in an individual delivery). ABC uses these Despatch Notes to issue invoices that are sent back to XYZ. Upon receipt of the invoices, XYZ schedules the delivery of invoices to customers along with the normal delivery of products.

c) The **Warehouse Management Process (WMP)**. This process refers to XYZ's task to maintain an appropriate level of inventory in its warehouse to be able to efficiently fulfil customer orders. After receiving the authorised order (see the OTP above), XYZ check their inventory for the ordered items. If the order can be fulfilled, the despatch of the products to the customer is scheduled. A Despatch Note is issued and accompanies the shipped items. Otherwise, the order is fulfilled only partially (if this is possible) and a backorder is created for the remaining items. At regular intervals, XYZ compiles all backorders into a *Backorder List* that is forwarded to ABC warehouse. Upon receipt of the Backorder List, ABC warehouse employees schedule a shipment of products to

XYZ. Products are sent so that all backorders can be fulfilled and a regular replenishment of XYZ's warehouse is also performed.

2.3 Modelling Approach and Issues

Fig. 2 illustrates the approach that was followed for modelling and analysing the processes under consideration. Since the modelling work presented herein was part of a wider business process change project, the approach shown in Fig. 2 does not correspond to the actual phases of the project as a whole. Rather, it is indicative of the activities that were performed within the modelling and analysis part of the project. As such, the approach is consistent with a generic approach to simulation modelling, such as the one presented by Law and Kelton (Law, 1991).

We initially intended to use the approach as a sequential process from one phase to the next, much like to the waterfall paradigm of Information Systems development. Of course, there were provisions for minor steps back when necessary (for example, if the validation process dictated that changes were necessary in any of the preceding steps). However, when the approach was applied in practice, various problems emerged. Some of these problems necessitated that previous phases were revisited to accommodate for further requirements that were not initially envisaged. The most important of these non-planned iterations are depicted in Fig. 2 as dotted lines. These iterations were the result of some specific issues that arose during the modelling process. We believe that most of these issues are of a generic nature (i.e. they can be expected to arise in most practical modelling situations). We will therefore present them in the context of our case study and articulate some possible improvement measures.

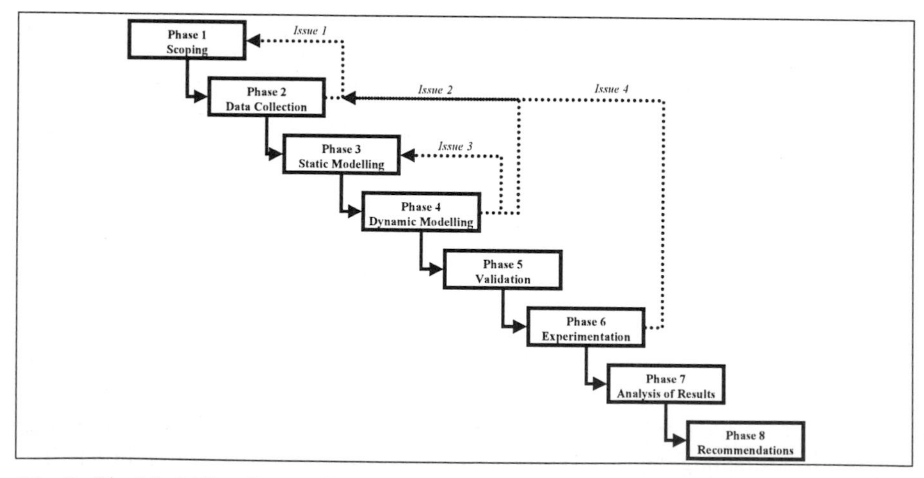

Fig. 2: The Modelling Approach

The first phase of our work (*Scoping*) was concerned with capturing the essence of the process and defining the boundaries of the process models. Interviews with key process participants (management and employees) of both companies were conducted. The aim was to obtain managers' and end-users' consent on the overall layout of the process and the individual activities it consists of. Moreover, the Key Performance Indicators (KPIs) for the analysis of the models were defined and agreed upon by the modellers and the decision-makers.

The next phase (*Data Collection*) involved the collection of the quantitative data needed for the development of the simulation model(s). Various categories of quantitative information had to be gathered, including:

a) *Activity durations*. It must be noted that these times are often stochastic (in other words, dependent on statistical function distributions to sample individual values during the simulation runs) resulting in increased complexity of the data collection exercise.

b) *Resource requirements*. For example, how many people are available and what other mechanisms are required to perform certain activities?

c) *Activity and resource schedules*. For example, what are the normal working hours for resources?

d) *Priority arrangements*. For example, what should happen when two or more activities compete for a single resource?

e) *Event frequencies*. For example, how often do customer orders arrive?

f) *Path routing mechanisms*. For example, what percentage of orders has to be placed in a backorder?

g) *Individual Activity costs* (based on time schedules and resource utilisation).

h) *Workplace constraints*. For example, queue or buffer sizes, maximum copies of activities allowed to run concurrently due to spatial or other limitations.

Data was collected via interviews, document and other observed evidence, as well as direct measurement in the workplace in both companies. Data collection was a complex and laborious endeavour, taking up a considerable amount of time in relation to the complexity of the overall project. The data collection was based on our perception of data requirements based on the conceptual model of the process that we had developed during the first phase (*Scoping*). However, we found out that some of the data we wanted was not available and could not be directly obtained by observation and measurement (at least not without an unacceptably high cost). We had therefore to revisit Phase 1 and re-scope the project by shifting the boundaries of our models in order to accommodate it to existing or easily accessible data. This problem was not unlike the problems reported by other researchers in similar situations (for example, Lee and Elcan 1996). Such problems can be detrimental for the success of a BPS project if the re-evaluation of model scope is significant. In such a case, the modellers could end up by studying the wrong problem or arriving at conclusions and recommendations that are not directly related to the initial questions asked.

Issue 1: *Data unavailability may lead to re-evaluation of model scope.*

Possible Remedy: If relevant and credible data cannot be obtained, it might be necessary for modellers and decision-makers to re-examine the desirability of proceeding with a quantitative process simulation model. It might be better to limit the modelling exercise to static and qualitative model development instead (for example, flowcharting).

Fortunately, in our case we did not need to modify the model scope significantly, so we were able to proceed with the third phase (*Static Modelling*) which was concerned with the development of a flowchart to depict the activities of the OFP in detail. A commercial off-the-shelf software package was selected as the basis for model development. The main requirement of the decision-makers was that the process modelling software should be easy to use in order to allow the companies to continue using the models after the end of the project. It should also encompass both static and dynamic modelling capabilities. An evaluation of different process modelling and simulation packages was carried out and *Process Charter* (Scitor, 1995) was selected for modelling, mainly because it combined ease of use with decomposition of the static and dynamic modelling phases. An initial static process model was developed in Process Charter to depict the activities within the OFP. Fig. 3 depicts this model.

The fourth phase (*Dynamic Modelling*) involved entering the quantitative data obtained earlier into the individual activities of the flowchart to provide all the information needed for dynamic simulation runs. Two problems arose during this phase:

a) The data obtained did not perfectly match the activities of the flowchart, so some manipulation of either the data or the flowchart had to be done.

b) It was realised that the simulation model would require the collection of additional data not initially envisaged during the initial Data Collection phase.

As a result, we had to revisit both Phase 2 (in order to collect more data as needed), and Phase 3 (in order to modify the flowchart to accommodate the existing data structures). This had various effects to the whole exercise. Firstly, the project as a whole was delayed as a result of having to abandon work on the current phase and switch to previous phases that were considered complete. Furthermore, the new static model had to be shown once more to customers in both companies to obtain their approval that it constitutes an adequate representation of the existing processes. Customers did not initially understand the need for rework on an already approved model. This resulted in customer complaints and dissatisfaction with the whole process. Additional time had to be spent to explain the need for the proposed route until customers were convinced.

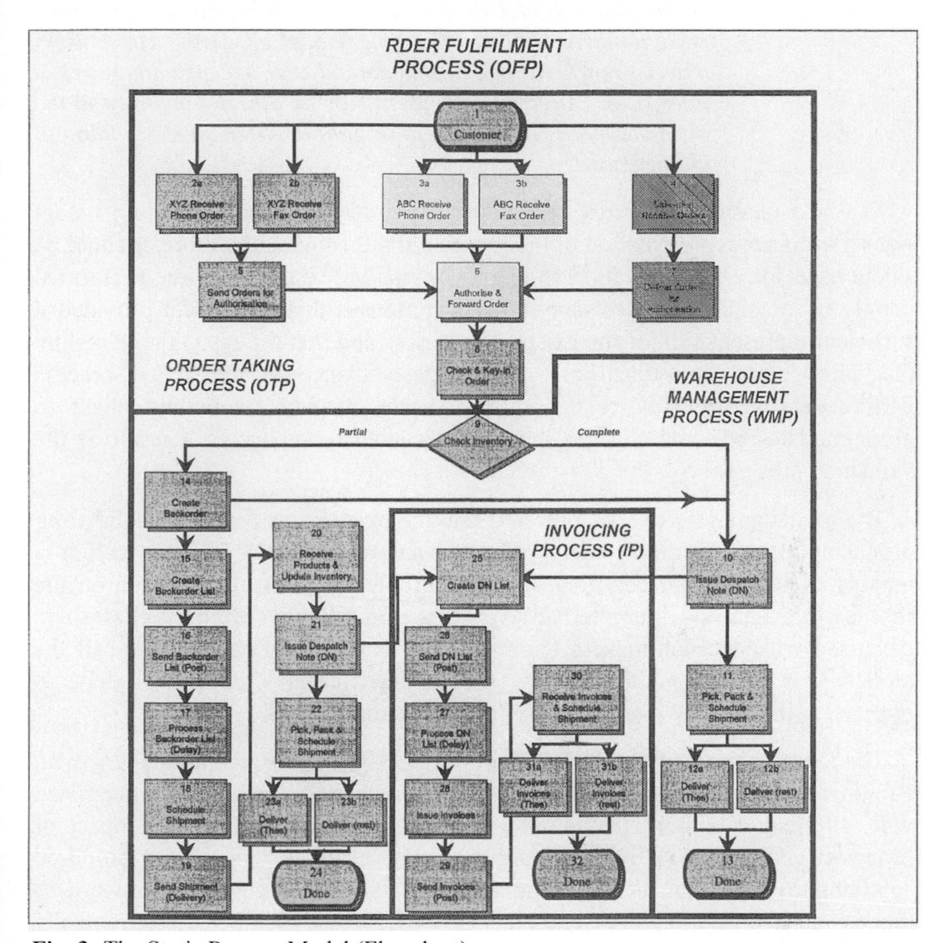

Fig. 3: The Static Process Model (Flowchart)

Issue 2: *Further data collection requirements may arise during the course of the modelling process.*

Possible Remedy: *It may be helpful if data collection is carried out only* ***after*** *the development of the static model. This way the collection of quantitative information can be planned to directly match the requirements imposed by the structure of the static model already developed and the dynamic model that will follow.*

Issue 3: *Dynamic model development may impose changes on the static model structure.*

Possible Remedy: *Users should be warned as early as possible, that dynamic modelling and data collection needs may influence the layout of the static model to a structure not perfectly matching the most intuitive way of arranging model elements. Thus, users may be more willing to understand and accept changes at a later stage. Alternatively, users may be directly involved in the dynamic model development to gain a better understanding of change needs.*

The next phase (*Validation*) involved making sure that the simulation model was an adequate representation of the real-world system and, therefore, it could be safely used for experimentation and decision-making. Validation was performed mainly by obtaining the decision-makers' consensus that the model provided a sufficient representation of the existing processes and that the results obtained by the initial simulation runs bore a close resemblance to the actual process performance. Of course, we had to re-visit the dynamic modelling phase to accommodate for some changes that were deemed necessary as a result of the validation process.

The model analysis clearly indicated that the existing processes were far from producing results within the stated management targets. It also brought to light a number of reasons that contribute to the inefficiencies identified. Based on the results of the analysis, four alternative process configurations were developed and discussed with decision-makers for acceptance and feasibility. We will call the models that correspond to the new process layout, the 'TO-BE' models, as opposed to the 'AS-IS' model that depicts the existing process.

The development of the TO-BE process designs presented modellers with additional problems and challenges. The TO-BE models needed to be populated with different data from the initial simulation model to reflect the impact of proposed changes on activity performance. For example, one of the proposed solutions involved that the two companies are linked in a computer network so that customer orders, invoices, and other information are exchanged electronically instead of in paper form. It was expected that such a change would significantly reduce the duration of certain activities, thereby removing many bottlenecks in the process. However, the TO-BE models had to be populated with data referring to the new activity durations. But how can, for example, the average time needed for a database query or a file transfer over the network be estimated since neither the IT applications nor the network are in place? Such information is critical to developing a valid and credible TO-BE model in order to arrive at informed decisions about the effect of the proposed changes on the model KPIs, yet it is very difficult to obtain in practice.

Issue 4: *Experimentation will lead to further data collection requirements that cannot be accommodated by direct measurement at the workplace.*

Possible Remedy: *Modellers will have to make detailed and justified assumptions about data they will use to populate the TO-BE models. Such data can be obtained for example, by observing similar settings in other real-world systems. Due to the additional assumptions made, sensitivity analysis of models becomes of paramount importance for informed analysis and decision-making.*
Another potential solution might be to develop further, detailed simulation models, to study how the proposed changes will influence the activities in the TO-BE models and incorporate the results of such simulations in the TO-BE BPS models.

In the next section, we will elaborate this final suggestion in more detail, by presenting a novel approach to BPS experimentation and analysis.

3 Towards the Integration of Business and Network Simulation Models

In this section we will draw from the experience gained in our case study and present a novel approach to the identified problems of experimentation and analysis in BPS models (see issue 4 above). We have mentioned that one of the major problems when experimenting with proposed changes on the business level will be to populate the TO-BE BPS models with valid and credible data on the impact of proposed changes on activity parameters. Such data cannot be easily collected with traditional data collection techniques, as they relate to a non-existing system and cannot therefore be obtained by direct measurement.

We propose that when the process changes involve the application of Information Technology (IT) applications and Computer Networks (CN), such data can be obtained by simulating the proposed changes at the detailed level of the network infrastructure that will support the redesigned processes. More specifically, modellers can develop *Computer Network Simulation (CNS)* models, experiment with alternative network configurations and IT application structures, and feed the CNS results into the TO-BE BPS models to support further experimentation and analysis at the business level.

Under our proposed scenario, simulation should start from the detailed level of CN and IT, with the development of models that depict the network infrastructure of proposed changes. Business domain data and IT application constraints will be the input data to CNS and will be used to drive the experiments performed on the models. CNS will be used to arrive in specific, detailed proposals for changes at the infrastructure level. The simulation output data from the TO-BE CNS models will be used to drive the development and operationalisation of high-level BPS models. These models will in turn be used as experimentation vehicles to assess alternative process design structures. Fig. 4 illustrates this approach.

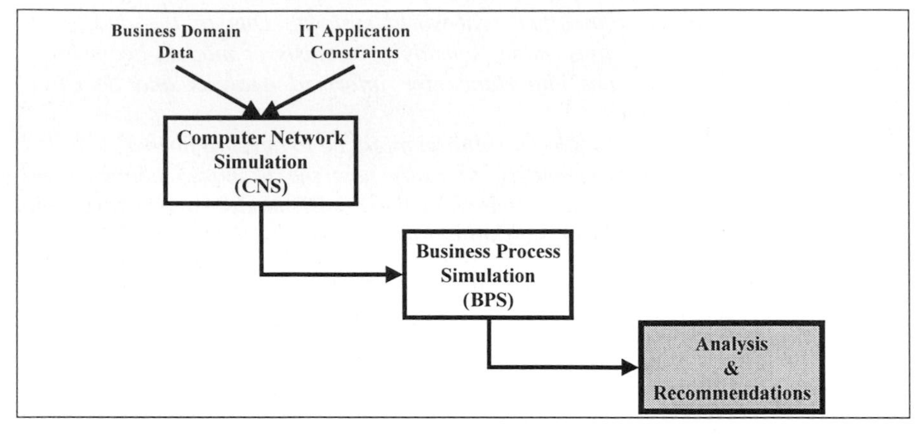

Fig. 4: Integrating BPS and CNS

The rationale behind this argument is simple. Due to the complementary nature of Business Processes (BP) and Information Technology (IT), one could assume that improvement and design efforts on these two organisational facets should be performed in parallel. However, this rarely seems to be the case in practice.

Discrete-event simulation is an example that clearly demonstrates this gap. A plethora of specialised simulation software environments have been developed to support organisations in modelling their business processes and IT infrastructure. On the one hand, there is a category of products, collectively referred to as Business Process Simulators (BPS), that allow for business process modelling and analysis. Examples of such software include *Simprocess* (by CACI Products Company), *Process Charter* (by Scitor Corporation), *Bonapart* (by UBIS GmbH), and *iThink* (by High Performance Systems Inc.). On the other hand, there is a distinctly different category of tools, namely Computer Network Simulators (CNS) that allow for modelling of and experimentation with the underlying computer network infrastructure. Examples of such software include the *Comnet* family of products (by CACI Products Company), *BONeS* (by Systems and Networks), and *Opnet* (by Mil 3).

However, despite the imminent interrelationship between business processes and Information Technology, there do not exist any simulation environments that allow organisations to model ***both*** elements and identify the impacts that changes on one will have on the other (Painter, 1996; Bradley, 1995). Although BPS and CNS products have proved themselves successful in their own application areas, there is a growing demand for a new generation of simulators that will unite BP and IT simulation under a single umbrella and facilitate parallel design of business processes and IT systems. Van Meel and Sol (Meel, 1996) have used the term

'*Business Engineering*' to refer to this dual design effort. To be a beneficial facilitator of *Business Engineering*, simulation should be able to address both the BP design and IT design problems.

Painter et al (Painter, 1996) have addressed this problem and proposed that a 'middle' layer is introduced between Business Processes and Computer Networks. The middle layer consists of models that depict the IT applications that run on the Computer Networks and support the Business Processes. Such a layer introduces a medium abstraction level as a mechanism for bridging the gap between BPS and CNS. Such a hierarchical view implies that the relationships are also hierarchical. However, the interrelationships between BP, IT, and CN are much more complex and may have different implications for the design of successful integrated simulation environments. Fig. 5 depicts these relationships as a tightly coupled dependency web. There is a need for further investigation of the nature of these dependencies in practice. Such a research effort could drive the development of integrated BPS and CNS environments that will be suitable for the needs of enterprise modelling. We envisage that interface mechanisms can be built between BPS and CNS models so that the required data can be automatically exchanged.

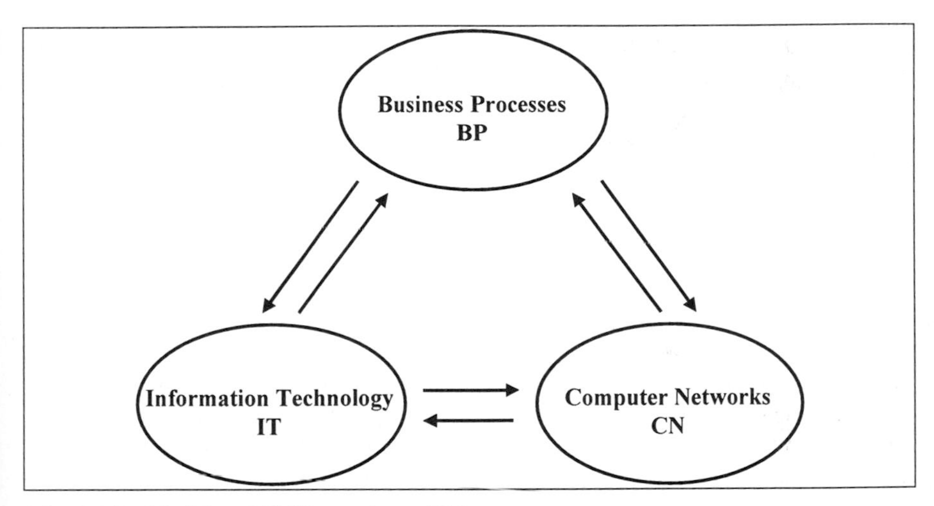

Fig. 5: The BP, IT, and CN Dependency Web

4 Conclusions and Further Work

In this paper we investigated the issues that may arise when transforming static flowchart models into dynamic simulations. Based on a real-world case study, we articulated four such issues:

a) Data unavailability may lead to re-evaluation of model scope.

b) Further data collection requirements may arise during the course of the modelling process.

c) Dynamic model development may impose changes on the static model structure.

d) Experimentation will lead to further data collection requirements that cannot be accommodated by direct measurement at the workplace.

We have also proposed a number of mechanisms by which such issues can be accounted for before and during the course of a modelling project. These mechanisms can be summarised as follows:

a) The desirability and feasibility of dynamic modelling should be re-examined in the light of data collection problems.

b) It may prove more efficient to carry out data collection after the development of an initial model layout, in contrary to what is proposed in most modelling methodologies.

c) User training and awareness of modelling issues and potential problems is necessary to ensure project success.

d) Integration of Business and Network simulation models may prove an efficient mechanism for alleviating problems relating to unavailability of data for simulation experimentation.

Of course, our main proposition (that experimentation problems may be overcome by combining BPS and CNS) is not applicable to every business modelling situation. Rather, it is mainly targeted to business change scenarios where IT applications and computer networks play an integral part.

Another potential limitation has to do with the complexity of the proposed approach. It may be argued that combining different simulation models may prove to be a complex and laborious endeavour that will need extensive business and technical skills to accomplish. After all, not many analysts or consultants exist that can effectively combine business and network simulation skills, let alone also master investment appraisal techniques.

On the subject of integrating BPS and CNS models, we envisage the development of automated interface mechanisms that will allow for the seamless exchange of information and data between different simulation models. On a more advanced scenario, further research could investigate the incorporation of the business investment appraisal process itself in the simulation models. This could be achieved by enriching simulation software with Cost-Benefit Analysis (CBA) capabilities that will allow decision-makers to compare the benefit of each proposed investment with the cost associated with its implementation. Similar mechanisms have already been implemented to incorporate Activity Based Costing (ABC) mechanisms in BPS software packages.

A number of further research avenues can be envisaged to validate and enhance the findings of our study. Real-life case studies of investment evaluation by simulation need to be pursued in order that will test our hypotheses and generate further implications of our approach in practical settings. Such a research could lead to further generalisations and, ultimately, towards a theory to explain the dynamics of BP, IT, and CN change. Such a theory could drive the development of automated tools to support the evaluation process and alleviate the problems of complexity and skills required for execution.

A major research project is currently under way that will investigate the above issues. The project is jointly undertaken by Brunel University and a number of industrial collaborators in the UK and is funded by the UK Government's *Engineering and Physical Sciences Research Council* (EPSRC) under the umbrella of the *Systems Engineering for Business Process Change* (SEBPC) programme. The project's research objectives are:

a) To understand the opportunities and obstacles in using discrete simulation to model business processes.

b) To develop theories and techniques for quantifying the effects of IT on business performance and translate them into specific simulation software requirements and design structures.

c) To integrate business, IT, and network modelling by Simulation. We envisage the development of a combined BPS and CNS Design Theory, as well as an accompanying set of methodologies, techniques, and software tools to address investment evaluation by simulation.

d) To generate a rationale and design guidelines for a new generation of simulation environments that will allow for the integrated modelling of Business Processes and IT infrastructure.

e) To demonstrating the validity of the approach in real-life business settings in order to obtain more information on industrial relevance and potential improvements.

References

Bradley, P.; Browne, J.; Jackson, S. and Jagdev, H. (1995): Business Process Re-engineering (BPR): A Study of the Software Tools Currently Available, *Computers in Industry, 25, 3*, pp. 309-330

Curtis, W.; Kellner, M.I. and Over, J. (1992): Process Modelling, *Communications of the ACM, 35, 9*, pp. 75-90

Giaglis, G.M. (1996): Modelling Electronic Data Interchange Through Simulation: An Industry-Wide Perspective. In the *Proceedings of the 8th European Simulation Symposium*, Genoa, Italy, October, pp. 199-203

Giaglis, G.M.; Paul, R.J. and Doukidis, G.I. (1998): Dynamic Modelling to Assess the Business Value of Electronic Commerce. In the *Proceedings of the 11th International Electronic Commerce Conference, vol. 1: Research*, Bled, Slovenia, June, pp. 57-73

Gladwin, B. and Tumay, K. (1994): Modeling Business Processes With Simulation Tools. In Tew, J.D., Manivannan, S., Sadowski, D.A. and Seila, A.F. (Eds.), *Proceedings of the 1994 Winter Simulation Conference*, Lake Buena Vista, FL, December, pp. 114-121

Law, A.M. and Kelton, D.W. (1991): *Simulation Modelling and Analysis*, 2nd ed., McGraw-Hill, New York

Lee, Y. and Elcan, A. (1996): Simulation Modelling for Process Reengineering in the Telecommunications Industry, *Interfaces, 26, 3*, pp. 1-9

van Meel, J.W. and Sol, H.G. (1996): Business Engineering: Dynamic Modelling Instruments for a Dynamic World, *Simulation and Gaming, 27, 4*, pp. 440-461

Mylonopoulos, N.A.; Doukidis, G.I. and Giaglis, G.M. (1995): Assessing the expected benefits of Electronic Data Interchange through Simulation Modelling Techniques, In the *Proceedings of the 3rd European Conference on Information Systems*, Athens, Greece, pp.931-943

Ninios, P.; Vlahos, K. and Bunn, D.W. (1995): Industry Simulation: System Modelling With an Object Oriented / DEVS Technology, *European Journal of Operational Research, 81*, pp. 521-534

Painter, M.K.; Fernades, R.; Padmanaban, N. and Mayer, R.J. (1996): A Methodology for Integrating Business Process and Information Infrastructure Models. In Charnes, J.M., Morrice, D.J., Brunner, D.T. and Swain, J.J. (Eds.), *Proceedings of the 1996 Winter Simulation Conference*, San Diego, California, December, pp. 1305-1312

Scitor Corporation (1995): *Process Charter User's Guide*, Scitor Corporation, Menlo Park, CA

Process Modelling during the Determining of Offers for Castings

Sylke Krötzsch[1], Ines Hofmann[1], Georg Paul[2] and Eberhard Ambos[1]

[1] Institute of Production Technology and Quality Management,
[2] Institute of Technical and Business Information Systems,
Otto-von-Guericke-University Magdeburg, P.O. Box 4120, D-39016 Magdeburg
{kroetz|paul}@iti.cs.uni-magdeburg.de, {ines.hofmann|eberhard.ambos}@masch-bau.uni-magdeburg.de

Abstract. *In modern companies, the use of large information systems is increasing, in order to support the phases of the product development process. The Institute of Production Technology and Quality Management and the Institute of Technical and Business Information Systems have worked together on an integration of software systems at various levels, in order to support the engineers at the development of products. The aim of this is the support and automation of the process of preparation of offers for castings with focus on modelling and execution of the identified processes. This article describes a process sequencial control, based on the PACO-Integration framework, which is explain ad using the example of the work scheduling in foundries and the submission of proposals of a dynamic reaction of the process control on varying conditions.*

Keywords. *determining of offers for castings, integration architecture, engineering systems, process sequencial control, process modelling*

1 Motivation

In companies more and more software systems are used in product development. These are either complex monolithic systems, so called general solutions, for instance CAD/CAM, or special solutions which support designated subtasks. In the area of the product development process, applications such as CAD solutions, systems for numeric control, EDM- or PDM-Systems are used intensively. Hence, the introduction of new systems into the existing hard- and software environment of the companies is difficult and often realisable only through considerable expense in time and costs. Furthermore a subdivision in several work was is to observed, whereby the cooperation is often difficult.

During the process of completing documents for offers and manufacturing for castings, a lot of different information (geometry, material, process engineering, quality, deadline) has to be considered. The data effectively has to be gathered, to archived, and systematised. Thereby, the cooperation of the several software systems used in foundries is necessary. Under real conditions, the documents for a casting has to pass through several departments (work scheduling, construction, sale, manufacturing, ...), which are often locally separated (Fig. 1).

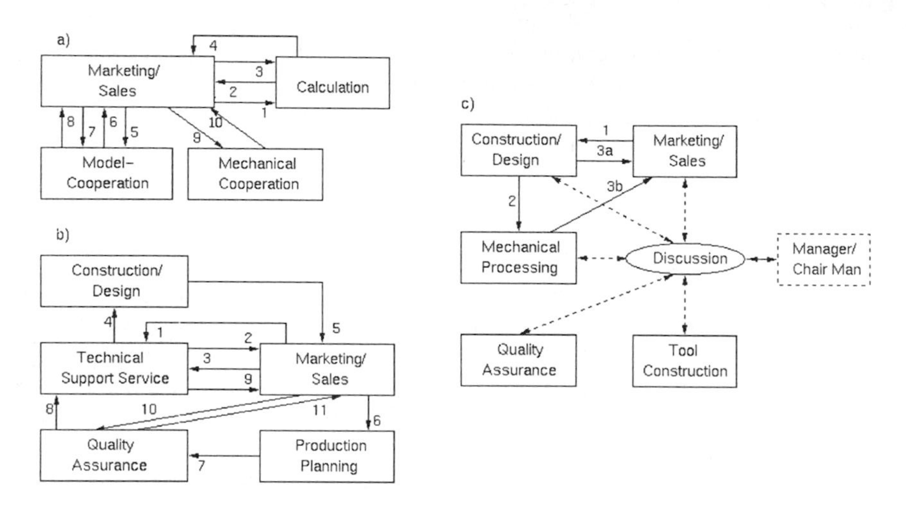

Fig. 1: Routes of the Documents in Different Companies

The consequence of this procedure is a loss of information and time. In order to increase the efficiency of the offer completion, it is necessary to implement new concepts for the product data management, process modelling, and process controlling. Therefore it has to be considered that in the past few years a lot of software systems for the development and manufacturing of castings have been developed, which have to be implemented expediently in a process sequential control and management. Consequently, it is necessary to model processes, in order to reengineer and to rationalise the process sequences in companies.

The aim of the integration framework is the realisation for the composition of different components of large software systems to a task-specific engineering system. In the Process Layer of the integration, it is attempted to realise a dynamic control dependent on the product development, in order to support the process of completing offers in foundries through modelling and automation and to transform concepts and methods for component oriented integration.

2 Integration Framework for Engineering Systems

*COR*BA (Common Object Request Broker Architecture) by OMG or (D)COM ((Distributed) Component Object Model) by Microsoft are examples for architectures, which provide methods and concepts to support the integration of different components of software systems relatively independent of computer and process boundaries. Thereby modern integration architectures for computer aided engineering systems must provide an integration on different levels (Paul, 1996). An integration framework (PACO[1]), which has been developed at the Otto-von-Guericke-University Magdeburg, is in response to adequate requirements. The framework should be understood as a concept for a system, in order to use furthermore the company's own software. In the following diagrams the PACO-Integration framework is explained (Fig. 2).

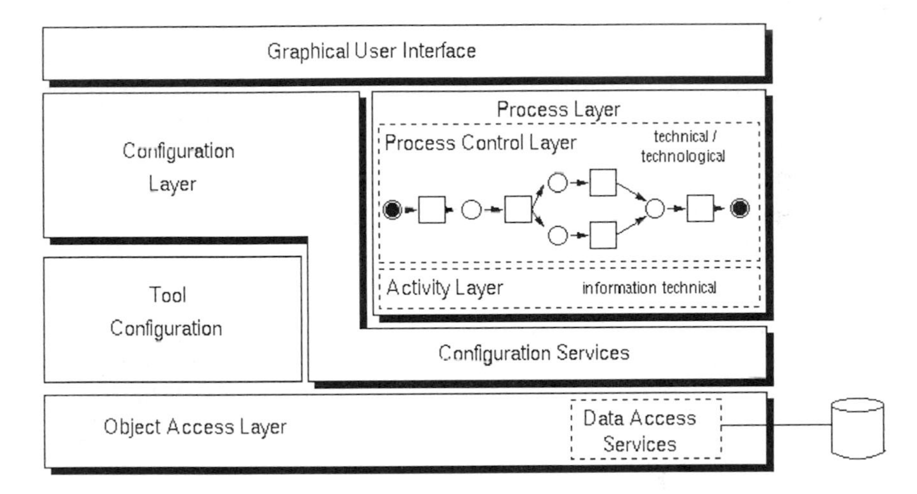

Fig. 2: Integration framework (PACO)

In order to achieve an integration of software systems, four integration layers can be considered. In the *Object Access Layer,* the necessary data from all the systems are collected and managed, for instance in a federated data model. The objects, e.g. product describing data or documents, are stored directly in the database or indirectly through references to the actual data objects. In order to use the functionalities of software systems better, in the *Configuration Layer,* components of the systems are assembled to desired configurations (Sattler, 1997). Components are modules with defined interfaces, where definitions, which are necessary, are made for the communication between the different software

[1] PACO stand for layers Process-, Activity-, Configuration- and Object Access Layer

systems. In this way dependencies respectivly interactions between the applications can be defined. The *Process Layer* is responsible for the modelling of process sequences which support the product development process. Todays engineering systems offer only limited possibilities for the modelling of sequences, so in this paper the realisation of the Process Layer is the main point. The *User Layer* provides all the functionalities of the total system and includes the graphical user interface, and the functionalities within the framework, in order to support the user on the work with the different hard- and software.

The four layers (Object-, Configuration-, Process- and User Layer) are based upon each other. The Process Layer consists of two layers, the *Process Control Layer* and the *Activity Layer*. The Process Layer realises a process sequential control, in order to flexibly coordinate the services provided by the Configuration Layer. A complete description of the total process sequences during the product development process shall not be achieved in the Process Control Layer. The goal is a partial automation of possible process sequences.

Further, more it is possible to implement the integration framework only partially or gradually. Through the possibility of using components of different software systems, systems it can be composed dependent on the desires of the users, thereby user-specific engineering systems are developed, which are tailor to a special branch of industry. The configuration is realised both, oriented on the components (functionality) and their cooperation and oriented on the process sequences and the single process steps. By this means the sequences can be modelled dependent on the necessary system components.

Based on the introduced integration framework, the Process Control Layer is presented in the following selection, in order to realise a process sequential control oriented on the determining of offers.

3 Proposal for a Process Integration

Engineering systems based on the PACO-Integration framework shall flexibly support the total product development process (Paul, 1997). For this, it is necessary to include the software available in the company into the engineering systems.

In todays companies cooperative work is recognisable. In order to realise this, it is necessary to model the executed sequences. Therefore, information about the tasks, the participants, the structure of the company, etc. must be provided. The engineering system must provide components for process modelling. The PACO-Integration framework realises these components through the Process Layer, in order to integrate only the components needed during the execution of the modelled sequences.

In order to support the product development process, it must operate with complex and interlocked processes. So parallel, iterative, or other structures in the sequences can be developed (Krötzsch, 1997). Another attribute of the product development process is the high, process-oriented information volume and the associated complex tasks for the project control. It is attempted to develop concepts for the reduction of product development times, while simultaneously assuring quality. The product development process is supported by a process sequential control, a dynamic reaction that some situations require.

Since the Process Layer of the PACO-Integration framework is responsible for the flexible control of sequences, particularly the execution, the formation of the execution and the aspects of the processes (what, how, with what, who) are important. It is assumed that the control is independent of the data management, the product structure management (EDM/PDM), or the project management. The functionalities are provided by the available systems and the components are integrated on demand.

3.1 Process Control Layer

In order to develop a process sequence control, which works with static sequences and also reacts dynamically on varying conditions (depending on the product structure), the Process Layer of the PACO-Integration framework is split into the *Process Control Layer* and the *Activity Layer*. By this means, the fact that the user works with technical/technological processes is achieved. The processes are specified as far as they are technical/technological processes. The Activity Layer realises the transfer into information-technical processes (*activities*) and therefore represents the connection between the technical/technological and the information-technical layer. The user itself cannot influence the execution of the activities of the activity net on the Activity Layer unless the intervention of the user is required by the system. That means each process step on the Process Control Layer can refer to several activities, which are executed on the Activity Layer. The Activity Layer combines abstract services from the Configuration Layer (e.g. „openCADFile(aCADDocument)“) to high-order services (e.g. „open existing CAD-File with name casting A“). The Process Control Layer uses these high-order services. The connection between Process Control Layer and Activity Layer is realised through a meta-description of the processes.

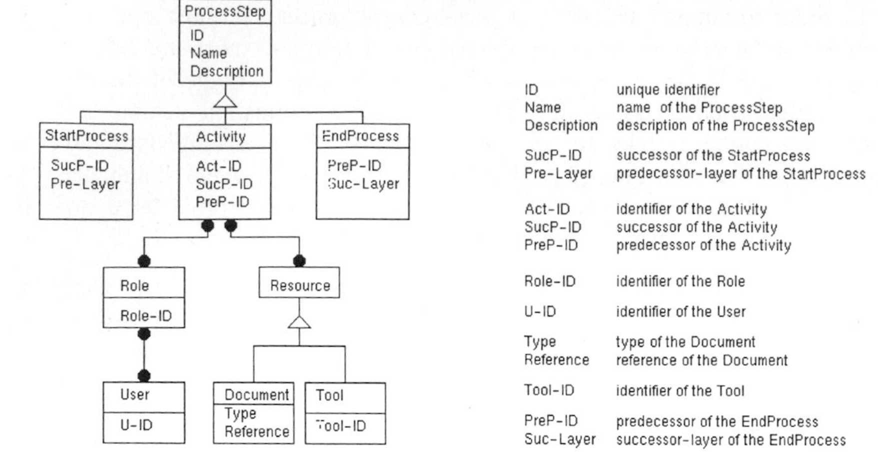

Fig. 3: Object Model for the Process Step

The modelling of the sequences is based on the object model in Fig. 3. Sequences that are modelled by the user, and pre-defined sequences are distinguishable. Pre-defined sequences are already modelled and provided by the system. That means these pre-defined sequences contain the calls of activities that are often executed by the engineer or the draftsman during the product development process. The other sequences, which are not pre-defined are engineering-technical sequences of tasks that can be used by the engineer during the modelling of new sequences. An example for a recurring sequence of tasks is the scheduling of times and operations for the manufacturing of castings. Sequences, including pre-defined sequences, consist of processes. A process can be a sequence or a process step. A process step is the smallest, atomic unit on the Process Control Layer. That means a process step is indivisible. There are three possible types of process steps (Fig. 3): *StartProcess*, *EndProcess*, and the call of an *Activity to* the Activity Layer.

The latter means, that the process step refers to the *Activity Layer* through an API and starts an appropriate activity. The Activity Layer is responsible for the execution of the activities. At *build time* the user must be able to access the repository that contains the available activities. Furthermore at runtime it must be verified whether the executing user is authorised to execute the process steps respectively the activities, they are running. Further, more it is possible to execute methods in order to verify the necessary roles and authorisations. The methods, as well as the possible executable activities, are provided by the Activity Layer.

Roles, users, tools, and documents must be assigned to the process step. The difference to conventional systems is the assignment of activities to the respective process steps. The activities are dependent on the used tools and can be selected from a list of available activities. In order to create the connection to the Activity Layer, the realisation of the API access and the specification of the processes will have to be implemented. A user can be assigned to one or more roles and a role can be assigned to one or more users. In Fig. 3, an activity with name "make pre-definitions" and identifier "Act-ID A8" could be available on the Activity Layer, which allows it to work after this scheme in order to make the pre-definitions. By this means, the user is supported as far as possible. The *Pre_Layer* of a StartProcess is „zero", it is the highest layer and the whole sequence starts there. In order to execute an activity on the Activity Layer, it is necessary to provide information about the user and the appropriate documents. The other information of the single process steps are used at the Process Control Layer only.

To support a user by the development of new products, it is required to model a coarse process scheme with the most important process steps or process sequences. It is possible to create processes, to connect processes to a sequence or simple constructions (e.g. sequences or loops), to assign pre-defined sequences to processes, and to set-up users. In order to make an assignment it is necessary that a pre-defined process sequence starts with a StartProcess and ends with an EndProcess. The total pre-defined sequence is assigned to a process as a sublayer.

The example in Fig. 4 shows the connection between Process Control Layer and Activity Layer. When a user models a coarse sequence, he defines a process. The user can then decide, whether the process shall be refined or activities, roles, user, tools, and documents will be assigned to the process. In the example of Fig. 4, the users are not considered, because one role can be assigned to several users and one user can be assigned to several roles.

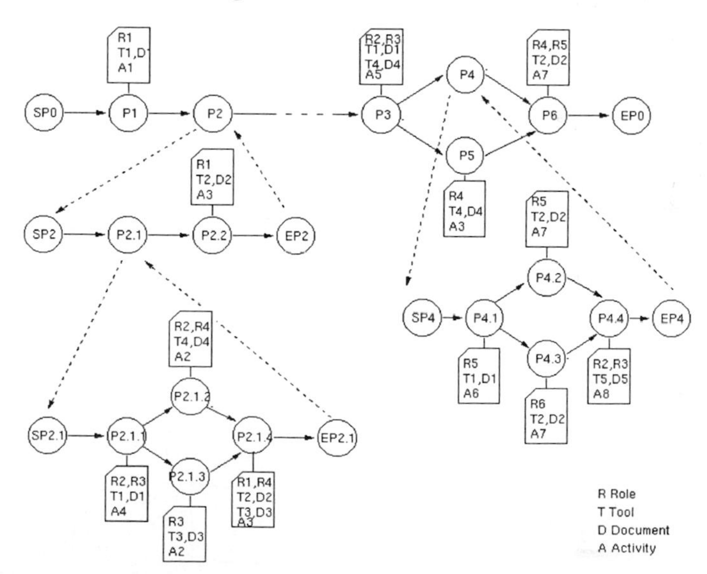

Fig. 4: Example for Process Modelling

The Process Control Layer accesses the API. With the available functions it is possible to get a list with the available activities and the user can the choose the desired activity for his process. Hence, it is possible that an activity is assigned to several processes, which are ordered on different layers because the execution is dependant on the results of the processes and activities that were finished before. The activities can consist of activity nets, because they can be refined on the Activity Layer. With help of the activities assigned to the processes, the execution is carried out on the Activity Layer.

3.2 Example: Determining of Offers for Castings

In order to realise a control on the Process Control Layer, a type of Petri-Net has been proposed. According to Petri-Net Theory, the modelling and execution of static sequences is possible, but not the dynamics of the actual net. The Process Layer is a link between the engineer and the engineering system and consequently it must be possible to react on dynamic situations.

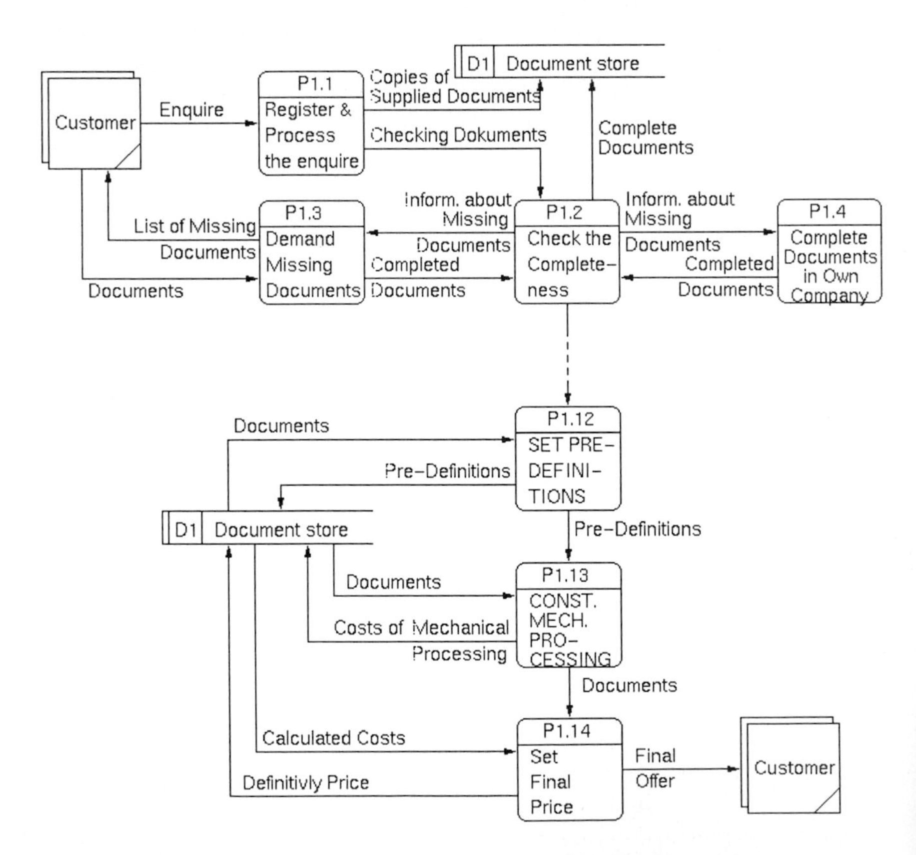

Fig. 5: Part of the Sequence “to Determine an Offer“ (Data Flow Diagram)

That means, not every possible sequence must be known at the beginning and therefore it is not necessary to model the sequences at that time. During runtime the sequences are dynamically defined. In order to differentiate from todays projects basic, recurrent sequences of the product development, process are created in cooperation with engineers. Examples are sequences for construction or design, for the generation of part lists, or for the determining of offers. Fig. 5 shows part of a possible pre-defined sequence for the determining of offers for castings. In the praxis the draftsman or engineer attemps to adapt available solutions or partial-solutions and only in 15 % of all cases a new construction has to be made (Krötzsch, 1997).

In order to use the created sequences it is necessary to cooperate with engineers at the beginning of development and modelling. The main processes of the sequences are the technical/technological processes. The assignment of activities in this case is not performed by the engineer at build time, but some pre-defined sequences are provided by the engineering system. Such static, pre-defined sequences are possible because, at the determining of offers in foundries, recurrent work-steps do exist. Following the sequence, which is shown in Fig. 5, is transferred into another sequence (Fig. 6). During the analysis phase, data flow diagram modelling was necessary because only unstructured and incomplete information exist in foundries. Very often the decisions are based upon experience. With the help of data flow diagrams it is possible to clearly present the necessary information.

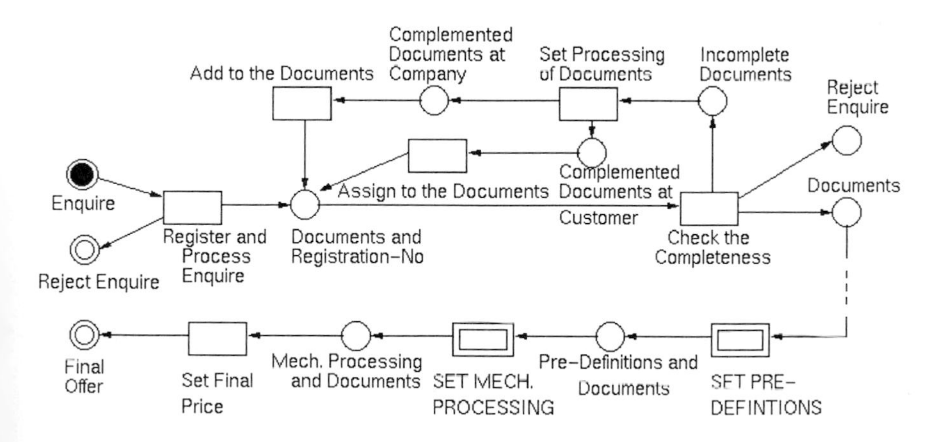

Fig. 6: Sequence for the Process "DETERMINE THE OFFER" (Workflow)

The modelled sequence for the determining of offers does not contain the distinction of the determining of offers for new, similar, or equal parts. This distinction will be realised in a next, refined sublayer. Due to complexity of the sequences, it is practical to further refine the processes. Thus the users are offered modelled, recurrent sequences (e.g. “make decisions“), which contain the realisation of the distinction of new, similar, or equal parts.

These are still static sequences, but a dynamic reaction is required and so the control will be split. On one hand a modelling tool will be developed, enabling the engineer to model sequences as accustomed. This method will be expanded such in a manner that the user is able to assign the activities from an appropriate list to the processes. The user can also use predefined sequences, which support parts of the product development process. In this way, the user is supported with the modelling of sequences. On the other hand, a control will be developed that reacts dynamically to situations. The investigations showed that no complete dynamic control, has been developed. In order to develop an engineering system with a part-dynamic control a rule-based solution is proprosed.

The demand for a dynamic control is from companies, because it is not possible to model all sequences during the product development process in advance. The product development process is a repetitive development process which must be controlled dynamically. Problems occur, for example, if the sequence depends on the product structure. In order to realise a control depending on the product structure, it is necessary to use the information and functionalities of the EDM/PDM-Systems. For this propose the product PDV, by BIM Consulting GmbH, was chosen because it offers the possibility to realise access to the data.

It is attempted to dynamically use prearranged recurrent processes. That means a set of rules is defined that can be evaluated. The execution of a rule depending sequence can be realised as follows. The first task is the identification of the processes. That includes the recurrent processes of the determining of offers. Based on an analysis, a set of rules must be defined. The set of rules contains the conditions which must be fulfilled, in order to execute a certain process. That means, all the processes have certain conditions. At runtime, these rules determine which process is executed next, considering the product structure and whether the rules are completely, partially, or not fulfilled. Due to the fact, that the fulfillment of conditions depends on the results of the executed processes a sequence is created and executed dynamically.

In order to distinguish this approach from the available process controls the focus is the product structure-driven (dynamic) control. In this paper the workflows of the Workflow Managment Systems are not the crucial point, instead, the technical processes of the product development are considered. The process is controlled depending on the information contained in the PDV-System. The existing prototype PACO has to be expanded, in order to realise a dynamic control, which depends on the executed user tasks, the product structure, and the requirements. That is, based on the product structure (e.g. information for the offers) the system chooses the process steps which are offered to the engineer as the next process. For instance, during the determining of offers the user works with the PDV-System and searches for documents for the offer. Depending on this information and possible supplements the pre-definitions are set. The evaluation of the results of the executed processes leads to automated decisions about the further sequence (e.g. pre-definitions for new, similar, or equal parts). Depending on the decision, the user is offered a list of additional pre-defined sequences (e.g. „set postprocessing" or „set assurance control"), which he can choose from.

4 Development of the Prototype

In coorporation with the Fraunhofer Institute for Factory Operation and Automation IFF in Magdeburg, a prototype has been developed at the Otto-von-Guericke-University Magdeburg. The PACO-Integration framework is the base for the development of a WWW-based client-server-system. The process sequences are described by a meta-language, in order to make it possible to specify roles, documents, and activities. Using a compiler, this meta-description is mapped onto the data model of the sequence control component so that, at the runtime of the system, an instance of the process model is created and executed.

The communication between client and server is realised through the HTTP Web Protocol. The user interface of the prototype, i.e. the client, is realised by a web-browser in connection with dynamically generated HTML-Pages. The functionalities of the web-browser is expanded by functions for the support and execution of activities. A central server provides services, e.g. for data management, for document management, or for process control. It is also possible that additional tools communicate with the environment through other protocols (e.g. DBMS-own protocols). The required service is activated at the server-side and the results are available as dynamically generated documents. The tools can either be available locally on client-side or can be loaded from the server. The prototype realises the role concept only in connection with the assignment of documents and tools.

As soon as a user logs in through his web-browser he receives, depending on his roles, an activity list (ToDo-List) with the activities that can be executed. After an activity has been chosen the selected activity script is loaded from the server and executed by the client. The activity script is stored on the PACO-Server and will be loaded when the activity has to be executed. Therefore, it is guaranteed that the client works with the current version. After an activity has been processed sucessfully, the user is informed by a user dialog and the process state of the runtime component is updated, so that the user is able to execute the next process step. In this manner a pre-defined system consisting of a set of cooperating tools is activated and standard operations are automated. Further, more the prototype offers the parallel execution of tasks.

5 Conclusion

The number of the systems that are used for the product development process in companies is increasing constantly. Due to the complexity of current systems, deficiencies in handling can occur. The aim of the introduced PACO-Integration framework is an integrated engineer environment, which supports the design and development tasks of the product development process. In contrast to other research areas the advantage is the capability to configure the engineering systems on different layers. It is possible to compose a software system using pre-defined components. Also, a configuration of severel process sequences is possible. With the help of the PACO-approach it is possible to decompose software systems into the actual main components, in order to compose specific engineering systems for special tasks.

The realisation of the integration framework will be done by the development of integration methods for the different layers. The developed concepts are provided as general tools for the modelling and executing of dynamic sequences. The development of the described process integration will be continued. Additional work concentrates on the adaptation of the available software for tool integration and the development of general concepts for the integration on the user layer.

References

Krötzsch, S.; Hofmann, I.; Kreutzmann, F.; Ambos, E. and Paul, G. (1997): Economic Supply of Information During the Determining of Offers for Castings. *Giesserei* 84(10):15-21, 1997 (in german)

Paul, G.; Kreutzmann, F. and Sattler, K.-U. (1996): An Integration architecture for Systems in the Field of CAD/CAM. *CAD/CAM-Report* 3, pages 130ff, 1996 (in german)

Paul, G.; Sattler, K.-U.; Endig, M.; Krötzsch, S.;Kreutzmann, F. (1997): An Integration model for Engineering Systems. In *3. Magdeburger Maschinenbau-Tage, Tagungsband I*, Logos Verlag Berlin, September 1997 (in german).

Sattler, K.-U. (1997): A Framework for Component Oriented Tool Integration. In. *4th International Conference on Object-Oriented Information Systems (OOIS'97)*, Brisbane, Australia, Nov. 10-12, Springer-Verlag 1997, pages 455-465.

Potential of Business Process Modelling with Regard to Available Workflow Management Systems

Rainer Endl and Martin Meyer

Institute of Information Systems, Research Unit Information Engineering, University of Berne, Engehaldenstrasse 8, CH-3012 Bern, Switzerland

Abstract. *Even though many workflow management systems (WfMS) are commercially available, only few existing systems are supporting business processes which are of main interest for the enterprise. This paper deals with the requirements of business process modelling and workflow modelling. These activities are not yet sufficiently supported by common methods. Based on the requirements, problem areas are identified and illustrated by SAP's Business Workflow (SAP BWF). Finally, we suggest some solutions to meet the outlined problems.*

Keywords. *Business Process Modelling, Workflow Modelling, Business Rules, Modelling, Requirements, Ad hoc-Workflows.*

1 Introduction

1.1 Motivation

The increasing dynamics and the continuous change of the market places forces the enterprises to meet thoroughly the needs and requirements of the customer. Many enterprises are trying to improve their competitiveness by designing their organisations to be process oriented, flexible, and adaptable. While the traditional theory of organisation focuses primarily on the structure, the trend to a process oriented organisation of an enterprise is significant. (Nippa, 1995)

In the context of this paper, business process- (BPM) and workflow-management (WFM) are treated as different views on the processes within an enterprise. While BPM is assigned to the conceptional level, WFM deals in the first place with the automation and the management of a business process or instances of them.

To model, analyse, and optimise business processes and building workflows, several tools and methods are commercially avalaible (Kurbel, 1997). Meanwhile, some of them have achieved a broad acceptance. Nevertheless, some problem areas still exist which restrict the potential benefits of the use of BPM/WFM, e.g.

- *Hetereogeneous modelling tools with different methodology and functionality:* Commercially available products are commonly based on different modelling approaches and are supporting different methods. The set of constructs provided by the modelling methods determines the power of the modelling methods and therefore the power of the tool supporting this method. (Endl, 1997) It is difficult for the user to find the most appropriate method for his needs.
- *The absence of proven directions and guidelines supporting the whole process of developing and modelling business processes and their implementation as workflows in an automated environment:* Most of the commercially available tools are not associated to such guidelines supporting the developing process. Thus, it is up to the user to apply the method and the tools more or less systematically.

1.2 Aim of this Paper

This paper investigates the modelling of business processes with respect to their implementation in workflow management systems. In the next chapter, the goals of modelling business processes and workflows and their requirements are explained in more detail. Based on these requirements, areas of problems are identified and illustrated by SAP's Business Workflow (SAP BWF). Finally, we suggest in chapter four some solutions to meet the outlined problems.

2 Modelling Business Processes and Workflows

2.1 Goals of Business Process Modelling

Modelling business processes can be divided into a strategic and a tactical/operational development task. The modelling on the strategic level is the starting point and determines the modelling on the succeeding level. In principle, the following goals of process modelling are achieved:

- *To help both, analysts and users, to gain a strong, commonly accepted understanding of the coherence of activities to achieve a business objective:* By using appropriate methods and tools, transparency about elements needed within a business process and their relationships can be achieved (Krcmar, 1994). On one hand, the model can reflect the actual state of the business, on the other hand the method can be used to develop a state which is desired for the future. The models foster the understanding of the relevant aspects of the business and their relationships and therefore facilitate the communication and discussion about issues of improvements. This goal can be achieved on the condition that the models are readable and understandable by different user groups.

- *Specification and configuration of packaged software:* Some vendors of packaged software provide their systems in conjunction with an enterprise model. (Buck-Emden, 1995) These so-called reference models often provide a process oriented view on the enterprise model and can be very useful for the evaluation and customisation of packed software: The reference model can support the identification and the choice of components and functions of the packaged software relevant to the enterprise. Given that the business processes planned for the future are modelled in the same way as the processes in the reference model of the packaged software, one can compare these models, determine the fundamental suitability of the packed software and furthermore derive the amount of work required for customisation. In the whole, the comparison of process models with reference models enables the evaluation and assessment of alternatives with respect to process structures and resources. (Scheer, 1998; S. 61 ff.) For that purpose, several reference models are commercially available today.
- *Basis for the development of application systems:* (Kurbel, 1997) The paradigm of process oriented modelling affects the methods for the development of application systems in the business environment, e.g. implementing a business process in a WfMS. Especially their implementation and therefore their automation with a WfMS requires the preceeding modelling of the business process.

2.2 Process- vs. Workflow-Modelling

As mentioned in the section above, one of the important prerequisites for workflow management is the process centred view. Workflow management learns from the business process modelling to take all aspects of an application system into account and not just to concentrate on certain partial aspects (Jablonski, 1996; S. 11f.). Thus, modelling business processes is a conceptional task while modelling workflows focuses primarily on the implementation aspects of a process and therefore provides a much more detailed view.

Given the different point of views, the following consequences occur:

- The modelling of business processes aims at the optimisation of the supply - customer relationship, while workflow modelling aims primarily at the operational aspects of the process and IT-related questions such as: Which role or office is responsible for the execution of an activity within the process? Which application system on which server is designated to support a given activity within the process? Which file formats are suitable to put the necessary information throughout the process?
- Analysing and documenting the business process is achieved with non-formal or semi-formal description languages in contrast to the specification of workflows, where formal description languages are strictly used. The implementation of the business process with a WfMS is primarily a task for IT-specialists, while modelling the processes is a task to be accomplished under the authority of the business departments. (Halter, 1998)
- Business process models are used as a communication instrument between the involved persons, e.g. IT-specialists, consultants, business department staff. By decreasing the level of abstraction more and more, operational and IT-related know-how is required to understand the models. Thus, a workflow model on the most detailed level serves as the formal specification of the workflow to be implemented in a WfMS.
- In both business process- and workflow modelling the same information types are needed. But stepping down the abstraction level, the information types become more and more concrete, e.g., the specification of functions leads to activities, the specification of entity types leads to data stores, and the specification of departments responsible for a function leads to roles responsible for some activities. In addition, according to the level of abstraction, additional modelling constructs are required.

2.3 Important Requirements

In the preceeding chapter we introduced the different levels of process and workflow modelling. Based upon this, several requirements can be derived. These are often listed in requirements catalogues for the evaluation of WfMS (Derungs, 1995; Endl, 1997; Heimig, 1997). In this chapter, we explain some selected requirements, whereas later in this paper ideas for their solution are suggested:

- *Generality:* The process of detailing the business process model must be accomplished with the same modelling instruments and without changing the paradigm.
- *Availability of directions and guidelines supporting the whole process of developing and modelling business processes and workflows:* Methods for business process- and workflow modelling must be supported by modelling guidelines and directions. These guidelines and directions are represented as

rules that determine how the whole modelling process from the first model to its implementation has to be accomplished. These rules must be enforced by a modelling tool to ensure the integrity of the models (Knolmayer, 1997).

- *Flexible Organisation model:* The method of modelling the organisational aspects of a process must support a broad range of organisational concepts. At least the following three concepts must be provided by a modelling method:

  - Role: Depending on the concrete definition, a role can be either a role with respect to a designated qualification or with respect to a specific competence- or knowledge-run activity within the process.
  - Users or persons involved in a specific activity within the process.
  - Organisational units and their relationhips.

  The modelling of roles, users and organisational units and its stepwise refinement must be possible as well. Furthermore, a flexible assignment of roles, users and organisational units has to be provided (Endl, 1997; Rosemann, 1998).

- *Support of ad hoc workflows* (Georgakopoulos, 1995; S. 125 ff.): In practice it is often not convenient - or even impossible - to model all the situations which potentially can occur at runtime. The user should rather have the possibility to modify the workflow instance at runtime if any exceptional situation occurs. Thus, at least the following operations for the treatment of ad hoc-workflows must be supported:

  - To skip an activity defined in the workflow model.
  - To repeat an activity or a sequence of activities, e.g., if any errors are encountered.
  - To abort an activity, e.g., a customer has cancelled his order which is already in execution. Depending on the activity, the abort operation triggers either a rollback of activities already executed or the activation of another workflow to reverse the completed activities.

  Concerning the modelling aspects, a method must support

  - The definition of possible exit points in activities.
  - The specification of events or event classes which lead to an abnormal termination of the activity.
  - The definition of the operation types mentioned above which are triggered by the exception events.

The mentioned ad hoc operations executed at run-time do not affect the workflow model. But some ad hoc modifications could be recognised as universally valid, i.e., for all subsequent workflow instances. In this case, the changes must be available on all levels of abstractions, not only on the level of the workflow specification. To avoid repeated top-down modelling caused by changes on the level of workflow specification, a bidirectional approach is required which makes persistent changes on all levels of abstractions available.

3 Problems for Modelling Processes and Workflows Illustrated by SAP BWF

In this chapter, the problems derived from the requirements outlined in the preceeding chapter are discussed and illustrated by SAP's BWF. This system represents the class of so-called *integrated WfMS*, because it is built upon a packaged software (Becker, 1997; S. 58.).

We chose SAP BWF because it is a part of SAP R/3 and therefore one of the most common WfMS (Meyer, 1998). Furthermore, some evaluations have considered SAP BWF as one of the most mature WfMS, with respect to the modelling component. (Petrovic, 1998, S. 8; Joos, 1997)

The following problems may occur when modelling business processes and workflows using SAP R/3:

- *Generality:* Most of the common WfMS are provided with modelling tools. These modelling tools are proprietary and can not interact with tools designed for process modelling. Thus, a gap arises between business process- and workflow modelling, leading to additional activities for workflow specification. Moreover, there is no way to ensure the integrity of the workflow model with respect to the business process model automatically or at least supported by a computer. These disadvantages may occur with SAP BWF as well. It contains a graphical editor, which allows the specification of workflows by using *Event driven Process Chains (EPC).* One can model the business processes directly with these EPC's. But nevertheless it is not possible to import business process models designed with another tool using a method other than EPC. It is the same reason why a feedback from the level of the workflow model to the level of business process models is also impossible. But this is an important requirement from the perspective of the management of the whole workflow life cycle (Galler, 1997; Kurbel, 1997, S. 75).
- *The absence of directions and guidelines supporting the process of developing and modelling business processes and workflows:* Common guidelines and directions are specific either to an enterprise or only to a part of the developing and modelling process. No solutions exist, with respect to the transformation of business processes into workflow models, which are sufficient for the practical use (Joos, 1997). In SAP BWF, for example, no sufficient guidelines

exist for the support of the development and the use of the WfMS. On one hand, the available guidelines distinguish the development phases roughly between planning, realisation and implementation, on the other hand, the guidelines focus mainly on the implementation phases (SAP, 1997a; SAP, 1997b).

- *Modelling the organisation:*
  - *Insufficient possibilities to model the organisation structure:* The modelling of organisational aspects are insufficiently treated by common methods for business process and workflow modelling. Many systems support only parts of the constructs required to model the organisation. Given an example, temporarily existing organisation units, e.g. projects, may not be integrated into the process model if they establish a matrix organisation structure. In common, the possibilities for modelling multi-dimensional organisations are very restricted (Rosemann, 1998). In SAP BWF there exists constructs for modelling occupied jobs, vacant jobs, organisation unit, roles and employees. The method is quite convenient to model hierarchical organisation structures (Meyer, 1997, S. 21). Using both modelling constructs, role and vacancy, it is partly possible to model temporary organisation units (Karl, 1997, S. 36). But of course this is not sufficient to model multi-dimensional organisation structures often found in enterprises.
  - *Insufficient possibilities to relate employees to jobs and substitutes:* Modelling the substitution of employees is insufficient in the sense that often only one person can act as a substitute to an employee. Further, defining a person as a substitute employee implies that he has the necessary qualifications to fullfil the job. We would prefer role based modelling of substitute employees instead of person based modelling. With role based modelling, a set of employees will be available, which contains all persons who potentially can act as a substitute. SAP BWF, for example, allows the definition of a substitute employee only at run-time. The role based *definition* of substitute employees is not supported. In this context, the insufficient possibilities to relate jobs to employees must be mentioned. Most of the common methods do not consider that a job in the same organisational unit may be owned by several employees at the same time, e.g., in case of shift work (Rosemann, 1998).
- *Insufficient flexibility with respect to ad hoc-workflows:* The requirements with respect to ad hoc-workflow support as stated in chapter 2.3 are either not or insufficiently supported by common modelling methods (Bürgi, 1998). For example, SAP BWF supports only some user initiated ad hoc-operations at run-time. Constructs for modelling aspects of ad hoc-workflows are not avaliable.

4 Solution Approaches

4.1 Standards Provided by the Workflow Management Coalition (WfMC)

The *Workflow Management Coalition (WfMC)* was founded in 1993 as an international non-profit making organisation by several vendors developing workflow management systems. In september 1998, the WfMC had 200 members, including consultancy companies, universities, WfMS-users and -vendors[1]. The WfMC states its mission as follows (WfMC, 1994):

- Increase the value of customer investment in workflow technology.
- Decrease the risk of using workflow products.
- Expand the workflow market through increasing awareness of workflows.

The WfMC developed a WfMS-reference model which specifies a framework for workflow systems as well as identification of their characteristics, functions and interfaces (WfMC, 1994; Eckert, 1995; Weiß, 1996) (cp. Fig. 1).

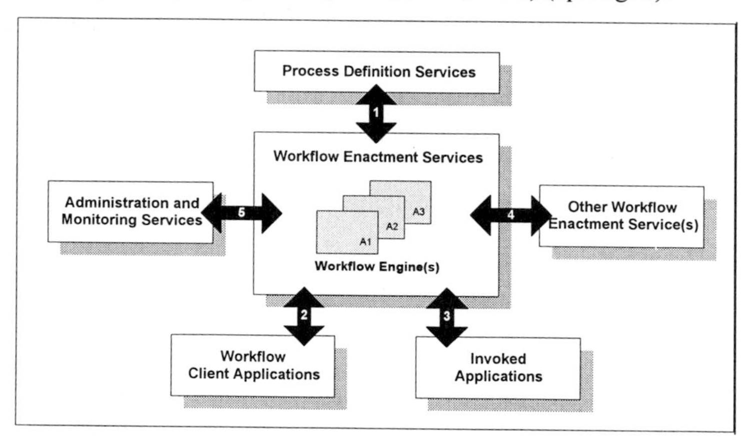

Fig. 1: The WfMC-reference model

The WfMC considered the five interfaces in the reference model as WAPIs (Workflow Application Programming Interfaces and Interchange Formats). At WfMC five working groups are established to work on one of the following interfaces (cp. Fig. 1):

[1] Home page of the Workflow Management Coalition: http://www.aiim.org/wfmc/.

- Process Definition Tools
- Workflow Client Applications
- Invoked Applications
- (Other) Workflow-Enactment Services (Interoperability)
- Administration and Monitoring Tools.

With respect to business process and workflow modelling, the interface is one of particular importance. It defines a standard interface between a process definition tool and the workflow engine(s). To achieve this, a standard is developed for the import and export of workflow specifications. The interchange format contains, among other things, the following information:

- Conditions under which a workflow starts and terminates.
- Identification of activities within a process.
- Identification of data types and access paths, definition of transition conditions, and rules to control the flow.
- Information concerning the allocation of resources.

Basically, the standardisation of the interface enables one the automatic transformation of the business process model into the WfMS and its use for workflow specification. But, currently, the functionality of the interfaces considered in the reference model is not defined in detail (Jablonski, 1997, S. 80). Furthermore, the question whether the interface is one that universally remains valid, because the WfMC has defined just the kernel of the interface model which may be enhanced by vendors with proprietary features (Derszteler, 1996, S. 591).

4.2 Approaches of the SWORDIES-Project

Within the project SWORDIES[2] (Swiss WORkflow Management in DIstributed EnvironmentS), which is partly sponsored by the Swiss National Science Foundation, the department of information engineering at the University of Berne researches business rule based modelling of business processes and workflows. The aim of this work is to develop a rule based method to analyse and model business processes and their specification in different target systems. Besides the method, general guidelines and directions will be provided to support the whole workflow life cycle. These guidelines are formulated as rules as well and result in a set of meta-rules.

---

[2] http://www.ifi.unizh.ch/groups/dbtg/SWORDIES/

Another goal is the consideration of aspects of modelling ad hoc-workflows which includes mechanisms to transform changes made on the level of workflow specification to a higher level of abstraction.

Business rules can be defined as statements about how the business is done, i.e., about guidelines and restrictions with respect to states and processes in an organisation (Herbst, 1997). Business rules trigger an action of an information system, send an alerter to a human factor or define the feasible space for human action. The rule based modelling method therefore builds on ECA-rules (ECA: *Event, Condition, Action*) well known in active database systems.The ECA-Rules were extended by constructs representing specific features of business and workflow modelling, e.g., to model the organisational structure (Knolmayer, 1997). In addition, a construct to model alternative actions (similarily to the *if... then... else...*-statements in programming language) were introduced which enhances ECA to *ECAA* (Knolmayer, 1997). Depending on the level of abstraction, the content of the rule components contains colloquial or formal statements. Every rule within the business process or workflow is identified by a unique name. The components of a rule are described as follows:

- *Event:* An elementary or complex event which triggers the processing of the rule. It contains the information when a business rule has to be processed (Herbst, 1995).
- *Condition:* Contains the conditions or circumstances which have to be checked before the action-part will be performed. (Herbst, 1997)
- *Action*: Contains information about what has to be done if the condition is true.
- *Alternative Action:* Contains information about what has to be done if the condition is false (for example Herbst, 1995).

Note, that while performing an action part of a business rule, events may be raised that trigger further business rules. In addition, the condition and the alternative action part of a rule are optional. Fig. 2 gives an example of a sequence of actions modelled by business rules containing only the parts event and action (Endl, 1998)

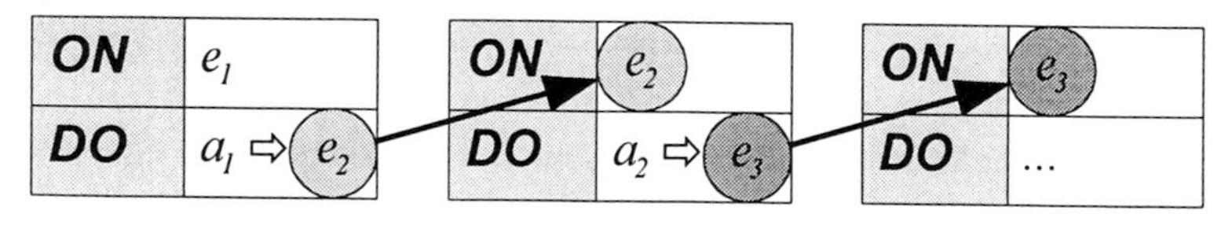

Fig. 2: Modelling a sequence of actions

Several case studies have shown that common modelling methods, e.g., EPC, fun soft-nets, and petri nets may be transformed into business rules at every level of abstraction (Knolmayer, 1997). In the same way, one can transform processes modelled by business rules to different target systems, e.g., in a workflow description language like TRAMS (Kradolfer,1997) or in database systems (Gatziu, 1995). Furthermore, business rules can be used to generate the parts of the logic of an application system (Mallens, 1997). So one can think of business rules as a standardised representation of business processes. The process models eventually obtained by employing different process modelling tools (by decentralised or virtual enterprises or along a supply chain) may be transformed to a rule based description of the business processes. This business rule model may be refined stepwise until the system is represented by elementary business rules (Fig. 3).

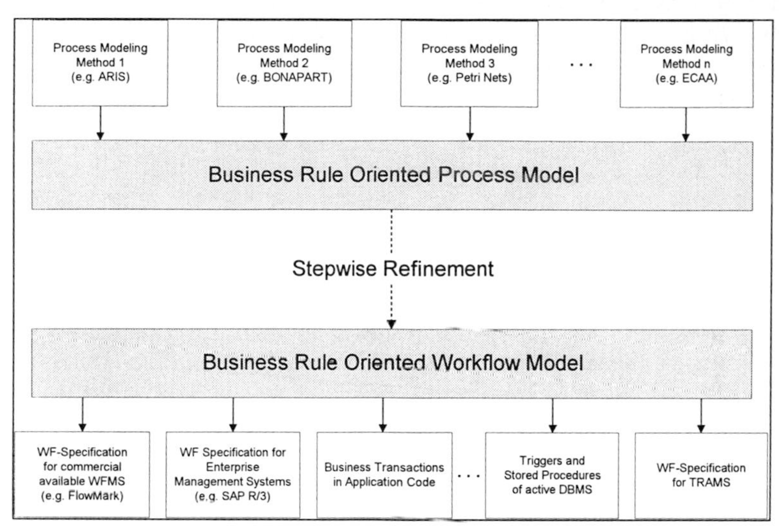

Fig. 3: Business rules as integration layer between process and workflow models

During the refinement of business rules, the relationship between the different levels is preserved (Endl, 1998). This is one of the most important prerequisites for bidirectional modelling and a potential advantage with respect to other modelling methods (e.g. EPC). Following the path of abstraction in reverse direction allows one to make changes on a very formal level, e.g. on the level of workflow specification, and transform the changes to the other abstraction levels in a controlled manner. Thus, ad hoc-modelling of workflows can be supported on all levels of abstraction.

5 Outlook

Currently, the considerations made up in the preceding chapter are tested for their practicability. In parallel, we are working on the specification of a repository supporting the business process and workflow life cycle. Furthermore, we started a project to develop guidelines and directions to support the business rule oriented process modelling.

Following the vision of a learning WfMS, we are looking at a bidirectional modelling method that can build its knowledge base. A learning WfMS registers all activities, to which in the past ad hoc-workflows were invoked, and evaluates and stores the relevant environment parameters. Based on this information, the WfMS can support the decision of the user about the continuation of the process if any situation occurs which is similar to an ad hoc-workflow initiated in the past. The user then can view the suggested alternatives at any level of abstraction and modify or accept one of them. Such a learning WfMS could be a valuable contribution to the knowledge management in an enterprise, in particular with respect to the organisational know-how.

References

Becker, M.; Vogler, P. (1997): Workflow-Management in betriebswirtschaftlicher Standardsoftware - Konzepte, Architekturen, Lösungen, Arbeitsbericht IM HSG/CC PSI/9, Version 1.0, Institut für Wirtschaftsinformatik, Universität St. Gallen 1997

Buck-Emden, R.; Galimow, J. (1995): Die Client/Server-Technologie des System R/3, Bonn et al.: Addison Wesley 1995

Bürgi, J.-M. (1998): Möglichkeiten und Grenzen für das Management von Ad-hoc-Workflows am Beispiel von SAP Business Workflow, Lizentiatsarbeit, Institut für Wirtschaftsinformatik, Universität Bern 1998

Carlsen, S. (1995): Organizational Perspectives of Workflow Technology, Working Paper, The Norwegian Institute of Technology, University of Trondheim 1995

Derszteler, G. (1996): Workflow Management Cycle, in: Wirtschaftsinformatik 38 (1996) 6, S. 591 - 600

Donatelli, S.; Simone, C.; Trentin, D. (1998): Combining abstraction and context: a challenge in formal apporaches to workflow management, in: van der Aalst, W. (Ed.), Proceedings of the Workshop on Workflow Management: Net-based Concepts, Models, Techniques and Tools (WFM'98), Computing Science Reports, Eindhoven University of Technology 1998, pp. 194 – 209

Derungs, M.; Vogler, P.; Österle, H. (1995): Kriterienkatalog Workflow-Systeme, Arbeitsbericht IM HSG/CC PSI/1, Version 1.0, Institut für Wirtschaftsinformatik, Universität St. Gallen 1995

Eckert, H. (1995): Die Workflow Management Coalition, Zielsetzungen, Arbeitsgebiete und erste Arbeitsergebnisse, in: Office Management 43 (1995) 6, S. 26 - 32

Endl, R.; Knolmayer, G.; Pfahrer, M. (1998): Modelling Processes and Workflows by Business Rules, Swordies Report, Institut für Wirtschaftsinformatik, Universität Bern 1998

Endl, R.; Duedal, L.; Fritz, B.; Joos, B. (1997): Anforderungen an Workflowmanagementsysteme aus anwendungsorientierter Sicht, Arbeitsbericht Nr. 92, Institut für Wirtschaftsinformatik, Universität Bern 1997

Galler, J. (1997): Vom Geschäftsprozeßmodell zum Workflow-Modell, Wiesbaden: Gabler 1997

Gatziu, S. (1995): Events in an Active, Object-Oriented Database System, Hamburg: Dr. Kovac Verlag 1995

Georgakopoulos, D.; Hornick, M.; Sheth, A. (1995): An Overview of Workflow-Management: From Process Modelling to Workflow Automation Infrastructure, in: Distributed and Parallel Databases 3 (1995) 2, S. 119 - 153

Halter, U. (1998): Auswahl und Konzeption von Workflow-Anwendungen, in: Informatik 5 (1998) 2, S. 18 - 22

Heimig, I.; Borowsky, R. (1997): Geschäftsprozeßgestaltung mit integrierten Prozeß- und Produktmodellen (GiPP), Ergebnisbericht aus den Projektgemeinschaften, Evaluierungskatalog zur Bewertung von Workflow-Systemen, Institut für Wirtschaftsinformatik, Universität Saarbrücken 1997

Herbst, H.; Knolmayer, G. (1995): Ansätze zur Klassifikation von Geschäftsregeln, in: Wirtschaftsinformatik 37 (1995) 2, S. 149 - 159

Herbst, H. (1997): Business Rule-Oriented Conceptual Modelling, Heidelberg: Physica 1997

Högl, M.; Derszteler G. (1997): Vom Business Process Reengineering zum Workflow-Management, Ein Vorgehensmodell zur Einführung effizienter, DV-gestützter Geschäftsprozesse, in: DV-Management 7 (1997) 1, S. 28 - 33

Jablonski, S.; Bussler, C. (1996): Workflow Management. Modelling Concepts, Architecture and Implementation. London et al.: International Thomson 1996

Jablonski, S. (1995): Workflow-Management-Systeme, Modellierung und Architektur, Bonn: Thomson 1995

Jablonski, S. (1996): Anforderungen an die Modellierung von Workflows, in: Österle, H., Vogler, P. (Hrsg.), Praxis des Workflow-Managements, Grundlagen, Vorgehen, Beispiele, Braunschweig - Wiesbaden: Vieweg 1996, S. 65 - 81

Jablonski, S. (1997): Architektur von Workflow-Management-Systemen, in: Informatik, Forschung und Entwicklung 12 (1997) 2, S. 72 - 81

Joos, B.; Endl, R.; Tombros, D. (1997): Evaluation von Workflow-Management-Systemen, SWORDIES Report Nr. 3, Institut für Wirtschaftsinformatik, Universität Bern 1997

Joosten, S. (1994): Trigger Modelling for Workflow-Analysis, in: Chroust, G., Benczur, A. (Hrsg.), Workflow Management: Challenges, Paradigms and Products, Conference

Proceedings of CONnectivity '94, Linz, October 19-21, München: Oldenbourg 1994, S. 236 - 247

Kappel, G.; Lang, P.; Rausch-Schott, S.; Retschitzegger, W. (1995): Workflow Management based on Objects, Rules and Roles, in: IEEE Data Engineering Bulletin 18 (1995) 1, S. 11 - 18

Karl, R.; Deiters, W. (1997): Workflow Management, Groupware Computing, Studie über SAP Business Workflow, Release 3.1, 2. Aufl., Pfaffenhofen: dsk 1997

Keller, G.; Nüttgens, M.; Scheer, A.-W. (1992): Semantische Prozessmodellierung auf der Grundlage ereignisgesteuerter Prozessketten (EPK), in: Veröffentlichungen des Instituts für Wirtschaftsinformatik, Heft 89, Universität Saarbrücken 1992

Kilov, H.; Ross, J. (1994): Information Modelling. An Object-Oriented Approach, Englewood Cliffs: Prentice Hall 1994

Knolmayer, G.; Endl, R.; Pfahrer, M.; Schlesinger, M. (1997): Geschäftsregeln als Instrument zur Modellierung von Geschäftsprozessen und Workflows, Arbeitsbericht Nr. 105, Institut für Wirtschaftsinformatik, Universität Bern 1997

Kradolfer, M.; Geppert, A. (1997): Modelling Concepts for Workflow Specification, Arbeitsbericht, Institut für Informatik, Universität Zürich 1997

Krcmar, H.; Schwarzer, B. (1994): Prozessorientierte Unternehmensmodellierung - Gründe, Anforderungen an Werkzeuge und Folgen für die Organisation, in: Scheer, A.-W. (Hrsg.), Prozessorientierte Unternehmensmodellierung, Gabler: Wiesbaden 1994

Kurbel, K.; Nenoglu, G.; Schwarz, C. (1997): Von der Geschäftsprozeßmodellierung zur Workflowspezifikation - Zur Kompatibilität von Modellen und Werkzeugen, in: HMD - Theorie und Praxis der Wirtschaftsinformatik 34 (1997) 198, S. 66 - 82

Mallens, P. (1997): Business Rule Automation, Naarden: USoft 1997

Meyer, M.; Pfahrer, M. (1997): Erfahrungen beim Einsatz von SAP Business Workflow und IBM FlowMark, Arbeitsbericht Nr. 93, Institut für Wirtschaftsinformatik, Universität Bern 1997

Meyer, M.; Wimmer, F. (1998): Bedeutung und Einsatz von SAP Business Workflow in der Schweiz, Arbeitsbericht Nr. 108, Institut für Wirtschaftsinformatik, Universität Bern 1998

Nippa, M.; Picot, A. (Hrsg.) (1995): Prozeßmanagement und Reengineering: Die Praxis im deutschsprachigen Raum, Frankfurt a. M. - New York: Campus 1995

Rosemann, M.; zur Mühlen, M. (1998): Modellierung der Aufbauorganisation in Workflow-Management-Systemen: Kritische Bestandsaufnahme und Gestaltungsvorschläge, in: EMISA Forum, Mitteilungen der GI-Fachgruppe 2.5.2 "Entwicklungsmethoden für Informationssysteme und deren Anwendung" 7 (1998) 1, S. 78 - 86

SAP AG (Hrsg.) (1997a): R/3 System Release 3.1G, Online Documentation, Compact Disk, Walldorf 1997

SAP AG (Hrsg.) (1997b): SAP Business Workflow - Einsatz und Konfiguration, Kursunterlagen zum Kurs BC 085, Regensdorf 1997

Scheer, A.-W. (1998): ARIS - Vom Geschäftsprozeß zum Anwendungssystem, 3. Aufl., Berlin et al.: Springer 1998

Siebert, R. (1998): Anpassungsfähige Workflows zur Unterstützung unstrukturierter Vorgänge, in: EMISA Forum, Mitteilungen der GI-Fachgruppe 2.5.2 "Entwicklungsmethoden für Informationssysteme und deren Anwendung" 7 (1998) 1, S. 87 - 90

Teufel, S.; Sauter, C.; Mühlherr, T.; Bauknecht, K. (1995): Computerunterstützung für die Gruppenarbeit, Bonn et al.: Addison-Wesley 1995

Vogler, P.; Jablonski, S. (1998): Editorial, Workflow-Management, in: Informatik 5 (1998) 2, S. 2

Weiß, D.; Krcmar, H. (1996): Workflow-Management: Herkunft und Klassifikation, in: Wirtschaftsinformatik 38 (1996) 5, S. 503 - 513

Weske, M. (1998): Überlegungen zur Flexibilisierung von Workflow-Management-Systemen, in: EMISA Forum, Mitteilungen der GI-Fachgruppe 2.5.2 "Entwicklungsmethoden für Informationssysteme und deren Anwendung" 7 (1998) 1, S. 91 - 95

WfMC (Hrsg.) (1998): The Workflow Reference Model, TC00-1003, Issue 1.1, 1994, http://www.aiim.org/wfmc/DOCS/refmodel/rmv1-16.html [Stand: 1998-04-16]

Zarri, G.P.; Azzam, S. (1997): Building up and making use of corporate knowledge repositories, in: Plaza, E., Benjamins, R. (Eds.), Knowledge acquisition, Modelling and Management, 10th European workshop, EKAW'97, Berlin et al.: Springer, pp. 301 – 316

Flexible Process Management

Matthias Ort and Uwe Pape

Technical University Berlin, Department of Computer Science, Office FR 5-5, Franklinstrasse 28-29, 10587 Berlin, Germany

Abstract. *Proceeding from the requirements placed on business process models and the evaluation of present-day methods and tools, this contribution presents a procedure for flexible process management. The prime focus is on the flexibility and especially the practicability of such a procedure. In order to master the complexity of operational systems, these aspects have been examined in various models in accordance with diverse criteria. In the concept presented here, these models received the necessary integration to allow flexible administration of both allocated tasks and the use of process information.*

Keywords. *ad hoc process, business process, flexible process management, process evolution, process modelling, process reengineering, task migration*

1 Introduction

Any analysis in the field of research and the area of business process modelling and workflow management systems brings to light a number of methods. In their functionality, these methods range from extended graphic tools to integrated systems to the analysis and simulation of processes and even include complex CASE tools as well as complete workflow management systems.

Almost every tool relies on its own methods and notation and generally follows a special procedure. Furthermore, one can observe a tendency for the complexity of tools to increase, as can be seen, for example, in the case of vast graphic notations and the growing number of diagrams they require. As a result, although specialists are able to work with these systems, it is very difficult, if not impossible to make them understandable or accessible to outsiders, even where these systems have a sound methodological basis.

Practical experience from science, instruction and immediate practice suggests that such tools do not always reduce the existing complexity of operational systems. On the contrary, special applications tend to be inflexible. Inadequacies in structures often become greater when they are electrified using a system.

Furthermore, there is a very strong link to the system or so-called reference processes already chosen. Thus, prefabricated patterns often block the creativity required to optimise existing structures.

To summarise: there is an urgent requirement for practicable systems that do not restrict flexibility, but nevertheless reflect the diverse perspectives within the observed segment of a real business environment.

2 Requirement Definition / Status Quo

It is becoming increasingly important for both the private sector and public-sector administrations to find ways of creating new organisational structures. Intensified competition and growing globalisation call for ever faster and more flexible reactions to changing conditions. In order to meet this requirement, an ever growing number of tools, techniques and methods are being used that focus on process orientation. This is expected to provide the necessary interdisciplinary and trans-functional perspectives for reorganisation methods and technical support of the goals in order to attain the necessary business processes by means of workflow management.

In many areas, such as the responsible allocation of tasks, the payment of salaries and organisation development, to name but a few examples, it is necessary to create a dual orientation for organisational structures in accordance with the criteria of development and processes. However, projects to optimise existing structures demand an integrated analysis of the tasks and of those working in a related business environment. Business processes are one way of representing the results. One can define them as a number of tasks logically linked to one another and executed to attain an operating result. The focus is no longer on organisational and functional tasks but on processes, thus abstracting from a structure-oriented and, consequently, static perspective. Business processes now tend to be concentrated far more on dealing with quite specific and, under certain circumstances, complex tasks in a company or in the public-sector administration and are thus oriented towards a dynamic, behaviour-oriented analysis of the systems.

In addition, there is enormous unused potential for taking measures to increase effectiveness and efficiency in existing information systems and databases. The procurement, processing, evaluation and forwarding of information constitutes the key productivity factor and main resource for management in both the private sector and in public-sector administrations. The targeted employment of information technology thus offers enormous potential for boosting effectiveness in the long term.

If one could show the semantic link between the business processes of the operational system and the related flow of information in a consistent model, it would represent an intuitively comprehensible and logically accurate representation of reality. Hence, models of operational business processes ensure both central access to information systems and the uniform administration of information (or rather: of their information systems).

A comparison of the elaborated requirement definition for an ideal system with the functionality of existing procedures and tools reveals considerable differences. Thus, it is not possible to take over conventional methods and techniques directly, if one wants to apply process modelling effectively and optimise processes by performing process modelling. This requires a comprehensive process management. The ideal solution would be a co-operation medium based on modelled processes. By collecting additional information and tips, business process models, along with the presentation of the information content of the process flow, can make further data available in the environment (information retrieval). Process optimisation is also supported. The subjective impression and individual skills of the user are still decisive for the desired result. The available information is, however, no longer limited to the relationships presented by each of the models, but can be extended as desired and made available at a decentralised level.

As mentioned at the beginning of this paper, there is a problem involved in transferring such a diversified functionality to a system, since it must allow mastery of the entire complexity of a function without limiting its flexibility. The fundamental idea is based on the fact that in process management interest is focused on the business process. The key factors are thus the process structure and its limitations - i.e. the results of the process - and not the individual work share. Experience and knowledge reappear in the business processes. Consequently, the system presented reflects these factors inasmuch as all the models not only cohere at the level of contents, but also at the level of semantics corresponding to the process perspective.

The basic criticisms directed at systems used so far are as follows:

- the divergence with respect to comprehensive employment of the supporting tool and the goals of conventional systems,
- the non-reproducibility of processes that are difficult to structure,
- the lack of links between models and partial processes,
- the lack of comprehensive simulation and of sensitivity analysis,
- the lack of process-configuration management and
- the considerable complexity of the methods and notations used.

Process optimisation is characterised by rationalisation measures, client-care and the enhancement of both transparency and quality. As staff know-how is frequently the most productive source of ideas for process innovation (members of staff know various ways of solving problems, sometimes have experience of rival companies and also develop their own ideas) the staff should be integrated as much as possible into the analysis and modelling as well as into the re-engineering and working phases.

This goal cannot really be achieved in large companies by establishing small innovation teams within the enterprise or by calling in external consulting companies. Processes and the extensive related information should be transferred to the system and maintained by the members of staff themselves. Only in this way is it possible to ensure that system data remains correct, complete and up to date.

This approach contrasts with that used in conventional systems. They support the activities of one or a small group of process analysts. For this reason, existing systems limit themselves to presenting graphic models of business processes. Furthermore, these models can also be simulated by stating identification parameters on the base of (prescribed) objective functions. As they originate from software engineering, and owing to the historical development of methods and techniques and their intended employment, conventional systems have a highly complex notation for describing processes. Consequently they are suitable for use as CASE tools, for example, although difficulties may arise in their utilisation (on a day-to-day basis) as organisation instruments among very diverse users.

Depending on the method used, the processes aiming at reducing existing complexity are considered in the light of various criteria (e.g. from the point of view of the description of an organisation as well as of data, function and control). Any link between the partial models in order to represent the actual business processes will thus take place in a causal or operation-oriented process. This procedure, which aims to reduce complexity, runs counter to the necessarily extensive notation required for process modelling. The corresponding software tools only inadequately support the logical and content links of information in the individual diagrams and partial models. Creating these links remains the task of the user.

In order to represent business processes, which are frequently very extensive, it is necessary to link the diverse (partial) process models to create an overall model. This also reveals a further shortcoming in traditional systems, since links have to be made manually, and distributed data storage (process storage) is not supported conceptually.

In view of the above, a process configuration management is needed which permits the administration of the partial processes and any versions of these processes that might be produced. It is thus possible to calculate and simulate the processes. Process variants could be simulated in diverse versions and analysed for their limitations and effects before being compared to one another, as in benchmarking. A system-supported sensitivity analysis would allow the user to simulate processes with respect to specific target figures. However, when performing the analysis, steps must be taken to ensure that the system does not have a restrictive impact on the simulation function and its parameters. Previous systems merely provide for detail simulation, i.e. the processes have to be described right down to the last detail and the simulation parameters fixed.

The key point in establishing a requirement definition for an ideal system lies in synthesising specific identification parameters for workflows which are difficult to structure. One particular aspect is characterised by the expression "management thinks and directs". There are a multitude of processes dealing with information whose clients cannot be clearly localised, or whose recipients can only be determined after a certain run-time. These informal, so-called ad-hoc processes cannot be described or modelled a priori. The optional process operation results from spontaneously arising structures, i.e. ad hoc.

Knowledge of the structure of processes is, on its own, an unsatisfactory basis for optimising them. Diverse pieces of marginal information not necessary for collecting process data constitute regulators defining the degree to which changes can be effected; they thus contain a high level of information. The potential for innovations and restructuring can often be found in marginal information.

The characteristic feature of all processes is the great volume of information necessary to begin processing. Criticisms have frequently been raised about the difficulties and amount of energy involved in procuring information. The pressure of deadlines, the dependence on external services and the necessary monitoring performed by superiors can have a very negative effect on members of staff in certain cases. The degree to which it is necessary to acquire knowledge of the details, or to gain an overview, as well as to present information can be established in relation to the process hierarchy.

The process classification can be used to grade the levels of detail. The resultant semantic loss can be compensated for by performing a drill-down between the levels of granularity. A clerk, for example, needs specific data to do his work, whereas the management and external parties are frequently interested in statistics or processed information that provide them with an overview.

The situation is markedly different in a large number of information systems. However, as in other areas, these systems are not accepted by users and are therefore not utilised in many cases. This is generally due to the fact that end-users find these 'exposed' information systems too complicated to operate in comparison with systems and tools with which they are accustomed. Owing to the information structure of these processes, the required information cannot be accessed by means of a workflow-management system either, since information is required before tasks can be performed. Hence, we wish to underline our demand that business process models be used to access information as a kind of „information recycling“ procedure.

The following criteria reflect the requirements placed on a flexible process management system:

- user-oriented, hierarchical process designs as well as control and monitoring
- intuitive and practice-related process modelling (document-centred modelling)
- logical and content-based links between the individual models for the purpose of making process descriptions
- the generation of a logical network structure by uniting (partial) processes in an overall process (process abstraction)
- the possibility of reproducing and generating ad hoc processes (information structures, showing an optimal procedure)
- process-configuration management and process-version management
- comprehensive hierarchical as well as allocated process simulation
- detailed information input and information processing methods, which can also come from the wider environment
- the integration of benchmarking functions.

In this way it is possible to bridge the existing gap between business process-modelling tools, workflow-management systems and information systems. Modelled data is not only available for the models for which it is designed, but it also becomes possible to assign data items to diverse processes and tasks. Thus, for example, the representation of forms is very helpful. During the process analysis they are assigned to the corresponding processes and later used by a workflow-management system or by an integrated application. In addition, information systems can be made available to them if required (see Fig. 1).

Whilst the workflow-management system takes over the control of the (specialised) application programmes, the process-management system contains the necessary process information and controls the information systems. Hence in all phases, and particularly in the planning phase, techniques are available to support communication and co-operation and may, indeed, even make them possible in the first place. New forms of organisation and work can be created on the basis of process analysis. The integrated system does not only provide a unique tool for modelling data and processes as well as for data and process control, but it is also to be seen as an instrument supporting the continuous administration and optimisation of processes.

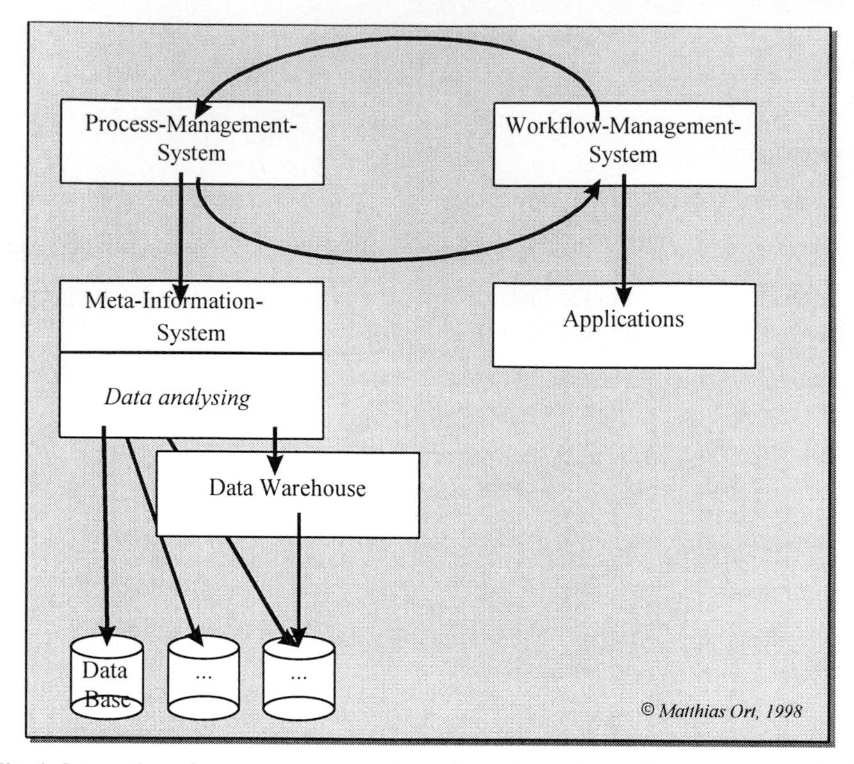

Fig. 1: Integration of process management as information and workflow management in the form of application control

3 Systems Planning

Up to now, business process-modelling tools have generally only been used within the framework of organisation analysis. However, the use of the acquired information even after a project has been concluded, is becoming increasingly important. Thus, an ever greater number of commercial systems are supplying interfaces for CASE tools and workflow-management systems.

The problems facing us nowadays are not primarily to be found in the structure of actual business operations, but in the structure of their processes. The key problems arise in procuring information and in passing on the knowledge and results gained. This interface problem can only be countered by planning, integrating and co-ordinating all process activities as early as possible. This also implies continuous concentration on the corresponding business processes (see Fig. 2).

To ensure that the data integrated into the system always remain up to date, correct and complete, the staff will have to maintain the data they themselves use in their field of work. As pointed out above, there is no lack of will to do so. However, a system of this nature is very difficult to operate. Understanding and modelling business process must not be left to the experts alone.

Consequently, it is necessary to have a dual-level user-surface. On the one hand, it must be possible for users to operate models intuitively and find them easy to comprehend, on the other hand, highly complex calling-in mechanisms are required in order to supplement an instrument of this type with the required functionality. (see Fig. 3).

The problem of modelling nevertheless remains. If modelling is left to the staff, it is possible that they will not describe abstract processes but only specific workflows. Furthermore, the present scale of the notation of existing methods renders simple modelling impossible. At present, the possibility of performing document-centred modelling is being analysed. The documents and forms being processed are enriched by process information and operational steps. Subsequently, the processes can be automatically "generated" from these data. In order to do so, the system must ensure the integrity of the data used. Moreover, the necessary notation is reduced to a minimum. Additional information can be inserted below the graphic surface if required.

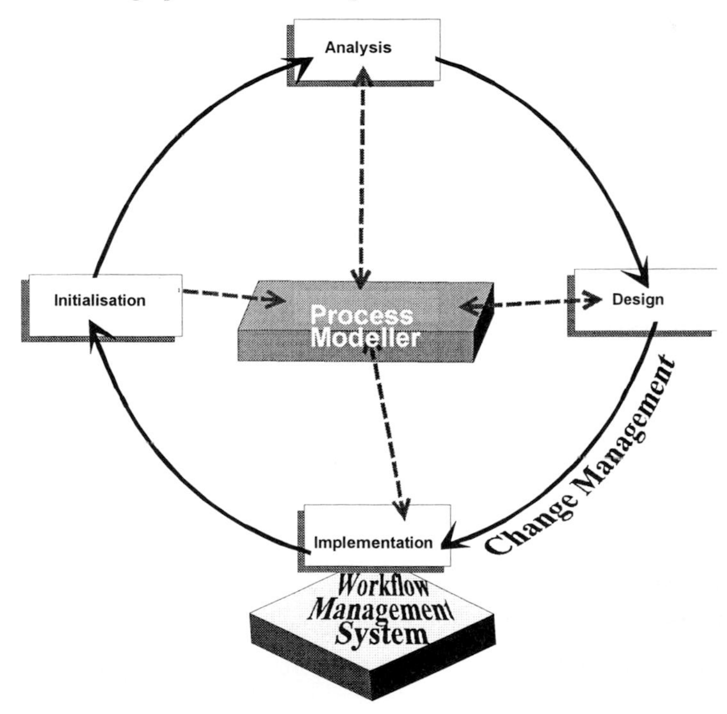

Fig. 2: Use of tools in process life-cycle

Another way of integrating processes into the system is to transform tasks into processes. To this end, a causal model has been developed which ensures the consistency of the contents of both tasks and processes. During the initial phases, the current tasks are analysed and categorised. Partial processes are then identified by creating so-called task pools which can be combined to form integrated process models by defining interfaces.

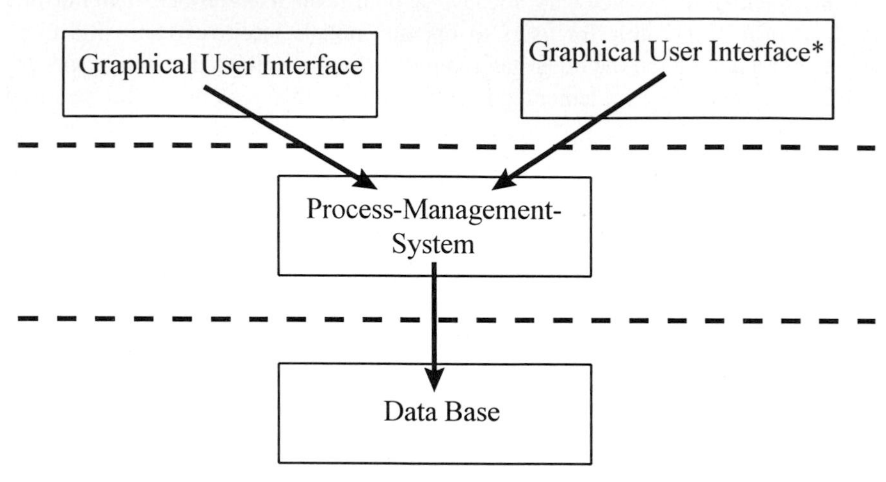

Fig. 3: The three-level architecture of the process-management system

During the implementation phase, use was made of diverse tool systems and of the available instruments of Internet and Intranet technology. Apart from the ensuring the possible allocation of components, they also ensure both platform independence and a high-level of user acceptance.

4 Technical Realisation

In order to involve as many employees as possible in the operation of the system, the planning of the process-management system makes use of a three-level software architecture. The heart of the application consists of the middleware components that integrate all of the essential functionalities. Dual-level access to the system satisfies the requirement of providing staff with an intuitive tool and gives the company and administrative organisation a powerful instrument supporting their work. As the available data are to be integrated as well, the databases are located in the lowest level of the architecture.

The partial models are linked to a meta-information concept based on the data level.

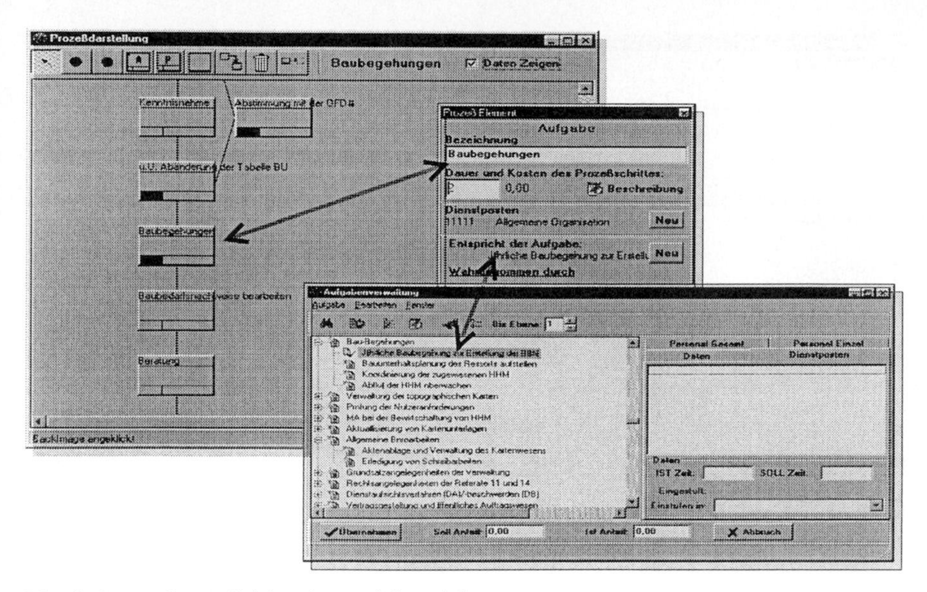

Fig. 4: Screendump: linking the partial models

Implementation was carried out using JAVA, thus ensuring a relatively high degree of platform independence. For reasons of performance, this mode of implementation is to be avoided for complex system surfaces. Delphi 3 was used as a development platform in this case.

The ad hoc processes only arise during run-time, thus making it possible to describe them. Consequently, these data are taken over from the workflow-management system. The processes are stored in a meta-database on a roll-basis and are only linked with the current operative data when they are called up by the user. In this way, the process knowledge is not implicitly stored in the database, but fills it with the current workflow data and the knowledge of the user operating the system. Storing this data also ensures that it can be used as information for evaluations and to aid decisions at a later date.

The adoption of diverse perspectives ensures that access to the system is context-related. Thus a distinction has been made between a user-specific, roll-based and process-oriented perspective. Information modelling is carried out in relation to the processes. To this end, information on the current process stages is stored in repositories. The concept of data analysis and the distribution of the information system was implemented using JAVA and CORBA-Standard.

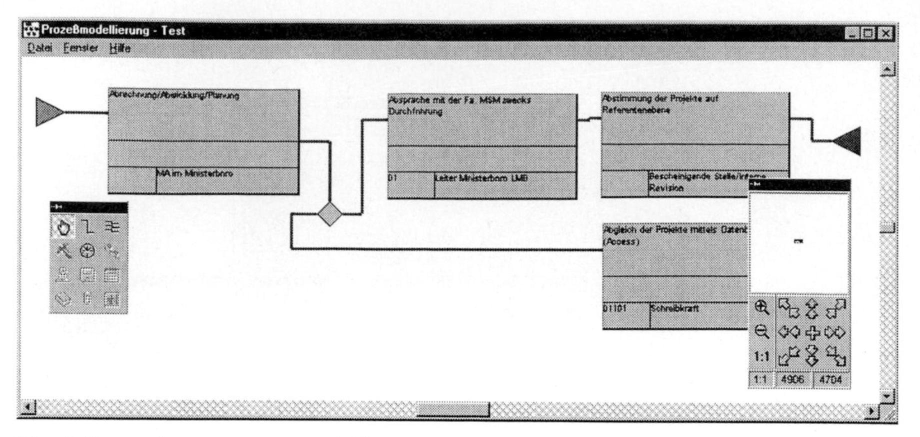

Fig. 5: Screendump: process modelling

5 Outlook

The wealth of information can be mastered by hierarchically modelling the processes. Specific information was represented at every level of detailing. The information at this level of detail can be represented using Virtual Reality Mega Language (VRML). If one „zooms" into an abstract process, further process details and additional information become apparent. The relevant information can be obtained by entering freely chosen concepts by means of a data-mining procedure. This procedure not only permits the use of data warehouses, but also allows the integration of distributed databases and gives the user a certain degree of freedom in the way he calls in data.

Future-oriented activities serve the effective integration of workflow-management systems. Internet technologies can be expected to play a key role in this respect. The Internet information-pulling principle was supplemented by an active, dynamic information pushing procedure.

Taking over business processes as workflow models, and vice versa, entails an implicit loss of information. Analysis must still be carried out to see how a process-management system and a workflow management system can operate with the same database.

References

Chrobok, R.; Tiemeyer, E. (1996): *Geschäftsprozeß. Vorgehensweise und unterstützende Tools*, in: zfo – Zeitschrift Führung + Organisation, 65 (1996) 3, S. 165-172

Davenport, T. (1993): *Process Innovation. Reengineering Work Through Information Technology*, Harvard Business School Press, Boston 1993

Gaitanides, M. et al. (1994): *Geschäftsprozeßmanagement*, Physica-Verlag, Heidelberg 1994

Jablonski, S. (1995): *Workflow-Management-Systeme: Modellierung und Architektur*, International Thomson Publishing, Reihe TAT, Nr. 9, Bonn 1995

Ort, M.; Hemmerling, B.; Wiedemann, T. (1998): *Intranet-based workflow management system*, Preprint 002, Potsdam 1998

Rosemann, M. (1996): *Komplexitätsmanagement in Prozeßmodellen*, Gabler-Verlag, Wiesbaden 1996

Scheer, A.-W. (1998): *Vom Geschäftsprozeß zum Anwendungssystem*, 3. Auflage, Springer Verlag, Berlin/Heidelberg/New York 1998

Stahlknecht, P.; Hasenkamp, U. (1997): *Einführung in die Wirtschaftsinformatik*, 8. Auflage, Springer Verlag, Berlin/Heidelberg/New York 1997

Travis, B.; Waldt D. (1996): *The SGML Implementation Guide*, Springer Verlag, New York 1996

W3C (1996): *Cascading Style Sheets, level 1*, W3C Recommendation, HTTP:\\www.w3c.org, 17 Dec 1996

W3C (1998): *HTML 4.0 Specification*, W3C Recommendation, HTTP:\\www.w3c.org, 24-Apr-1998

Workflow Management Coalition (1994): *The Workflow Reference Model*, November 1994, Technical Report WFMC-TC-1003, 1994

Workflow Management Coalition (1996a): I*nterface 1: Process Definition Interchange*, Document Number WFMC-TC-1016 Version 1.0Beta, Brüssel, Mai 1996

Workflow Management Coalition (1996b): *Workflow Management Coalition Audit Data Specification*, Document Number WFMC-TC-1015 Version 1.0, Brüssel, November 1996

Workflow Management Coalition (1998): I*nterface 1: Process Definition Interchange Process Model*, Document Number WFMC-TC-1016-P Document Status - 7.05 beta, August 1998

Wirtschaftsinformatik/AEDV, *Geschäftsprozeßmodellierung*, Skript, Technische Universität Berlin, Fachgebiet Wirtschaftsinformatik/AEDV, Berlin 1998

Benchmark-based Dynamic Process Management with an Example of a Decentralised Organised Service Industry

Dieter Specht, Jürgen Heina and Robert Kirchhof

Chair of Industrial Engineering, Brandenburg Technical University of Cottbus, Universitätsplatz 3/4, D-03044 Cottbus,
specht@prodwi.tu-cottbus.de

Abstract. *This paper describes a method of process benchmarking to support the dynamic modelling of business processes using a decentralised organised service industry of the insurance industry as an example. The way in which the integration of business process management and service system formation contribute to complexity control and reduction is shown. One of the possibilities for complexity management in service companies is the approach taken by the benchmark-based dynamic business process management. Through the integrated approach of service systems formations, the company is able to offer competitive customer-oriented services.*

Keywords. *Business Process Modelling, Process Benchmarking, Dynamic Business Process Management, Service System Formation, Complexity Control*

1 Introduction

For several years the development of "the services and information society" has been promoted as the new model for economic development. With the conclusion of the 90's in sight many businesses are facing a number of the following great challenges: Competition on a world wide scale, increased pressure on pricing levels, growth of individual consumer demands, the globalisation of both sales and input markets, a demand increase in quality, a continuous growth in the technology concerning the production of goods and services. The above has resulted in organisational structural changes within business, and also a reduction of the product life cycle process due to conditions in the market place in many areas.

There are many ways of reacting to the above paradigm changes. (Österle, 1995) In addition to the optimisation of company structures and internal processes, new products and services are required as wen as new forms of co-operation and co-ordination of business are increasingly required. (Specht, 1998) This complexity and dynamism must be recognised, planned and actioned by means of suitable methods and instruments. (Specht, 1996)

This paper is based on the application of process benchmarking which supports the modelling of business process within an insurance company. The company is characterised by a heterogencous organisational structure and consists of several branches differing in size adjoined to a central division. The spectrum of insurance services offered by all branches is the same. However, considerable differences in efficiency and quality of the services are recognised.

It will be shown that this procedure in business process management is suitable for service system formation, complexity control and reduction. The insurance industry is particularly affected by the changes as described above, and the increasing dynamism and complexity. From the customer's point of view a few years ago, this was a national business. Nowadays, insurance companies operate on an international basis. This development has consequences for the relationship between insurer and customer, which is no longer characterised by relatively inelastic prices and long-standing relationship based on trust, but by the comparison of services offered. With the increasing globalisation and the concurrent individualisation of customer preferences, a decline in the long-term commitment to the company has occurred, resulting in both price and quality of the service becoming more important. Due to the previous policy of "profit based on turnover" in the insurance business strategy, the price and quality war has led to an extension of the services offered. As these measures have gradually destroyed the economies-of-scale, views had to be changed. The insurers have reacted to these new challenges through world-wide agreements, a restructuring of both sales markets and product and service programs as well as that of company and process configuration.

As far as a consumer-oriented service is concerned, insurance companies may choose from a great range of services. Because services are only generally realised by the customer, service providers operating nationally must be present at numerous locations. Thus the provision of services is often realised decentrally, resulting in increased co-ordination requirements. This business concept required a hierarchical division of territories, which led to the development of virtually independent branches with, in some cases, different EDP support, different cost structures and heterogeneous staff qualifications. The separation of the branches was magnified by the use of the organisational units as independent companies or profit centres.

2 Complexity in Terms of Both Description and Design-Parameters of the Business Processes

A large number of new forms of services are created through reforming, repackaging services and networks and through the grouping of products and services into system services. The above being caused by increasing and more differentiating consumer needs. However, as a consequence of these measures, the complexity of both the services and the processes causing them, increases. The companies tread a path which can lead to higher complexity costs caused by the increase in variety service variant. (Belz, 1997) To respond to this it is necessary to change policies via a comprehensive examination of the company and from variable and dynamic strategies. The analysis of the internal and external company structure is to effectively support the business process management and provide starting-points for the creation of service systems.

Companies are complex systems consisting of a large number of sub-systems between which exist numerous, multi layered connections. The subsystems are involved in the exchange of services, information, material, energy and people. As a rule they usually interact by constantly changing configurations. Complex systems can take on many different states within a short period of time. (Servatius, 1985)

The complexity increases with the number of subsystems, the variety of interactions, the number of possible configurations, and the speed of the change of the system's conditions. With increasing complexity, the behaviour of the system itself becomes more indeterminable and more unpredictable.

The descriptive dimensions of the company's structure are the market or demander complexity, the product and process complexity. In turn, each of these objects can be described with regards to the multitude, ambiguity, variety and speed of change (see Fig.1).

Increasing complexity is linked to an increasing technical and economic success risk and a restricting of the company's planning possibilities. A large number of subsystems and stakeholders coupled with relations, relationships and conditions of the subsystems results in structuring, co-ordination and transparency problems.

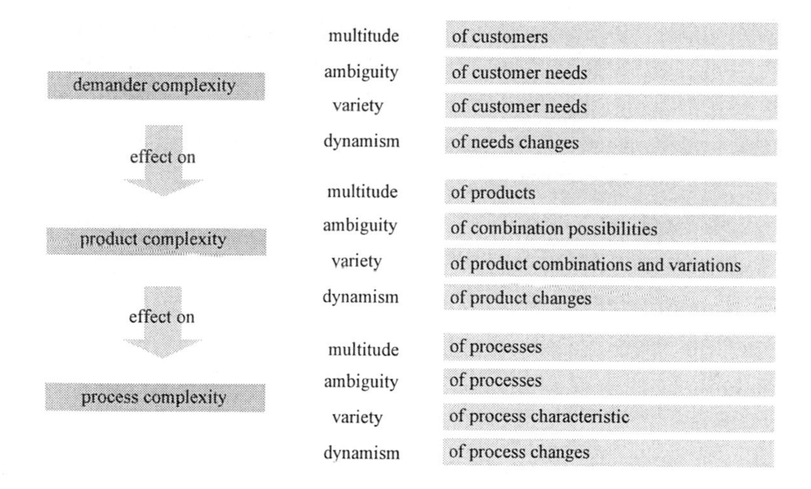

Fig. 1: Dimensions of complexity

A company can respond to the problems occurring through changes in its environment and through increasing complexity by means of the business process management. In terms of an internal application of the business process management this means:

- the development of process models as well as indicators for their dynamic adaptation or alteration,
- the dynamism of business process via the formation of flexible and combinable semi processes which can be integrated into service system and which are capable of being adapted to different consumer needs when combined,
- establishing a process control using code numbers.

The business process management contributes to the transparency of the company's complexity, through its orientation towards the business process via the value adding chain. The business process optimisation and the business process redesign are also, almost automatically, directed at controlling or reducing the complexity. A quantitative or qualitative appreciation of this interrelation has, until now, not been clearly formulated in either scientific literature or management approaches. With this in mind, it can be emphasised that complexity within the company gives a sufficient, but not a compulsory reason for the introduction of business process management. However, with regards to process modelling, when it is combined with business process management, it becomes clear that process models are a sufficient and necessary requirement to control or to reduce the complexity of the company system or to make it more easily understood.

When combined with an integrated complexity examination to provide the description and design parameters for process modelling, benchmark-based dynamic business process management can be directed towards:

- an increase in consumer orientation, plus a strong orientation towards the success factors of the company,
- the creation of a service system to enhance consumer benefits, to reduce product-independent complexity and to reduce costs,
- the reduction of process complexity through a standardisation of flexible and combinable semi processes,
- the establishment of a continuous-process-improvement process by repeated benchmarking through use of process parameters,
- the improvisation of core competence management and the establishment of knowledge management,
- self-control and self-organisation as well as organisational learning,
- more flexible organisational structures,
- the design of process interfaces which overlap departments and companies.

When carrying out business process optimisation and process benchmarking a service company cannot be compared to the above or the solutions for production company. Services are characterised by: (Bullinger, 1995)

- a general intangibility,
- a high individuality and consumer orientation,
- the simultaneous preparation and consumption of an immediate interaction by the consumer,
- participation of the customer in the service preparation,
- services cannot be stocked up,

- they must be presented to the customer and often, cannot be improved after presentation,
- often they cannot be transported (with the exception of telecommunication services).

In the service sector, products and processes are closely linked to each other. For this reason, in business process optimisation, product and process models always have to be considered as a whole. But existing tool-based and established approaches to both business process analysis and modelling show only a few combination possibilities between products and processes. Another requirement by business process modelling in the service sector is due to the consideration of dynamic changes of the processes. Due to the fact that the service is only realised through the process and that this takes place immediately at or even during the customer's participation, the actual product configuration of a service is not static, but dynamic. Thus the process causing the service is always subject to a dynamic process configuration which is comparable to a product configuration. Services are not only products, but at the same time are also the processes causing them.

This view leads to the conclusion that, both during business process optimisation and during business process redesign within the service sector, not only are the processes within the company changed, but also the final product offered. This ought to be taken into consideration during process analysis and modelling.

3 Business Process Modelling

Processes are defined as objective, temporal and spatial interconnections of activities of a value adding chain. (Striening, 1988; Gaitanides, 1983) Each process can be thought of as a chain of semi processes and considered separately. For a description of the processes and their process parts as well as the hierarchical classification of different literature terms in use. (Davenport, 1993; Sommerlatte, 1990; Striening, 1988; Nippa, 1995; Hammer, 1994)

The processes can be arranged under two suitable classifications: The process system for *business and support processes* and the reproducibility for *creative and repetitive processes*.

- *Business processes* illustrate the core competence of the company, they correspond to the company's chosen path, have direct consumer relationships, add value and usually occur in more than one area. (Gaitanides, 1996)
- *Support processes* support business processes and are a necessity for their implementation. Usually, they span more than one business process and do not add value.

- *Creative processes* are linked to different customer requirements and normally occur only once in a specific way.
- *Repetitive processes* show a high level of repetition and always follow similar or identical patterns. The processes are characterised by a relatively clear apportion of responsibility and funds.

When looking at process modelling and process benchmarking the sphere of examination must be both directed towards business and support processes as well as repetitive. Creative processes show strong procedural funds and responsibility variations making modelling and comparative consideration difficult. Consequently, process modelling for creative and processes with little repetition must take place at a highly abstract level. Modelling can be conducted for repetitive processes. Reference models can be defined and specific process parameters can be assigned, which should facilitate subsequent comparisons. The spectrum of process modelling and benchmarking applications is shown in Fig. 2.

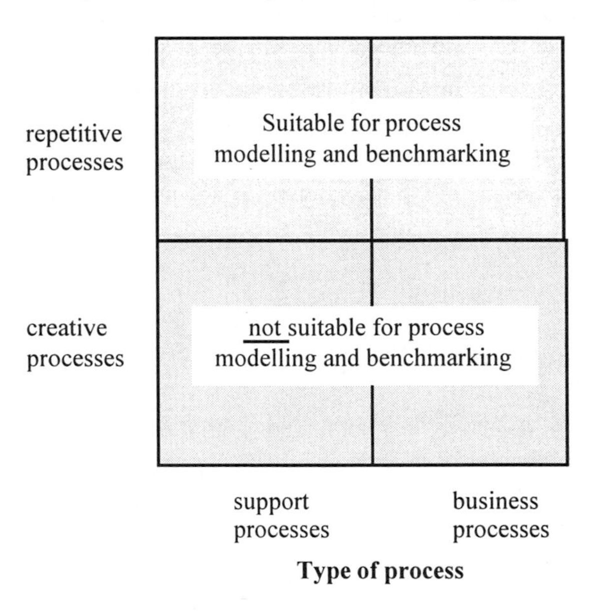

Fig. 2: Process Classification

Process modelling is presented below in the form of a flow diagram. The model displays the linking of phases and can be adjusted to fit specific company-related requirements. The flow diagram showing the description dimensions for complexity (cf. Fig. 1) must be considered before the process modelling so that process benchmarking can be carried out within the framework of business process optimisation.

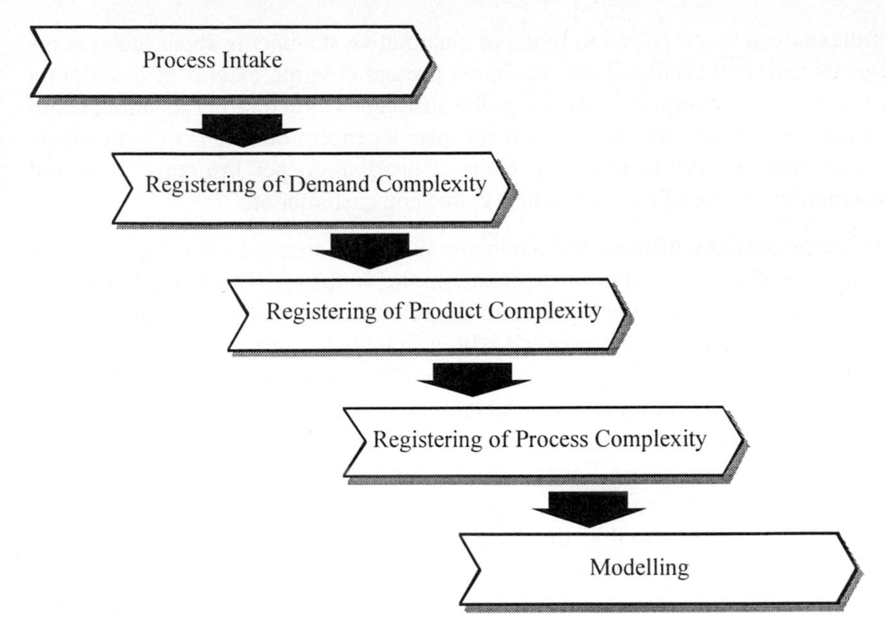

Fig. 3: Flow Diagram Depicting Process Modelling

Process Intake

The insurance company's success factors, which are critical for competition, have to be realised and separated from each other. In this way hypotheses concerning existing business and support processes were developed through the cooperation of selected experts within the company. In certain locations, interviews and expert workshops were carried out in order to identify individual tasks, responsibilities, media and aids within the business and support processes, as well as the way they interrelate and check and refine the hypotheses. As a result, eight business processes and six support processes were ascertained. The results were presented in flow charts, which considers the differentiations amongst the reproduction capability of the process. The differentiation with mainly creative and repetitive processes was carried out with expert support. The repetitive processes were registered and documented in more detail and the process recording expenditure for creative processes was kept to a minimum. In this phase the company's branches were considered in sequence. In terms of a cybernetic closed loop, the process model to be designed was analysed and redefined in a repetitive way. Through being location-specific, peculiarities were made transparent.

Registration of demander complexity

During the registration of the complexity the question of the underlying extent of description arose. If one follows an object-related, system-oriented complexity term, as during the formation of the complexity's descriptive dimensions, the complexity can be described in terms of quantitative statements about the systems elements and their relationships etc. In the present case the extents of description for the demander complexity were e.g. the customer's structure per location, result or turnover figures of the consumers per location, duration of consumers relationships, industry distribution, contract distribution per insurance class and consumer, incidence of loss per industry, line and customer etc.

With the services offered, the company aims at a specific consumer segment which represents the premium segment in the industry. The annual turnover figures of the individual consumers are the relevant criteria for this. In addition, for traditional reasons, some so-called VIP-customers are served. For a transparent representation of the demander complexity in all branches, an ABC analysis of all consumers related to turnover is conducted. As a preparation for this, all contracts concerning respective consumers from different lines (liability, legal protection etc.) have to be summarised. Several size classifications are formed. The branches only recruit a small number of their consumers from the desired premium segment. In comparison to this, the largest part of the turnover is achieved through many small customers. Moreover, the share for these respective customers varies strongly in terms of size classifications among the branches. In order to get a clear view of the consumer structure, the customer's share per the various size classifications and the average number of contracts per customer and the average turnover per customer are ascertained for each of the different locations.

Registration of product complexity

The registration of product complexity considers the heterogeneous product range that can be observed particularly within service companies. It is dependent upon the region and matches with the corresponding customer preferences. The parameters to be determined per location are, among others, the number of product types, the different types of service packaging and the different kinds of services accompanying the product of service package with regards to insurance benefits.

The company offers the consumer a service in various forms. The problem is, that almost every service is tailor made for the consumer. In addition, various services can be combined, making a more comprehensive service offer. This leads to a very heterogeneous product range. The product complexity is a direct cause of both the high process complexity and process variant variety (see Section 2). It is the cause of inefficient service provision and makes the modelling and optimisation of the processes more difficult.

In the area of insurance, benefits are traditionally classified as lines, e.g. property, liability, health, automobile, transportation insurance etc. This classification is based on the risk to be covered. The functional areas in the company including EDP systems etc. are oriented towards this classification. This historically developed line classification has to be taken into consideration. For the registration and representation of product complexity, a line classification has to be introduced, which in addition to marking off as the facts and content, combines the lines with respect to the comparative complexity criterion of the activity.

Registration of process complexity

Process complexity shows a high correlation with product complexity. Its description serves as the assessment of the service provision process. In order to show the process complexity for each business process modelled, the processes are arranged according to the product relationships (here insurance lines) and fragmented into semi processes. In addition, the individual activities of each semi process are differentiated with regards to their contribution, in adding value to the core and supporting activities. A core activity of the business process "acquisition" is, for example, visiting consumers. It adds value and costs arise. An example of a supporting activity, as a process-internal support service, is, in this process, key data administration. Despite the increase in costs there is no value added since the value of the service remains constant for the consumer.

Modelling

Within the framework of modelling, the processes are described according to the methods of system analysis. The results of the preceding complexity examinations were registered according to the relevant process activities, their knock on effects and results, combinations, branchings and loops, responsibilities and executives as well as means for implementation, especially EDP systems programs. Important conditions that determine the size of the models are the process hierarchies and the details of modelling. The hierarchy of the processes was elaborated in interviews and workshops conducted by process role players and experts. As far as details are concerned, concurring aggregation levels were found which sufficiently describe the sequence of activities that led to a defined process result. For the comparative assessment of the processes, process parameters were registered and assigned to the business process and, when possible, to the partial processes. In principle, quantitative and qualitative parameters can be distinguished. Quantitative data is the number of consumers per staff, the number of contracts per staff, the number of staff per process etc. This data was extracted from the operative systems. It was more difficult to determine the cost data for service processes of high complexity. The inclusion of the process times required for this, as well as their allocation to the large number of process variants represented expenditures, could not be justified any more. For this reason the staff were made familiar with the process models and asked for a percentage estimate of the expenditure for each process or partial process structured according to lines, sometimes even according to consumers. The qualitative data found in this way was evaluated in specialist talks and sample surveys, the statistical error rate was low, sometimes further investigations were necessary. Finally, the frequency of implementation was assigned to the processes completely and in proportion to the number of processing operations. The process models were elaborated using the ARIS tool set.

Plausibility considerations must form the end of process modelling. The expenditure necessary for this should not be underestimated since it is decisive for the quality of the process models.

The advantages of the described action model already showed at this stage. The modelling of the business processes was carried out in the following way:

- a fragmenting and grading of the business processes were carried out according to certain kinds of services, i.e. product and process models were combined and thus the subsequent design of the service system was simplified,
- qualitative and quantitative market, customer and product data were immediately assigned to the business processes, as well as the product-specific semi processes in the model,
- right from the beginning, a subsequent comparability of the processes was guaranteed beyond branch boundaries, which made the allocating of the corresponding process parameters easier,
- the modelling expenditure was reduced by separating relevant noteworthy objects from less relevant ones.

4 Process Benchmarking in Dynamic Business Process Management

Result-orientated action in business process optimisation requires the deduction of informative reference numbers and the consequent orientation towards top-level services. For this, benchmarking is a suitable management instrument. This method contains the systematic comparison of one's own products, services, processes and procedures with those of other industries, other companies within the same field or internal organisational units. (Camp, 1989; Camp, 1995; Mertins, 1995) According to this, depending on the choice of partner for the comparison purposes, internal and external benchmarking is distinguished. (Siebert, 1997) From benchmarking results, hints for using the available success potential in one's own company, and for existing differences in services compared to competing companies, are deducted.

The aim of the above course of action is the establishment of a benchmark-based dynamic design of business processes based on a systematic and integrated complexity registration, modelling, and parameter registration. During implementation, first and continuous applications of the method must be distinguished.

First process benchmarking

During the first application, the processes to be assessed must be transformed on a common comparative basis. This took place during modelling to a great extent. The process models of the individual branches were brought into line during structuring. As a next step, the process parts, which were to be compared, were selected. The parameters of the demander and product complexities, which influence the process parts, were then assigned to them. This too, happened to a large extent during modelling. For the efficiency assessment of the processes, residual efficiency factors were additionally formed from the process and complexity parameters according to lines, e.g. number of contracts per man month and line etc.

On the basis of the comparison of the process and efficiency parameters, process benchmarking was carried out. For this, the individual process parts and parameters were discussed in workshops and the best process was identified. However, a direct comparison of the processes is generally not possible, as here, the differences in the demander and product structures which cannot be influenced by the branches must be taken into consideration. The result was subjected to a plausibility test, which, as far as possible, arithmetically eliminated the identifiable structural differences. The processes in the order branches which were made in a comparable way, were compared with the "best practice process", and weaknesses-strengths profiles were drawn up.

The branches of the company were traditionally compared by means of global parameters. But in this way the heterogeneous structures, customers and products could not be taken into account. The individual branches developed in different ways and created their own processing standards, which could not be standardised through the use of centrally controlled organisational instructions.

Through the methods described above and the accompanying analysis, a perceptive view of the individual company parts, as well as the service and customer spectrum, was conceived. With reference to the individual processes of the branches, in being able to compare the benchmarks, short-term recommendations for action can be drawn up in the form of saving, changing and simplification possibilities for processes. In addition, long-term possibilities for improving the process can be immediately generated.

Dynamic process benchmarking

However, the aim should be to carry out this benchmarking not only once, but to establish a continuous process comparison. In the long run, the process modelling expenditure decreases. The methodology described here is characterised by the dynamism connected to it. Since the processes of all the branches are transparent, an active interchange regarding product and process solutions takes place amongst the process experts, account representatives and staff, therefore going beyond branch and department boundaries. The resulting effect of the changes can be directly seen using permanent parameter monitoring. A continuous and dynamic redesign of the processes developed makes the company not only more efficient, as regards process security and cost efficiency, but above all it gains flexibility and strength of innovation concerning the market through the establishment of the transfer of knowledge concerning products and processes.

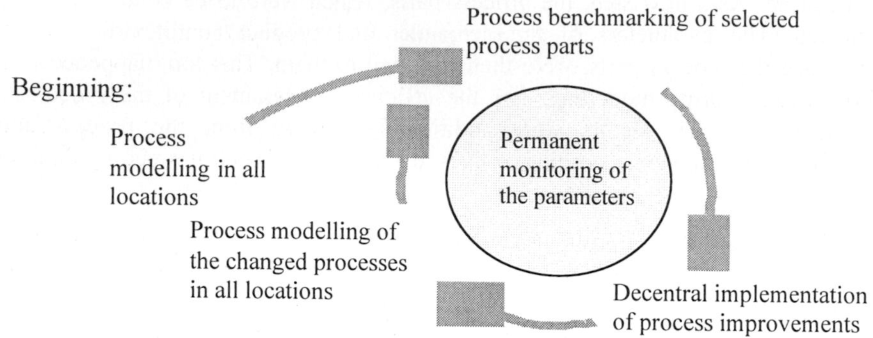

Fig. 4: Dynamic process modelling

5 Dynamic Business Process Management and Service System Formation for Complexity Control

The company had a high demander, product and process complexity before the project was carried out, but now the redesign or the improving of the processes resulted in a reduction of the variant of the products. This resulted almost by itself from the close relationship between product and process during services due to process simplification, process parallelisation, process elimination etc.

Since the interrelations between business process optimisation and complexity reduction or control have previously been insufficiently described, the question that must be answered is to what extent business process optimisation influences the structure of the services offered and thus influences the competitiveness positively or negatively. In the action model described in Section 3, a certain demander complexity was taken as a basis and this was gradually broken down to product and process complexity. After the change in process complexity, the reverse path must be taken. This means that the change of the process complexity must be investigated, looking at the effects on the structure of the services offered and the service variants for the four description dimensions multitude, variety, ambiguity and speed of change, and these changes in turn must be reproduced on the demanders. As a consequence it may happen that previously important customers cannot be handled in the way they used to be.

In order to avoid this dilemma, the optimisations must be carried out in an integrated way, in the direction of the processes and in the direction of the demanders. During the integrated structure of the service systems, innovative service-market combinations are looked for and developed with the consumer. For the control and reduction of the complexity in the company, it is necessary to combine the benchmark-based dynamic business process management with the construction of the service system. The combination of both methods serves to avoid a service variant variety which has been caused by the adaptation of individual customer requirements and the establishment of a service system. Service systems are characterised by:

- a clear structuring of core products, business processes and partial processes.
- a flexible process design combining standardised and separate process services with individual process services to form consumer-tailored total solutions,
- high transparency of the problem solution for the consumer and thus a high degree of consumer orientation,
- clear competitive advantages created through customer-tailored products with a high level of process standardisation.

The structure of service systems is shown in Fig. 5. Around the core products of the company, types or packages of services are grouped which become more and more customer-specific the further they are away from the core products themselves.

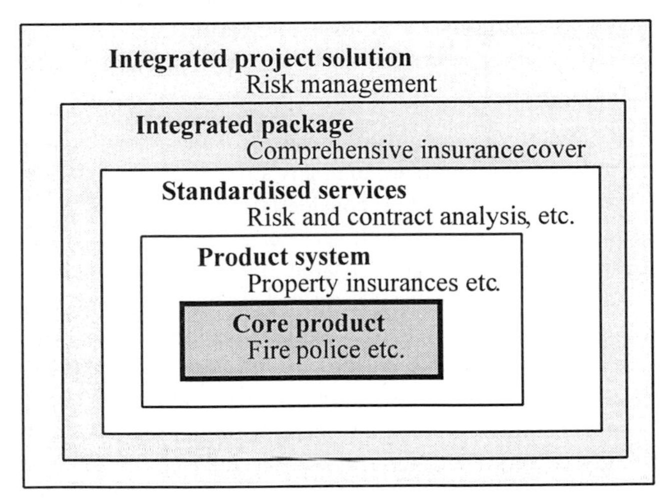

Fig. 5: Service system (following (Belz, 1997))

The service system of a service provider is closely connected with its process system. In the service sector the concept of service systems follows the approach of the standardisation of services and processes as well as the shift of the creation of the customer-tailored service variant to the latest time possible within the whole process. This approach is known through variant management.

Through the permanent process redesign, an innovatory process of finding product ranges, which are connected with dynamic process benchmarking, is established. Consequent monitoring and analysis of the customer structure show possibilities of segmenting customers or need characterising during product formation. In addition, starting points for an active influence on one's own suppliers, are created. Changes in the organisation with regards to hierarchy, the self-management on the basis of parameters, as well as the decentralisation or centralisation of functions and processes have been proposed. If this is done, the resulting effects of the measures can be demonstrated immediately.

6 Conclusion

The aim of this paper is to explain the methods of process benchmarking to support the modelling of business processes in a clear way using a decentralised organised service provider of the insurance industry as an example. It was described above, the way in which integration of business process management and service system formation contribute to complexity control and reduction.

Process modelling is a sufficient and necessary requirement to be able to make the complexity of the company system transparent plus, to control and reduce it. A peculiarity during the provision of services results from the similarity of the product caused with the process generated.

The action model for the modelling of company processes is described using a service provider as an example A heterogeneous company structure with different, regionally characterised locations was taken into consideration. The basis was the identification of the business processes as well as the graphical and verbal description of the processes. Creative, standardisable, repeatable as well as supporting processes were identified. The processes mainly taking place in a creative way showed strong variations as regards to processes and times. Thus, reference model formation and benchmarking on the basis of parameters was not useful. For standardisable and repeatable processes, process parameters were defined on the basis of the quantity and value framework, which was also established. These process parameters were subjected to plausibility examinations and registered for all locations and relevant processes. An almost complete image of the company was created. Due to the complexities of the products and processes there were difficulties when the processes were compared with regards to the different service repackaging, service provision linked to the products, the level of creativity in the processes, and the different support of these processes by means of EDP and administration. In order to be able to compare the processes these effects were eliminated to the greatest possible extent.

On the basis of these results, internal benchmarking was carried out, during which the company processes in all locations were aligned with those processes that showed the best process parameters ("best practice"). Thus, for each process a reference model was created giving recommendations for the form of the design of the processes for each location and revealing potentials for saving and qualification. In certain cases a savings potential of more than 10% was proven with this approach.

Through the inherent possibility of dynamic process improvement, through continuous monitoring of the process parameters and subsequent benchmarking, the company gained higher flexibility and efficiency. A special advantage resulted from the improved exchange of company-specific know-how among the various locations, so that due to such prudent action, a permanent exchange of experience regarding process and product knowledge was initiated. In consequence, benchmarking companies with a decentralised form of organisation are able to model suitable business processes for all company units. This will make the high product and process complexity manageable and will adapt itself dynamically to changing environmental conditions.

So that benchmarking can consider the optimal product mix per market segment for the company, an accompanying build-up of a service systems is necessary. For complexity control and reduction within the company, the build-up of a service must be combined with benchmark-based dynamic business process management.

The approach taken by the benchmark-based dynamic business process management presented here, is one of the possibilities for complexity management in service companies. Through the integrated approach of service systems formations, the company is able to offer competitive customer-oriented services. The methodology presented can be transferred to all service companies.

References

Belz, C.; Schuh, G.; Groos, S. A.; Reinecke, S. (1997): Erfolgreiche Leistungssysteme in der Industrie in: Belz, C. et al. (Hrsg.): Industrie als Dienstleister. (1997), St. Gallen.

Bleicher, K. (1995): Das Konzept Integriertes Management. 3. Aufl., Frankfurt a.M. / New York 1995.

Bullinger, H. J. Dienstleistungsmärkte im Wandel – Herausforderung und Perspektiven" in: Dienstleistung der Zukunft. Wiesbaden, 1995, S. 45-49.

Camp, R. (1989): Benchmarking: The Search for Industry Best practices that lead to superior Performance. ASQC Quality Press. In Deutschland 1994 erschienen unter dem Titel Benchmarking. Hanser Verlag, 1994.

Camp, R. (1995): Business Process Benchmarking. Finding and Implementing Best Practices, ASQC Quality Press.

Chrobok, R. (1996): Geschäftsprozeßorganisation in zfo (1996) Heft 3.

Davenport, Th. (1993): Process Innovation - Reengineering Work Through Information Technology. Boston 1993.

Gaitandies, M. (1983): Prozeßorganisation. München 1993.

Hammer, M.; Champy, J. (1994): Business Reengineering - Die Radikalkur für das Unternehmen. Frankfurt 1994.

Mertins, K.; Kempf, S.; Siebert, G.: Benchmarking - Praxis in deutschen Unternehmen. Springer Verlag.

Nippa, M.; Picot, A. (1995): Prozeßmanagement und Reengineering. Frankfurt a.M. / New York 1995.

Österle, H. (1995): Business Engineering - Prozeß- und Systementwicklung. Bd. 1, Berlin / Heidelberg / New York 1995.

Servatius, H. G., Methodik des strategischen Technologie-Managements, in: Technological economics; Band 13; Berlin 1985.

Siebert, G. (1997): Prozeß-Benchmarking - Methode zum branchenunabhängigen Vergleich von Prozessen, Diss. IPK Berlin.

Sommerlatte, T.; Wedekind, E. (1990): Leistungsprozesse und Organisationskultur. In: Little A.D. (Hrsg.) (1990): Management in Hochleistungsorganisation. 2. Aufl., Wiesbaden 1990, S. 23-36.

Specht, D. (1996): Dynamische Produktionsstrukturen, in: Wildemann, H. (Hrsg.) (1996), Produktions- und Zuliefernetzwerke, München 1996, S. 147-168.

Specht, D.; Siegler, O.; Kahmann, J. (1998): Innovationspotentiale in Kooperationsverbunden, in: Wildemann, H. (Hrsg.) (1998), Innovationen in der Produktionswirtschaft: Produkte, Prozesse, Planung und Steuerung, München 1998 (noch nicht erschienen).

Striening, H. D. (1988): Prozeßmanagement - Ein Weg zur Hebung der Produktivitätsreserven im indirekten Bereich. In: Technologie & Management (1988) 3, S. 16-26.

Computer Integrated Redesign and Visualisation of Business Processes

Jürgen Gausemeier, Oskar v. Bohuszewicz,
Andreas Lewandowski, Andreas Siebe and Martin Soetebeer

HEINZ NIXDORF INSTITUT, Fürstenalle 11, 33102 Paderborn

Abstract. *Globalisation and rapidly changing business environments require enterprises which are able to identify success potentials of the future quickly and to develop them on time. »Strategic Production Management« is a comprehensive management methodology developed to overcome these obvious challenges. The method OMEGA has been developed at the Heinz Nixdorf Institut for designing business processes. It is possible to transfer process and product data out of the engineered modelling-tool PRESTIGE into EDM- or Workflow Management Systems. Furthermore, a new approach for the visualisation of business processes is presented. Virtual Reality Technology is used to experience a simultanous view on the physical model of the factory as well as the business process model.*

Keywords. *Business process reengineering, Strategic production management, EDM-systems, BPR-tools, Virtual Reality*

1 Introduction

The current marketplace requires flexibility and a high rate of productivity. To meet these challenges business managers have to reengineer and optimise their business processes by making extensive use of information technology (IT) and sophisticated production technology.

However, many companies have only minimal knowledge of production internal efficiency and the potential that Information Technology offers. Using the approach of Strategic Production Management, a way in which the chances and trading options can be seen and recognised by the implementation of Information Technology was found. One deciding factor of the Strategic Production Management is the Business Process Optimisation.

The method OMEGA and the software tool PRESTIGE have been developed in the Heinz Nixdorf Institute to support the process of Business Process Optimisation and are efficient aids to the rational modelling of process organisation. Where optimised business processes are introduced, it is important that the new processes and process organisation are communicated to the employees in a graphical method that ensures the acceptance of all involved.

A new approach is the visualisation of the business processes in a three dimensional, computer generated model of an industrial company. By using an intuitive user-interface based on Virtual Reality Technology, the employees have the opportunity to view the newly designed business in advance. Here alongside the more object orientated processes in the areas of fabrication and assembly, the abstract processes such as Order Transactions and Purchase can be displayed. At the Heinz Nixdorf Institute, the Cyberbikes AG a fictive average-sized company, which has been planned and realised as at Virtual Model Enterprise.

2 Strategic Production Management

The structure of the comprehensive management methodology "Strategic Production Management" consists of four management methods: scenario management, strategic management, process management, and information technology management. The methodology of "Strategic Production Management" guarantees a continuous transition between the different levels.

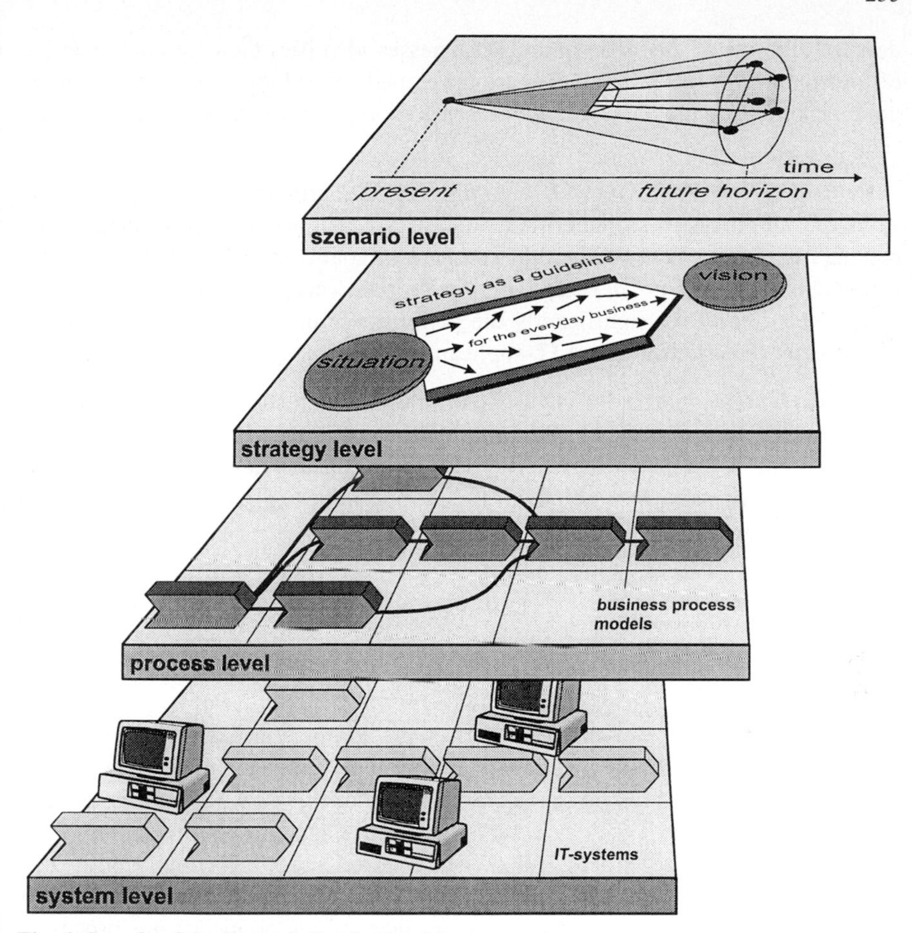

Fig. 1: Levels of the Strategic Production Management

Scenario level: In order to secure sustainable competitiveness, enterprises must principally deal with the products and markets of the future. The scenario management, derived by the Heinz Nixdorf Institut, is the appropriate method to deal with future developments and to derive opportunities and risks resulting from the different scenarios (Gausemeier, 1996). Scenario Management is based on the principles of system-thinking and "multiple futures". From an interconnected number of influence-factors on a certain topic, key factor and their multiple future projection have to be developed and combined to complex scenarios. At the Heinz Nixdorf Institut, a programmed software the "SCENARIO MANAGER", is supporting this group-based work, leading to visionary views, and future opportunities.

Strategy level: Based on the scenarios, visions are developed, convincing mission-statements are postulated and future opportunities are identified. A detailed analysis of the strength and weaknesses identifies the core competencies of the company. Finally, strategies are developed to lead the way to the strategic goals according to the overall strategic orientation given by the mission statement (Pümpin, 1992).

Process level: At this level according to the principle "structure follows strategy", all processes are identified, analysed, redesigned, and implemented in a manner to align them with the company's goals, to strengthen the core competencies, and to develop an appropriate corporate culture. Process Management improves significantly the performance of the company and creates the basis for the efficient use of IT-applications.

System level: Concepts for the usage of information technology (IT), i.e. EDM, CAD or MRP, are developed which includes computer systems, communication systems, and software systems. The IT is applied to support efficiently the company's processes and to take the opportunities offered by the future development of IT.

3 Optimising and Implementing Business Processes

The business processes of most enterprises are characterised by the division of labour combined with a functional hierarchical structure of the company. Furthermore, business processes have mutated over years forced by historic changes of the company. Today it seems clear, that the success of a company is mainly influenced by process- and customer-oriented behaviour. With this background, many enterprises take the challenge in redesigning or optimising their business processes according to customers demands. Optimising the processes finally leads to adapting principles of the new or modified processes to the hierarchical structure of the company as well as to the optimised technology, e.g. EDM-systems.

The principle of redesigning or optimising business processes follows the 5 phases shown in *Fig. 2*. After reviewing the strategic targets, which are worked out on the strategy level, a business process reengineering project is carried out on process level.

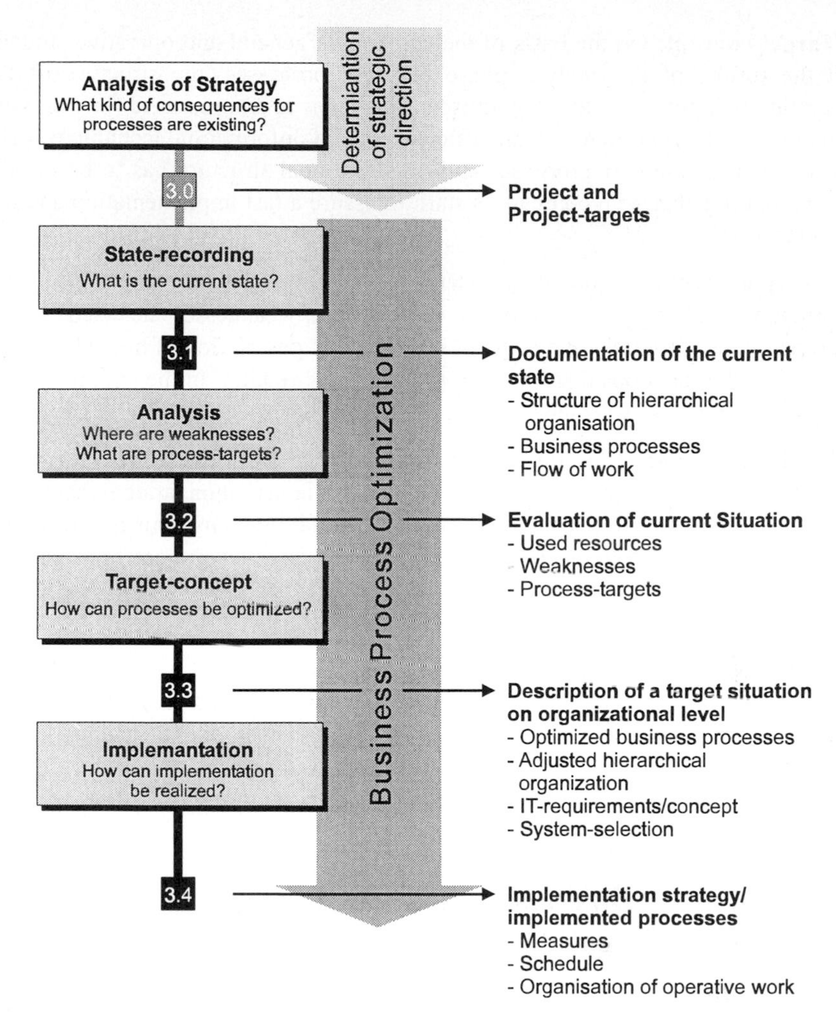

Fig.2: Structure of project-phases

Current state recording: First of all, it is important to get a complete overview of the hierarchical structure and over all business processes taking place in the company. As basis of the analysis-phase a model of the flow of work is designed combining process-information and information of organisational units.

Analysis: Throughout the analysis phase the process-team has to identify the weaknesses and problems existing in the flow of work by examinations of the recorded data. Certain spots for action and process targets can be developed. The question of what is going wrong and why something is going wrong has to be answered.

Target concept: On the basis of the company's general and operative strategy and the results of the analysis-phase business processes, the structure of the organisational units have to be optimised. This leads to an optimised flow of work constituting the conditions of an efficient use of information technology. The development of the new processes and organisational structure has to be closely worked out together with company's staff, to assure a fast implementation and full acceptance.

An IT-concept is worked out tending to complete all requirements on information technology in order to remove the weaknesses detected in the analysis-phase. Generally, the concept is a semantic description of how IT-support is allowing the functions that business-processes are require. In the end, it leads to a system selection (Gausemeier, 1997).

Implementation: The implementation-phase is constitutes the inter-phase between the process and the system-level. The implementation strategy contains a plan of measures and a concrete schedule. Regarding customising of EDM-systems the implementation phase requires itself targeted project management. Therefore similar structured phase-models exist for the system level. The transition is fluent, depending on the range of the project.

On the system level it is now possible to work in project groups to implement, for example, an EDM-system according to general- and process strategy targets. All operative efforts are focussed on the effectiveness and efficiency of the company's business.

4 Modelling Business Processes Using OMEGA/Prestige

Considering the described phase model (*Fig. 2*), a main emphasis lays on the design of the business processes. This is supported by OMEGA an object-orientated method to model and analyse complex business processes in the way they exist in a company's reality.

The object orientation is based on used principles such as class building, aggregation and navigating in certain model hierarchies. The identified objects are characterised by attributes and existing relations between different objects are determined. With OMEGA, it is possible to record all necessary information about organisational units and their structure, technical resources (e.g. IT-applications) and all input- and output-objects of the business processes (comp. *Fig. 3*). Particularly the flow of information and material can be demonstrated using the 11 design-elements (comp. *Fig. 4*).

Importance is put on an easy understandable, semantic model founded on a very semantic visualisation of what business a company is doing. Nearly all 11 design elements are to be recognise intuitively. The small number of element helps to simplify the model and guarantees a good overview for all participants of the optimising project (consultants as well as company's personnel). OMEGA-models therefore, offer the adequate basis for discussion, workshops and development of target-concepts for the whole optimising-team.

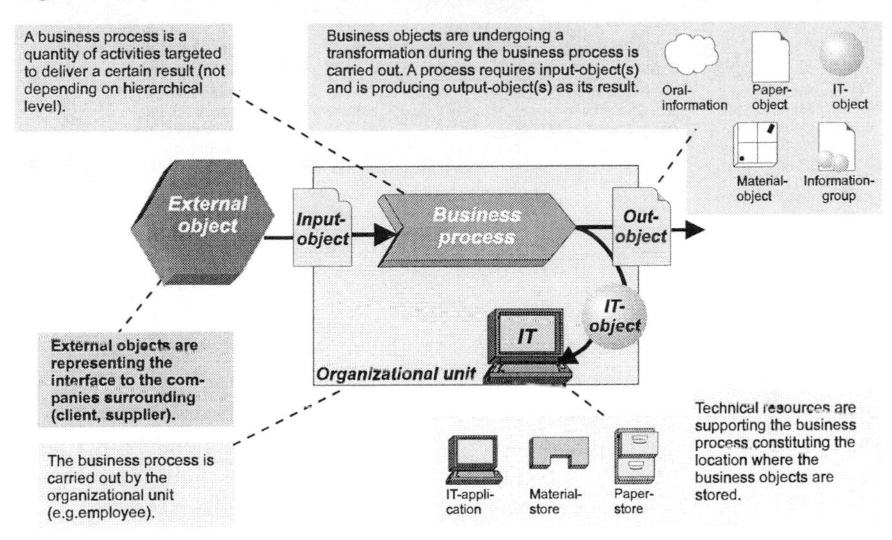

Fig. 3: Principles and basic elements of OMEGA

In OMEGA, as *Fig 3 and 4* shows, a business process is represented by a broad arrow inscribed with its name. Every new business-object is originated in a process and put into the flow as output-object. The input-objects run through the process in the arrow direction and will either be transformed or just needed as additional information leaving the process as an output-object.

The organisational units are modelled as a rectangle surrounding the process, representing the employee and location where a certain task is done. Necessary resources to carry out the process are modelled into the organisational unit. The process receives the business objects from these resources. They are transferred to the next process or are stored in a technical resource if they are not needed to continue the workflow.

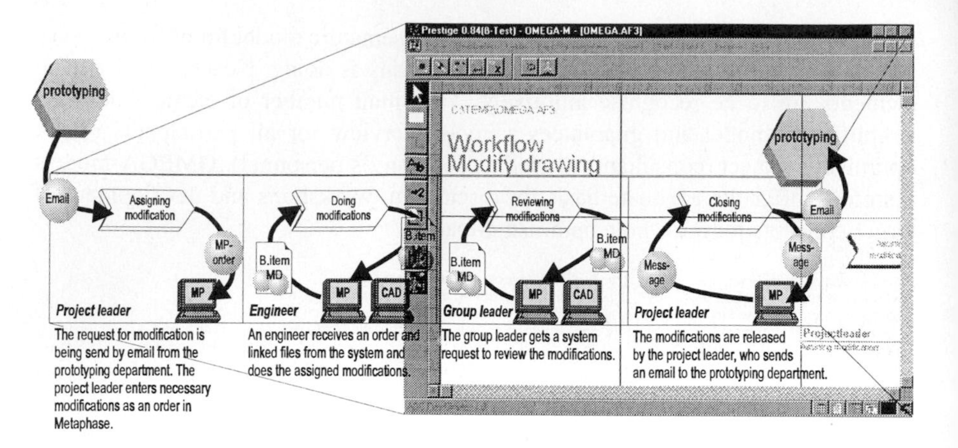

Fig. 4: Flow of work modelled using OMEGA and Prestige

The relations between process-process and process-resource are symbolised via arrows without inscriptions connecting these elements (direct connection) and transferring the business-objects. Communication between processes can also be indirect, when business-object are stored in a technical resource and further on in the model a following process receives these business-objects from the technical resource appointed to the organisational unit of that process (comp. *Fig.4*).

OMEGA as a semantic design method and PRESTIGE as its software-tool support the phase of process redesign as well as all three steps of customising EDM-systems. In applying some rules concerning the consistency of the formal models, it enables the user to work with only one method and one tool for modelling business processes and product data.

OMEGA has been successfully used in phases of redesigning business processes in many projects with industrial partners. Exporting processes and their activities of workflow management components of EDM-systems have been successfully tested for METAPHASE™. The next step will be to extend the exporting module according to the requirements of the WFMC (Workflow-Management Coalition). This enables the import and export of WfMC compatible workflow models.

5 Visualisation of the Business Processes in a Virtual Model Enterprise

In order to communicate with those employees that are not directly involved in the Business Optimisation Process, the enterprise can be shown in a computer generated Virtual Enviroment with its design and functions. Apart from the almost realistic visualisation of the design, the business processes themselves are made visible in a generally understandable way. Obviously, this is only possible in a computer, as the processes cannot be seen in reality. Through the simultaneous and consistant display of the „real" enterprise and the procedures running within, both the processes and their meaning to the success of the enterprise can be made understandable.

The Virtual Model Enterprise is therefore built up as a Prototype, a true to life design orientated simulation model of an exemplary industrial production enterprise. With the help of VR-Technology, the visitor can make an interactive excursion through the simulation model.

The simultaneous and interconnected presentation of material objects, such as maschinery and workplaces, and immaterial objects, like IT- Applications and dataflow in a virtual enviroment, offer a perspective never previously seen, which make the complex procedures of an industrial enterprise intuitively understandable to all participants. The Virtual Model Enterprise is especially suitable for the training and further education of employees, the projection and validation of futuristic enterprise concepts and technology transfer.

6 The Concept of Visualisation

6.1 Visualisation of the Information- and Dataflow

Visitors to the Model Enterprise should always have a freedom of choice in which area they visit during their tour of the Virtual Model Enterprise, without losing the general idea of the whole establishment. Earlier methods of developing a presentation, scripts and storyboards are more ideally suited for planning linear series of events, and unsuitable for the possibilities offered by a Model Enterprise. The information which the visitors receive is made available in a number of logical rooms, which can be entire rooms in the building or just locations where certain circumstances are explained.

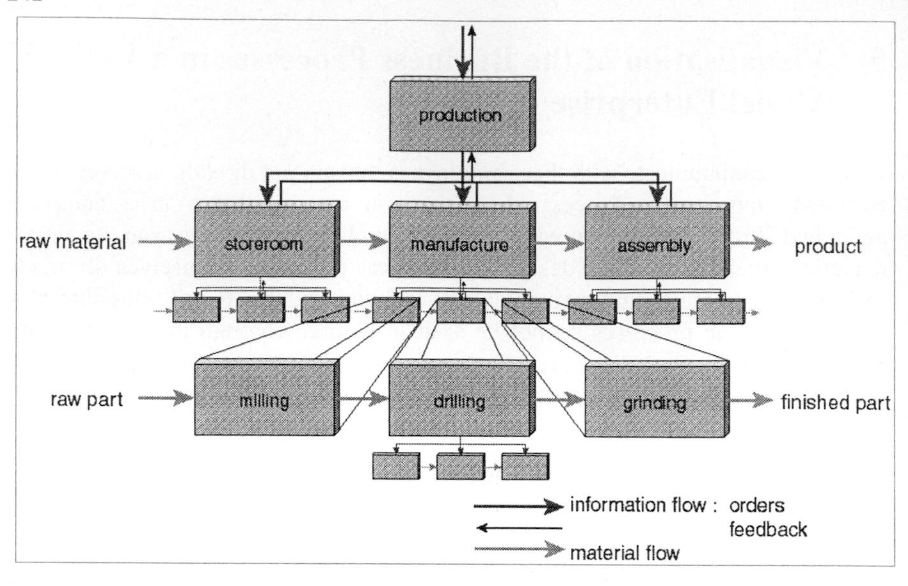

Fig. 5: The logical structure of information about Material and Information Flow

As the visitor enters each room, which explains a self-contained section of the enterprise, the connections between each room or section of room can be made. The rooms are laid out in a logical tree structure, where the horizontal branches show the timing of the procedures, and the vertical degree of detail in which the information is displayed. In each room the visitor can decide whether he will continue to observe the process, receive detailed information or a larger overall view. Using this method, the visitor can follow the process directly. When a stage is more or less of interest the visitor can, at any time, switch to more detail or move to a higher level.

A large part of the information is contained in the model itself. For example, the visitor can observe the flow of material directly, by simply following the item through the production process. In an area where the visitor has had an overall view of the production process, he can obtain more detail, by switching to a view of the individual machines.

This structure is also used for the Business Process Model, which would not be visible in reality. Data flowing through the computer network of the enterprise would be shown as a material representation, as shown in Fig. 6. The visitor can then follow the now visible flow of information in the same way as the flow of material.

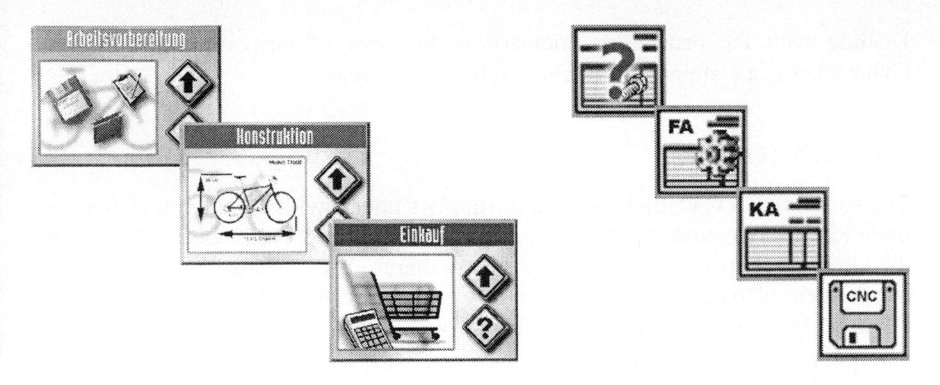

Fig. 6: Graphic Representation of Organisation Units and Information Objects.

Because some of the data is exchanged between Organisation Units which are separated by long distances, for example different departments, distant departments are represented by signs that are attached to each workplace. In this way, the visitor is able to see the procedures at work in one office, without having to follow the data through the whole enterprise. Later, the visitor can look up the department involved and observe the same procedure from another point of view. This avoids the visitor having to follow a document through the whole enterprise and being confused by the numerous departments involved.

By touching the control panel on the righthand side of the screen (diagram 6) the visitor can either receive information about the department or be taken there directly. This function accelerates the passage through the enterprise, although the visitor might become disorientated when switching his position.

When showing the procedures, it is always important to bear in mind that the visitors should have the impression that they are moving through a real industrial enterprise. Certain abstract methods of visualisation may make it possible to show more detail, they are however, due to the lack of relation to reality, definitely harder to interpret.

7 Realisation of the Prototype: The Cyberbikes Inc.

For this purpose, a fictitious enterprise was planned which is ideally suitable for presentations, and at the same time fullfills all requirements of real enterprise concepts. The result, Cyberbikes Inc. is an enterprise in cyberspace that differentiates itself from its competitors through its modern production techniques, high quality, enviromentally consciousness and strong customer orientation.

Dealing with the production industry in the area of mechanical engineering, Cyberbikes is a company producing high-class bicycles.

7.1 Objectives

The basic idea of CYBERBIKES is to supply students with a system which allows them to explore a modern industrial enterprise. An appropriate system should give the user the impression of „really being there“. The clearest solution to this requirement is to design the core of the system as a Virtual Environment, based on realistic looking 3-D models representing all objects. A typical industrial enterprise consists of, the plant building, the various departments of the production plant with its machine tools, robots and transportation systems, the offices of engineering, administration and management, and perhaps even a cafeteria.

Fig. 7: A user immersed in the CYBERBIKES virtual Environment

Although a realistic looking 3-D model is the most important part of such a virtual environment there are other issues which have to be addressed. A simulation of an industrial enterprise must also represent the data to be processed within the production. For example machine tools and robots have to operate as they do in reality. Thus, there is a need for special simulation models.

The challenge is to implement CYBERBIKES as a virtual environment. Graphics rendering complex 3-D models, 3-D sound, simulation of machine tools and robots, clear simulation of material flow and the flow of information have to operate in real-time. Furthermore, there is a need to integrate software which enables a person to control the various business processes. The most realistic solution would be to use real IT applications -- i.e., software which is used in the actual world of an industrial enterprise -- e.g. manufacturing control systems (MCS). Real IT applications should be integrated whenever considered reasonable. At the same time simplifications are often unavoidable for reasons of feasibility.

7.2 General Approach

CYBERBIKES is a virtual environment representing a medium-sized enterprise which produces high-quality bicycles. The bike was chosen because it is a well-known product to everybody, atthough the idea behind the system is not limited to bicycles.

CYBERBIKES was designed from scratch, starting with the basic design of the production processes. These are based on actual production processes which can be found in the modern bike manufacturing industry. The basic design has been reworked in several different forms resulting in a detailed design. At this stage the space requirements for the plant building were defined as well as how many and what types of machine tools and production systems were needed. At the end of the detailed design stage all the information necessary to start geometric and non-geometric modelling were available (see section 3).

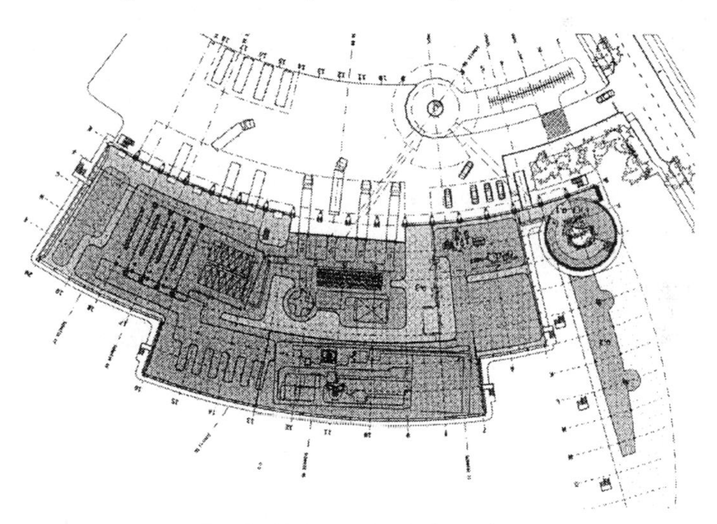

Fig. 8: CYBERBIKES floor plan

Fig. 9: Bird´s-eye view of CYBERBIKES

7.2.1 System Overview

To explore the CYBERBIKES virtual environment the user puts on a head mounted display (HMD) enabling him/her to move through the model and interact with objects within the virtual environment, using a special kind of 3-D input device called a „3-D-mouse“. During the walkthrough of the scene, standard magnetic tracking devices track the user´s head and hand. The virtual environment itself is generated and controlled by a four processor graphics workstation (SGI Onyx). The workstation is equipped with two separate graphic subsystems to generate stereoscopic images. Acoustic rendering is carried out by an entry level workstation (SGI Indy) which drives an 4-speaker ambisonic audio subsystem (Bamford, 1995].

The system is installed in the „Software Theatre“, a cinema-like auditorium equipped with 30 seats and a large-scale stereoscopic projection screen (Fig. 4 and Fig. 5). This facility, located in the Heinz Nixdorf MuseumsForum next to our institute, was designed by our group for the presentation of CYBERBIKES and other virtual reality applications.

In the Software Theatre trained instructors offer visitors guided tours around the CYBERBIKES virtual environment. The audience follows the instructors view presented through the HMD he wears on a large-scale projection screen. Interested users who wish to explore CYBERBIKES themselves can use the HMD while the instructor can lead them to points of interest.

Fig. 10: The Software Theatre used for presentation of the Cyberbikes virtual environment

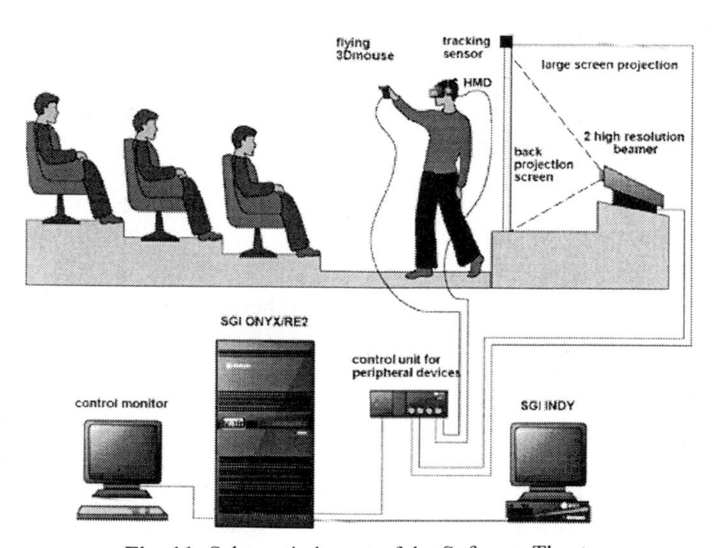

Fig. 11: Schematic layout of the Software Theatre

7.2.2 Training Procedure

After the user has entered the CYBERBIKES production, he is able to start a production process using a manufacturing control system (MCS) and watch the working numerical controlled machine tools and industrial robots. The results of his/her interaction are immediately visualised by the manufacturing control system. The user is also able to modify the production process currently running.

The manufacturing control system (MCS) is an example of software that can be found in a real world enterprise. We have integrated such a system into CYBERBIKES (Gausemeier, 1996a). The graphical user interface of the MCS is displayed on a virtual monitor. The user immersed in CYBERBIKES interacts with the system by means of a touch screen metaphor using hand, which is represented in the virtual environment by a 3-D hand icon.

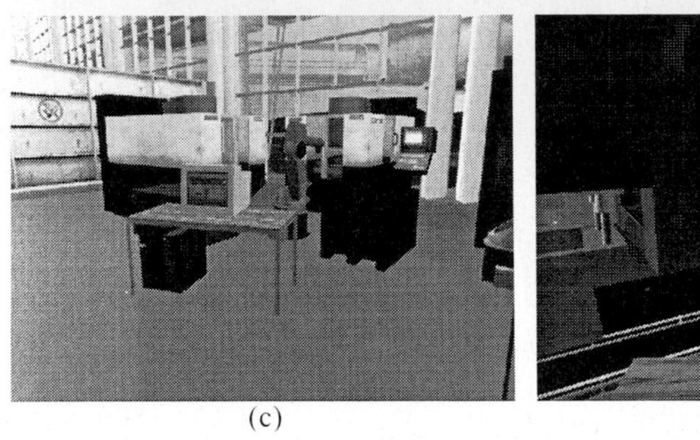

(c)

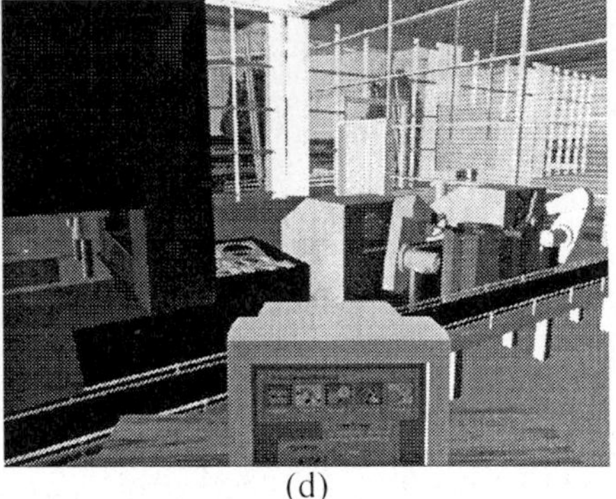

(d)

Fig. 12: Terminals in the manufacturing area

7.3 System Design and Implementation

7.3.1 Geometric Modelling

All geometric modelling was done with 3-D CAD tools. The plant building itself and the interior of the offices were designed by the department of architecture at the University of Fine Arts in Berlin, Germany. All other models were created at our institute. The machine tools and industrial robots were modelled with the help of drawings and photographs of real machine tools used for bicycle production. Realism of the models was enhanced by using realistic materials and texture maps. Furthermore, a lot of time was spent optimising the models for the use in a real-time environment. The whole geometric data set of CYBERBIKES consists of 1,357,000 polygons. 129 megabytes are required to store it in a compact binary data format (108 MB geometry files and 21 MB texture files). Fig. 7 shows a selection of different inner and outer views from CYBERBIKES. They underline the complexity and the level of realism of the CYBERBIKES virtual environment.

(a) (b) (c) (d)

Fig. 13: Views from the CYBERBIKES virtual environment

Real industrial enterprises are operated by actual human labour. Thus, human beings should also be included in a realistic simulation of an industrial enterprise. Realistic models of human beings for the use in real-time environments are difficult to create (Durlach, 1995), especially when numerous representations are needed simultaneously. In most existing VR-applications humans look and behave like toys. We decided to make light of this situation by specifically modelling humans as toys (Fig. 8). We believe this is an acceptable solution for our application.

Fig. 14: Toy models representing workers in the pedal crank production

7.3.2 Non-Geometric Modelling

Data-processing in modern industrial enterprises is hierarchically structured into four management layers, ranging from issuing production orders with declared delivery dates to the control of particular machine components, and where each layer processes specific tasks.

Planning and Disposition Layer: The management of primary data and material management occurs at this layer. Customer orders are managed and handed on to the next layer.

Manufacturing Control Layer: At this layer, the customer orders received from the planning and disposition layer, are transformed into manufacturing orders. At this stage, planning of details is carried out -- i.e., staffing schedules of particular machines are generated by considering the predetermined delivery dates taken from customer orders. As a part of the manufacturing layer, NC-programs are passed to machines on demand. In addition, the manufacturing layer ensures the efficiency of the production run. Reports of disturbances are collected and forwarded to subsequent applications.

Process Control Layer: This layer is responsible for the coordination of the production equipment and the specific transport systems. Records of material flow in the production process are kept here and processing data of the individual machines is registered for quality control.
Machine Control Layer: This is the lowest management layer. It contains the controls for all automated manufacturing machines.

We have included the most important parts of each layer into the non-geometric model of the CYBERBIKES virtual environment. It is a simplified model of production control in an industrial enterprise. The user of CYBERBIKES is able to process customer orders. These orders generate manufacturing instructions which are subsequently forwarded to manufacturing units consisting of machine tools and industrial robots. The flow of information associated with these processes is invisible in the real world. In the CYBERBIKES virtual environment however, users can see and thus understand the flow of information. Information such as data is visualised by means of metaphors as shown in Fig. 9.

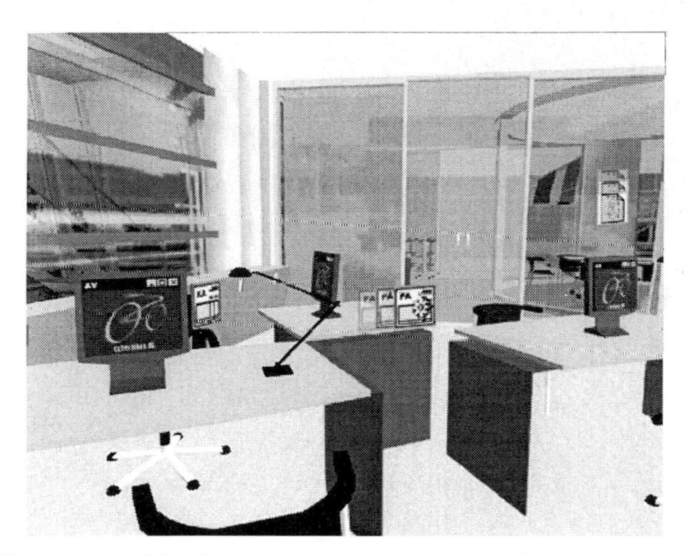

Fig. 15: Visualisation of the flow of information within CYBERBIKES. (Data packages moving from one host to another)

7.3.3 Real-Time Environment

The simulation model for the CYBERBIKES virtual environment comprises of all geometric and non-geometric models, as shown in Fig. 10.

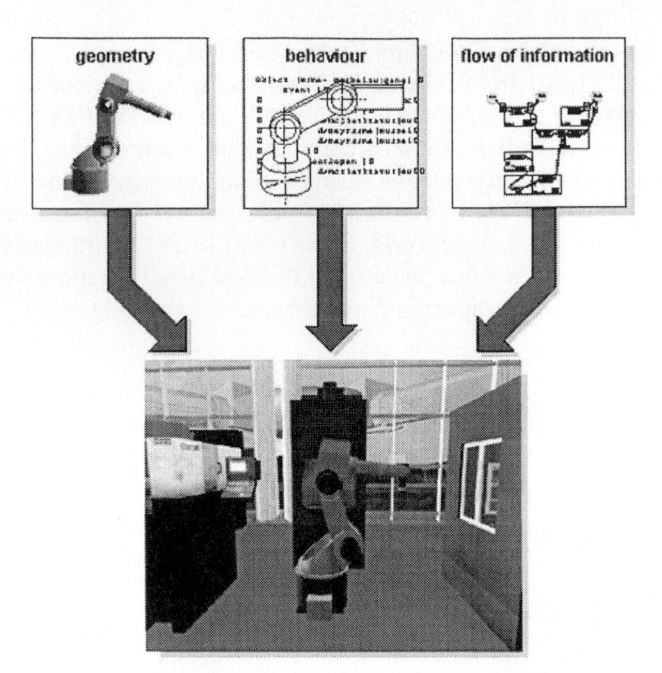

Fig. 16: The parts of the simulation model

The simulation model is executed through the real-time environment. The most important requirement of the system is to deliver real-time frame rates (more than 15 frames/sec). This is crucial to maintain the illusion of a interactive virtual world. The hardware resources needed to process the complete CYBERBIKES data set in real-time is nearly a magnitude greater than the state of the art graphics hardware delivers today.

7.4 Evaluation of the System

A first version of CYBERBIKES was completed in October 1996. This version consisted of the plant building and 20% of the production systems. At that time no offices were included. Since then the virtual environment has been continuously updated. Several production systems as well as office areas have been added. CYBERBIKES is scheduled to be completed in December 1997.

Since the initial version of the system was available CYBERBIKES has been presented and tested in the Software Theatre mentioned above. Since then more than 50,000 people have seen these presentations. 10% of the visitors used the HMD themselves. The majority of those who have immersed themselves in CYBERBIKES got a positive impression. There where a number of complaints about the low resolution of the HMD. Several users had difficulties finding their way around when being immersed in the CYBERBIKES virtual environment. It is obviously necessary to have efficient navigational aids in virtual environments of such complexity. A detailed study on the usability of the system in order to find out whether it meets its design goals will be done shortly after its completion.

References

Abramovici, M.; Gerhard, D. (1997):Engeneering Daten Management (EDM)-Anspruch, Wirklichkeit und Zukunftsperspektiven; in: Industrie Management Spezial 1997; S.E11-E15

Bamford, Jeffrey S. (1995): An Analysis of Amibisonic Sound Systems of First and Second Order; Master's Thesis; University of Waterloo; Waterloo; 1995

Durlach, N. I.; Mavor, A. S. (Editors) (1995): Virtual Reality: Sientific and Technological Challenges; National Research Concil; National Academy Press; Washington DC; 1995

Ebbesmeyer, P. (1997): Dynamische Texturwände - Ein Verfahren zur echtzeitorientierten Bildgenerierung für Virtuelle Umgebungen technischer Objekte; Dissertation; Heinz Nixdorf Institut; Universität-GH Paderborn; 1997

Fahrwinkel, Uta (1995): Methode zur Modellierung und Analyse von Geschäftsprozessen zur Unterstützung des Business Process Reengineering; HNI-Verlagsschriftenreihe, J.Gausemeier (Hrsg.); Bd.1; Paderborn; 1995

Gausemeier, J.; Lewandowski, A.; Siebe, A.; Fink, A. (1997): Rechnerunterstützte Methode zur Optimierung von Geschäftsprozessen; in: ZWF 7-8; 1997; S. 347-350

Gausemeier, J.; Hahn, A.; Schneider, W.; Kespohl, H. (1997a): Integrated Process and Product Data modeling - Ein Ansatz zur schnelleren Einführung von Engineering Daten Management Systemen; in: VDI-Berichte 1352; VDI-Verlag; Düsseldorf; 1997

Gausemeier, J.; Fink, A.; Schlake, O. (1996): Szenario-Management - Planen und Führen mit Szenarien; 2. bearb. Aufl.; Carl Hansa Verlag; München; 1996

Gausemeier, J.; von Bohuszewicz, O.; Gehnen, G.; Grafe, M. (1996a): Cyberbikes: An Immersive Virtual Environment controlled by real CIM-Applications; Proceedings of the FIVE 1996 Conference; Pisa; December 1996

Kaplan Robert, Norton David (1996): The Balanced Scorecard. Translating Strategy into Action; Harvard Business School Press; Boston; 1996

Pümpin, C. (1992): Strategische Erfolgspositionen. Methodik der dynamischen strategischen Unternehmensführung; Verlag Paul Haupt; Bern; 1992

Part 3 Logistic Processes

Componentware for Supporting Transport Logistics

Jörg Rieckmann

Computer Science Department, Databases and Information Systems Group,
Brandenburg Technical University Cottbus, Germany

Abstract. *This paper describes the development of a system for the support of end users in making complex decisions. After surveying the according application area, stable variants (as processes, SQL statements, and programmes) are modelled in terms of components in the spirit of componentware. In addition, robust mechanisms for combining components are offered to the end user. Thus, the user may compile (new) solution strategies while being guided by the system. The flexible exception handling is guaranteed by the concepts of the componentware and by combining and adapting given components by the user at run time.*

Keywords. *Decision Support, Process Modelling, Workflow, Exception Handling, Patterns, Templates, Components, Parameterisation, Case Based Reasoning*

1 Introduction

In particular in the area of logistics, the increasing globalisation of markets and the resulting competition raise high requirements for the participating companies and their executives. By the introduction of flat company hierarchies, responsibility increasingly moves towards the single employee. In particular in the operational level, specific areas suffer from a more than proportional rise of their complexity.

In this case, *decision support systems* as commercial *reporting tools* do not offer much help. Here, user information queries are based on predefined SQL statements. Without knowing the underlying database structure, no adaptations of queries are executable, and furthermore, unexperienced users are unable to make adaptations at all. SQL statements are usually set on top of *star* and *snow flake schemes*, and thus, determine the use of *Data Warehouses* (Inmon, 1996). Additionally, they do not offer interfaces for integrating external programs. For guiding the user, opportunities concerning solution strategies and corresponding realisations by solution processes do not exist. The use of *workflow management systems* (Jablonski, 1997) (WFMS) is impossible, since the complete modelling of all business processes and according process control is not feasible (Pohl, 1995). Current rudimentarily existing capabilities for handling *exceptions* finally exclude WFMS for our purpose.

This paper describes the design of a decision support system (*reporting platform*) for supporting material requirements (MR) planning in the area of logistics in transport. The system equips the MR planner (*dispatcher*) with components of different granularity (as *process modules*, *SQL statements*, and *extern programs*). Before, during the modelling phase, an IT specialist had to develop scenarios for the problems including corresponding solution strategies in terms of the mentioned components. To equip the dispatcher with the ability to adapt or combine components, necessary mechanisms for the cooperation between components are offered by coupling mechanisms and predefined interfaces.

Through the modelled components, the reporting platform offers the dispatcher a flexible adaptation of differently complex SQL statements (Abiteboul, 1995) by simultaneously hiding the details of SQL. Furthermore, the reporting platform guides the user through decision making and solution processes which are in turn adaptable by the user. Including the additional opportunity of incorporating extern programs, it provides it with flexibility and functionality missing in the above mentioned tools.

1.1 System Requirements

The area of MR planning in transport companies distinguishes itself by increasing requirements in daily business as well as in an increasing amount of information to be processed (often by a just-in-time requirement). Because of realities, a variety of information has to be dynamically combined. Individually developed application programs are usually too static, and therefore, unable to cover the various information needs.

Typical cases occuring in a truck MR planning, for example, are:

- Assigning transport goods to resources (as drivers and trucks).
- How to procced in case of an accident?
- How to proceed in case of technical problems with trucks?
- Loss of drivers or trucks and subsequent replacement.
- Necessary re routing of trucks.

A variety of processes would need to be predefined to support these kinds of processes. However, this is not completely realisable, since the variants might alter constantly. Therefore, we introduce a different approach. We extract stable subprocesses from the system, make them available via components, and offer the possibility of combining these components with each other. Combining components needs to be realised through an user friendly interface. In addition, the system should support the user in searching for components which fit into the current context. The user may adapt found solutions to her/his needs. Afterwards, the solution processes are executed while the user is optimally guided.

However, the variety of extracted components is too manifold and too unstructured, and thus, not practicable as a tool. Therefore, the reporting platform should use as a basis for proposing solution strategies the problem specification given by the dispatcher in her/his terminology. All solution strategies are based on the components designed in the modelling phase.

At any time, the dispatcher needs to be able to adapt and combine found components before using them. This also holds for exceptions occuring at run time which, for example, determine whether to include new components into the solution path. It should be possible to save composed and adapted components, so that they may be applied in future problems.

2 Realisation

2.1 Survey

Decision support systems shall support the user in solving complex, application specific problems. In just a few cases, these "decision making processes" can be modelled in terms of *business processes* (Scheer, 1994; Ferstl, 1995) for specific application areas, which are transformed into workflow processes for controlling the problem solution at run time. In the following, we synonymously use the notions of business process, decision making process, and workflow process, since they all possess the same semantics in our reporting platform. The transformation of business and decision making processes into workflow processes is guaranteed by the system, and therefore, transparent to the user.

The concept of the developed system is diagrammatically illustrated in Fig. 1.

The dispatcher communicates with the reporting platform on a technical level (in terms of action sequences). To solve the problems, she/he executes several actions (as assigning goods to transport resources, information queries) which she/he initiates and specifies by point and click. The mapping of action sequences into processes is guaranteed by the systems architecture. Single actions determine a number of process steps and their included functions (illustrated by rectangles in the process level, analogous to the EPC syntax[1]).

---

[1] ECP – Event Controlled Process Chains, (Scheer, 1994)

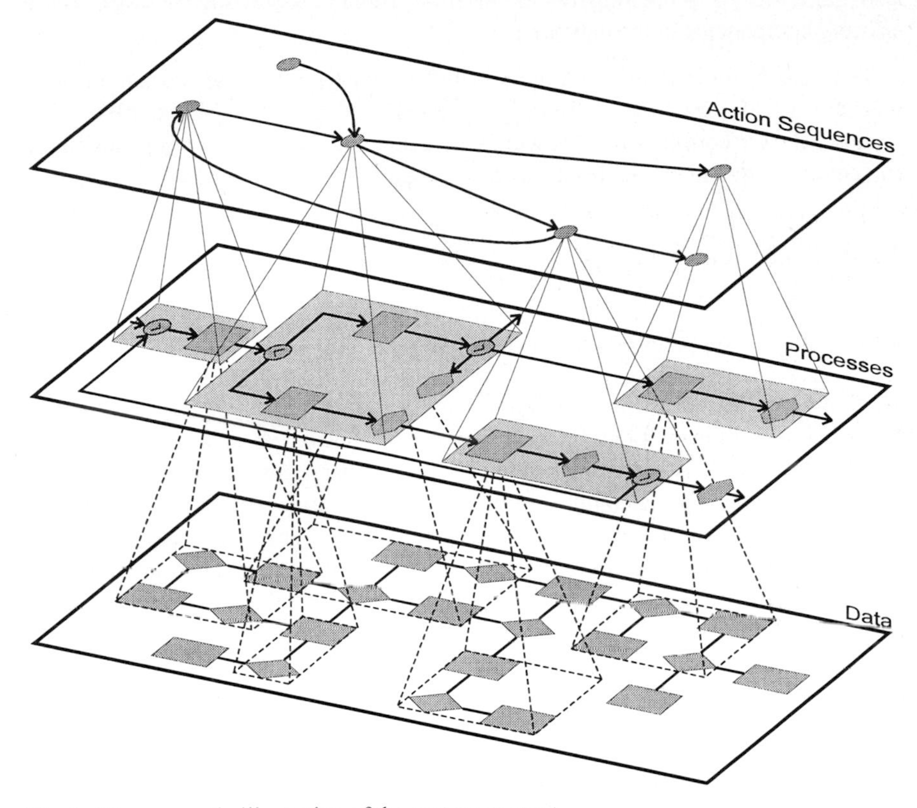

Fig. 1: Diagrammatic illustration of the system concept

Besides functions, processes may contain further subprocesses, SQL statements, and extern programs. By including subprocesses, additional levels arise between the process and data level. Through the processes and the included SQL statements, the dispatcher gets varying views on the data structures (data level). The data access is realised by accordingly adapted components at the process level. Changes at the level of action sequences cause the automatic and transparent adaptation at the process level including the accordingly new mapping onto a data view. The basis for the mapping of information requirements of the user from the technical concept level onto the implementation level is a structured co design of data structures, functions and user interfaces during the design of the information system (Thalheim, 1997). Thereby, a consistency preserving transformation throughout each layer of abstraction is guaranteed.

The system is realised by a *framework based software architecture* (Lewis, 1995; Pree, 1997). The use of concepts in the field of componentware (Nierstrasz, 1996; Schmid, 1996) achieves the required flexibility for the adaptation of components initiated by the dispatcher. Through this opportunity, the IT specialist also gains during modelling the application domain, since she/he might reuse existing components in the new area.

The search for possible solutions will be supported by the system through mechanisms of *case based reasoning (CBR)* (Kolodner, 1993; Wess, 1995). The system offers according modules which are evaluated due to their relevance to the current problem and are expressed as percentage.

2.2 Technical View

2.2.1 Use of the Componentware Concept

The notion of *componentware* comprises application systems which are generated by the composition of predefined software components of different granularity (Nierstrasz, 1995). A framework controls the cooperations of components by a *composition language*. Flexible instances of composition languages are created through the use of *script languages* (Nierstrasz, 1997). These languages offer the necessary "*glue*" to connect components with each other. The implementation of these concepts in the reporting platform enables to fulfil the strong requirements concerning the adaptivity. The varying process modules and SQL statements are represented as components and may be connected by combination, aggregation, and specialisation.

During the domain modeling, the IT specialist creates the application independent components as a basis for future use (*frozen spots*) (Pree, 1997). Their transformation into application specific components (*hot spots*) follows during run time through *parameterisation* as a technique of specialisation (Guntermann, 1998). Parameterisation includes the choice of implementation variants of an application system. It is realised by assigning predefined ranges to specific parameters .

The structures of components are modelled by *templates* (Gamma, 1995). Their input parameters initiate an adaptation of the templates, because different modules and components itself will be included into the skeleton of the algorithm. Besides parameters, *meta data* created during the domain modelling acts as a basis for the component adaptation.

The delivery of the components (stored in source code) to the run time system is deferred until the invocation of the component. Thus, it is possible to react in the case of unpredicted exceptions, since the user process is not wholly pre-instantiated, and pre-instantiation of SQL commands is not necessary. *Control constructs* (Jablonski, 1996) as *conditions* and *iterations* determine the behaviour of the processes.

Therefore, a completely specified process template can be altered during run time, since the components are not entirely joined to pre-instantiate the whole process, but they are adapted and interpreted at invocation time only.

In Fig. 2, the compilation of single components demonstrates that templates may be built upon templates themselves. For future requirements, a location transparent cooperation, for example via the Internet, should be enabled. For this reason, accordingly middle ware concepts such as CORBA (OMG, 1995) have to be integrated .

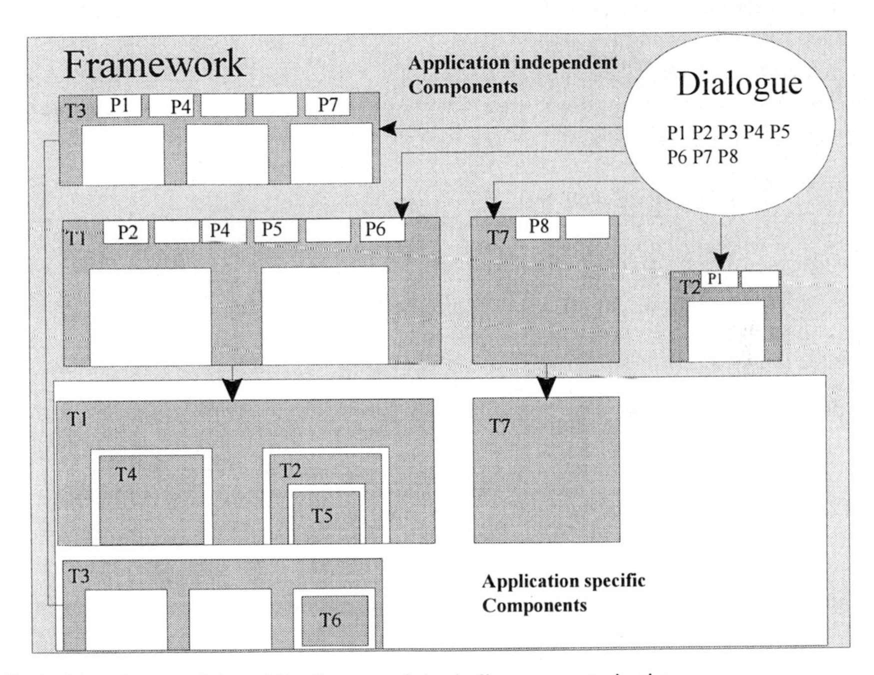

Fig. 2: Template modules of the framework including parameterisation

Template modules may only be used as a unit, if they do not offer the possibility of integrating other modules (see T4 to T7 in Fig. 2). These are process modules and predefined SQL statements with an atomar granularity (as, for example, an insert operation in SQL or a function call in a process) as well as processes which do not allow further specialisation.

The adaption of the framework modules and templates is realised by parameterisation. The parameters are generated through the problem specification and further desires for adaptations from the dialogues with the dispatcher. The parameter input of templates yields the desired alteration and composition by simultaneously taking meta data into account. Concerning the process modules, meta data comprise, for example, the modelling of the data flow in terms of an interface definition to enable the data transfer between the single components, as well as rules for combining process modules with the according consistency check.

The framework monitors the entire process as a control component. The predefined, application independent components modelled as process modules and SQL statements will be transformed into application specific components by the parameterisation.

Process modules do not always possess a start and an end event, but they might contain alternative process paths which do not join again. In general, these paths branch and join through AND/OR/XOR conditions. If process modules need to be concatenated, a consistent transition from one process module to the next has to be guaranteed by *integrity constraints*. In the case of several alternatives, the user needs to be involved through dialogue boxes.

2.2.2 The Template Concept

By the use of templates as an internal structure of the varying components, a high flexibility concerning the exception handling at run time is achievable. Input parameters of the components determine the according context and initiate the adaptation of the templates.

In the following, we describe the template concept in several examples. For this reason, we first define a specification language for the process control. A straightforward grammar borrowed from *communicating sequential processes* (CSP) (Hoare, 1985) will be sufficient for our examples.

Sequential execution:	;	B_1 is executed before B_2
Parallel execution:	||	$B_1 \| B_2$ is finished only if both B_1 and B_2 are finished. In a sequential execution either sequence is admitted.
Non-deterministic execution:	[]	The process is finished as soon as one of the part processes is finished.
Iterative execution:	*	
Conditional instruction:	$\alpha \rightarrow p$	As long as α is true, p will be executed.
Optional instruction:	{}	The process may optionally be executed.
Input:	?	
Output:	!	

The general template structure follows:

Template name	T_x	
Parameters	[Type]	Process / SQL
	$[P_1]$, $[P_2]$, [...], $[P_n]$	Components, Data Objects, etc.
Modules	$[B_i]$, $[B_{i+1}]$, [...], $[B_k]$,	Predefined
	$[B_m]$, $[B_{m+1}]$, [...], $[B_q]$	Variable
Conditions	$[C_i]$, $[C_{i+1}]$, [...], $[C_k]$,	Predefined
	$[C_m]$, $[C_{m+1}]$, [...], $[C_q]$	Variable

The parameter "Type" determines further specialisations during the adaptation of the template. These may be either process or SQL templates which are differently translated. The other parameters (P_1 to P_n) control the adaptation and consist of components, data objects, and meta data as, for example, additional conditions.

Predefined modules and *predefined conditions* are parts of templates which must not be altered or omitted, since they have been specified during the domain modelling of this specific area.

The dynamic aspect of the adaptation is covered by the *variable modules* and *variable conditions*. The modules themselves may be templates, so that, process templates may cascadate. Furthermore, a user may enrich process nodes by additional SQL statements in terms of SQL templates, which support her/him in solving the current problem. While adapting and inserting the varying modules, it is normally necessary to include additional conditions and modules which are determined by the system itself.

In this paper, we exemplarily demonstrate the application of the template mechanism using process templates in the area of MR planning[2]. Process templates admit the following adaptations:

- Sequential concatenation of processes.
- Insertion of subprocesses.
- Deletion of subprocesses.
- Insertion of iterations.

---

[2] See (Rieckmann, 1998) for the transformation of SQL templates into SQL statements.

- Inclusion of additional programs or SQL statements into single process nodes.
- Deletion of queries and programs from single process nodes.

The process "Disposition" (MR planning) can simplifiedly be described by the EPC represented in Fig. 3:

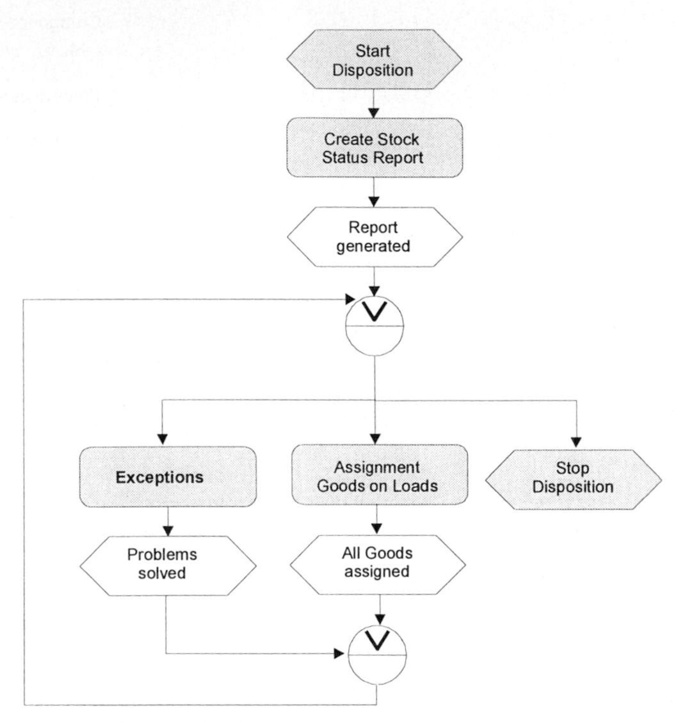

Fig. 3: Main process "Disposition"

Before assigning goods on Loads/Trucks, the dispatcher first informs about the freight volume. Afterwards, the planning task starts, which usually continues throughout the rest of the day. Nevertheless, a significant part of the day is used for specific problems as, for example, *loss of drivers*, *loss of trucks*, or *missing capacities*. Processing these exceptions needs to be done parallelly, and thus, the planning will often be interrupted. Fig. 3 illustrates these exceptions by an according function "Exceptions".

The following two process templates specify the "Disposition" process. Template-1 comprises the function for generating a survey and the subsequent iteration of a module B_2, until the user terminates the process. To specify the "Disposition" process, "Template-2" is chosen through the parameter P_2, and is used as module B_2. The place holders P_i, B_i, and C_i shall emphasise that additional input parameters may enable further adaptations - beside the parameter used for integrating Template-2. Thus, a user might, for example, include another SQL statement into a process node via drag & drop. The according SQL template would be inserted as an additional module B_i, and the conditions would be extended by the sequence condition (';' template-y ';').

The following template specifies the process "Disposition":

Template-1 *Disposition*

Parameters [Type: Process]

[P_1:Process_create_stock_status_report],

[P_2:Process_assignment_goods_on_loads], ..., [Pi]

Modules [B_1:Process_create_stock_status_report],
[B_2:*Template*-2], ..., [B_i]

Conditions [C_1:(B_1 ; *({EXIT ? → Stop_Disposition}
[] B_2))], ..., [C_i]

Template-2 specifies the "Assignment Goods on Loads" while permitting exceptions:

Template-2 *Assignment Goods on Loads*

Parameters [Type: Process]

[P_1:Process_assignment_goods_on_loads],
[P_2:Variable], ..., [P_i]

Modules [B_1:Process_assignment_goods_on_loads],
[B_2:Variable], ..., [B_i]

Conditions [C_1:(B_1 [] B_2:Variable)], ..., [C_i]

In this example, module 2 in Template-1 will be replaced by Template-2 which enables the handling of exceptions. A parameter specifies the process module that will be inserted in the case of an exception. Thus, the problem solution is realised by specialising the main process "Disposition".

The template which results by composing "Template-1" and "Template-2" possesses the following structure:

Template-1 + Template-2 *Disposition + Assignment Goods on Loads*

Parameters [Type: Process]

[P_1:Process_create_stock_status_report],
[P_2:Process_assignment_goods_on_loads],
[P_3:Variable], ..., [P_i]

Modules [B_1:Process_stock_status_report],
[B_2:Process_assignment_goods_on_loads],
[B_3:Variable],..., [B_i]

Conditions [C_1: (B_1 ; *({EXIT ? → Stop_disposition}
[] (B_2 [] B_3:Variable)))],

..., [C_i]

Fig. 4 represents a subset of possible exceptions whose corresponding process modules have to be inserted depending on the current event.

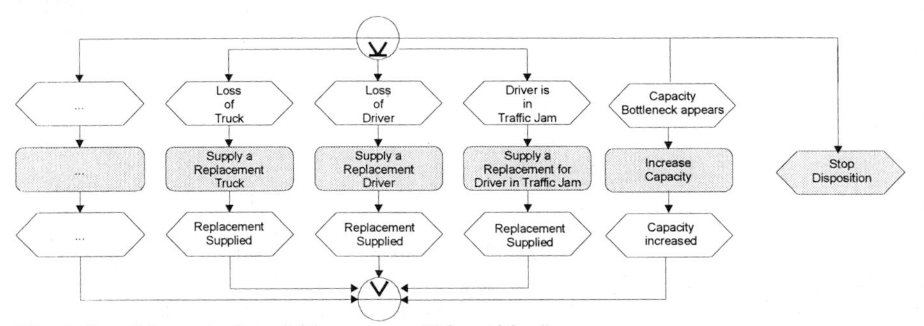

Fig. 4: Possible exceptions in the process "Disposition"

In the following example, we assume the loss of a truck which causes a more sophisticated solution process. For readability, we represent this process as a single process module, even if it may in turn consist of several templates. However, the process of mechanism of combining varying templates and modules is generic, so far, that arbitrary nesting is permitted.

The specification of the executable decision making process looks as follows:

Template-1 + *Disposition + Assignment Goods on Loads + Loss of truck*
Template-2 +
Loss of Truck

Parameters [Type:Process]
[P_1:Process_create_stock_status_report],
[P_2:Process_Assignment_goods_on_loads],
[P_3:Loss_of_truck],..., [P_i]

Modules [B_1:Process_create_stock_status_report],
[B_2:Process_assignment_goods_on_loads],
[B_3:Loss_of_truck],
..., [B_i]

Conditions [C1: (B_1 ; *({EXIT ? → Stop_disposition}
[] B_2 [] B_3))],

..., [Ci]

2.2.3 Combining Process Components

To ensure that the combination of process modules yields a consistent total process several integrity constraints need to be checked. While concatenating two modules, the following connection variants are possible:

1. The data flow is modelled, and a data transfer exists between both processes.
2. There is no data flow, or the data flow is not necessary.
3. The outgoing process paths match the incoming process paths.
4. The processes belong to the same domain, or they stem from different domains.

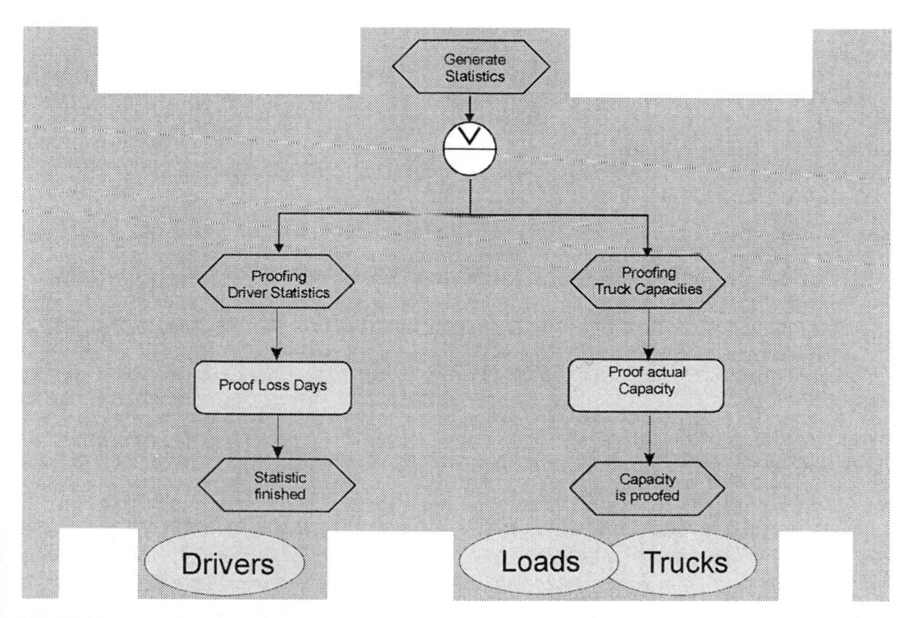

Fig. 5: Process Component-1

The mechanism of "putting together" the single components is guarded by *generic integrity constraints*. Thus, it is not forced to manually define integrity constraints for each transition between components. The integrity check uses these specified constraints as well as the according meta data. Fig. 5 and Fig. 6 exemplarily illustrate a conflict situation that has to be solved at run time.

The notches at the bottom shall indicate the opportunity to connect other components. As an interface, the "Component-1" in Figure offers three data object drivers, loads and trucks (represented as ellipses) which are defined by SQL statements. These data objects serve as input parameters for a process module to be connected. Fig. 6 presents a second component that shall be connected to "Component-1".

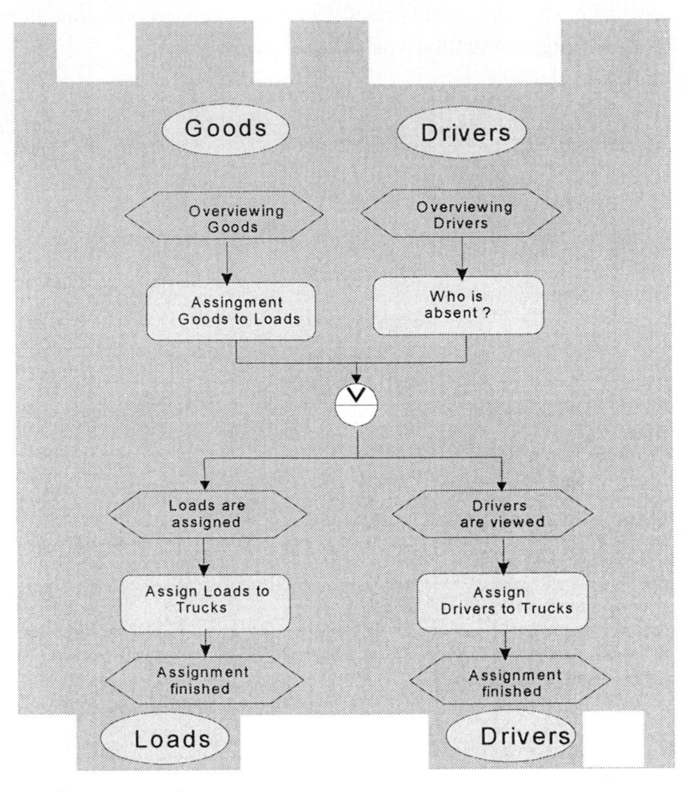

Fig. 6: Process Component-2

In this case, it is practical to permit the possibility of connecting both components, even if there is no unambigous compliance between their interfaces. If the user has chosen to execute the left path in "Component-1" to learn about driver statistics, then a *partial compliance* exists between the output interface of "Component-1" and the input interface of "Component-2" which might already meet the users needs. Maybe, in this particular situation, the user is not interested in following the goods, but he wishes further investigations in current driver assignments or specific drivers.

If the user executed queries in "Component-1" about the loading capacity, the compliance concerning the interface of "Component-2" will be lost. However, this situation should not yield an error message, but a dialogue box to solve the problem. Some alternatives are imaginable:

- A possible iteration over the left path in "Component-1".
- A complete data transfer is not necessary or not possible; further considerations of these data are not wished.
- "Component-2" is executed as a separate process, so that, previous data are not relevant in the new context.

Through the freedom in the cooperation of varying process components, the user is enabled to individually compile solution stategies. Since it is impossible to anticipate each connection, manifold possibilities for combinations are thereby offered.

Fig. 7 diagrammatically represents a possible connection of both processes.

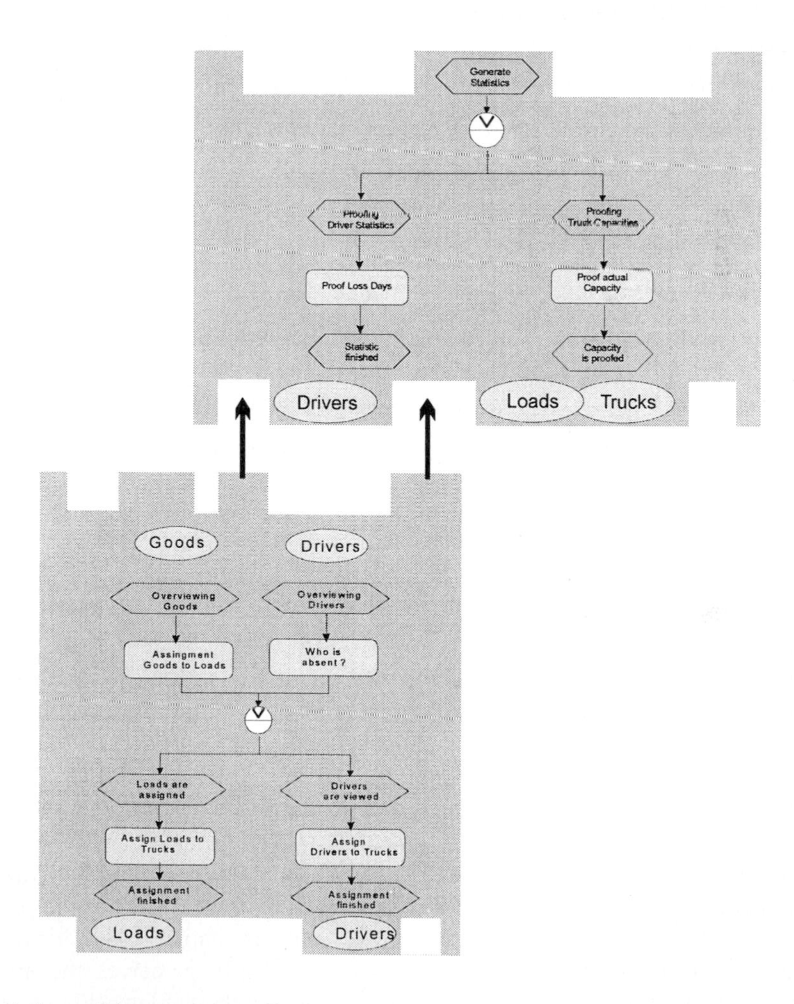

Fig. 7: A possible connection of both processes.

2.2.4 Case Based Reasoning

Case Based Reasoning (*CBR*) became known by several publications in the United States at the end of the eighties. It is a combination of ideas and methods in the areas of statistics, information retrieval, machine learning, and the development of expert systems (Bartsch-Spörl, 1996). The CBR method was chosen for the realisation of the reporting platform, since the criteria of similarity is qualified for finding solutions for similar problems within a case base, and thus, for the reasoning for SQL templates and processes. Within the SQL templates, the problem area is described by attributes (= designators for the columns in the SQL template). Within the processes, the problem specification is realised by the included SQL templates, subprocesses, and external programmes.

The problem description of the dispatcher will be exploited for the search for components (cases) which possess a specific similarity to the current problem. Found components do not necessarily yield the desired solution, so that, the user is free to perform further adaptations. Therefore, the user connects different components via drag & drop which determines a solution strategy adapted for her/his needs. A further and important fact in the area of case based reasoning concerns the storage of new solutions which successively yields a higher number of reuseable solutions.

The CBR mechanisms fulfil the following three aims:

1. Given information queries described by specific attributes, i.e., the problem to be solved. Search for a similar problem by searching for according SQL templates in the case base. Create a corresponding SQL statement.

2. Search for processes in the case base whose input interfaces are similar to the given one.

3. Store new solutions for use in future problems.

2.3 Implementation

In the following, we describe two masks of the user interface for the information retrieval and the adaptation of the solution. It surveys the realisation of the concepts introduced sofar.

2.3.1 The Information Supply

In the left window in Fig. 8 titled by "Verfügbare Informationen" (available information), the existing system information is graphically arranged in the corresponding *domains*. Each node (folder) contains similar cases having an adequate level of abstraction. Thus, the user efficiently perceives the area of interest. By an entry in the *domain box* in the search window (Fig. 8, right window), the current domain area may be extended or reduced at any time (within the assigned permissions).

The dispatcher may also navigate through the tree to find the information of interest. Within the hierarchy, the degree of information details can be specialised as well as generalised via mouse click, so that, the complexity of information becomes evident. With this technique, subprocesses and information queries are perceivable at single nodes.

2.3.2 Information Search

The search window is arranged similarly to general index search techniques as in help systems. In a glossary, all information concerning the current information query and the chosen information element (as attributes, information queries, processes) are offered in an alphabetical order. Besides the information within the window containing available information, the glossary comprises of descriptors which serve as an additional description at a more abstract level. Each entry in the glossary references n ($n \geq 1$) is entered into the window containing available information. By a combo box, the information supply is restrictive to specific information elements.

The *problem box* (underneath the glossary) enables the user to describe the problem by combining varying information elements. Furthermore, descriptors may be included into the description of the problem. This problem description will be transformed into a starting base which is suitable for the search structure of the case base. It enables to search the solution space by an index structure. With the chosen attributes, corresponding SQL templates can be found. With these SQL templates, processes can be found. And finally, with SQL templates and processes, superior processes can be found. Because of the input set, similar cases will be determined through CBR.

In Fig. 8, three entries were made to describe the problem (problem box). The entry "Fahrerausfall" (loss of a driver) is a descriptor which references two processes ("Ausfallgrund des Fahrers ermitteln" (determine the reason for losing the driver), "Ausfall eines Fahrers" (loss of a driver) which are represented in window "Verfügbare Informationen" (available information) and marked by a special colour). For this moment, both processes possess an equal probability concerning the problem solution.

An additional description of the problem with information queries "Ermittlung eines Subunternehmers" (determination of a subcontractor) and "Standort des Lastzuges ermitteln" (localize transported goods) results in a new solution set.

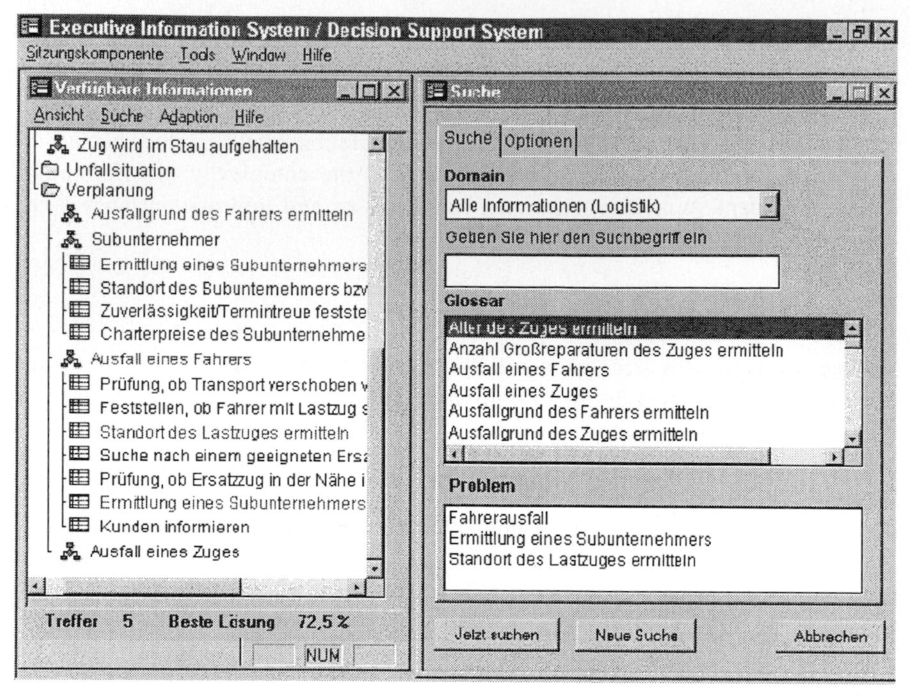

Fig. 8: Search result computed by case based reasoning

The analysis of the hits "reveals" that the process "Ausfall eines Fahrers" (loss of a driver) should get a higher probability, since it contains both information queries which were chosen to describe the problem. By a specific weighting, i.e., information queries detected within a found process get a higher weight, this process is finally the best solution concerning the problem description.

2.3.3 Information Adaption

In the window "Verfügbare Informationen" (available information -Fig. 9, left window), the search result is represented by a special colour. In addition, each of the proposed solutions is attached by a probability which indicates howfar this solution solves the described problem. In a status line at the bottom of the window, the number of found solutions as well as the percentage of the best solution are presented.

Via drag and drop, the search result can be moved into the window "Informationspool" (Fig. 9, right window) and can be further processed there. As an opportunity, all found cases can automatically be moved into the window "Informationspool" (menu item "Adaption"). Thereby, found cases will be combined and automatically be compiled to a new case. The case with the highest probability is represented first (in the example: "Ausfall eines Fahrers" (loss of a driver)).

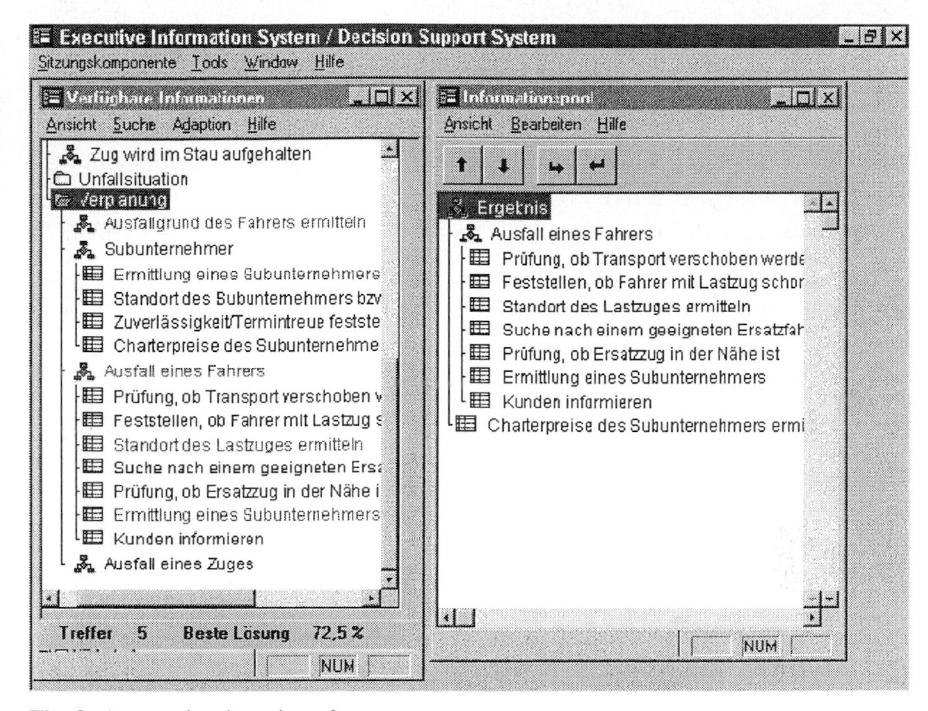

Fig. 9: Automatic adaptation of a new process

Subsequently, this new case may be further processed by the user. In Fig. 9, the information query "Charterpreise des Subunternehmers ermitteln" (determine costs of the subcontractor) from the process "Subunternehmer" (subcontractor) was additionally moved into the resulting process via drag and drop. This adapted decision process may now be executed via a single button press. The execution is realised by a workflow management tool which is part of the reporting platform.

3 Conclusion

The designed program enables the user to solve the distinguishing problems. Given a problem specification, the user "finds" solution strategies, which she/he may adapt to her/his desires. By the component oriented program, the exceptional handling and the dynamic adaptation to new realities is made possible at run time. In addition, no specific system knowledge is needed, since the user communicates with the system in her/his technical language.

The possibility of connecting and adapting components to new solutions strongly supports the reuse of already modelled modules. Storing new solutions yields a "learning system".

The restriction of structurable decision making problems through the use of processes as solution strategies has not proven to be a disadvantage. By the possible combination of the distinguishing modules and their loose coupling, the user may compile the desired information and solution strategies. The more granular the modules are, the more flexible can be reacted. On the other hand, this flexibility yields a higher effort concerning the choosing and combining of the components.

The use of templates has also proven to be an advantageous concept concerning the possibility of combining components. This flexibility of course yields more time for modelling.

References

Abiteboul, S.; Hull, R.; Vianu, V. (1995): Foundations of Databases, Addison-Wesley

Bartsch-Spörl, B.; Wess, S. (1996): Editorial. In: Künstliche Intelligenz, Heft 1, Interdata

Ferstl, O.K.; Sinz, E.J. (1995): Der Ansatz des Semantischen Objektmodells (SOM) zur Modellierung von Geschäftsprozessen, in: Wirtschafts- informatik 37. Jg., Vieweg

Gamma, E.; Helm, R.; Johnson, R.; Vlissides, J. (1995): Design Patterns, Addison-Wesley

Guntermann, T.; Popp, K. (1998): Betrachtungen zur Anpassung objektorientierter Standard-anwendungssysteme, in: Rundbrief GI-Fachausschuß 5.2, 5.Jg., Heft 1

Hoare, C.A.R. (1985): Communication Sequential Processes, Prentice-Hall International

Inmon, W.H. (1996): Building the Data Warehouse, 2nd Ed., New York

Jablonski, S.; Bussler, C. (1996): Workflow Management, Thomsen Computer Press

Jablonski, S.; Böhm, M.; Schulze, W. (Hrsg.) (1997): Workflow-Management Entwicklung von Anwendungen und Systemen, dpunkt.verlag

Kolodner, J. (1993): Case-Based-Reasoning, Morgan-Kaufmann

Lewis, T. et. al. (Hrsg.) (1995): Object Oriented Application Frameworks, Manning Publications

Nierstrasz, O.; Meijler, T. (1995): Requirements for a Composition Language, ECOOP 94,LNCS 924, Springer

Nierstrasz, O.; Schneyder, J.-G.; Lumpe, M. (1996): Formalizing Composable Software Systems - A Research Agenda, in: Proceedings 1st IFIP Workshop on Formal Methods for Open Object-based Distributed Systems FMOODS `96, Paris

Nierstrasz, O.; Schneyder, J.-G./Lumpe, M./Achermann, F. (1997):Towards a formal composition language, in: Proceedings of ESEC ´97 Workshop on Foundations of Component-Based Systems, Zürich

Object Management Group (1995): CORBAfacilities: Common Facilities Architecture, Revision 4.0, November

Pohl, K.; Jarke, M.; Dömges, R. (1995): Unterstützung schwach strukturierter Geschäftsprozesse. In: Informationssystem Architektur, Rundbrief GI-Fachausschuß 5.2, 2. Jg, Heft 2

Pree, W. (1997): Komponentenbasierte Softwareentwicklung mit Frameworks, dpunkt.verlag,

Rieckmann, J. (1998): Die Unterstützung von Benutzerprozessen beim Einsatz von DSS/EIS durch Workflowmanagementsysteme, in: Hummeltenberg, W. (1998): Information Management for Business and Competitive Intelligence and Excellence ,Vieweg

Scheer, A.-W. (1994): Wirtschaftsinformatik, Referenzmodelle für industrielle Geschäftsprozesse, 5. Aufl., Springer

Schmid, H. A. (1996): Creating Applications From Components: A Manufacturing Framework Design, IEEE Software

Thalheim, B. (1997): Codesign von Struktur, Funktionen und Oberflächen von Datenbanken, Informatik-Bericht I-05/1997, BTU Cottbus

Wess, S. (1995): Fallbasiertes Problemlösen in wissensbasierten Systemen zur Entscheidungs-unterstützung und Diagnostik, Dissertation , Universität Kaiserslautern

Modelling of Track Bound Line Operation in Public Transport Systems

Kai Frederik Zastrow

IFB Institut für Bahntechnik GmbH, Carnotstr. 1, D-10587 Berlin, Germany

Abstract. *This paper describes a model to demonstrate a track bound line operation in public transport systems. After a short overview on the basic elements of the track bound line operation, the mathematical model is presented. The following simulation shows some typical scenarios of disturbances. Finally the essential conclusions for transport operators are drawn.*

Keywords. *track bound line operation, public transportation system, delay, headway, basic interval timetable, waiting time*

1 Introduction

You have probably encountered the following situation yourself: you reach an underground station in a large town where a lot of passengers have already gathered on the platform. Although according to the timetable a three minute interval between trains is scheduled you seem to have to wait endlessly for the next train. At last a full train arrives at the station. As the train is already overloaded you decide to wait for the next train which follows immediately.

Such a situation, which often arises after an interruption to train movements, presents an unsatisfactory condition for the operator as well as the passengers.

A research project concerned with implementing automatic operation for the Berlin underground offered the chance to examine this phenomenon. This data serves as a basis to establish strategies for terminating operative disturbances.

2 Elements of Track Bound Line Operation

Let us examine a route (or part of a route within a network) which is operated as a line. A certain amount of stations are situated on this line where passengers can board and disembark including a starting point and an endpoint. A public timetable determines the departure times of the trains for every station, possibly also the arrival times. A train operates its route from starting point to endpoint of the line stopping at every station and subsequently passing through the terminal loop. After passing through the terminal loop the train begins its route in the reverse direction and reaches its point of departure after another loop.

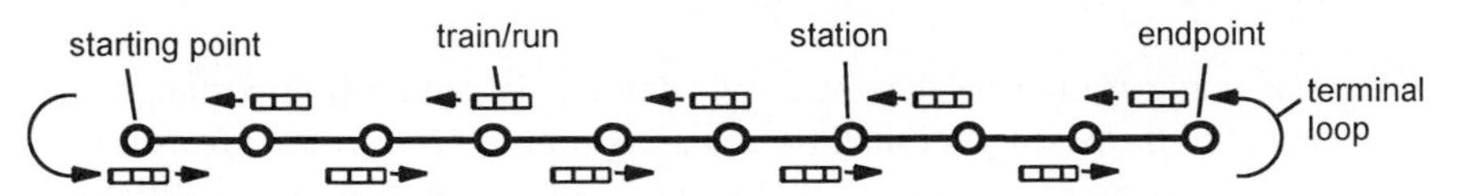

Fig. 1: Operation of a line

In addition to trains, runs are also a part of line operation. A train journey according to the timetable begins at the starting point and ends at the endpoint without executing a loop. This means that a run in the line operation usually effects several consecutive train journeys. The sequence of a run is determined in the allocation list. Train headways (interval between two consecutive trains) of the same length are called a Basic Interval Timetable. The Basic Interval describes the duration of the constant headways. For Light Rail Transit, underground and tramway operation a flexible Basic Interval Timetable is usually applied, as the Basic Intervals change with the time of day.

The adherence to the timetable of an observed train (meaning punctuality) can be influenced by technology, environment, passengers and other trains. A feedback of the observed train towards other trains and towards passengers arises if for example the cancellation of a train causes a higher passenger flow at short notice and the journey of the succeeding train is delayed.

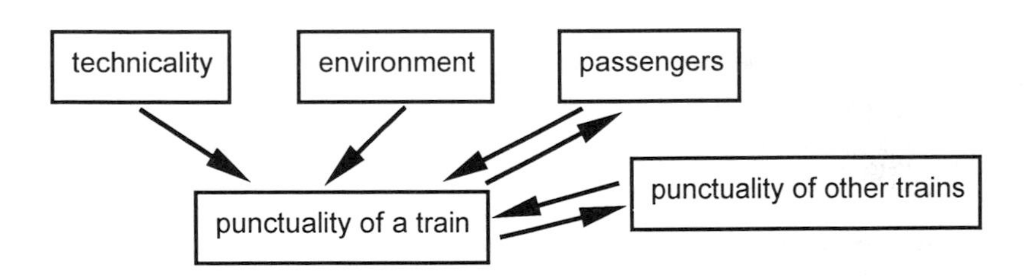

Fig. 2: Influences upon the timetable

It is obvious that the influence of passengers and other trains is of special importance as an oscillating effect of disturbances can arise by the existing feedback. This phenomenon will be analysed in the following text.

With a Basic Interval Timetable with a headway of under ten minutes it can be assumed that the passenger flow will be almost constant, or in other words, that the arrival of passengers at the station is unrelated to the timetable. This situation, which mostly arises in large towns, is the pre-condition for the following examined model.

3 The Opinion of Passengers

As the passenger should be the focus of attention in public transportation, the requirements of the passengers shall be examined first. These requirements do not always have to correspond with the requirements of the operator as the following examples indicate:

The cancellation of a train in a 20 minute Basic Interval will have more serious consequences for the passenger than the same situation within a three minute Basic Interval. From the view of the operator both cases constitute a cancellation of a train without significant differences.

If according to a Basic Interval of five minutes, every scheduled train journey starts with a delay of exactly 20 minutes the passenger is not inconvenienced after the first delayed train, even if the line does not correspond with the train allocation list and therefore an operative disturbance exists.

In order to examine line operation from the viewpoint of all passengers *(amount m)* two specific characteristic values apply:

the average waiting time *E(W)* on the platform for a passenger

$$E(W) = \frac{\sum_{k=1}^{m} W_k}{m}$$

W_k = *waiting time for passenger k*

and the average delay *E(D)* during the journey for a passenger

$$E(D) = \frac{\sum_{k=1}^{m} R_k - S_k}{m}.$$

R_k = *real journey time for passenger k*

S_k = *scheduled journey time for passenger k*

The last one is needed in order to account for delays of a train journey which arise after the passenger has boarded.

These characteristic values constitute a problem as they are difficult to measure without a closed ticketing system (turnstiles or optical sensors in the stations in order to record passenger flows) and are also dependent on the features of the timetable. The question arises if a rigid timetable is sensible.

The average waiting time $E(W)$ on the platform for a passenger depends on the headway H and the arrival time of the passenger. The basis here is a random arrival time of the passenger on the platform. This can be assumed for short headways (see above).

During the observed time span the headways are numberd from H_1 to H_n. The average waiting time for a passenger, who arrives within the headway H_k on the platform, is the expectation value $E(W_k)$:

$$E(W_k) = \frac{H_k}{2}\,. \tag{1}$$

The probability that a passenger arrives on the platform within headway H_k, corresponds to headway H_k related to overall time:

$$P_k = \frac{H_k}{\sum_{i=1}^{n} Z_i}\,. \tag{2}$$

The expectation value of waiting time W regarding overall time is

$$E(W) = \sum_{k=1}^{n} P_k E(W_k)\,. \tag{3}$$

Inserting equations (1) and (2) results in

$$E(W) = \frac{\sum_{k=1}^{n} H_k^2}{2\sum_{i=1}^{n} H_i} = \frac{E(H^2)}{2E(H)}, \tag{4}$$

with $E(H^2) = \frac{\sum_{k=1}^{n} H_k^2}{n}$ and $E(H) = \frac{\sum_{i=1}^{n} H_i}{n}$.

Inserting the variance $var(H) = \sigma^2 = E(H^2) - (E(H))^2$ results in

$$E(W) = \frac{1}{2}(E(H) + \frac{var(H)}{E(H)})\,. \tag{5}$$

The first term of sum $\frac{E(H)}{2}$ corresponds with the average waiting time, if there is no deviation from the average headway (meaning an unchanging interval without delayed or premature trains). The second term of sum grows in proportion to the individual trains deviation from the average headway. Furthermore it grows in proportion to the frequency of headways (with constant deviation). Thus with long headways referring to the deviation this influence remains small.

The following example explains the meaning of equation (5): A line has a headway according to an exact five minute interval timetable. The average waiting time for a passenger with a random arrival time is therefore two minutes and 30 seconds. Operative disturbances cause alternating headways of three and seven minutes. Even though the average headway still amounts to five minutes, the average waiting time of a passenger will rise by 16 %, or increase to two minutes and 54 seconds.

This is the reason why the operator must aim to avoid variations of headways.

4 Model of Regularities in Train Delays

Delays must be distinguished between primary and consequent delays. Primary delays relate directly to a train concerned. Among other factors the situation and duration of the disturbance are important. Consequent delays result from primary delays. These are train journeys which can not be effected according to timetable because they are encumbered by other trains. A special consequent delay arises through the phenomenon of the oscillating effect which is analysed as follows.

As mentioned earlier, the arrival of the passengers on the platform is regarded as unrelated to the timetable. If only one direction of traffic is examined, all the passengers waiting on the platform will try to board an arriving train. The resulting equation is:

$$\textit{Amount of boarding passengers} = F \cdot H = C \cdot T. \tag{6}$$

F = *passenger flow to the station (persons per second)*

H = *headway (seconds) (respectively at train departure)*

C = *efficiency of doors (persons per second)*

T = *door opening time (seconds)*

If F and C are constant, this results in

$$\frac{C}{F} = \frac{H}{T} = \frac{H_s}{T_s}. \tag{7}$$

H_s = *scheduled headway*

T_s = *scheduled door opening time*

In the case of a delay this results in

$$H = H_s + D_0 + T - T_s \,. \tag{8}$$

D_0 = *delay (seconds) of a train arriving at the station*

If equation (8) is inserted in (7) and resolved after the door opening time T this results in

$$T = T_s(1 + \frac{D_0}{H_s - T_s}) \,. \tag{9}$$

The additional delay of a train at departure, which arises through longer door opening time results in

$$\Delta T - T - T_s = D_0 \frac{T_s}{H_s - T_s} \,. \tag{10}$$

The sum of delay at train departure therefore amounts to

$$D_1 = D_0 + \Delta T = D_0(1 + \frac{T_s}{H_s - T_s}) = D_0 \frac{H_s}{H_s - T_s} \,. \tag{11}$$

If the train continues its journey and the passenger flow at the next station is the same as at the first station, the total delay at departure from the next station is

$$D_2 = D_1 \frac{H_s}{H_s - T_s} = D_0(\frac{H_s}{H_s - T_s})^2 \tag{12}$$

and at departure from station m

$$D_m = D_0(\frac{H_s}{H_s - T_s})^m \,. \tag{13}$$

The total delay grows exponentially with every further station, so that an initially small delay D_0 can quickly result in a very big delay without any other existing disturbance if passenger flow to the stations remains constant.

Let us examine the following train. The following index is applied here: The first index indicates the station the second index the train. The time D_{12} indicates the delay of train 2 at departure from station 1.

The headway for the following train results in

$$H = H_s + D_{02} - D_{11} + T - T_s \,. \tag{14}$$

D_{02} = *delay of the following train (train 2) on arrival at station 1*

D_{11} = *delay of the train 1 at departure from station 1*

The additional delay of the following train at departure from station 1, which arises through longer door opening time is therefore analogous to the equations (10) and (11)

$$\Delta T = T - T_s = (D_{02} - D_{11})\frac{T_s}{H_s - T_s} = (D_{02} - D_{01}\frac{H_s}{H_s - T_s})\frac{T_s}{H_s - T_s}\,. \tag{15}$$

It becomes obvious, that in the special case $D_{02} = D_{11}$ no additional delay for the following train arises.

The total delay of the following train at departure from station 1 is

$$\begin{aligned} D_{12} &= D_{02} + (D_{02} - D_{01}\frac{H_s}{H_s - T_s})\frac{T_s}{H_s - T_s} \\ &= D_{02}\frac{H_s}{H_s - T_s} - D_{01}\frac{H_s T_s}{(H_s - T_s)^2}\,. \end{aligned} \tag{16}$$

The additional delay of the third train at the first station is

$$\Delta T = (D_{03} - D_{12})\frac{T_{plan}}{H_{plan} - T_{plan}}\,. \tag{17}$$

So by inserting (16) the total delay of the third train reaches

$$\begin{aligned} D_{13} &= D_{03} + (D_{03} - D_{02}\frac{H_s}{H_s - T_s} + D_{01}\frac{H_s T_s}{(H_s - T_s)^2})\frac{T_s}{H_s - T_s} \\ &= D_{03}\frac{H_s}{H_s - T_s} - D_{02}\frac{H_s T_s}{(H_s - T_s)^2} + D_{01}\frac{H_s T_s^2}{(H_s - T_s)^3} \end{aligned} \tag{18}$$

or the total delay of train n at station 1

$$D_{1n} = \sum_{i=1}^{n} (-1)^{n-i} \frac{H_s T_s^{n-i}}{(H_s - T_s)^{1+n-i}} D_{0i}\,. \tag{19}$$

Now the equation for the delay of a train n at departure from station m is generalized:

$$D_{mn} = \sum_{i=1}^{n} (-1)^{n-i} \binom{m+n-i-1}{m-1} \frac{H_s^m T_s^{n-i}}{(H_s - T_s)^{m+n-i}} D_{0i} \,. \tag{20}$$

H_s = scheduled headway (seconds) (respective departure of trains)

T_s = scheduled door opening time (seconds)

D_{0i} = delay of train i on arrival at the first station (seconds)

This is a recursive relation, as the value is dependant on the delay of the train n on arrival at the observed station $D_{(m-1)n}$ as well as the delay of the preceding train (n-1) at departure from the observed station $D_{m(n-1)}$. By using these two parameters the equation (20) can be described in a recursive form:

$$D_{mn} = \frac{H_s}{H_s - T_s} D_{(m-1)n} - \frac{T_s}{H_s - T_s} D_{m(n-1)} \,. \tag{21}$$

In order to calculate this equation for all $m, n \in \mathrm{IN}$, an anchorage is needed. This requires the marginal condition that train operation before the examined train runs according to schedule:

$$D_{m0} = 0 \ \text{ for all } \ m \in \mathrm{IN} \,. \tag{22}$$

Furthermore constant values have to be implemented for the initial delays of every train:

$$D_{0n} = \text{constant} \ \text{ for all } \ n \in \mathrm{IN} \,. \tag{23}$$

This mathematical model provides the opportunity to simulate line operation.

This phenomenon does not only occur in public transport but also in buildings with several lifts. Here passengers will also board the first lift to arrive and thus delay its journey. But in this case the problem loses its importance because the following lift can make up for lost time and overtake the full lift. This is also the case with overcrowded buses in line operation. In track bound transport systems with one track per direction the possibility of trains overtaking each other during operation does not usually exist.

5 Simulating Scenarios of Disturbances

Microsoft Excel® offers the possibility of calculating and demonstrating scenarios of disturbances during line operation. For this reason the train delays are presented in a table. One column presents the delays of a train and one line the delays for a station. Apart from the delays of the individual trains the average waiting time for a passenger waiting in a station is also calculated.

Hs (s) 180		Ts (s) 30		Hmin (s) 60			
	train 0	train 1	train 2	train 3	train 4	train 5	train 6	E(W) (s)
initial delay	0	120	0	0	0	0	0	
station 1	0	144	24	0	0	0	0	110
station 2	0	173	53	0	0	0	0	116
station 3	0	207	87	0	0	0	0	126
station 4	0	249	129	9	0	0	0	140
station 5	0	299	179	59	0	0	0	157
station 6	0	358	238	118	0	0	0	185
station 7	0	430	310	190	70	0	0	219
station 8	0	516	396	276	156	36	0	271
station 9	0	619	499	379	259	139	19	340
station 10	0	743	623	503	383	263	143	406
station 11	0	892	772	652	532	412	292	487
station 12	0	1070	950	830	710	590	470	586

Table 1: Delays of all trains for one primary delay of two minutes

For calculating the individual delay values the recursive equation (21) is applied. The mathematical anchorage is reached by an initial train (train 0) which operates punctually as well as the given initial delays for each examined train.

Further conditions form the basis for the simulation:

- every value of delay is bigger than 0, meaning there are no early trains as every train only departs when its scheduled departure time is reached,
- as no two trains can arrive at one station at the same time, a minimum headway has to be respected. This is also necessary for safety reasons in order to guarantee a minimum distance between trains. This distance results from the characteristics of the train control system.

The minimum value of train delays corresponds to the "delay of the preceding train at the same station" minus "scheduled headway" plus "minimum headway".

The simulation model programmed accordingly can now be used to describe different scenarios of disturbances.

In every case the basis of the valid model is peak operating time in a large town. This means the scheduled headways are around two to six minutes. The door opening time is between 10 and 50 seconds and the minimum headway between one and three minutes.

5.1 Scenario 1

Typical variables that apply to this scenario are a scheduled three minute Basic Interval, a scheduled door opening time of 30 seconds and a minimum headway of two minutes.

Through a technical disturbance (i. e. a faulty track occupation notice) one train receives a unique primary delay of one minute. All the other trains operate punctually up to this point. Because of the additional passengers on the platform due to the delayed train, the delay grows with every station.

Fig. 3 describes explicitly that after only a few stations the two following trains are affected and after some time all the trains accumulate behind the delayed train. At the departure of the twelfth station the delay of the train which was held up through a technical disturbance for only one minute already has a total delay of around nine minutes. A time reserve of nine minutes for the terminal loop is therefore necessary so the train can start its return journey according to schedule after passing through the loop. The bigger the time reserve for the terminal loop, the more trains are necessary to provide the same transport efficiency in train kilometers. Therefore a compromise is needed for economic reasons in calculating the time reserve for the loop. This can be optimised with the aid of such a simulation.

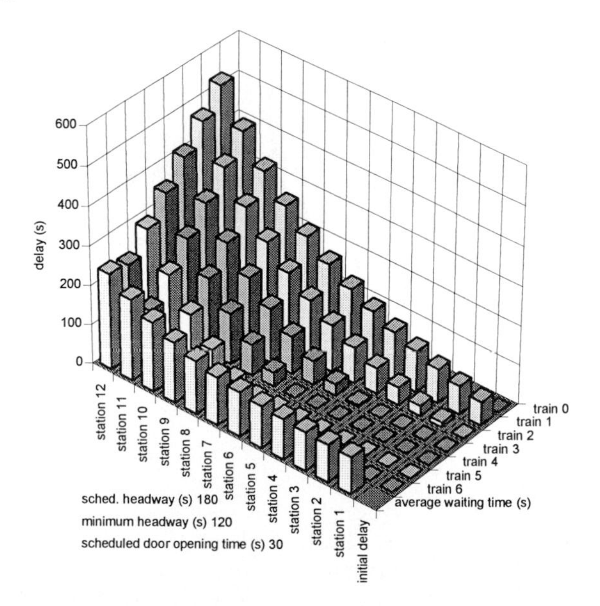

Fig. 3: Train delays with an initial delay of one minute

5.2 Scenario 2

The basis is the same as in scenario 1 except that due to the operation of a new vehicle type with more efficient passenger doors a reduction of the planned door opening time of 10 seconds can be reached.

Fig. 4 shows that with an identical primary delay of one minute after 12 stations (according to scenario 1) a delay of only four minutes is reached. Less than half of the scenario 1 results. This example shows that certain conditions, like purchasing new vehicles with more efficient doors can reduce the need for time reserves in the terminal loop and therefore the amount of vehicles can also be reduced.

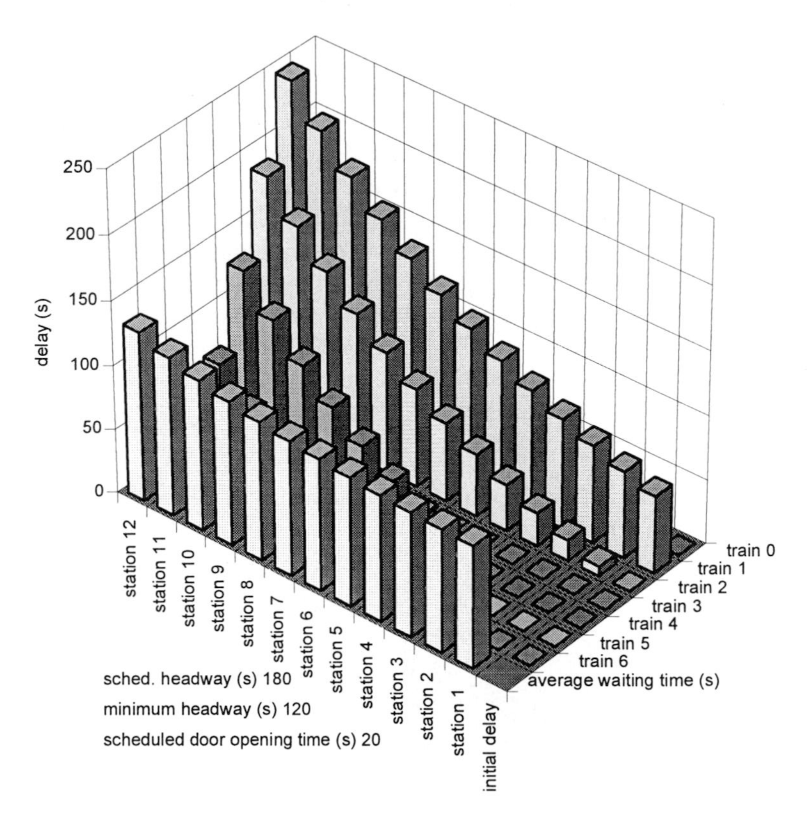

Fig. 4: Train delays with shorter door opening times

5.3 Scenario 3

In this scenario a minimum headway of only one minute is calculated in order to show the effect of time made up by the trains. Apart from this the basis is also a scheduled three minute Basic Interval and a planned door opening time of 30 seconds.

Because of a disturbance of the track all the trains of one line have a primary delay of 40 seconds.

In Fig. 5 the oscillating effect of the delays becomes obvious. While the trains 1, 3 and 5 add to their delays continually at every station, the following trains 2, 4 and 6 can even diminish their delays at the first stations. Through the growing delays of the preceding trains the headways to these trains become shorter as well as the amount of waiting passengers at the stations. Only when the minimum headway is reached does the delay increase for the trains 2, 4 and 6.

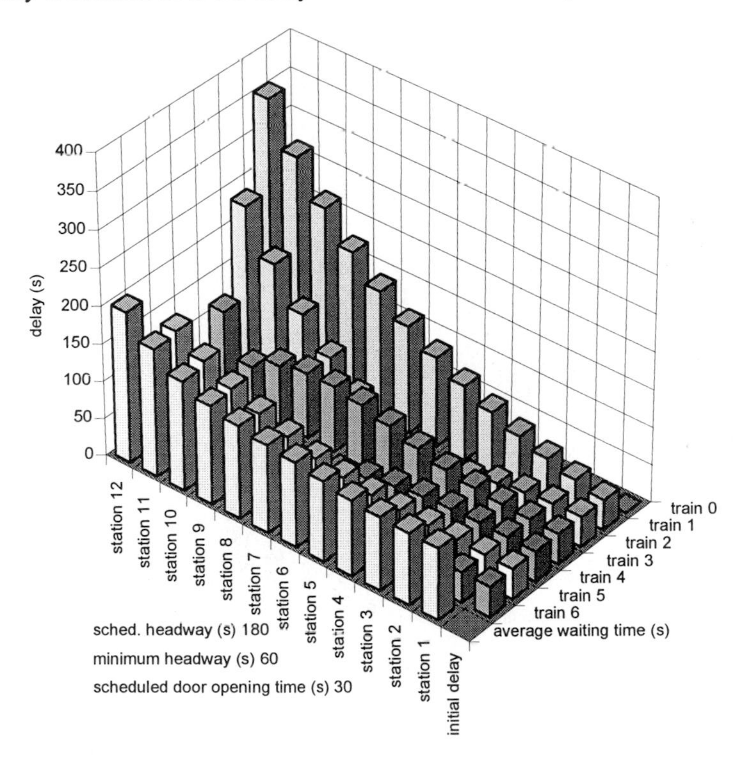

Fig. 5: Train delay with a primary delay for all trains

5.4 Scenario 4

In this scenario the aim is to show how even very small primary delays can have an effect on the waiting time of passengers on the platform.

Using the same basis as scenario 3 a primary delay of all trains of only 20 seconds is calculated. This has the effect that a few trains can diminish their initial delay after a few stations completely and operate punctually (see fig. 6). In spite of the successful making up of time the average waiting time at the stations rises continually.

Even if a small amount of delayed trains can be noticed in the operation statistics as a positive effect for the passengers, this does not necessarily mean a shortening of the average waiting time for passengers. Here such a simulation can be used to optimise operational strategies in terminating disturbances and so the service of the transport operator is made more attractive to the passengers.

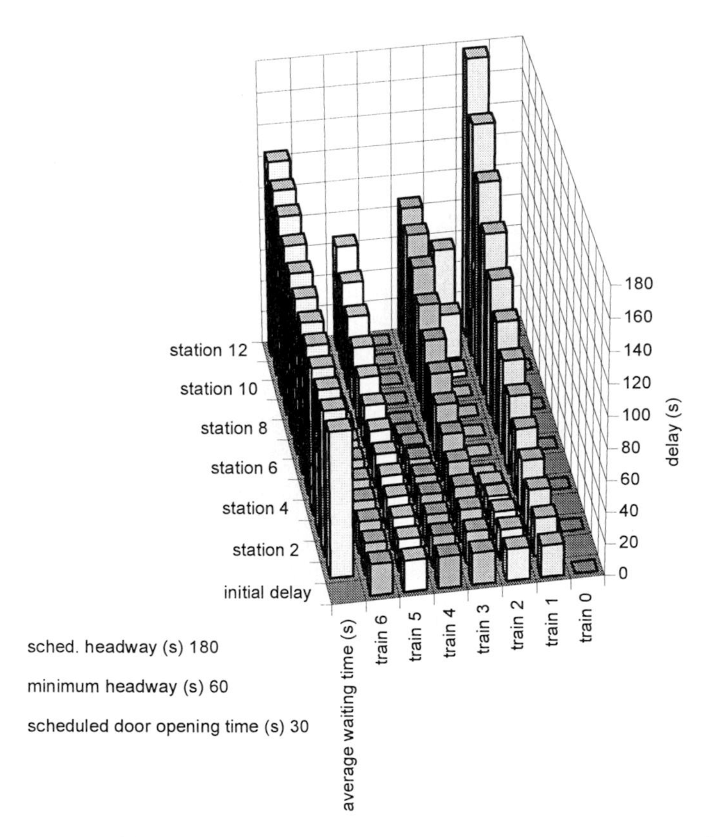

Fig. 6: Rise of the average waiting time in spite of the increasing amount of punctual trains

6 Limits of the Model

Of course track bound line operation cannot be entirely covered by a mathematical model with relatively simple basic conditions as made above.

The developed equations require a constant passenger flow F to the stations. This can be assumed with a scheduled headway of under 10 minutes, meaning that passengers do not plan their arrival time according to the timetable. For which headway's such a behaviour can be assumed depends on the timetable knowledge that the passengers have. Furthermore fluctuations in passenger flow are also possible with small headway's. For example it is possible at major events that passenger flow F may occasionally reach door efficiency C. Furthermore passenger flow varies from station to station.

The efficiency of the doors C is not constant in practice. It depends on the occupation grade of the train (i. e. heavily occupied trains with longer door opening time can still have a very low passenger exchange).

Initially, in the model only the boarding passengers are observed, as they increase through a delayed train. The influence of the disembarking passengers becomes apparent later when the additional boarding passengers leave the train and so delay it again after the average duration of a passenger journey.

How the accumulation of delays affects operation even after the end of the line depends mainly on the time reserve for the terminal loop. Here the question also arises if reserve trains are ready at the end of the line.

A consideration of all these aspects which restrict the model would complicate the mathematical model considerably. This is the reason why the model can only aim at examining some relevant characteristics of line operation.

7 Prospect

In spite of the restrictions listed above the developed model is suitable to demonstrate important aspects of line operation as they become clear in the chosen scenarios.

The aim of the simulation of disturbance scenarios is to optimise operation by means of terminating disturbances either through operation personnel or by implementing corresponding automatic systems.

This can have the effect of

- a facilitation for personnel,
- an increase of efficiency of operation and
- an increased attraction for the passenger.

The opportunity is provided to consider the possibilities of terminating disturbances especially while implementing new technologies, i. e. for automatic driverless operation.

References

Berg, W. (1982): Innerbetriebliche Gesetzmäßigkeiten des öffentlichen Linienbetriebes, Dissertation, ETH Zürich, Institut für Verkehrsplanung und Transporttechnik, IVT-Bericht 82/2

Newell, G. F. and R. B. Potts (1964): Maintaining a Bus Schedule, Proceedings Second Conference Australian Road Research Board, Vol. II Part 1, p. 388-393

Newell, G. F. (1974): Control of Pairing of Vehicles on a Public Transportation Route, Two Vehicles, One Control Point, Transportation Science, 74/3, p. 248-264

Osuna, E. E. and G. F. Newell (1972): Control Strategies for an Idealised Public Transportation System, Transportation Science, 72/1, p. 52-72

Part 4 Engineering Processes

The Application of the Hamiltonian Formalism by Means of Tensor Calculus to an Electrical Circuit

Ute Diemar

Technical University of Ilmenau, Division of Theoretical and Experimental Electrical Engineering, P.O. Box 100565, D-98684 Ilmenau, Germany

Abstract. *The contribution demonstrates the possibility to establish the Hamiltonian of an electrical system in two forms, dependent on a covariant- and contravariant, description of these systems, respectively to establish the canonical equations of motion. A basis system will be created for an electrical circuit, which is embedded in the Euclidean space.*

Keywords. *Hamiltonian, Lagrangian, dissipation function, electrical circuit, covariant contravariant coordinates, covariant momentum, contravariant momentum, canonical equations of motion, Euclidean space, Riemannian space, metric tensor, basis vectors.*

1 Introduction

The increasing interconnection between most of the various fields of sciences demands more general, as well as more interdisciplinary, describing and calculation methods, respectively. Specific methods of separate areas are not sufficient to describe all system components and their interactions completely.

In today's time of the miniaturisation of the most varied systems (microelectronic, nanoelectronic, micromechanic, mechatronic), the mutual interaction between system components cannot be neglected any more. Therefore, a theory is necessary, which on the one hand, covers different areas of sciences under common points of view and, on the other hand, produces the demanded equations of motion. The Lagrangian and the Hamiltonian formalism offers this possibility. The basis of this formalism is variational calculus. The energy balance of the regarded system constitutes the starting point of the examination and not the parameter of states. Systems of most different natures (electrical, electromechanical, mechanical, thermal, optical) and also their coupling are described by means of the Lagrangian and Hamiltonian formalism independent of the type of their energy structure.

Therefore, the problem to be examined has to be formulated as a variational function to use the Euler-Lagrangian differential equation to establish the equations of motion.

The application of the Lagrangian and Hamiltonian formalism offers a comprehensive possibility to handle technical systems under generalised points of view in its entirity. The classical methods are bound to the Euclidean space. The transfer of the Lagrangian and Hamiltonian formalism into the Riemannian space (it contains the Euclidean space as a special case) is possible without restrictions of the generality. It is necessary to change to the Riemannian space if the elements of the metric tensor are not constant but they are functions of the co-ordinates of the space. For this, it is necessary to go on to a co- and contra-variant way of looking at the parameters of state. The connection between the metric of the space, the energy balance, the Lagrangian and Hamiltonian formalism and the equations of motion are clearly explained in (Süße, 1994; Süße, 1996; Süße, 1997). The applicability of this extended theory is guaranteed by its generality. The universality of the formalism used will be demonstrated in this contribution. The metric will be determined on the basis of this formalism and a basis system will be derived for this metric. From this, it is apparent in which kind of space the mapping occurs (Euclidean or Riemannian space). The equations of motion are derived from the Hamiltonian of this electrical system and the solution is also calculated.

2 Basics

All technical calculations are based explicitly or implicitly on a particular geometry. In the majority of the cases, it is the Euclidean geometry. But it is not always advantageous to use the Euclidean space for all calculations. Special problems simplify considerably if they are calculated with the calculation in a suitable space. Now it is necessary to introduce a metric to make quantitative statements independent of the used one. A metric exists for the Euclidean space, which is valid for the whole space.[1] The Riemannian space is determined by the introduction of a manifold. The nonlinear metric is defined on a determined area of a manifold. A differentiable manifold has euclidean behaviour in the infinitesimally small limits. That means, a linearisation occurs in this way, which an Euclidean metric is given in every local vector space, so that the „law of the Pythagoras" is valid. Further more, that means that the Euclidean space is a special case of the Riemannian space. It is embedded in it. We do not want to go into the

---

[1] The metric is labelled by the so-called metric tensor, which elements are linear in the Euclidean space.

theory of the Euclidean and Riemannian space and, so refer to (Süße, 1997), p. 23-38.

As all calculus is bound to special spaces, so is the calculus of electrical systems, too. An electrical circuit is also bound to a defined space. If this space is measurable by a metric, a base $\boldsymbol{B}(\mathbf{g}_1, \mathbf{g}_2, \ldots, \mathbf{g}_n)$ exists and also a base $\boldsymbol{B}`(\mathbf{g}^1, \mathbf{g}^2, \ldots, \mathbf{g}^n)$. The base vector $\mathbf{g}^i$ is orthogonal to the base vector $\mathbf{g}_j$ of the original base vector with $j = 1, 2, \ldots, n$ and $i \neq j$ and the scalar product of $\mathbf{g}_i$ and $\mathbf{g}_i$ is one. The $\mathbf{g}_i$ are called co-variant base vectors of the co-variant base $\boldsymbol{B}$ and the $\mathbf{g}^i$ are called contra-variant base vectors of the contra-variant base $\boldsymbol{B}'$.

This will be proven for the circuit shown in Fig. 1. If a basis system ***(B, B')*** exists, it is determined by an electrical circuit embedded in the Euclidean space. In the case of non-linear elements of the metric tensor, the calculus is based on the facts of the Riemannian space. In the following consideration it is necessary to consider the parameters of state of the electrical circuit in co- and contra-variant forms. Two possible descriptions of the generalised Hamiltonian ***H*** ((Süße, 1997) pp. 54-57) follow from this fact. The relation between the generalised momenta are:

$$p_k = p_k(\dot{q}^\mu, q^\nu, t) = \frac{\partial L}{\partial \dot{q}^k} \quad and \quad p^k = p^k\,(\dot{q}_\mu\,, q_\nu\,, t) = \frac{\partial L}{\partial \dot{q}_k} \tag{1}$$

respectively, and the Hamiltonians are given by

$$\begin{aligned} H^+ &= H^+(p_k, \dot{q}^k, t) = p_i \dot{q}^i - L(\dot{q}^k, q^k, t) \\ H^- &= H^-(p^k, \dot{q}_k, t) = p^i \dot{q}_i - L(\dot{q}_k, q_k, t) \qquad i,k = 1,\ldots,f \quad . \end{aligned} \tag{2}$$

The classical Hamiltonian is a pure conservative function. The losses occuring in real systems are considered in the canonical equations of motion according to (Süße, 1996). A Lagrangian and dissipation function is established for each system element, which terms are reflected in the equations of motion. It is referred to (Süße, 1994; Süße, 1996; Süße, 1997) for the theory of the extended Hamiltonian formalism and the losses contained in it. Both forms of the Hamiltonian and the resulting equations of motion are established for the circuit in Fig. 1 in this contribution. Furthermore, the transfer of the Hamiltonian H^+ into the Hamiltonian H^- is demonstrated and assumes a regular metric.

3 Tensor Examination of the Electrical Circuit

A base in co- and contra-variant form will be established for the electrical circuit shown in Fig. 1 on the basis of the tensor calculus. The Hamiltonian in its two forms will be built and the canonical equations of motion are derived (Süße, 1997). The following circuit is given (Fig.1):

3.1 Determination of a Co- and Contra-Variant Base

The metric tensor of a system, in this case of an electrical circuit, is the starting point for setting up a base (Süße, 1997). For this, it is necessary to analyse the structure of the system.

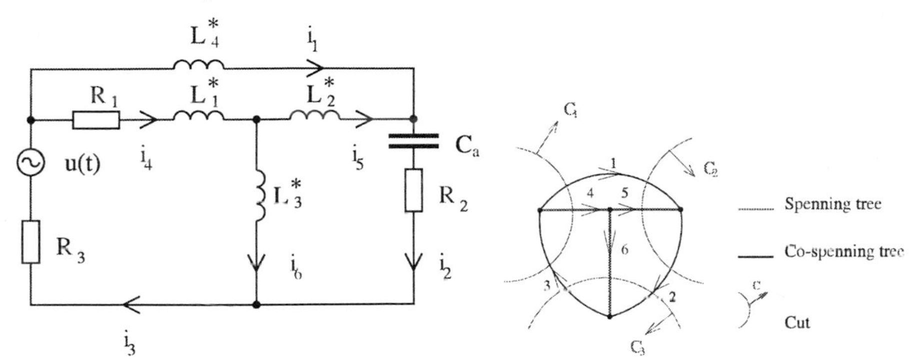

Fig. 1: RLC – circuit

Fig. 2: Graph of the circuit

The oriented graph of the circuit follows from Fig. 1 in Fig. 2. The incidence matrix of cutsets is

$$
\begin{array}{l} \text{branches} \quad 1 \quad 2 \quad 3 \quad 4 \quad 5 \quad 6 \\
\begin{pmatrix} 1 & 0 & -1 & 1 & 0 & 0 \\ 1 & -1 & 0 & 0 & 1 & 0 \\ 0 & 1 & -1 & 0 & 0 & 1 \end{pmatrix} \rightarrow \begin{pmatrix} i_4 = i_3 - i_1 \\ i_5 = i_2 - i_1 \\ i_6 = i_3 - i_2 \end{pmatrix} \end{array}
$$

It is possible to build the Lagrangian and the dissipation function for each system, which represents the so-called {L, D}-model. The dissipation function contains the losses of the system. The generalised momenta then result from the Lagrangian. The {L, D}-model is in the contravariant form:

$$
\begin{aligned}
L &= \frac{L_1^*}{2}(\dot{q}^4)^2 + \frac{L_2^*}{2}(\dot{q}^5)^2 + \frac{L_3^*}{2}(\dot{q}^6)^2 + \frac{L_4^*}{2}(\dot{q}^1)^2 - \frac{1}{2C_a}(q^2)^2 - u(t)q^3 \\
&= \frac{L_1^*}{2}(\dot{q}^3 - \dot{q}^1)^2 + \frac{L_2^*}{2}(\dot{q}^2 - \dot{q}^1)^2 + \frac{L_3^*}{2}(\dot{q}^3 - \dot{q}^2)^2 + \frac{L_4^*}{2}(\dot{q}^1)^2 - \frac{1}{2C_a}(q^2)^2 - u(t)q^3 \quad (3) \\
D &= \frac{R_1}{2}(\dot{q}^4)^2 + \frac{R_2}{2}(\dot{q}^2)^2 + \frac{R_3}{2}(\dot{q}^3)^2 = \frac{R_1}{2}(\dot{q}^3 - \dot{q}^1)^2 + \frac{R_2}{2}(\dot{q}^2)^2 + \frac{R_3}{2}(\dot{q}^3)^2 .
\end{aligned}
$$

The co-variant generalised momenta are then

$$p_k = \frac{\partial L}{\partial \dot{q}^k} \quad \rightarrow \quad \begin{aligned} p_1 &= (L^*{}_1 + L^*{}_2 + L^*{}_4)\dot{q}^1 - L^*{}_2\dot{q}^2 - L^*{}_1\dot{q}^3, \\ p_2 &= -L^*{}_2\dot{q}^1 + (L^*{}_2 + L^*{}_3)\dot{q}^2 - L^*{}_3\dot{q}^3, \\ p_3 &= -L^*{}_1\dot{q}^1 - L^*{}_3\dot{q}^2 + (L^*{}_1 + L^*{}_3)\dot{q}^3. \end{aligned} \tag{4}$$

The metric tensor follows from the momenta with

$$[g_{ij}] = \left[\frac{1}{K}\frac{\partial p_i}{\partial \dot{q}^j}\right] = \frac{1}{K}\begin{pmatrix} L^*{}_1 + L^*{}_2 + L^*{}_4 & -L^*{}_2 & -L^*{}_1 \\ -L^*{}_2 & L^*{}_2 + L^*{}_3 & -L^*{}_3 \\ -L^*{}_1 & -L^*{}_3 & L^*{}_1 + L^*{}_3 \end{pmatrix}. \tag{5}$$

The connection between the base and the metric tensor is independent of every technical condition

$$g_{ij} = g_i \cdot g_{j,} \tag{6}$$

and it follows that:

$$\begin{aligned} g_{11} &= (g_{1x})^2 + (g_{1y})^2 + (g_{1z})^2; \\ g_{12} &= g_{1x}g_{2x} + g_{1y}g_{2y} + g_{1z}g_{2z} = g_{21}; \\ g_{22} &= (g_{2x})^2 + (g_{2y})^2 + (g_{2z})^2; \\ g_{13} &= g_{1x}g_{3x} + g_{1y}g_{3y} + g_{1z}g_{3z} = g_{31}; \\ g_{23} &= g_{2x}g_{3x} + g_{2y}g_{3y} + g_{2z}g_{3z} = g_{32}; \\ g_{33} &= (g_{3x})^2 + (g_{3y})^2 + (g_{3z})^2. \end{aligned} \tag{7}$$

The comparison of Eq. (7) and Eq. (5) yields:

$$(g_{1x})^2+(g_{1y})^2+(g_{1z})^2=\frac{1}{K}(L^*_1+L^*_2+L^*_4),$$
$$(g_{2x})^2+(g_{2y})^2+(g_{2z})^2=\frac{1}{K}(L^*_2+L^*_3),$$
$$(g_{3x})^2+(g_{3y})^2+(g_{3z})^2=\frac{1}{K}(L^*_1+L^*_3), \qquad (8)$$
$$g_{1x}g_{2x}+g_{1y}g_{2y}+g_{1z}g_{2z}=-\frac{L^*_2}{K},$$
$$g_{1x}g_{3x}+g_{1y}g_{3y}+g_{1z}g_{3z}=-\frac{L^*_1}{K},$$
$$g_{2x}g_{3x}+g_{2y}g_{3y}+g_{2z}g_{3z}=-\frac{L^*_3}{K}.$$

There is an equation system of six equations with nine unknowns. From it follows that three quantities are freely eligible for election. These quantities are set to zero. In this case, it was advantageous to set $g_{2z}=g_{3y}=g_{3z}=0$. The solutions for the co-variant base vectors are

$$\vec{g}_1=\begin{pmatrix}\frac{L^*_1}{\sqrt{K}\sqrt{L^*_1+L^*_3}}\\ \frac{\sqrt{C}}{\sqrt{K}\sqrt{L^*_1+L^*_3}}\\ \frac{\sqrt{L^*_4}}{\sqrt{K}}\end{pmatrix};\quad \vec{g}_2=\begin{pmatrix}\frac{L^*_3}{\sqrt{K}\sqrt{L^*_1+L^*_3}}\\ \frac{\sqrt{C}}{\sqrt{K}\sqrt{L^*_1+L^*_3}}\\ 0\end{pmatrix};\quad \vec{g}_3=\begin{pmatrix}\frac{-\sqrt{L^*_1+L^*_3}}{\sqrt{K}}\\ 0\\ 0\end{pmatrix} \qquad (9)$$

under the conditions $L^*_1L_2^*+L^*_1L_3^*+L_2^*L_3^*\neq 0$; $K\neq 0$ and $L^*_1+L_3^*\neq 0$ (with $L^*_1L_2^*+L^*_1L_3^*+L_2^*L_3^*=C$). Because of

$$\vec{g}^{\,i}=g^{ij}\vec{g}_j \qquad (10)$$

it is possible to evaluate the contra-variant base vectors. For this, it is necessary to evaluate the inverse of the matrix of the coefficients of the metric tensor g_{ij}. It is

with $\Delta=L_4^*(L^*_1L_2^*+L^*_1L_3^*+L_2^*L_3^*)=L_4^*C$.

$$[g_{ij}]^{-1}=[g^{ij}]=\frac{K}{\Delta}\begin{pmatrix}C & C & C\\ C & C+L^*_1L^*_4+L^*_3L^*_4 & C+L^*_3L^*_4\\ C & C+L^*_3L^*_4 & C+L^*_2L^*_4+L^*_3L^*_4\end{pmatrix} \qquad (11)$$

with $\Delta = L_4^*(L^*_1 L_2^* + L^*_1 L_3^* + L_2^* L_3^*) = L_4^* C$.
The contra-variant base vectors $\mathbf{g}^1$, $\mathbf{g}^2$, $\mathbf{g}^3$ can be evaluated to

$$\vec{g}^1 = \begin{pmatrix} 0 \\ 0 \\ \dfrac{\sqrt{K}}{\sqrt{L^*_4}} \end{pmatrix} \quad ; \quad \vec{g}^2 = \begin{pmatrix} 0 \\ \dfrac{\sqrt{K}\sqrt{L^*_1 + L^*_3}}{\sqrt{C}} \\ \dfrac{\sqrt{K}}{\sqrt{L^*_4}} \end{pmatrix} \quad ; \quad \vec{g}^3 = \begin{pmatrix} \dfrac{\sqrt{K}}{\sqrt{L^*_1 + L^*_3}} \\ \dfrac{L^*_3\sqrt{K}}{\sqrt{C}\sqrt{L^*_1 + L^*_3}} \\ \dfrac{\sqrt{K}}{\sqrt{L^*_4}} \end{pmatrix} \tag{12}$$

with the help of (10) and (11) and under consideration of the summation convention of Einstein.

In order to check the results, the laws of the tensor calculus are used.

1. The multiplication of the corresponding base vectors supplies the coordinates of the contra-variant metric tensor

$$\vec{g}^i \cdot \vec{g}^j = g^{ij}. \tag{13}$$

2. The contra-variant and co-variant base vectors are related by

$$\vec{g}^i \cdot \vec{g}_j = \delta^i{}_j \qquad ; \qquad i, j = 1,\ldots,3. \tag{14}$$

Both conditions are fulfilled.

3.2 Setting Up the Hamiltonian H^+

The Hamiltonian H^+ will now be established by means of tensor calculus. H^+ follows with Eq. (2)

$$H^+ = H^+(p_k, q^k) = p_1\dot{q}^1 + p_2\dot{q}^2 + p_3\dot{q}^3 - L(\dot{q}^k, q^k, t). \tag{15}$$

We obtain with eq. (4)

$$H^{+}=\frac{1}{2}\begin{bmatrix}(L^{*}{}_{1}+L^{*}{}_{2}+L^{*}{}_{4})(\dot{q}^{1})^{2}-L^{*}{}_{2}\dot{q}^{1}\dot{q}^{2}-L^{*}{}_{1}\dot{q}^{1}\dot{q}^{3}\\-L^{*}{}_{2}\dot{q}^{2}\dot{q}^{1}+(L^{*}{}_{2}+L^{*}{}_{3})(\dot{q}^{2})^{2}-L^{*}{}_{3}\dot{q}^{2}\dot{q}^{3}\\-L^{*}{}_{1}\dot{q}^{3}\dot{q}^{1}-L^{*}{}_{3}\dot{q}^{3}\dot{q}^{2}+(L^{*}{}_{1}+L^{*}{}_{3})(\dot{q}^{3})^{2}\end{bmatrix}$$

$$+\frac{1}{2C_{a}}(q^{2})^{2}+u(t)q^{3} \tag{16}$$

and

$$H^{+}=\frac{K}{2}g_{ij}\dot{q}^{i}\dot{q}^{j}+\frac{1}{2C_{a}}(q^{2})^{2}+u(t)q_{3,} \tag{17}$$

respectively. The Hamiltonian depends on the generalised momenta and coordinates and follows in accordance with the relation

$$\frac{K}{2}g_{ij}\dot{q}^{i}\dot{q}^{j}\equiv\frac{1}{2K}g^{lm}\dot{p}_{l}\dot{p}_{m} \tag{18}$$

now to

$$H=H^{+}(p_{k},q^{k},t)$$

$$=\frac{1}{2K}\frac{K}{\Delta}\begin{pmatrix}C & C & C\\ C & C+L_{1}^{*}L_{4}^{*}+L_{3}^{*}L_{4}^{*} & C+L_{3}^{*}L_{4}^{*}\\ C & C+L_{3}^{*}L_{4}^{*} & C+L_{2}^{*}L_{4}^{*}+L_{3}^{*}L_{4}^{*}\end{pmatrix}p_{i}p_{j}$$

$$+\frac{1}{2C_{a}}(q^{2})^{2}+u(t)q^{3}. \tag{19}$$

3.3 Setting Up the Canonical Equations of Motion by Means of the Variational Derivation

The canonical equations of motion can be derived from the Hamiltonian. The generalised equations are

$$\dot{q}^{i}=\frac{\partial H^{+}}{\partial p_{i}}\qquad;\qquad\dot{p}^{i}=-\frac{\partial H^{+}}{\partial q_{i}}-\frac{\partial D^{i}}{\partial\dot{q}} \tag{20}$$

and for the example of Fig. 1

$$
\begin{aligned}
\dot{q}^1 &= \frac{1}{L^*_4}(p_1 + p_2 + p_3), \\
\dot{q}^2 &= \frac{1}{\Delta}[(C + L^*_1L^*_4 + L^*_3L^*_4)p_2 + Cp_1 + (C + L^*_3L^*_4)p_3], \\
\dot{q}^3 &= \frac{1}{\Delta}[(C + L^*_2L^*_4 + L^*_3L^*_4)p_3 + Cp_1 + (C + L^*_3L^*_4)p_2]
\end{aligned}
\tag{21}
$$

and

$$
\begin{aligned}
\dot{p}_1 &= R_1(\dot{q}^3 - \dot{q}^1) = \frac{R_1}{\Delta}[L^*_3L^*_4p_2 + (L^*_2L^*_4 + L^*_3L^*_4)p_3], \\
\dot{p}_2 &= -\frac{1}{C_a}q^2 - R_2\dot{q}^2 \\
&= -\frac{1}{C_a}q^2 - \frac{R_2}{\Delta}[Cp_1 + (C + L^*_1L^*_4 + L^*_3L^*_4)p_2 + (C + L^*_3L^*_4)p_3, \\
\dot{p}_3 &= -R_1(\dot{q}^3 - \dot{q}^1) - R_3\dot{q}^3 - u(t) \\
&= -\frac{R_1}{\Delta}[L^*_3L^*_4p_2 + (C + L^*_2L^*_4 + L^*_3L^*_4)p_3] \\
&\quad - \frac{R_3}{\Delta}[Cp_1 + (C + L^*_3L^*_4)p_2 + (C + L^*_2L^*_4 + L^*_3L^*_4)p_3] - u(t).
\end{aligned}
\tag{22}
$$

3.4 Solution of the Equations of Motion and its Graphical Representation

Now it is possible to evaluate all branch currents and voltages of the circuit elements. Eq. (23) represents a system of ordinary differential equations of second order. It is solvable very quickly by means of known methods and computers, respectively (e.g. computer program „mathematica" (Wolfram, 1997). The simultaneous differential equations

$$\dot{p}_1 = (L^*{}_1 + L^*{}_2 + L^*{}_4)\ddot{q}^1 - L^*{}_2\ddot{q}^2 - L^*{}_1\ddot{q}^3 = R_1(\dot{q}^3 - \dot{q}^1),$$
$$\dot{p}_2 = -L^*{}_2\ddot{q}^1 + (L^*{}_2 + L^*{}_3)\ddot{q}^2 - L^*{}_3\ddot{q}^3 = -\frac{1}{C_a}q^2 - R_2\dot{q}^2, \qquad (23)$$
$$\dot{p}_3 = -L^*{}_1\ddot{q}^1 - L^*{}_3\ddot{q}^2 + (L^*{}_1 + L^*{}_3)\ddot{q}^3 = -R_1(\dot{q}^3 - \dot{q}^1) - R_3\dot{q}^3 + u(t)$$

offers as solution for the branch currents i_1, i_2, i_3 with $L^*_1 = 5.0$ mH, $L_2^* = 3.0$ mH $L_3^* = 2.0$ mH, $L_4^* = 6.0$ mH, $C = 30.0$ nF, $R_1 = 1.0$ kΩ, $R_2 = 500.0$ Ω, $R_3 = 100.0$ Ω, u(t) = 10.0 V sin(2 π 50 Hz t):

$$i_1 = -6.25 \cdot 10^{-4}\, mA - 2.5mAe^{2.810^{-7}\frac{1}{s}t} + 2.5mA\cos[100\pi\, Hzt]$$
$$+ 68.0mA\sin[100\pi\, Hzt],$$

$$i_2 = 4.85 \cdot 10^{-8}\, mA + 7.40 \cdot 10^{-7}\, mAe^{2.810^{-7}\frac{1}{s}t}$$
$$- 7.88 \cdot 10^{-7}\, mA\cos[100\pi\, Hzt] - 50.26nA\sin[100\pi\, Hzt], \qquad (24)$$

$$i_3 = 4.07\mu A - 2.3mAe^{2.810^{-7}\frac{1}{s}t} + 2.3mA\cos[100\pi\, Hzt]$$
$$+ 60.2mA\sin[100\pi\, Hzt].$$

The remaining currents and the voltages over the circuit elements are calculated by means of the Kirchhoff's laws. Fig. 3 shows the currents i_1 to i_6.

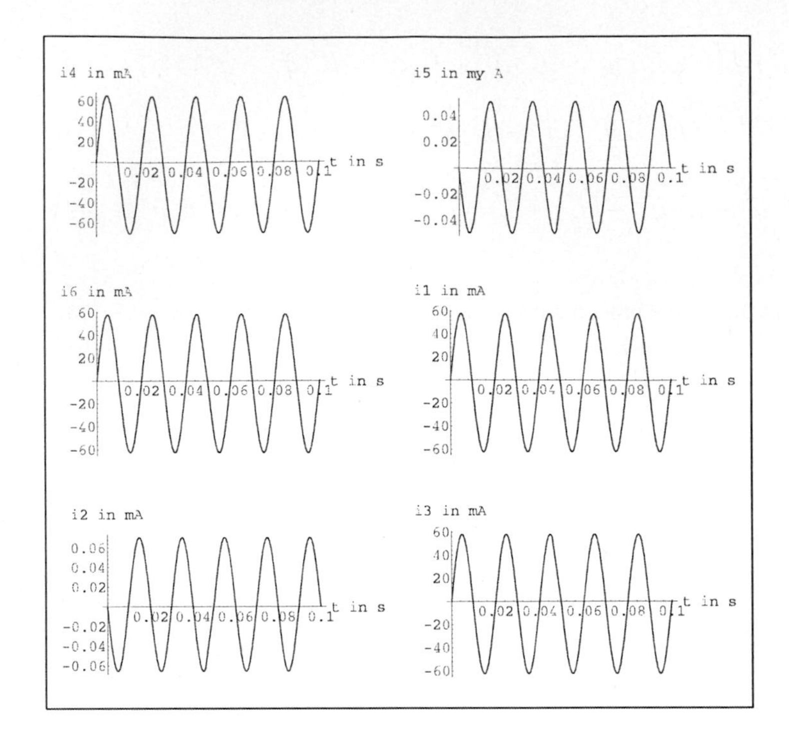

Fig. 3: Time dependence of the currents $i_1 - i_6$ established by means of the Hamiltonian H^+

3.5 Setting Up the Hamiltonian H^-

It is possibe to set up the second form of the Hamiltonian H^- by Eq. (2) in an elegant way by using the tensor calculus. The *{L, D}* - model is given by Eq. (3), dependant on contra-variant co-ordinates. The Lagrangian and the dissipation function are converted, dependant on co-variant co-ordinates, by means of the relation (3)

$$\dot{q}^i = g^{ij}\dot{q}_j. \tag{25}$$

For this, it is necessary to invert the coefficents of the metric tensor (to obtain g^{ij}). The Hamiltonian H^- is then:

$$H^{-} = H^{-}(\dot{q}_k, q_k, t) = \frac{K}{2} g^{ij} \dot{q}_i \dot{q}_j + \frac{1}{2C_a}(g^{21}q_1 + g^{22}q_2 + g^{23}q_3)^2$$
$$+ u(t)(g^{31}q_1 + g^{32}q_2 + g^{33}q_3) \ . \tag{26}$$

With the tensor relation

$$\frac{K}{2} g^{ij} \dot{q}_i \dot{q}_j = \frac{1}{2K} g_{lm} p^l p^m \tag{27}$$

we obtain the Hamiltonian H $^{-}$, dependant on the generalised momenta

$$H^{-} = H^{-}(p^k, q_k)$$
$$= \frac{1}{2K^2} \begin{bmatrix} L^*_1 + L^*_2 + L^*_4 & -L^*_2 & -L^*_1 \\ -L^*_2 & L^*_2 + L^*_3 & -L^*_3 \\ -L^*_1 & -L^*_3 & L^*_1 + L^*_3 \end{bmatrix} p^i p^j \tag{28}$$
$$+ \frac{1}{2C_a}(g^{21}q_1 + g^{22}q_2 + g^{23}q_3) + u(t)(g^{31}q_1 + g^{32}q_2 + g^{33}q_3).$$

The setting up of the equations of motion is carried out done by means of the canonical equation in the corresponding form

$$\dot{q}_i = \frac{\partial H^{-}}{\partial p^i} \quad , \quad \dot{p}^i = -\frac{\partial H^{-}}{\partial q_i} - \frac{\partial D}{\partial \dot{q}_i}. \tag{29}$$

4 Conclusion

The Hamiltonian formalism is applied to electrical circuits on the foundation of the tensor calculus. The advantages of the tensor and variational calculus are purposefully used. A distinction in co- and contra-variant quantities is necessary, because there is a non-cartesian base. The resulting two forms of the Hamiltonian are transferable onto each other by means of the metric tensor.

We obtain a Riemannian space from an Euclidean space, if a system is subject to defined constraints. That means, a system with n variable system quantities and z constraints is mapped on a $(n - z)$ – dimensional Riemannian space. The current-

and mesh-equations, especially of an electrical system, represents these constraints. They are connected directly with the degree of freedom of the system. The expenditure of the calculation by means of classical methods is increased with the increasing degree of freedom. The increasing degree of freedom does not cause problems regarding its applicability by the Lagrangian and Hamiltonian formalism. There is an Euclidean metric, as long as the metric tensor consists of constant co-ordinates. It is possible that a non-Euclidean metric can be devired, if the circuit contains non-linear elements. The consideration of co- and contra-variant quantities is then necessary, e.g. the created grid should be adapted to the geometry of the examined area for the calculus of fields. Then the Euclidean space is left. This is a further advantage of this explained method.

The Hamiltonian H^{+} is given preference over H^{-} by the consideration of generalised systems by means of the two forms of the Hamiltonian, because the set up of the energy and dissipation balance generally occurs in contravariant co-ordinates and therefore the Hamiltonian is dependant on contra-variant co-ordinates and furthermore, it is dependent on the co-variant momenta.

5 References

Calcin, M.G. (1996): Lagrangian and Hamiltonian Mechanics; World Scientific Publishing Co Pte Ltd; Singapore; 1996

Rolewicz, St. (1985): Metric Linear Spaces. Mathematics and its Application (East European Series); D. Reidel Publishing Company; Dordrecht/Boston/Lancaster; 1985

Seppälä, M.; Sorvali, T. (1992): Geometry of Riemann Surfaces and Teichmüller Spaces; North-Holland Mathematics Studies 169; North-Holland - Amsterdam London New York Tokyo.1992

Süße, R.; Marx, B. (1994): Theoretische Elektrotechnik, vol. 1: Variationsrechnung und Maxwellsche Gleichungen. B.I. Wissenschaftsverlag, Mannheim, 1994. Now Springer-Verlag, Heidelberg, 1997.

Süße, R.; Diemar, U.; Michel, G. (1996): Theoretische Elektrotechnik, vol. 2: Netzwerke und Elemente höherer Ordnung; VDI-Verlag GmbH; Düsseldorf; 1996; Now Springer-Verlag; Heidelberg; 1997

Süße, R.; Kallenbach, E.; Ströhla, T. (1997): Theoretische Elektrotechnik; vol. 3: Analyse und Synthese elekrotechnischer Systeme; Wissenschaftsverlag Ilmenau; Ilmenau; 1997

Wolfram, S. (1992): Mathematica, Ein System für Mathematik auf dem Computer; Addison-Wesley-Verlag GmbH; 1992

Determination of the Magnetic Stray Field with an Equivalent Source Model

Eugen Nolle

Blum GmbH, Erich-Blum-Str. 33, 71665 Vaihingen/Enz

Abstract. *Based on the equivalence of current-carrying circuit-loop and magnetic dipole density the field of a current-carrying air-core coil is approximated by the field of two magnetic equivalent charges which are to be attached to the winding axis at the distance of the coil ends. Thereafter the influence of a highly permeable core on the equivalent charges of the air-core coil are indicated with a known approximation equation, then the final equivalent source model of an EI-shaped small transformer is developed. In the final example the maximum stray field of a small transformer, which – as expected – occurs along the winding axis, is estimated and compared to concrete measured values.*

Keywords. *small transformer, stray field, equivalent source model*

1 Introduction

With the simulation systems available at present electromagnetic fields can be calculated with almost arbitrary exactness. The advantage of such a method is that complex structures can also be simulated by taking into account non-linear characteristics and connections.

In applying this method to magnetic circuits with a high percentage of iron, however, these are often grave differences between calculation and measurement results. In most cases attempts are made to improve the calculation by

– more numerical efforts and / or

– by 'suitable', i.e. measurement-adjusted magnetisation curvatures.

As a staff member of a firm which processes around 20 000 tons of various tranformer sheets per year and as a user of a commercial field calculation program I am observing with a certain astonishment how research institutes and universities proudly squeeze the seventh digit behind the decimal point out of their computers without even mentioning, let alone taking into account, the typical spread width of the magnetic characteristics of transformer sheets which is up to

30 percent. (In non-standardised magnetic materials, such as St 1203 or similar materials, there is even more scattering as a rule.)

In view of all this I am now introducing a model for the determination of the magnetic stray field of small transformers which yields the physical connections as formulas with appropriate exactness which are useful for engineers.

2 Equivalent Sources of a Current-Carrying Air-Core Coil

These connections can be shown particularly clearly with an EI-shaped small transformer in no-load operation, which is the operational status with the largest stray field.

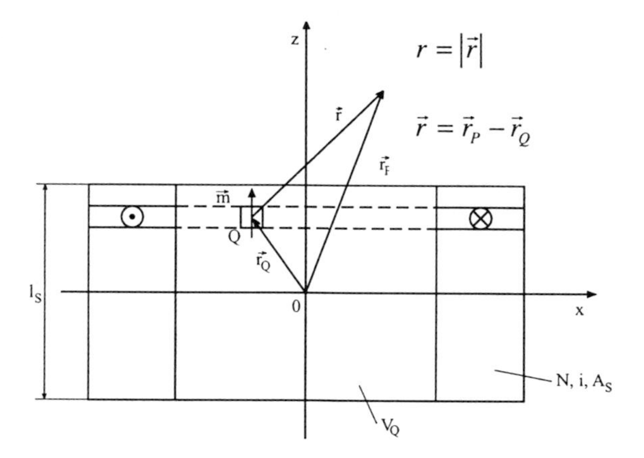

Fig. 1: Primary coil of a small transformer without core

Proceeding step by step, and, according to fig. 1, I only consider the primary coil with the magnetisation current i(t) with:

A_S – effective coil cross-section

l_S – coil length and

N – primary number of turns.

Imagine them symmetrically arranged in the origin of a coordinate system with the winding axis being identical with the z-axis. The primary coil usually consists of several windings out of which I symbolically marked one elementary winding. In 'electromagnetic theory' it is shown that the magnetic field of a current-carrying circuit loop can be presented as the magnetic field of a homogeneous double-layer which is distributed over an area of arbitrary position over the conducting contour, i.e. for instance on the marked plane for example. Taking the positive counter advance sense for the current into account, the primary coil as a whole can be replaced by a homogeneous dipole distribution inside of the coil (cross-section: A_S and length: l_S) of the density

$$\vec{m} = \vec{e}_z \frac{\mu_0 Ni}{l_S} \tag{1}$$

The marked volume element dV_Q then yields with the spacing vector

$$\vec{r} = \vec{r}_P - \vec{r}_Q \qquad \text{and} \qquad r = |\vec{r}|$$

for the magnetic scalar potential in the base P the contribution

$$dV_m(P) = \frac{\vec{m}}{4\pi\mu_0} \underbrace{\frac{\vec{r}}{r^3}}_{= -grad\frac{1}{r}} dV_Q = \frac{\vec{m}}{4\pi\mu_0} \overbrace{grad_Q \frac{1}{r}}^{= grad_Q \frac{1}{r}} dV_Q \tag{2}$$

Thus one can find the resulting potential by summing up the elementary contributions, i.e. by integration over the coil interior to

$$V_m(P) = \frac{\vec{m}}{4\pi\mu_0} \iiint_{V_Q} grad_Q \frac{1}{r} dV_Q = \frac{\vec{m}}{4\pi\mu_0} \oint\!\!\!\oint_{Rd\{V_Q\}} \frac{1}{r} d\vec{F}_Q \tag{3.1}$$

$$= \frac{m}{4\pi\mu_0} \left\{ \iint_{F_1} \frac{\vec{e}_z d\vec{F}_Q}{r} + \iint_{F_2} \frac{\vec{e}_z d\vec{F}_Q}{r} + \iint_{F_3} \frac{\vec{e}_z d\vec{F}_Q}{r} \right\} \tag{3.2}$$

$$= \frac{m}{4\pi\mu_0} \left\{ \iint_{F_1} \frac{1}{r} dF - \iint_{F_2} \frac{1}{r} dF \right\} \tag{3.3}$$

with

$$\vec{e}_z\, d\vec{F}_Q = dF_1 = dF \qquad \text{on } F_1$$

$$\vec{e}_z\, d\vec{F}_Q = -dF_2 = -dF \qquad \text{on } F_2$$

$$\vec{e}_z\, d\vec{F}_Q = 0 \qquad \text{on } F_3$$

where the volume integral, using the modified Gauss' divergence theorem, is at first re-transformed into an envelope surface integral and then evaluated taking the constant dipole direction into account. This intermediate result presents the potential of two homogeneous surface charges ±m, which according to figure 2 are distributed on the surfaces F_1 or F_2, respectively. With regard to multipole development (equivalent sources) mainly the charge sums and their centers are of interest. The first are yielded in an elementary way to $\pm Q_m{}^*$ with

$$Q_m{}^* = m\, A_S = \frac{\mu_0\, Ni\, A_S}{l_S}, \tag{4}$$

while the latter necessarily lies in F_1 or F_2, respectively, on the winding axis.

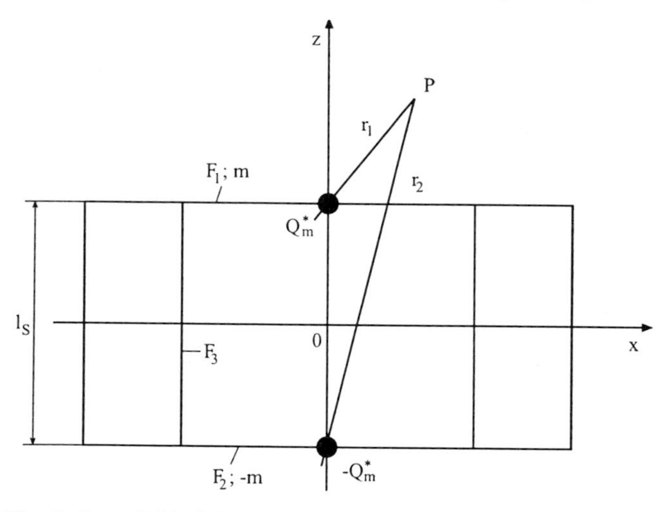

Fig. 2: Stray field of the primary coil, approximated by equivalent charges

As the result of the first step: The primary winding of a transformer without core behaves as two magnetic point charges $\pm Q_m{}^*$ which are to be attached to the winding axis at the distance of the coil ends $z_Q{}^* = \pm l_S/2$.

3 Equivalent Sources of an Iron-Core Coil

Let me now come to the second step which, other than the mainly mathematically exact first step, is based on an approximate relation which has been tested over many years. According to Meinke/Gundlach: 'Taschenbuch der Hochfrequenztechnik' the inductance of an air core coil increases by inserting a permeable core as in figure 3 by a factor of

$$\mu_w \approx 1 + \left(\frac{0{,}45 + \frac{l_K}{D_K}}{0{,}45 + \frac{l_K}{\mu_{rK} D_K}} - 1 \right) \cdot \left(\frac{D_K}{D_S} \right)^2 \approx 1 + 2 \frac{l_K \cdot \sqrt{A_{Fe}}}{A_S}, \tag{5}$$

where the second approximation is true for unsaturated transformer sheets with $\mu_{rK} >> 1$ and is generalised for arbitrary core and coil cross-sections.

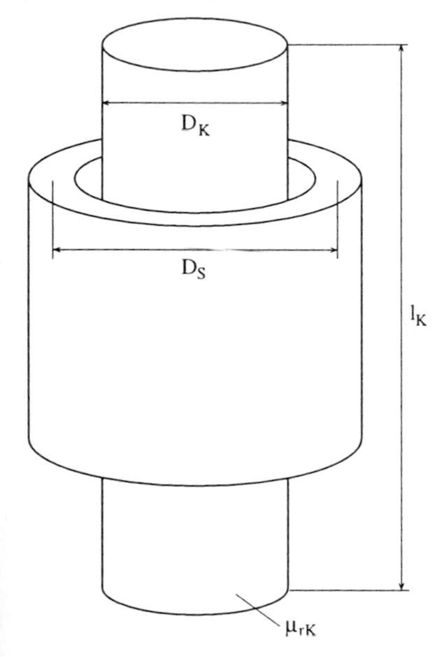

Fig. 3: Increase in the stray field of an air core coil dure to a permeable core.

$$Q_m = \mu_w Q_m^* \approx \left(1 + 2 \frac{l_K \sqrt{A_{Fe}}}{A_S} \right) \frac{\mu_0 N i A_S}{l_S}, \tag{6}$$

Because of the proportionality between the magnetic flux Φ and inductance L at a given current i in the point charges Q_m* of the primary coil, they increase due to the iron core, as well as by the factor μ_w which at the same time shift then along the winding axis to the core suface with $zQ = \pm l_K/2$.

$$V_m(P) = \frac{Q_m}{4\pi\mu_0}\left(\frac{1}{R_1} - \frac{1}{R_2}\right) \tag{7}$$

4 Equivalent Source Model of an EI-Shaped Small Transformer

Thus, figure 4 presents a simple model of the stray field of a small transformer which takes into account all relevant factors, such as winding data, coil and core dimensions, as quantities. Even the various types of transformer sheets and the precision in core fabrication (air gap, etc.) are integrated with the open-circuit current as company-specific empirical values.

Denoting the spacing between the equivalent charges $\pm Q_m$ and an arbitrary point P in the exterior space of the transformer by R_1 and R_2, its stray field can be approximated in a well-known way with the magntic scalar potential of two point charges or directly with their magentic flux density

$$\vec{B}_\sigma(P) = \mu_0 \vec{H}_\sigma = -\mu_0 \, grad \, V_m \tag{9}$$

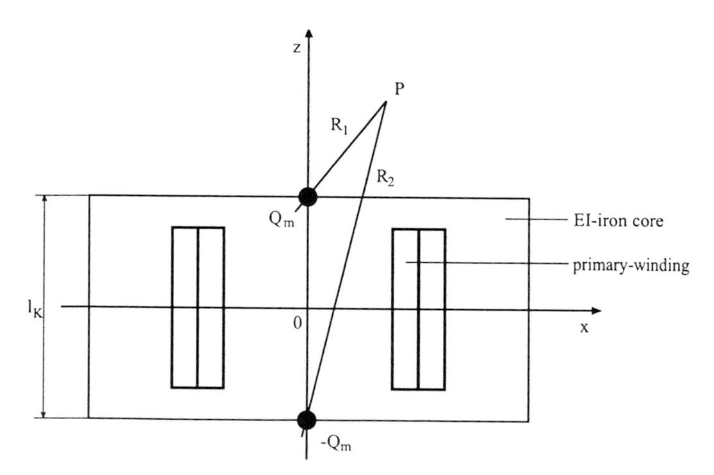

Fig. 4: Equivalent charges for the stray field of a small transformer

5 Axial Fields of an EI-Shaped Small Transformer in No-Load Operation

In such problems mainly the maximum stray field is interesting, f.e. for preventive health care or in order to avoid electromagnetic disturbances. In the EI small transformers this occurs in the extended winding axis, i.e. in the z-axis. With $x = y = 0$ and $|z| < l_K/2$ the general relations can be simplified as

$$V_m(z) = \frac{Q_m}{4\pi\,\mu_0}\left(\frac{1}{\left|z - \frac{l_K}{2}\right|} - \frac{1}{\left|z + \frac{l_K}{2}\right|}\right) \tag{10}$$

with V_m being an odd function and $B_{\sigma z}$ being an even function in z.

$$\vec{B}_\sigma(z) = \vec{e}_z\,\frac{Q_m}{4\pi}\cdot\left|\frac{1}{\left(z - \frac{l_K}{2}\right)^2} - \frac{1}{\left(z + \frac{l_K}{2}\right)^2}\right|, \tag{11}$$

The latter is shown in figure 5 for a medium-sized small transformer (EI 78 = 250 VA), where, in order to check the model, the calculation was made with the measured peak value of the open-circuit current. Moreover, measured values are indicated for two clearly-defined spacings, they were measured on the above mentioned transformer with a field measurement probe according to DIN 454110 (German Industrial Standard).

It is obvious that the introduced model yields useful calculation values with less than 10 percent deviation from the measured value at comparatively small spacings in a magnitude of the transformer dimensions.

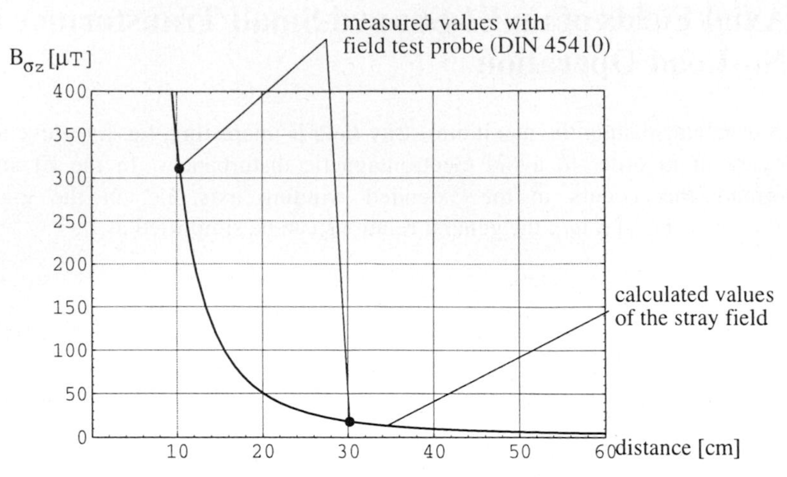

Fig. 5: Measured and calculated values of the stray field of a small transformer for points on the extended winding axis

The insecurity mentioned at the beginning due to the magnetic characteristics of the transformer sheets as well as due to fabricational influences mainly concerns the open-circuit current, however, it has a full effect on the length of the stray field: $B_\sigma \sim i_0$. Therefore such statements on the strayfield of small transformers in no-load operation can never be more exact than statements on the open-circuit current.

Everybody who knows the spread width of the open-circuit currents of small transformers in the series will understand my astonishment in view of the improper use of field calculation programmes for such problems and learn to value the models of the often complex reality.

6 Remarks

1) The stray field of small transformers follows the strongly distorted open circuit current and not the impressed, approximately sinusoidal main flux.

2) For other core types similarly simple equivalent charge configuration can be found, f.e. for UI cores as quadrupole or for 3UI cores as three time shifted, spaced dipoles.

3) Disturbances of the stray field by neighbouring permeable and / or conducting planes can be taken into account in approximation with the partial reflection method.

The Development of a Process Model in the Energy Supply System

Janet Nagel

RISA Sicherheitsanalysen GmbH, Krumme Str. 55, 10627 Berlin, Germany

Abstract. *Process models in the energy supply system are frequently used to mathematically delineate the interrelation between components (i.e. consumers, generating plants and heat distribution systems) within this system. Various process models can be deployed depending on the type of approach in solving contentious questions in order to explain by example, the system assumed in the energy distribution system. As far as plant upgrading is concerned, a specific model will have to be developed to illustrate on the one hand the flow of energy between individual components within the system, and one which allows for a definite regulation of the expenses incurred from the installation of generating equipment and distribution systems on the other. It is imperative that certain procedures be simplified, go that a computer-aided solution to inherent problems can be found. The process model described here allows us to simultaneously compare all possible energy supply options by computer therefore, the best economic and ecological option can be ascertained. This process model thus represents one alternative to the method deployed for embracing particular variants which allows for only one option to be computed at any one time.*

Keywords. *energy model, process model, mixed-integer-linear-optimization, calculation of dynamic economic profitability, sensitivity analysis, scenario, energy supply, daily load operation, operating company*

1 Introduction

Process models are deployed in the energy supply sector to delineate strategies for upgrading power stations as well as co-generation plants. We have to differentiate between dynamic and static models. In the case of dynamic models, any upgrading is effected in stages (Hamann, 1969) whereas static models only permit upgrades to be effected in a particular instance. (Egberts, 1979; Fendt, 1979; Hanfler, 1990; Readwin, 1985) Factors such as the technological status of a plant, its output efficiency and in some cases the actual site of a plant may play an important role. We can also provide answers to questions concerning the replacement of existing plants. Specific plans of action depict another area in which process models can be deployed. (Maubach, 1994; Pang, 1976; Riedel, 1988) In this case we can calculate how to deploy existing plants most cost-effectively within the energy supply system in order to keep up with the demand for electricity and/or heating requirements over a 24 hour period. In order to do so we must mathematically delineate start up and shut down procedures and the levels of energy required.

Process models in the energy sector generally depict the interrelation between the demands for energy by consumers in towns and cities as well as energy conversion plants including the provision of energy to consumers wherever they may be. The establishment of a model for regulating strategies to upgrade plants which enable them to use biomass as an energy source for a specific period can be seen below. Biomass is a special kind of energy source, as it can be produced by the consumer. Technologically, biomass is mostly deployed for producing thermal energy for economic and technological reasons. The heat produced can be deployed in locally situated heating plants as well as in central heating plants and district heating power stations. We will now go into this in more detail.

2 Problems and Difficulties

If we look at the economic potential of biomass in the production of energy we can see that there are various questions which aris. These can be answered using a process model which can be transformed into a system of mathematical equations and subsequently fed into the computer. These questions include:

1. Is the potential supply of biomass sufficient for any economic deployment?
2. Which technologies and/or energy hook-ups could deploy biomass as a source of energy in contrast to fossil fuels?
3. Which external political, ecological and economic conditions should be in force, e.g. in relation to fuel prices and subsidies?
4. What effect will the deployment of biomass have on the supply structures of a particular community?

The energy supply system must be considered with these points in mind. As we initially stated in our introduction, a system for supplying power is comprised of various components such as consumers, generating plants and distribution systems. These components are united by the thermal flow which is directed to every individual consumer (GB) from locally situated thermal stations (HA) and central heating plants (HW) at location S1 via supply lines NTHW, NKN and domestic hook-up points (Fig. 1). These heat flows are divided up at each of the nodal points KN and must be reproduced, not forgetting that the heating requirements of the consumer regulate the initial dimensions of the system.

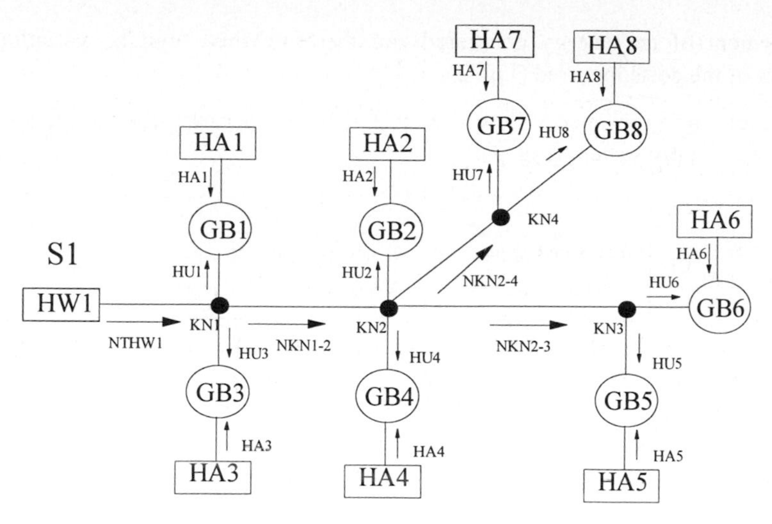

Fig. 1: Factual illustration of an energy supply system (excerpt)

Joint management operating companies will be created to operate central generating plants. These may take varying forms. On the one hand, a community may pool its resources and form an operating company (community TH), whereas, on the other, investors perhaps in the guise of farmers, for example, act as operating companies themselves (investor PA). Farmers can then produce biomass

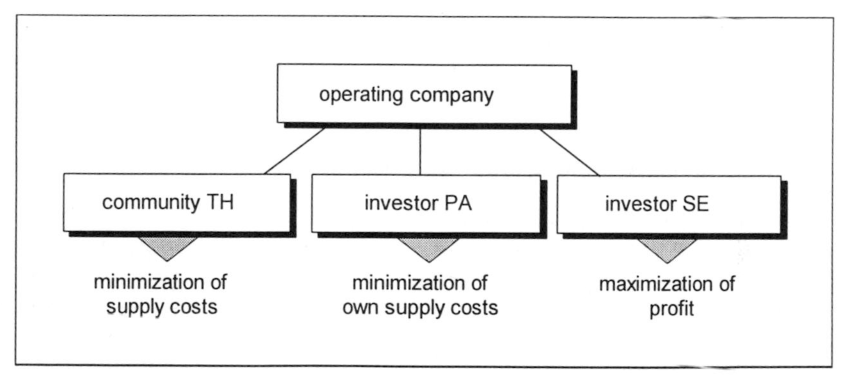

Fig. 2: Types of operating companies

themselves and deploy it to cover their own energy requirements; they can sell any surplus to local consumers if the economics of the operation are sound. Finally, investors who do not require any use of the energy generated (investor SE) may exist in the form of power-supply companies and their sole purpose would be to sell any energy produced. All three options have different aims as regards the deployment of the energy generated and therefore there will be variations in respect of the costs incurred (Fig. 2).

In addition to this energy flow, we must also consider cash-flow aspects within the power supply system (Fig. 3).

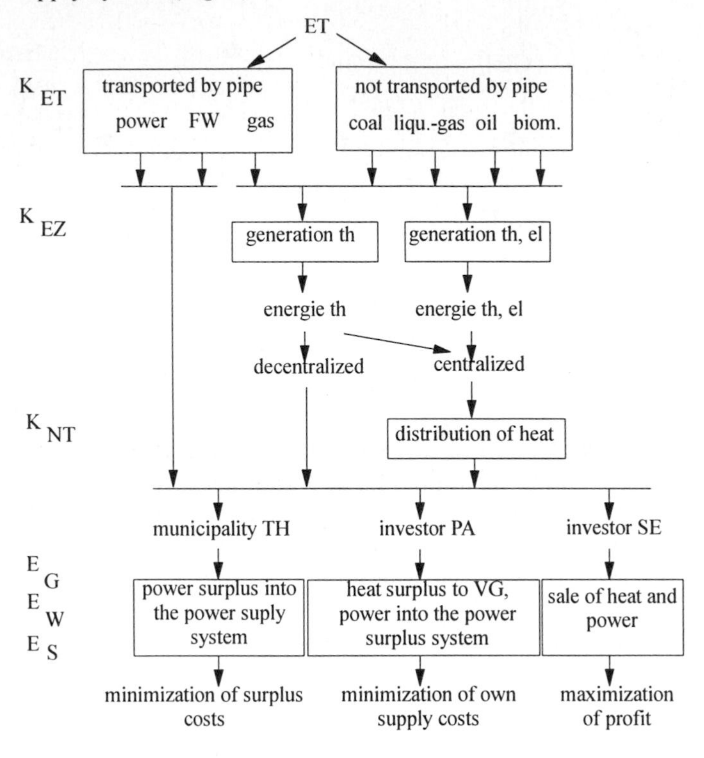

Fig. 3: Energy and cost flow in the energy supply system

On the one hand, costs (K_{ET}) are incurred by purchasing energy (ET). If any waste is produced, it may in fact allow for income to be generated. Energy supplies can be divided up into two categories: e.g. energy sources dependent on pipelines etc., i.e. long-distance heating systems (FW) and those which can be procured independently such as liquid gas (liqu.-gas). This means that price and tariff structures can differ. On the other hand, costs (K_{EZ}) are incurred from the installation and operation of processing plants (th.: production of thermal energy, el.: generating of electricity). If the plant in question is a central power supplier costs (K_{NT}) will be incurred from the distribution of heat. Depending on how the plant is operated and the energy produced (heat and/or electricity), additional income may be generated from the sale of energy (E_W), the transfer of surplus electricity into the national grid (E_S) as well as from issuing credit vouchers for any electricity not procured from the national grid (E_G).

In order to draw up plans for upgrading generating plants and at the same time to exactly define which costs will be incurred, it is essential that the efficiency of the plants be evaluated. This can be achived in a number of ways.

The output of a plant can either be assessed by a computer application for a particular twelve month period and then be used as a standard. Or, this can be determined from the amount of energy currently used by consumers. We must consider here that consumers who have to be supplied with energy from the centrel system, ought to be evaluated by means of our computer-aided procedure. If we are dealing with co-generation plants we should remember that the efficiency of the plant is especially important in economic terms and it would be wise to divide plants into two categories - normal and peak load types. (Klien, 1991; Suttor, 1991) Normal load plants cover basic requirements and thus operate at full efficiency over longer periods, whereas peak load plants merely take over when extremely high loads are required; they operate only at full peak efficiency over much shorter periods. Full peak efficiency levels are an indirect indicator for the usage full capacity of the technologies deployed. (Sauer, 1982) As far as economic factors are concerned, this division into categories has a number of large advantages as plants may be selected to cover basic energy supply requirements which require high investment, but which incur low fuel costs. Peak load plants deploy technology which requires lower investment but are operated with high fuel costs. In this manner a greater economically sustainable form of supplying energy can be established.

A process model must be developed based on the approach to problems referred to above and must depict the energy supply system in accordance with the questions raised.

3 Establishing of Process Models

The approach to problems and difficulties within the energy sector demands a mixed-integer-linear formulation, as it renders possible a decision as to whether a plant should be built or not by taking integer variables (1-0 condition) into account. Solving the system of equations signifies that this mathematical formulation will test the computer to its full capacity as the calculations are very intense. This means that the process model must be formulated in such a manner that on the one hand, it allows for an exact as possible technological and thus economic determination of the individual components within the system and, on the other hand, ensures that computers remain able to assist in answering questions. This would signify that Fig. 1 above would have to be simplified.

In order to do so, an actual supply zone will be split into categories depending on their physical structures (consumer groups VG) (Fig. 4). Similar consumer types, e.g. domestic households and catering establishments and similar types of buildings such as individual houses and blocks of flats, can be classed together within these categories if they show similar types of consume deploying similar energy levels over the course of a twelve month period.

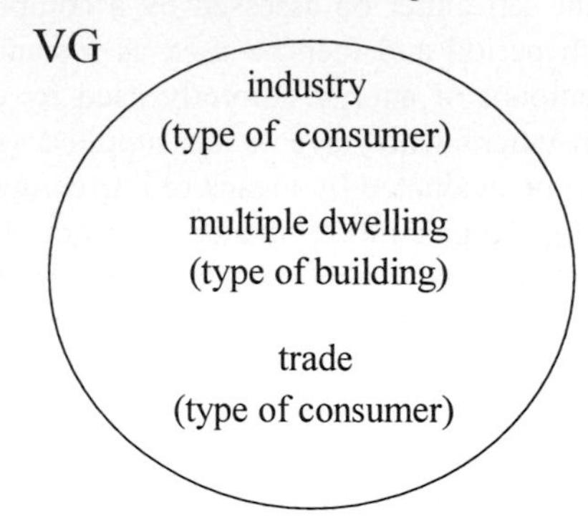

Fig. 4: Categorisation of consumers into specific consumer groups and types of building

The pipeline required for the central provision of long distance heat is not, as previously illustrated, laid from consumer to consumer, but is combined in a grid square over its entire length. We must consider that the diameters of district heating pipelines are becoming more and more insubstantial due to the perceived reduction in the number of consumers. Therefore, an average diameter will be calculated using the following equation: (Glück, 1985; Hakansson, 1986)

$$d = \sqrt{\frac{4 \cdot \dot{Q}_{NT}}{c_p \cdot \Delta T \cdot \rho \cdot \pi \cdot c}} \tag{1}$$

The speed c at which the heat is provided, is at a level of between 1 and 3 m/s depending on the diameter of the pipeline. (Sauer, 1982; Winkens, 1994) The temperature spread ΔT between offset and return is of the utmost importance in dimensioning the diameter of the pipe and reduces this diameter when the spread is substantial. $\dot{Q}_{NT}$ shows the total output in a grid square.

In order to ascertain the exact level of the costs incurred, the efficiency of domestic hook-up points (HU), the manner in which they, and domestic heat distribution systems function (HA) must be determined for each individual consumer. This is done by equating the requirements of the various types of consumer groups with the congruent amounts pertaining to HU and HA and by multiplying them by the number of consumers within a specific consumer type category. As a result, we have a simplified energy supply system which is shown in Fig. 5.

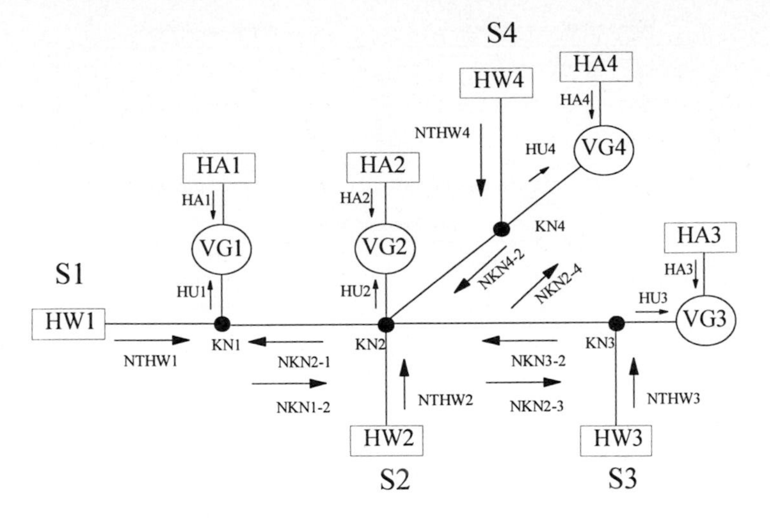

Fig. 5: Functional model of an energy supply system

On this basis, the energy requirements of each individual consumer can be provided for by all types of technology currently available on the market. As far as the task at hand is concerned, it would be sufficient to represent the generating plants as black boxes; offset and return processes are paid less attention to in this case. We may easily make use of this simplification due to the fact that the determination of upgrade strategies is based on preliminary planning stages.

As described above, the efficiency level of the plants is required for assessing how to equip generating plants and distribution systems. As those consumers who have to be supplied from the central system, still have to be determined using computer procedures, it is essential that actual periods of energy demand be calculated in the form of typical daily load operations as shown in Fig. 6. Three typical load operations can be drawn from the examples illustrated in the literature (Dehli, 1984; Gernhardt, 1993; Rouvel, 1985; Rouvel, 1983) and differentiated between:

- buildings partially heated at certain times,
- buildings completely heated,
- public institutions/facilities.

In the case of buildings which are only heated for part of the day, rooms are heated according to need. Heating is reduced overnight. In the case of completely heated buildings, rooms are heated all day and temperatures reduced at night. In public buildings, temperatures are reduced after the initial heat up phase in the mornings.

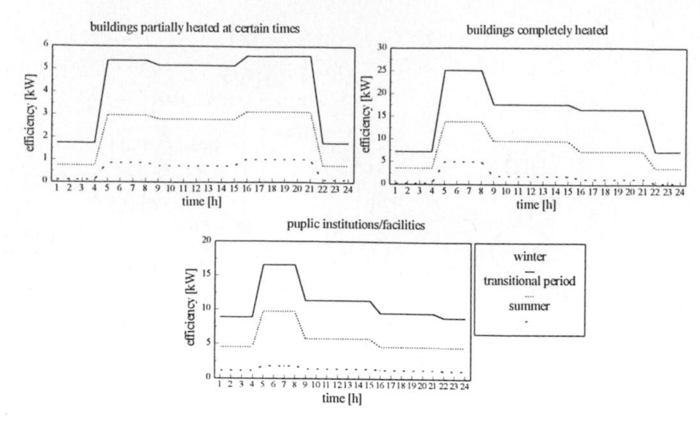

Fig. 6: Typical daily load performance

This method renders it possible for an accurate classification of the loads/requirements of each individual consumer within the system and the times when these requirements manifest themselves. This allows us to calculate exactly when peak loads will be reached. In order to further simplify the model, the load will not be recorded hourly, but rather each 24 hour period will be divided up into mornings, daytimes, evenings and nights and the average load for each period will be determined accordingly. Loads cannot be recorded for every single day of the year, so figures will be summarised for the summer period, the overlap into spring resp. autumn , and winter.

The delineation of upper and lower assessment limits will still play an important role in the establishing of process models, as we can use them to define energy flows and costing levels. Taking our three types of operating company into consideration, we can establish various assessment limits which can be seen in Fig. 7 and 8.

In the case of the operating company ‘community TH’ the assessment limit is applied to the whole of the area to be supplied with energy (Fig. 7). Consumers, both central and locall situated generating stations and distribution systems can be found within these limits. The fuel required for the heating of the plants is transported into the system from the outside, whereas with electricity, any emission of output gases and waste products are removed from the system. Investment, efficiency and performance-related costs re the generating and distribution components of the system must be expended. The direction taken by any flow of capital for primary resource material is dependent on whether it has any real value on the market and whether costs may arise which must be covered by the plant operator. Or whether we are dealing with waste products whose remediation would involve charges and would provide for additional income. The same applies to waste produced during thermal conversion such as ash and cinders. This could either be transported to a storage depot/landfill site which would also involve a handling charge. Or it could be sold on the market in the form of fertiliser, for example. If surplus electricity is fed into the grid, this would also signify that funding will be injected into the system.

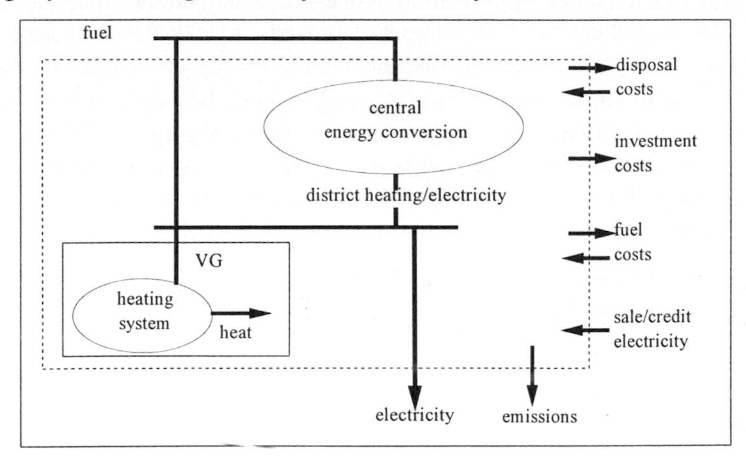

Fig 7: Assessment limits 'community TH'

The assessment limits for the type of operating company described as 'investor PA' below, function in the same manner (Fig. 8). Generating plants are included in these limits as is the distribution system in the case of long distance heating. We should note here that the assessment of 'investor PA' and 'investor SE' does not include charges for connecting consumers' property to the long distance heating system (link-up from the middle of the road across consumer property to individual houses and domestic hook-up points) as it is to be expected that consumers are obliged to pay these charges in the form of a connecting fee. These charges will therefore not be included in the model.

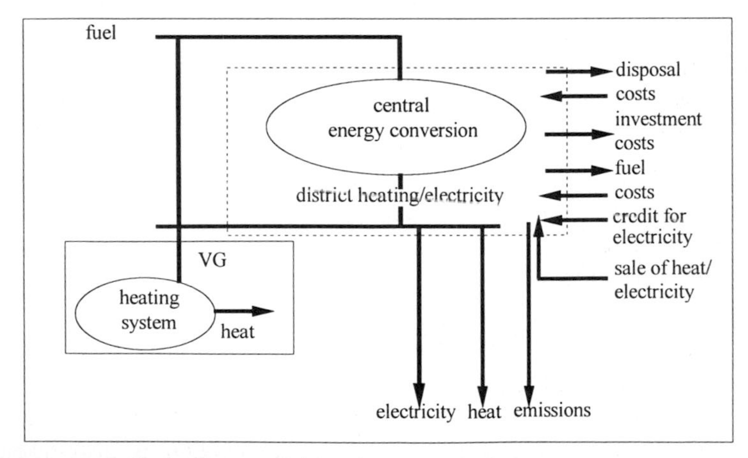

Fig. 8: Assessment limits 'investor PA' and 'investor SE'

4 Mathematical Model

During the mathematical optimisation process it is imperative that the correct variable for describing the technologies deployed, are calculated in accordance with the problem at hand. In the case of thermal energy this would be the heat output $\dot{Q}$. In addition to this there is also a requirement for a definition of decision variables (1-0 condition). The target function forms the main element of the mathematical process. The components and variables, as regards our approach to problems and difficulties, are contained in this target function and the stipulation for maximising or minimising of the target range will be calculated.

As regards to thermal energy production in the 'community TH' form of operating company, this target function must contain the following cost items:

- Investment costs for:
  - central and locally situated generating plants (*IHW, IHA*),
  - fuel tanks (*THW, THA*),
  - long distance heating networks (*NI*),
  - domestic hook-up points (*IHU*),
- efficiency related outlays for generating and distribution stations (*LHW, LHA, LHU*),
- performance related outlays for generating and distribution stations (*AHW, AHA*),
- charges for waste remediation from energy conversion processes (*AEK*),
- charges for supply of external long distance heating (*FWK*).

In order to take account of any differences in outlay and revenue, the calculation of dynamic economic profitability is based on the deployment of any annual payments on capital value, or is based on any annual payments on the cash value if there is no revenue being put into the system. The method used to calculate annual payments takes into account average capital or cash value over a twelve month period. (Winje, 1991)

If the operating company involves the generation of thermal energy only, any annual payments on cash value is assessed in relation to the stipulation for minimizing the target range, as there is no revenue entering the system. The target range and individual cost items can be seen as follows:

$$\begin{aligned} Min \overset{!}{=} Z = {} & IHW + LHW + AHW + BHW + THW + IHA + LHA + AHA + BH \\ & + THA + NI + IHU + LHU + AEK + FWK \end{aligned} \tag{2}$$

If combined electricity and thermal energy generation is also taken into account, the outlays for this type of plant (KWK stations) (*IKWK, BKWK*) will also be taken into consideration within the target function. At the same time, revenue will be generated by the sale of electricity (*ELS*) and thermal energy (*EFW*). If electricity is produced and utilised by the operator him herself, a credit will be issued for that amount. As revenue is present within the system in this case, the target function will be based on the annual payment on capital value. In order to retain the stipulation for minimisation, equation 2 will be multiplied by minus one. The target function will then read as follows:

$$\begin{aligned} Min \overset{!}{=} Z = {} & IHW + LHW + AHW + BHW + THW + IHA + LHA + AHA + BH \\ & + THA + NI + IHU + LHU + AEK + FWK + IKWK + BKWK \\ & - ELS - EG - EFW \end{aligned} \tag{3}$$

Costs can be calculated from the specific outlays which are incurred at levels depending on plant efficiency or performance related factors. These costs are calculated separately for either central or locally situated plants. Furthermore, there is a division of plants into those deploying fossil fuels and those using biological fuels, so various technologically specific costing factors can be modulated for different scenarios and sensitivity analyses. The mathematical model has been interpreted three dimensionally for purposes of the approach taken in this case, i.e. the majority of the variables is depicted in relation to the three following components (indices): technology (HW), location (ST) and timeframe (T). This would signify that the technological, ecological and economic parameters and variables can be assigned to the three components accordingly in the mathematical equations. The components are grouped together in categories which may also be subdivided into further subsets. Using this to attain a total formulation would signify that the values of the variables and parameters of the individual components can be accumulated within a specific group.

The target function thus comprises of efficiency and performance related costing items. The target function can be represented in an equation as follows:

$$\begin{aligned} Min \overset{!}{=} Z = {} & R \cdot \sum_{HW} \sum_{ST} k1_{HW} \cdot \dot{Q}_{HW,ST} + \sum_{HW} \sum_{ST} k2_{HW} \cdot \dot{Q}_{HW,ST} \\ & + \sum_{HW} \sum_{ST} \sum_{T} k3_{HW} \cdot Q_{HW,ST,T} + \ldots \end{aligned} \tag{4}$$

k1 to k3 represent specific outlays which can either be ascribed to the model as constants or can be interpolated within the upper and lower limits, if alongside, the degression effects as well as the increasing technological efficiency are taken into account of. R is a factor representing annual payment. $\dot{Q}$ (efficiency) and Q (performance) are positive variables which have to be calculated by computer, although there are several additional prerequisites, the most important of which being the operating company 'community TH':

- assaying re each component within the system,
- efficiency/output capping,
- linking of thermal efficiency and performance.

In accordance with Fig. 5, balance sheet equations must be established for each individual component so that either the total capacity limits can be deducted or that the total level of performance supplied must correspond to existing consumer requirements $RVA_{VG,T}$ including any losses made in conversion (VHU) at domestic hook-up points.

$$\sum_{HW} Q_{HW,ST,T} = \sum_{NTHW} Q_{NTHW,T} \tag{5}$$

$$\sum_{HA} Q_{HA,VG,T} + \sum_{HU} Q_{HU,VG,T} \cdot VHU = RVA_{VG,T} \tag{6}$$

In relation to the following additional preconditions

$$\dot{Q}_{HW,ST} \geq N1_{HW,ST} \cdot LL_{HW,ST} \tag{7}$$

$$\dot{Q}_{HW,ST} \leq N1_{HW,ST} \cdot LU_{HW,ST} \tag{8}$$

re central generating plants at each location ST, $\dot{Q}_{HW,ST}$ as a positive variable will be constrained within the range of upper and lower efficiency limits.

The product with binary decision variable $N1_{HW,ST}$ will determine whether or not technology will be manufactured. The same preconditions must also be established for locally situated generating stations, networks and domestic hook-up points for each type of consumer group with the corresponding binary decision variables $N2_{HU,VG}$, $N3_{HA,VG}$ and $N4_{NT}$.

So as to create a correlation between the performance of the generating stations in the assessment sheets and their output efficiency within the target function, the performance and efficiency factors of the generating plants must be as combined:

$$\frac{Q_{HW,ST,T}}{ZAB_T} \cdot VSF \leq \dot{Q}_{HW,ST} \cdot VF_{HW} \tag{9}$$

Thus, it can be established that the performance generated by every station at each location and time period $Q_{HW,ST,T}$ divided by the hours within these periods ZAB_T must be lower than output efficiency $\dot{Q}_{HW,ST}$ multiplied by availability VF_{HW}. This will show the exact level of thermal energy output attained whilst taking into account the respective technical and operating conditions. (VDWE, 1981) Efficiency can be increased by 25% via factor *VSF* in order to guarantee supplies. Thus, the fact that the actual peak load demand can be higher than the average values forecast must be considered, as well as the fact that a further general safety factor can be incorporated which, in practice, will be within a range of approx. 10% to 15% during the construction of technical equipment and generating stations. This additional precondition must be established for heating plants, networks, and generating stations to an equal degree by modifying the relevant indices.

5 Results and Points for Discussion

The optimisation procedure was applied to a typically rural community with a population of 600 and a main-road village infrastructure which corresponds to our 'community TH' model. With the help of this procedure we were able to establish the most economically effective energy supply option for each consumer and were therefore in a position to make a decision on whether or not technology should be manufactured. Under the conditions relevant at the present time, consumers within this community with a usage requirement of above 100 kW and high annual heating requirements could be supplied with energy from an automatically fired solid fuel burner using biological materials as fuel (H_FABio), which would be economical to run (Fig. 9). This would provide 22% of the total demand. Other consumers would be supplied from an oil-fired heating plant (H_OLGO).

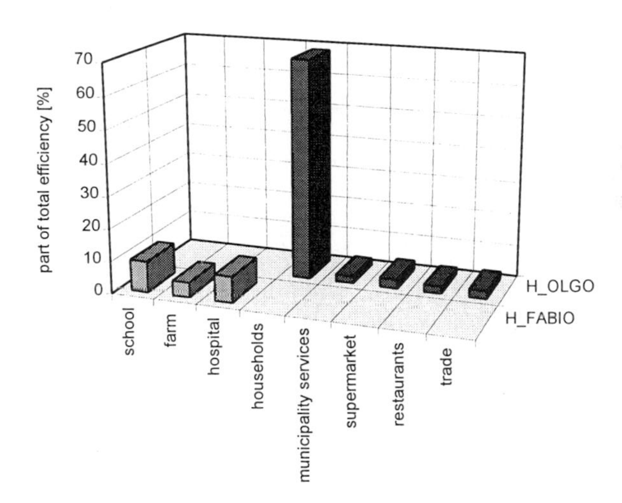

Fig. 9: Energy supply structure of a rural community

Using this as a basis, we were able to examine and assess the effects of supplying energy with biological material as a primary resource material, by modifying accompanying conditions and prerequisites for the cost-effective deployment of biological materials by modifying costing factors.

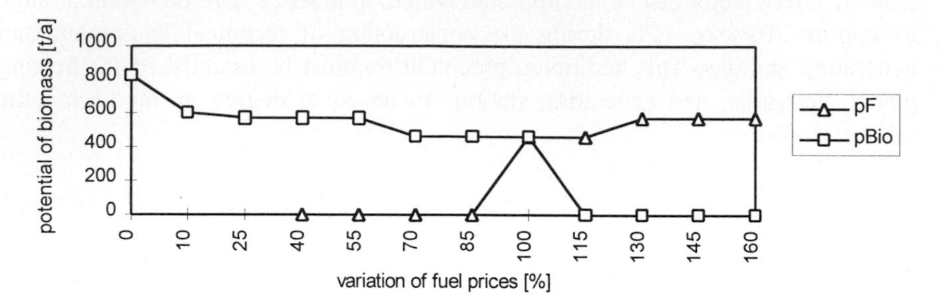

Fig. 10: Potential of biological material in correlation with prices for fossil fuels (pF) and biogenic energy sources (pBio)

Equally, the potential of biological material required for its cost-effective deployment can be assessed. Fig. 10 illustrates this in correlation with the price of biological materials in excess and below the current price (100%) of 0.03 DM per kWh.

A division of the plants into categories of normal and peak load stations and into seasonal periods S (summer), W (winter) and overlap (U) at various times during a 24 hour period mornings (M), daytimes (T), evenings (A) and overnight (N) is also possible using this model.

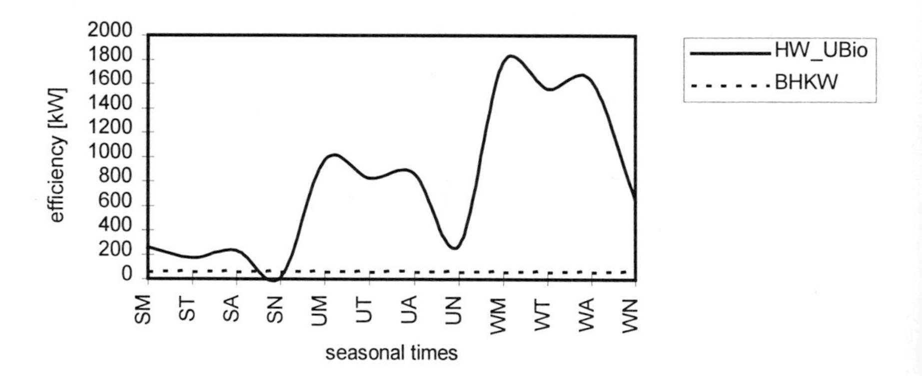

Fig. 11: Thermal loading procedure at thermal conversion station (HW_UBio) and co-generating plant (BHKW)

The results shown above illustrate that the process model established enables us to offer solutions to the problems initially confronted despite the simplifications we have made. In comparison to the variant calculations deployed to date, this process model offers an opportunity which takes into account the calculations for each variant within one model as well as establishing a premise for taking decisions based on existing demand, local resources and the technology available on the market. These results render it possible for suggestions for modifying any preconditions. The supply structures of the communities we looked at, the potential of deployable energy sources, conversion-related emissions and resulting costs can all be listed as results. Answers can also be found to specific questions as to the influence of particular factors on economic viability and the modulation of the energy supply structure, e.g. as regards to looking into the cost of transporting fuel sources such as those which are dealt with in the variant pertaining to source fuel prices. Comparative assessments for the appraisal of cost-saving exercises in relation to other energy supply potentials can also be carried out. If individual consumers have either no interest in seeing analyses of the opportunities re energy supply carried out or if there is no real requirement to upgrade generating plants in the immediate future, these will not be taken into account during any assessment. Customers/users can update or extend those parameters assigned to the process model at any time.

References

Dehli, M. et al. (1984): Überlegungen zur elektrischen Wärmeversorgung in hochwärmegedämmten Wohngebäuden. In: Elektrizitätswirtschaft 83, Nr. 3, S. 110-120

Egberts, G. (1979): Kostenoptimale Entwicklungsperspektiven des Raumheizungssektors im Energieversorgungssystem der Bundesrepublik Deutschland - ein Optimierungsmodell. Aachen, RWTH, Fakultät für Maschinenbau, Diss.

Fendt, H. (1979): Regionale Energieplanung. München, Ludwig-Maximilian-Universität, Diss.

Gernhardt, D. et al. (Bearb.) ; Ruhr-Universität Bochum, Lehrstuhl für Nukleare und Neue Energiesysteme (Hrsg.) (1993): Erstellung von Modellgemeinden sowie Darstellung elektrischer und thermischer Lastganglinien für Nordrhein-Westfalen. Bochum : Ruhr-Universität-Bochum. (Forschungsvorhaben IV B3-258 002). - 6. Technischer Forschungsbericht.

Glück, B. (1985): Heizwassernetze für Wohn- und Industriegebiete. Berlin : VEB Verlag für Bauwesen

Hakansson, K. ; Meyer-Clasen, B. (Mitarb.) (1986): Handbuch der Fernwärme Praxis. 3. Aufl. Essen : Vulkan-Verlag

Hamann, K. (1969): Beitrag zur komplexen Behandlung von Wärmeversorgungsproblemen mit Hilfe der dynamischen Optimierungsmethode. Dresden, TU, Sektion Energieumwandlung, Diss.

Hanfler, M. (1990): Methoden zur langfristigen Planung der Wärmeversorgung - dargestellt am Ausbau der flächendeckenden Fernwärmeversorgung der Städte Erfurt und Eisenach. Weimar, Hochschule für Architektur und Bauwesen, Fakultät Städtebau, Architektur, Gesellschaftswissenschaften, Diss.

Klien, J. (1991): Planungshilfe Blockheizkraftwerke : Ein Leitfaden für Planer und Betreiber. Karlsruhe : Verlag C.F. Müller. (Praxis Kraft-Wärme-Kopplung, Bd. 3)

Maubach, K.-D. (1994): Mittelfristige Energieeinsatzoptimierung in Versorgungssystemen mit Kraft-Wärme-Kopplung. Wuppertal, Bergische Universität-Gesamthochschule, Fachbereich Elekrotechnik, Diss.

Pang, C. K. ; Chen, H. C. (1976): Optimal short-term thermal unit commitment. In: IEEE Trans. on Power App. and Systems 95, Nr. 4, S. 1336-1346

Readwin, A. M. (1985): Ein energiewirtschaftliches Planungsmodell zur Bestimmung effizienter Versorgungsstrategien : Angewandt am Beispiel der Energieversorgung in Hamburg. Frankfurt/M. : Verlag Peter Lang. (Betriebswirtschaftliche Beiträge zu Energie-, Rohstoff- und Umweltfragen, Bd. 4)

Riedel, D. ; Schulz, M. (1988): Energie-Einsatzoptimierung in Industrie- und Heizkraftwerken. In: ETG Fachbericht 26. Mannheim, S. 43-50

Rouvel, L. (1985): Energiebilanz von Gebäuden. In: Ruhrgas AG (Hrsg.): Ruhrgas Handbuch : Haustechnische Planung. Stuttgart : Krämer, S. 222-235

Rouvel, L. (1983): Systemtechnik- Benutzerverhalten - Heizleistungsgänge. In: VDE et al. (Veranst.): Energieversorgungskonzepte : Möglichkeiten - Meinungen - Mißverständnisse (Tagung Schliersee, 1983). Berlin : Springer-Verlag, S. 63-82. (Schriftenreihe der Forschungsstelle für Energiewirtschaft, Bd. 16)

Sauer, E. ; Zeise, R. (1982): Energietransport, -speicherung und -verteilung. Köln : Technischer Verlag Resch, Verlag TÜV Rheinland. (Handbuchreihe Energie, Bd. 11)

Suttor, K.-H. ; Suttor, W. (Hrsg.) (1991): Handbuch Kraft-Wärme-Kopplung : für Planer, Betreiber, Ausbildung. Karlsruhe, Verlag C.F. Müller. (Praxis Kraft-Wärme-Kopplung, Bd. 1)

VDWE e.V. (Hrsg.) (1981): Begriffsbestimmungen in der Energiewirtschaft. Teil 2 : Begriffe der Fernwärmewirtschaft. 5. Ausgabe. Frankfurt/Main : VWEW

Winje, D. ; Witt, D. (1991): Energiewirtschaft. Berlin : Springer-Verlag, Köln : Verlag TÜV Rheinland. (Handbuchreihe Energieberatung / Energiemanagement, Bd. II)

Winkens, H.-P. (1994): Fernwärmespeicherung, -transport und -verteilung. In: Forschungszentrum Jülich, Programmgruppe Technologiefolgenforschung (Hrsg.): IKARUS : Elemente für Klimagas-Reduktionsstrategien. Jülich : KFA Jülich. - Studie ET9188A im Auftrag des Bundesministers für Forschung und Technologie. Teilprojekt 4 "Umwandlungssektor"

Different Techniques of the Modelling of Hydrodynamic Couplings

Peter Jaschke and Heinz Waller

Ruhr-Universität Bochum, 44780 Bochum, Germany

Abstract. *The following paper describes different methods of the investigation of numerical models for hydrodynamic couplings. First an overview of the methodology of modelling is presented. Analytic physical, black-box and hybrid modelling are explained in detail. The experimental identification of the models for the hydrodynamic coupling is presented. Different modelling methods and the appendant results are compared. A critical reconsideration of the methods and the results are presented at the end of the paper.*

Keywords. *black-box, hybrid, hydrodynamic coupling, identification, non-linear system*

1 Introduction

Hydrodynamic couplings possess a wide field of application in drive system design because of their property of vibration damping and isolation. The dynamic characteristic of a whole drive system is therefore substantially determined by these hydrodynamic components.

In Fig. 1, the general structure of a hydrodynamic coupling is presented. The essential parts are two vane wheels (the pump and the turbine) and the housing, enclosing the wheels. The power transmission from the pump to the turbine takes place with the aid of a fluid (mostly oil), which has been filled into the housing.

The stationary characteristic of hydrodynamic couplings has been analysed very precisely by experiments. Only very poor knowledge about the dynamic properties has been investigated till now. To guarantee safe operation it is not only necessary to adjust a motor to the working machine according to the stationary requirements but also the dynamic characteristic of the whole drive system has to be considered. Dynamic processes can be generated by periodic excitations (e.g. piston engines) or by impact excitation (e.g. the blocking of a shippropeller in ice, the fixing of a coal plane). Therefore, it is necessary to analyse the dynamic characteristics of hydrodynamic couplings. The investigations are performed on a test bench (Fig. 2). The coupling is assembled between two drive units – secondary controlled axial piston units, which are supplied by a constant pressure network. Because of the very low moments of inertia and the very high torque the hydrostatic drives are well suited to generate dynamic excitations for the coupling. The controlling of the drive units as well as the acquisition and the analysis of the measured data is done with the aid of a real time computer network.

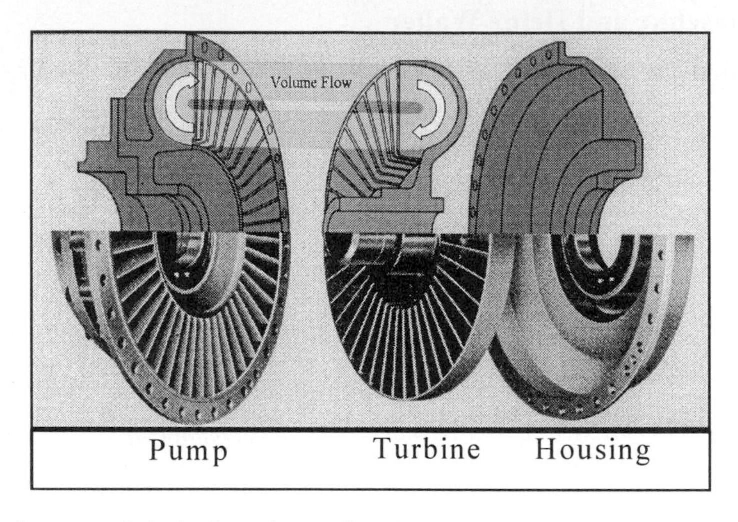

Fig. 1: Structure of a hydrodynamic coupling

Fig. 2: Test bench

The stationary and dynamic transfer characteristics of the hydrodynamic coupling are highly non-linear and a theoretic analysis seems to be nearly impossible or at least very extensive. Therefore, the modelling of the dynamic properties will be very sophisticated. In the following text different possibilities of modelling of the coupling are presented.

2 About Different Modelling Techniques

The purpose of modelling is to investigate mathematical models that represent the characteristics of a technical system. Different techniques can be used to solve this problem.

2.1 Analytic Physical Modelling

The analytic physical modelling is based on the assemblage of all physical a priori knowledge and on the subsequent setup of the differential equations specifying the system. This method enables a simulation of a system without any experiments. The other techniques of modelling systems – presented in section 2.2 and 2.3 – are based on experiments.

In the field of mechanics for example the

- kinematic relationships
- mechanic balance equations (static and dynamic equilibrium conditions, laws of conservation of energy and mass etc.)
- thermodynamic relations if necessary
- material laws

have to be assembled. But often it will be difficult to estimate, a priori, how accurate the different physical effects have to be considered, so that the complete model will supply reasonable results. Beyond this fact it will remain doubtful for different problems whether the accuracy, which is necessary to consider special effects, can be achieved. The effort for the derivation of an analytic physical model will increase immensely with the postulated accuracy. Therefore, it might happen that the intention of modelling cannot be fulfilled by this procedure respectively or that the realisation cannot be performed economically. Because of the complexity of the investigated models their benefit may be restricted (e.g. to long computing time).

2.2 Black-Box Modelling and Identification

For experimental modelling only prior defined input and output values of the system are considered. The inner state of the system is of no interest. Only the relation between the input and output is important. Therefore no or only few knowledge about physical connections is required (black-box modelling). This approach is based on generally formulated differential or integral equations (for static or stationary operation it may only be functions) which have to adjusted to the real system.

Fig. 3: Black-box model

There are many different possibilities to generate black-box models. A rough classification distinguishes linear and non-linear models. Moreover, one can differentiate between continuous time and discrete time systems. According to the applied model approach recursive or non-recursive models may result which have different properties. Finally, it should be mentioned that parametric and non-parametric models can be distinguished.

The characteristics of linear continuous models have been studied in most details. The superposition theorem is valid and single solutions are superposable and can be generalised.

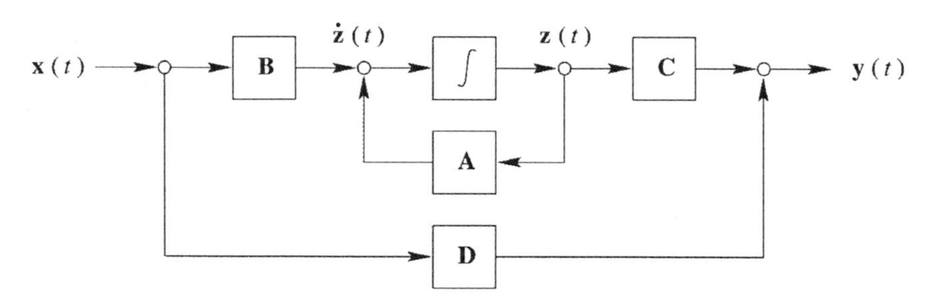

Fig. 4: State space model

Three different mathematical formulations can be used for the modelling of linear continuous time systems.

- The differential equation e.g. in the state space (Fig. 4)

$$\begin{aligned} \dot{\mathbf{z}}(t) &= \mathbf{A}\,\mathbf{z}(t) + \mathbf{B}\,\mathbf{x}(t) \\ \mathbf{y}(t) &= \mathbf{C}\,\mathbf{z}(t) + \mathbf{D}\,\mathbf{x}(t) \quad . \end{aligned} \tag{1}$$

- The weighting function (or the matrix of weighting functions)

$$\mathbf{h}(t) = e^{\mathbf{A}t}$$
$$h_{ij}(t) = \sum_l k_{ijl}\, e^{\lambda_l t} \quad . \tag{2}$$

λ_l are the eigenvalues of **A**.

- The frequency response

$$\mathbf{H}(\Omega) = \mathbf{C}(\mathrm{j}\Omega\mathbf{I} - \mathbf{A})^{-1}\,\mathbf{B} + \mathbf{D}$$
$$h_{ij}(\Omega) = \sum_l \frac{\left(\overline{\mathrm{A}}_{ij}\right)_l}{\mathrm{j}\Omega - \left(\lambda_{ij}\right)_l} = \frac{\sum_l \left(z_{ij}\right)_l (\mathrm{j}\Omega)^l}{\sum_k \left(n_{ij}\right)_k (\mathrm{j}\Omega)^k} \quad . \tag{3}$$

Frequency response and weighting functions are connected by the Fourier transformation. The differential equation may include physical information in an obvious form. In frequency responses and weighting functions, the physical information must no longer be evident, because they describe only input-output-relations.

The time history of a system can be evaluated with the convolution integral

$$\mathbf{y}(t) = \int_0^t \mathbf{h}(t-\tau)\,\mathbf{x}(\tau)\,\mathrm{d}\tau \quad . \tag{4}$$

By the Fourier transformation this equation is reduced to a simple product, which is the reason for its preferred application in the frequency domain

$$\mathbf{Y}(\Omega) = \mathbf{H}(\Omega)\,\mathbf{X}(\Omega) \quad . \tag{5}$$

Whereas the evaluation of the convolution theorem in the time domain is much more expendable. For non-linear systems no highly developed theory is known. As the superposition theorem is not valid, every problem has to be analysed individually. A general formulation for non-linear continuous systems (with continuous static and dynamic nonlinearity) is found in Volterra series

$$y(t) = \int_0^\infty h_1(\tau_1)x(t-\tau_1)\,\mathrm{d}\tau_1$$
$$+ \int_0^\infty\int_0^\infty h_2(\tau_1,\tau_2)x(t-\tau_1)x(t-\tau_2)\,\mathrm{d}\tau_1\,\mathrm{d}\tau_2 \tag{6}$$
$$+ \int_0^\infty\int_0^\infty\int_0^\infty h_3(\tau_1,\tau_2,\tau_3)x(t-\tau_1)x(t-\tau_2)x(t-\tau_3)\,\mathrm{d}\tau_1\,\mathrm{d}\tau_2\,\mathrm{d}\tau_3$$
$$+\cdots\;.$$

The Volterra kernels $h_1(\tau_1)$, $h_2(\tau_1,\tau_2)$, ... correspond to the order of the non-linearity of the analysed system. The handling of the Volterra series, however, is often too difficult in application. Therefore, simplified models e.g. Wiener or Hammerstein models are preferred (Fig. 5).

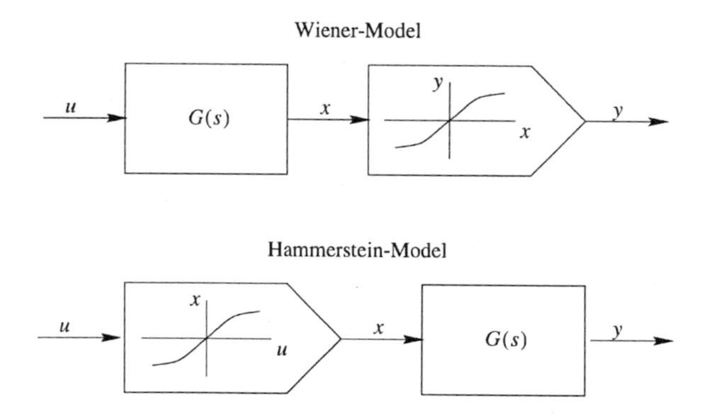

Fig. 5: Wiener and Hammerstein model

These models are composed of a linear dynamic and a non-linear static part. For this technique of modelling knowledge about the non-linearity is required. When the differential quotient is approximated by the difference quotient discrete models are derived.

For parametric identification it is necessary to assume a definite structure of the model for the analysed system. For this model structure, then an optimal set of parameters has to be determined by minimising the error between the response of the model and the real system (Fig. 6).

For the parameter identification, many different methods have been developed. The mean squares method, the method of instrumental variables and the maximum likelihood method should be mentioned. When an optimal set of parameters of the models has been determined then the model has to be verified, comparing data of simulations and experiments. The results of the model will only make sense if it is analysed for excitations which lie in the area of that used during identification. Very seldom the model will also be valid for areas beyond the used excitation range.

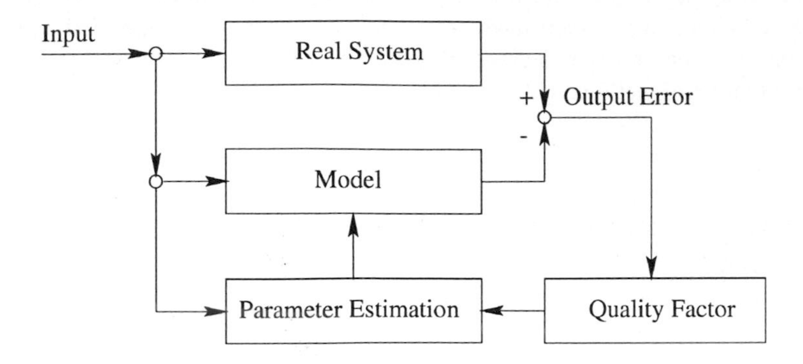

Fig 6: Block diagram of identification

2.3 Hybrid Modelling

An interconnection between analytic physical and black-box modelling is called a hybrid method. Here a restricted mapping of the physical phenomena into a relatively simple mathematical model is made that does not reflect a very accurate formulation but shows the rough structure of the physical process and the order of the nonlinearity. The structure of the mathematical model is adopted and the coefficients which result from physical consideration are collected. But the values of these collected coefficients are adapted to the real physical system by identification. This methodology combines the advantages and the disadvantages of the two other methods.

The following disadvantages are worth mentioning: In contrast to analytic physical modelling, experiments are necessary for hybrid modelling to investigate the mathematical model. The determination of the parameters is more expendable in comparison with most of the black-box modelling methods, because the parameters may depend in a non-linear way from the output error of the model (Fig. 6). This fact leads to non-linear optimisation problems that are computer time consuming. The necessary starting values for optimisation can mostly be determined from the values of the simplified physical model. But one has to pay attention that the parameters are identificable.

The advantage of hybrid modelling is the structure of the model. If black-box models are set up for non-linear systems, it might happen, that the model will consist of a structure, that will not match with that of the system but that will nevertheless characterise the dynamic properties for the identified area of operation. Outside of this region the model may loose its validity. However, the structure of hybrid models is adapted to the real system and the model may also characterise the system within certain limits even outside the identified area (at least the tendency). This fact may reduce the measuring effort. Moreover, the number of parameters which have to be determined may be essentially lower. Also, the number of inner degrees of freedom will be lower than it is for the analytical physical method.

3 Analytic Physical Modelling of Hydrodynamic Couplings

3.1 One-Dimensional Flow Theory

The first approach to characterise the dynamic properties of hydrodynamic couplings by physical consideration took place with the one-dimensional flow theory (Frömder, 1962). It was assumed that the fluid flow consists of constant characteristics over the whole cross-section between the blades. On the basis of the nonstationary laws of conservation of energy and momentum, the differential equations for the model have been derived. At first, only pipe frictions have been considered. Herbertz (Herbertz, 1973) developed a more complete model for the hydrodynamic torque converter. If this model is transferred to the coupling, the following equations result:

$$\begin{aligned} M_P &= a_{11}\,\dot{\omega}_P + a_{12}\,\ddot{V} + a_{13}\,\omega_P\,\dot{V} + a_{14}\,\omega_T\,\dot{V} + a_{15}\,\dot{V}^2 \\ M_T &= a_{21}\,\dot{\omega}_T + a_{22}\,\ddot{V} + a_{23}\,\omega_P\,\dot{V} + a_{24}\,\omega_T\,\dot{V} + a_{25}\,\dot{V}^2 \\ 0 &= a_{31}\,\dot{\omega}_P + a_{32}\,\dot{\omega}_T + a_{33}\,\ddot{V} + a_{34}\,\omega_P\,\dot{V} + a_{35}\,\omega_T\,\dot{V} \\ &\quad + a_{36}\,\omega_P^2 + a_{37}\,\omega_T^2 + a_{38}\,\dot{V}^2 \quad . \end{aligned} \tag{7}$$

The parameters a_{ij} are composed of geometric data of the coupling, of the density of the fluid and of flow loss factors. The latter can only be estimated, because no secure laws about the losses exist. Depending on the simplifying assumptions about the fluid flow, special terms of Equations 7 can be neglected.

These models frequently show only insufficient accurate results. This fact is based partly on the estimation of the loss factors and partly on the assumption that the fluid flow consists of constant characteristics over the whole cross-section between the blades. Measurements performed within the scope of the SFB 278 (a special research centre at the Ruhr-University Bochum sponsored by the *Deutsche Forschungsgemeinschaft*) have proved that the flow consists of many vortices so that the assumption of a one-dimensional flow theory is no longer valid. Hence, a more accurate method described in the following text is necessary to find an accurate analytic physical model.

3.2 Finite Volume Method

The basic equations (Reynolds averaged conservation equations with coriolis modified k-ε-model) for the characterisation of a three-dimensional nonstationary turbulent fluid flow as assumed for the hydrodynamic coupling form a non-linear partial differential equation system, that can be formulated in dimensionless form. As no analytic solution can be found, numerical methods have to be applied to find an approximate solution. For the modelling of the coupling, a finite volume method with a non-orthogonal grid fitted to the contour (Formanski, 1996) was used. A picture of the grid used for the analysis together with the stationary characteristic curve is shown in Fig. 7. The simulation of a dynamic process is reproduced in Fig. 8.

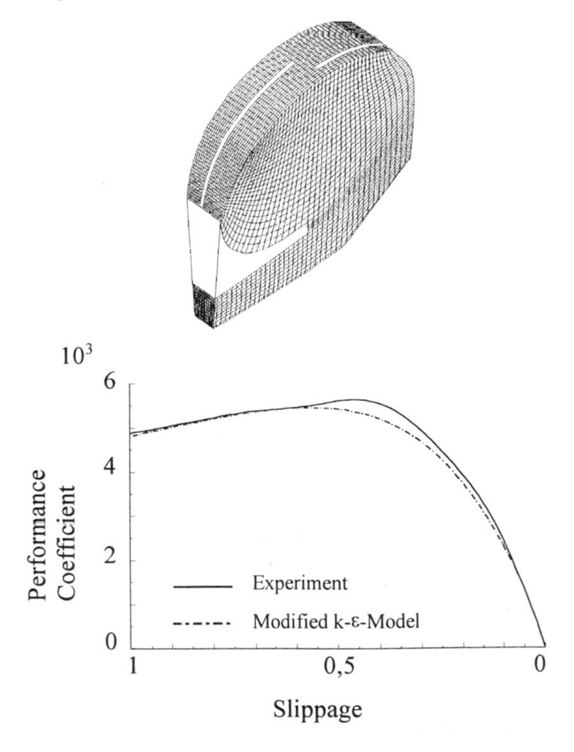

Fig. 7: Computation grid

The deviations between experiments and simulations result not only from the numerical approximation but also from the assumptions utilised for the modelling. Only a one phase flow has been analysed, that is to say, a completely filled coupling with incompressible oil with constant material parameters was supposed. Moreover, the same number of blades for the pump and the turbine has been assumed in contrast to the experimentally analysed coupling because it was necessary to reduce the number of grid points by taking advantage of symmetry. Thereby the amount of computer time and storage is also reduced. Nevertheless 40000-60000 cells were necessary for the numerical calculations. A computer time of between 20-50 hours was needed for the determination of only one nonstationary operating point. Because of the discretisation in circumferential direction, the number of blades, the speed of the pump and the slippage, a certain time discretisation Δt is necessary for dynamic simulations. Then for the simulation of only one period of the dynamic process a computer time of 40 days was needed on a modern workstation. This effort is much greater compared to the simulation effort with experimentally identified models .

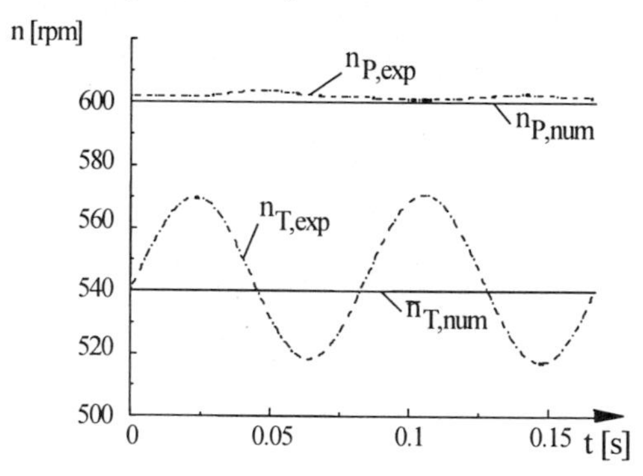

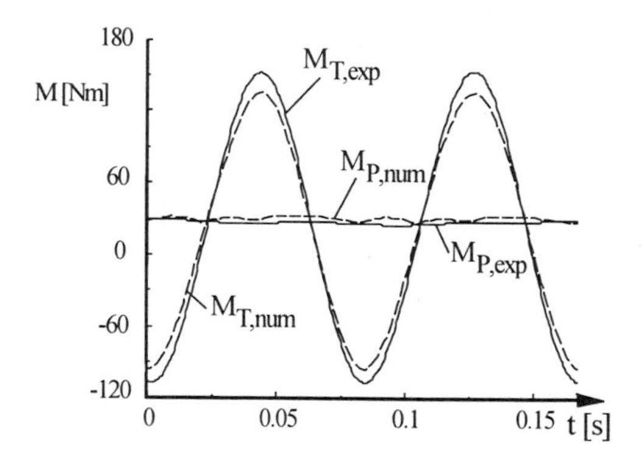

Fig. 8: Simulations with the finite volume method

4 Experimental Modelling of Hydrodynamic Couplings

For the experimental modelling of a hydrodynamic coupling it is necessary to measure its dynamic characteristic. The excitation of the coupling is realised with a newly constructed test stand. The characteristics of the hydrostatic drives used for excitation – driving and braking – is by no means ideal. The absence of retroaction necessary for state space formulations cannot be assumed. Always there is an interaction between the drive, the coupling and the brake. Therefore, the subsystem to be modelled and identified has to be cut free by introducing a subsystem boundary. The physical variables passing through the system boundary are the values by which interaction with other parts of the whole system is performed.

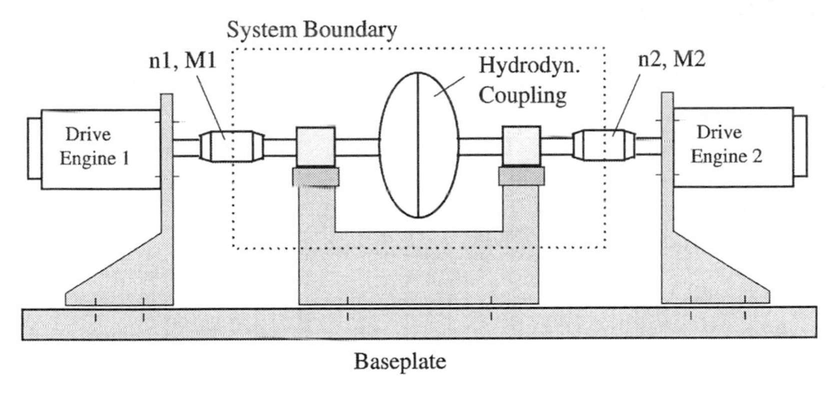

Fig. 9: Subsystem boundary

For the definition of the mathematical formulation the physical input-output relations of the system have to be considered. For the couplings in consideration, it is essential that both the number of revolution and the torque may be input, as well as output, values.

4.1 Black-Box Models

4.1.1 Linear Submodels

The linear modelling possesses the advantage that it is possible to establish models in the time as well as in the frequency domain, moreover a different assignment of input and output variables can be easily formulated. A linear model with two input and output variables, as used here, is called in terms of electrical engineering two port or four pole model.

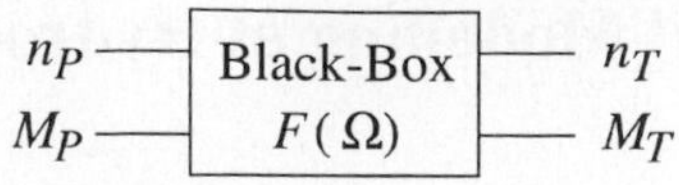

Fig. 10: Dynamic model

The Matrix $\mathbf{D}(\Omega)$, which characterises the transfer properties between the deviations of the number of revolutions and the torque from a certain operating point of the subsystem, here is called the dynamic damping matrix

$$\begin{bmatrix} M_P(\Omega) \\ M_T(\Omega) \end{bmatrix} = \begin{bmatrix} d_{11}(\Omega) & d_{12}(\Omega) \\ d_{21}(\Omega) & d_{22}(\Omega) \end{bmatrix} \begin{bmatrix} n_P(\Omega) \\ n_T(\Omega) \end{bmatrix}$$
$$\mathbf{M}(\Omega) = \mathbf{D}(\Omega)\mathbf{n}(\Omega) \quad . \tag{8}$$

Two independent experiments are necessary to determine the four frequency functions $d_{ij}(\Omega)$ for one frequency point. For each of the experiments the hydrodynamic system is excited monofrequently. The dynamic damping matrix can be evaluated by determining the amplitudes of the four state variables from experimental data by Fourier analysis. The indices 1 and 2 in the following equation distinguish the two different experiments

$$\begin{bmatrix} d_{11} & d_{12} \\ d_{21} & d_{22} \end{bmatrix}_{(\Omega)} = \begin{bmatrix} M_{P1} & M_{P2} \\ M_{T1} & M_{T2} \end{bmatrix}_{(\Omega)} \begin{bmatrix} n_{P1} & n_{P2} \\ n_{T1} & n_{T2} \end{bmatrix}_{(\Omega)}^{-1}$$
$$\mathbf{D}(\Omega) = \mathbf{M}(\Omega)\mathbf{n}(\Omega)^{-1} \quad . \tag{9}$$

In the presented form, the dynamic damping matrix is a non-parametric model for the dynamic characteristic of the coupling. For the investigation of a parametric model the four elements of $\mathbf{D}(\Omega)$ have to be approximated by complex broken rational functions (linear system)

$$d_{ij}(\Omega) = \frac{\sum_{k=0}^{m} (z_{ij})_k (\mathrm{j}\Omega)^k}{\sum_{l=0}^{n} (n_{ij})_l (\mathrm{j}\Omega)^l} \quad . \tag{10}$$

Restrictions (stability, causality) as postulated by linear system theory have to be considered but shall not be discussed here anymore. When the functions of all four elements are available it becomes possible to analyse the dynamic characteristic of the whole drive system in the frequency domain. The damping matrix for the coupling has only to be introduced into a computer program for the elastic elements of a drive system.

By inverse Fourier transform one can obtain linear differential equations in the time domain. The transformation becomes especially simple if a common denominator polynomial is used for all four elements $d_{ij}(\Omega)$

$$\begin{aligned} M_P(\Omega)\sum_{l=0}^{n} n_l(\mathrm{j}\Omega)^l &= n_P(\Omega)\sum_{k=0}^{m}(z_{11})_k(\mathrm{j}\Omega)^k + n_T(\Omega)\sum_{k=0}^{m}(z_{12})_k(\mathrm{j}\Omega)^k \\ M_T(\Omega)\sum_{l=0}^{n} n_l(\mathrm{j}\Omega)^l &= n_P(\Omega)\sum_{k=0}^{m}(z_{21})_k(\mathrm{j}\Omega)^k + n_T(\Omega)\sum_{k=0}^{m}(z_{22})_k(\mathrm{j}\Omega)^k \quad . \end{aligned} \tag{11}$$

The models obtained this way are only valid in a small area around the underlying operating point. If operating areas of a larger size have to be considered several linear submodels have to be determined which are valid for different operating points.

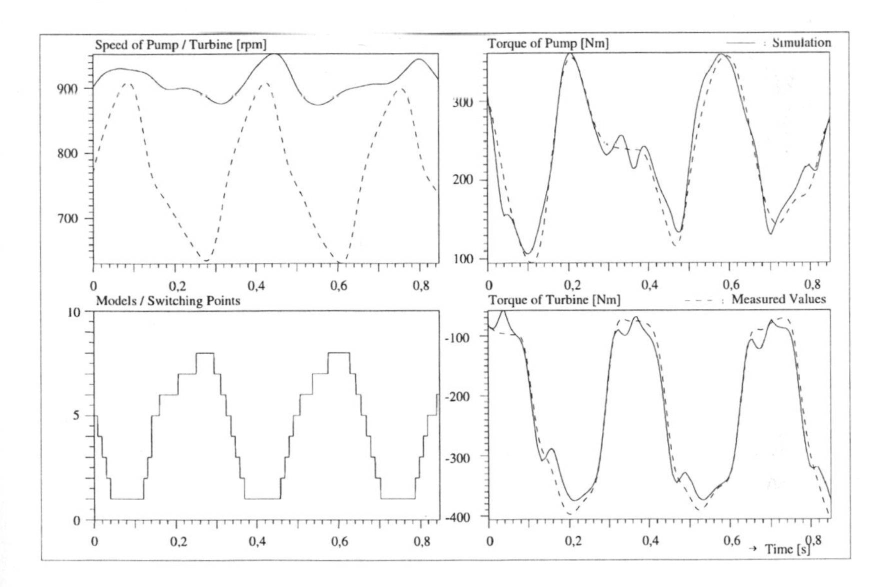

Fig. 11: Simulation with linear submodels

If the scopes of application of two submodels are adjacent to one another then it can be switched between these submodels to simulate enlarged areas of the non-linear system. If several subareas are crossed with high acceleration in simulations transient processes can result at the switching points. These transient processes may cause some trouble for the simulation. This fact can be overcome by non-linear models which will be valid for an enlarged area of operation.

4.1.2 Non-linear (Time) Discrete Model Formulations

If one wants to investigate models which can be elementarily realised and identified by a digital computer, often discrete difference equations or systems of equations are favoured. The advantage of these models is the fast identification of the parameters because they are linearly dependant on the output error of the model (Fig. 6). The mathematical models for the characterisation of the coupling will be derived from the Kolmogoroff-Gabor-polynomial (recursive model). At this point it should be mentioned that non-recursive models of this kind can be interpreted as a discrete formulation of the Volterra series. In the following, a model is presented by which the dynamic properties of the coupling can be described sufficiently accurately

$$\begin{aligned}
M_P(k) = \overline{M}_P &+ a^1_{1,1} M_P(k-1) + \sum_{i=0}^{4} b^1_{i,1} n_P(k-i) + \sum_{i=0}^{4} b^1_{i,2} n_T(k-i) \\
&+ \sum_{i=0}^{4}\sum_{j=i}^{4} b^1_{ij,11} n_P(k-i) n_P(k-j) + \sum_{i=0}^{4}\sum_{j=0}^{4} b^1_{ij,12} n_P(k-i) n_T(k-j) \\
&+ \sum_{i=0}^{4}\sum_{j=i}^{4} b^1_{ij,22} n_T(k-i) n_T(k-j) \\
&+ b^1_{000,111} n_P(k) n_P(k) n_P(k) + b^1_{000,112} n_P(k) n_P(k) n_T(k) \\
&+ b^1_{000,122} n_P(k) n_T(k) n_T(k) + b^1_{000,222} n_T(k) n_T(k) n_T(k)
\end{aligned}$$

$$\begin{aligned}
M_T(k) = \overline{M}_T &+ a^2_{1,2} M_T(k-1) + \sum_{i=0}^{4} b^2_{i,1} n_P(k-i) + \sum_{i=0}^{4} b^2_{i,2} n_T(k-i) \\
&+ \sum_{i=0}^{4}\sum_{j=i}^{4} b^2_{ij,11} n_P(k-i) n_P(k-j) + \sum_{i=0}^{4}\sum_{j=0}^{4} b^2_{ij,12} n_P(k-i) n_T(k-j) \\
&+ \sum_{i=0}^{4}\sum_{j=i}^{4} b^2_{ij,22} n_T(k-i) n_T(k-j) \qquad (12) \\
&+ b^2_{000,111} n_P(k) n_P(k) n_P(k) + b^2_{000,112} n_P(k) n_P(k) n_T(k) \\
&+ b^2_{000,122} n_P(k) n_T(k) n_T(k) + b^2_{000,222} n_T(k) n_T(k) n_T(k) \quad .
\end{aligned}$$

To define the structure of this model (number of time displacements and order of the non-linearity) several models with different structures have been identified and the quality of the models has been tested by simulations. No physical knowledge was necessary. To determine the parameters of the model the real system is excited with multivalued pseudo-random noise. To characterise the system, recursive models should be favoured as the system possesses memory and, therefore, it can be sufficiently approximated by a discretised differential equation. In fact it was possible to cover large areas of operation with such models, even when they have been crossed with high accelerations. Deviations could only be observed when a stationary operating had been recalculated. This fact had to be recognised for the stationary areas in Fig. 12 (dotted line for simulation results).

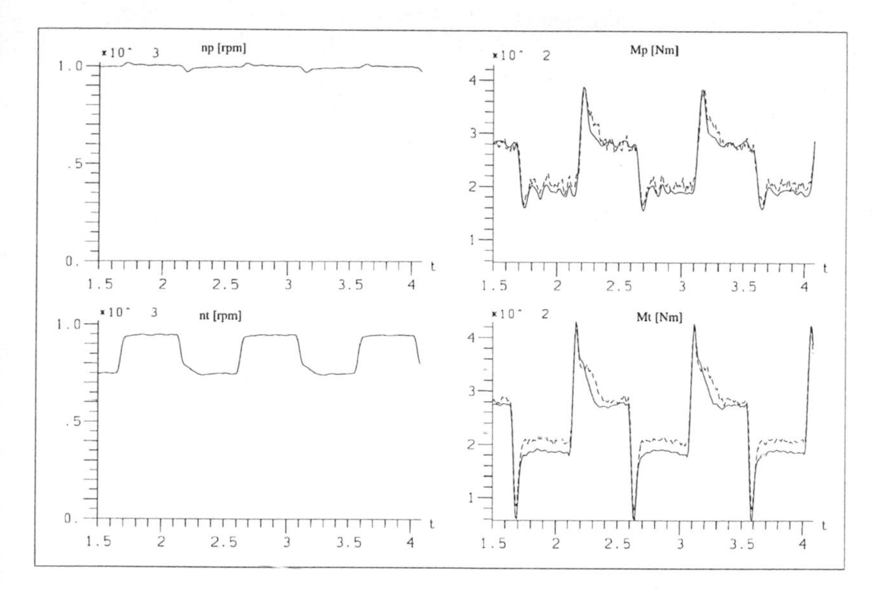

Fig. 12: Simulation with non-linear difference model

Moreover, it should be mentioned that simulations of whole drive systems will be restricted by discrete models, because the sample time is implicitly enclosed in the parameters of the model of the coupling and so all components have to use the same sample time. In certain circumstances identification problems can result if for different components of the drive system, which have very different eigenfrequencies, difference models with the same sample time shall be evaluated.

4.2 Characteristic Values

If the physical connections are so complex that no physical model can be constructed one can try to find characteristic values that determine the process (by theoretical considerations or experiments). The number of these characteristic values is reduced by introduction of dimensionless characteristic values (dimension analysis). It is known that the system can be characterised by a function of these dimensionless values. If it is not possible to specify this connection on the basis of theoretical considerations (because of complexity of the problem) then it can be found from measurements of models. This connection can be transformed to real devices by the law of similarity.

So far for the specification of the stationary operating properties of the coupling, the characteristic values of flow engines have been used.

Slippage: $$s = 1 - \frac{n_T}{n_P}$$

Performance coefficient: $$\lambda = \frac{M_P}{\rho D^5 \omega_P^2}$$

Conversion: $$\mu = \frac{M_T}{M_P}$$

Efficiency Factor: $$\eta = \mu(1 - s)$$

Neglecting the friction torque (this is only possible for the normal stationary points of operating with $1 \geq s \geq 0{,}03$) the conversion is $\mu = 1$ and hence $\eta = 1 - s$. For the description of the stationary characteristic of the coupling, a relation between the performance coefficient and the slippage is necessary. It is easy to realise different stationary operating conditions on the test stand to determine the performance coefficient. A function derived by regression analysis will be a mathematical description of the stationary operating characteristic for the area covered by measurements. For the modelling of the stationary performance this experimental methodology will suffice. But if projection has to be performed the physical modelling must be used.

If the methodology is extended to dynamic operating, additional characteristic values have to be introduced that consider time dependency

$$K_1 = \frac{\dot{\omega}_P}{\omega_P^2}$$
$$K_2 = \frac{\dot{\omega}_T}{\omega_P^2} \quad . \tag{13}$$

To characterise the dynamic performance of the coupling completely, a connection between the five characteristic values s, λ, μ, K_1 and K_2 has to be found. Due to the inner degrees of freedom of the fluid, it can only be realised in form of a differential equation. It will be difficult to determine the structure of the differential equation only from measured data because of the nonlinearities. Consequently more physical connections should be introduced (hybrid method).

5 Hybrid Modelling of Hydrodynamic Couplings

Hybrid modelling is based on a relatively simple model of analytic physical modelling. For the characterisation of the fluid flow a one-dimensional flow theory is used. It has been proved that for the description of the performance of the coupling this theory alone does not suffice. However, if this type of model is adapted to a real system by the methodology of hybrid modelling it becomes obvious that relatively simple types of models, e.g. the model of Hasselgruber (Hasselgruber, 1965), already suffice to characterise the dynamic performance of the coupling in an operating area that has been covered by non-linear black-box modelling, with discrete difference equations:

$$\begin{aligned} M_P &= a_{11}\,\dot{\omega}_P + a_{12}\,\omega_P\,\dot{V} + a_{13}\,\omega_T\,\dot{V} \\ M_T &= a_{21}\,\dot{\omega}_P + a_{22}\,\omega_P\,\dot{V} + a_{23}\,\omega_T\,\dot{V} \\ 0 &= a_{31}\,\omega_P^2 + a_{32}\,\omega_T^2 + a_{33}\,\dot{V}^2 + a_{34}\,\ddot{V} \quad . \end{aligned} \tag{14}$$

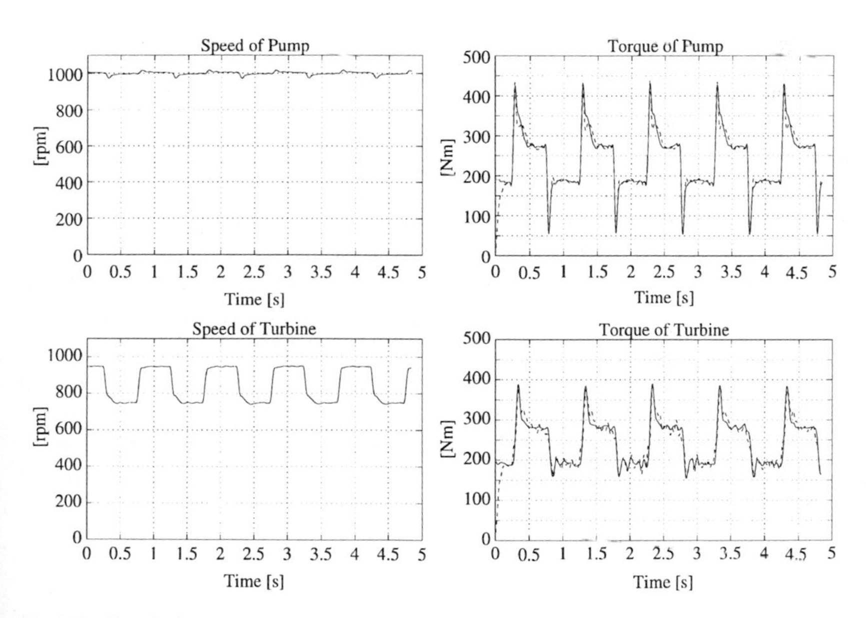

Fig. 13: Simulation with a hybrid model

For the model demonstrated in Equation (14) a linear dependency exists between 10 parameters. To find a unique identification, the third equation has to be divided by a certain parameter, (e.g. a_{34}), then the model consists only of 9 parameters which have to be estimated – in comparison to the non-linear discrete model approach where about 150 parameters are necessary. It can be recognised that the structure of the hybrid model is better than that of a black-box model. But the computational effort to estimate the 9 parameters of the hybrid model is very high because they are non-linearly dependant on the output error of the model.

In Fig. 13, a simulation with the hybrid model is shown (dotted line: simulation). The benefit of a hybrid model, in comparison with an expendable analytic physical model, is higher because simulations can be performed easier and faster. The reason for this is the number of inner degrees of freedom. The hybrid model only possesses one inner degree of freedom whereas 40 000 to 60 000 cells have been used in connection with the finite volume model – in doing so the advantage of symmetry has been already utilised.

6 Summary

Hydrodynamic couplings are highly non-linear systems which are difficult to be modelled. Different possibilities of modelling have been presented. All methods consist of advantages as well as disadvantages so that no method can be regarded as optimal. The choice of the method of modelling therefore depends on the postulations posed on the mathematical model. The disadvantages connected by this method then have to be accepted.

The analytic physical modelling of the coupling is advantageous because of the possibility of projection and of the interpretation of the flow but it is often not applicable because of the expensive grid generation and because of the computational effort. The description of the input-output-characteristic can be formulated with black-box models. For this type of model the rapid identification should be emphasised. Disadvantageous is the fact that the structure of the model is difficult to derive from experiments but, on the other hand, the time step size is implicitly determined by discrete model formulations. Thus, often the hybrid method of modelling should be favoured though the identification of the model parameters may be computer time consuming. But models can be derived, with which simulations can be performed fast, easily and accurately and the size of the time step can be varied as necessary.

References

Behrens, H. (1997): Nichtlineare Modellierung und Identifikation hydrodynamischer Kupplungen mit allgemeinen diskreten Modellansätzen (Ruhr-Universität Bochum 1997)

Folchert, U. (1994): Identifikation der dynamischen Eigenschaften hydrodynamischer Kupplungen (Ruhr-Universität Bochum 1994)

Formanski, Th. (1996): Numerische Untersuchung von dynamischen Betriebszuständen hydrodynamischer Kupplungen (Cuvillier Verlag Göttingen 1996)

Frömder, J. (1962): Das dynamische Verhalten der Föttinger-Kupplung (Technische Hochschule Hannover 1962)

Hasselgruber, H. (1965): Zum Drehschwingungsverhalten der hydrodynamischen Kupplung (Industrieanzeiger 87. Jg. 1965, Nr. 8, S. 121-123, Nr. 25, S. 436-438)

Herbertz, R. (1973): Untersuchung des dynamischen Verhaltens von Föttinger-Getrieben (Universität Hannover 1973)

Part 5 Production Processes

Simulation-based Production Control for Complex Industrial Processes by the Use of a Multilevel Concept

Matthias Thiel, Jochen Beyer, Roland Schulz and Peter Gmilkowsky

Technical University of Ilmenau, Faculty of Business Administration, Department of Business Related Computer Science I, PF 10 05 65, Helmholtzplatz 3, 98684 Ilmenau, GERMANY

Abstract. *Actually, there is a trend of integrating simulation based components in enterprise information and decision support systems. The available data in production planning and control software systems can be used to construct the models, the objective-oriented realisation of simulation experiments and the economic evaluation of the results. In this context, the conception and realisation of integrated systems is very important for the analysis, modelling, simulation and optimisation of business processes. The paper describes an approach of the production scheduling problem under consideration of high complex process conditions. The new control strategies are based on hierarchical concepts. Local and global mechanisms are used for the order scheduling. These methods are integrated in a software system for short time production planning and control.*

Keywords. *production control, simulation, interaction, coordination, semiconductor industry*

1 Introduction

The production planning and control in an enterprise means the use of computer aided systems for the organisational planning, short time controlling and monitoring of the production processes under consideration of aspects of quantity, quality, time and capacity.

The scheduling of waiting orders is the control problem of discrete production processes. In practice, simple heuristics like priority rules are used, but the results of this procedure are not optimal (Beyer, 1997). A generally selected rule is used for local decisions. But each waiting queue is another one - so the scheduling problem at each equipment is to be solved separately. Interdependencies between the equipment are ignored - the global view is lost. Therefore, the global objectives can not be achieved - the solution is not optimal. The problem is that the results, which a production planning and control decision will cause, are unknown at the decision time. When problems occur in realising the production process, you need the knowledge of the dispatchers to correct the order scheduling (Schwinn, 1992).

Complex industrial processes, such as in the semiconductor industry, are characterised by parallel equipment, aspects of load and setup and technological conditional times. Present solutions offer a regulation instrument including dispatch rules and additional load and setup rules, but these regulation instruments neither regard the coherence of the individual rules nor do they represent the complete production control problem. Up to now, there is a difference between practical demands and available solutions.

2 Simulation-based Production Control

Simulation can be conceived as examination of a process or system by the use of a substitute system called the simulation model. The simulation model represents the essential objects and behaviour of a real or hypothetical production system. This simulation model is the base of the concept of the simulation-based production control (see Fig. 1 (Weck, 1991)) to overcome the problems mentioned above.

The simulation model enables the user to test various schedules without interference to the real system. By integrating the simulation model into a complex information and control module an interconnection between existing software systems can be established (Beyer, 1995).

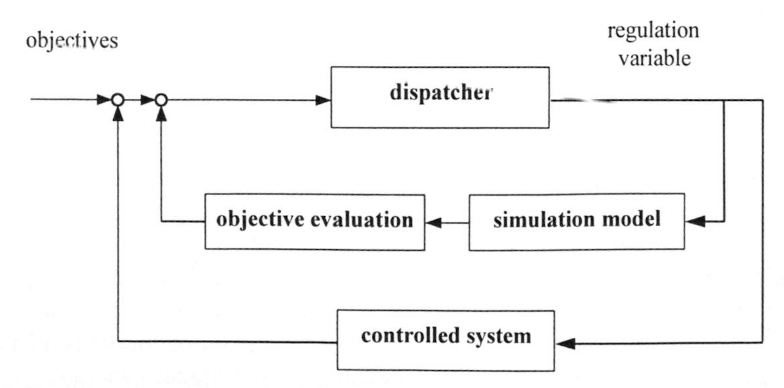

Fig. 1: Decision support by a simulation model

Generally, the order scheduling problem is represented as a discrete poly-optimisation problem. The following objective criteria are derived from the objective of the maximisation of the production efficiency:

- minimisation of cycle times
- minimisation of the transgressions of due dates
- minimisation of the amount of setup costs
- minimisation of capitalisation
- maximisation of capacity utilisation

A selected set of control strategies must be realised to detect the optimal control strategy. By the use of the process model, each single control strategy must be simulated. After that, you can choose the strategy which fulfils the above mentioned objectives best. In comparison with optimisation methods, this approach is more flexible and has a less computational effort.

3 Modelling Component

The modelling component, as a service provider, delivers the simulation functionality to the other components as a result of dynamically generated simulation models (Gmilkowsky, 1997).

The main problem of all modelling and simulation components is the flexibility needed to support short-term changes of structure, technique and organisation in the real process. Therefore, as it is not possible with feasible expenditure to manually change a static simulation model to short-term modifications, the component automatically generates context and granularity sensitive simulation models. Nevertheless, the modelling component itself has no implied simulation functionality but generates source code for an external simulation system which carries out the simulating. This reduces the development time and allows the usage of reliable external simulation systems.

The underlying principle of the process modelling component is the production process abstraction based on an expandable object-oriented metamodel. This metamodel serves as a common basis to represent the descriptors relevant to the whole process: structure, technique and organisation. In an object oriented model description, where no concrete model is specified but objects of the following class hierarchy are used for a description of the process that is closer to the real structure. The selection and modelling of a concrete simulation model is then done automatically, context sensitive and dynamically by the model generator. This is in contrast to classic component oriented systems, where a human modeller has to build the model by combining separate components with a fixed defined functionality and their parameterisation.

The architecture of the simulation and modelling component follows the client/server paradigm. At the centre of the whole system is the relational ORACLE-database which stores all description relevant information. Furthermore, by using a database it is possible to manage multiple class hierarchies and multiple instances within the system. This is important to insure reproducible simulation studies and is the foundation for exploring new knowledge about the process from the collected data of the history of this process.

4 Problem Solution by Means of a State Control System and a Multilevel Concept

4.1 Survey to the Principles of the Solution

Concerning the high complexity and dynamics of the production control, the state control system was selected as general conception (Kahlert, 1993). The following figure represents the structure of the state control system. The basis of this method is the conception of the production control as a regulator, which gives the strategies for controlling the production system as a regulation variable to the regulation system.

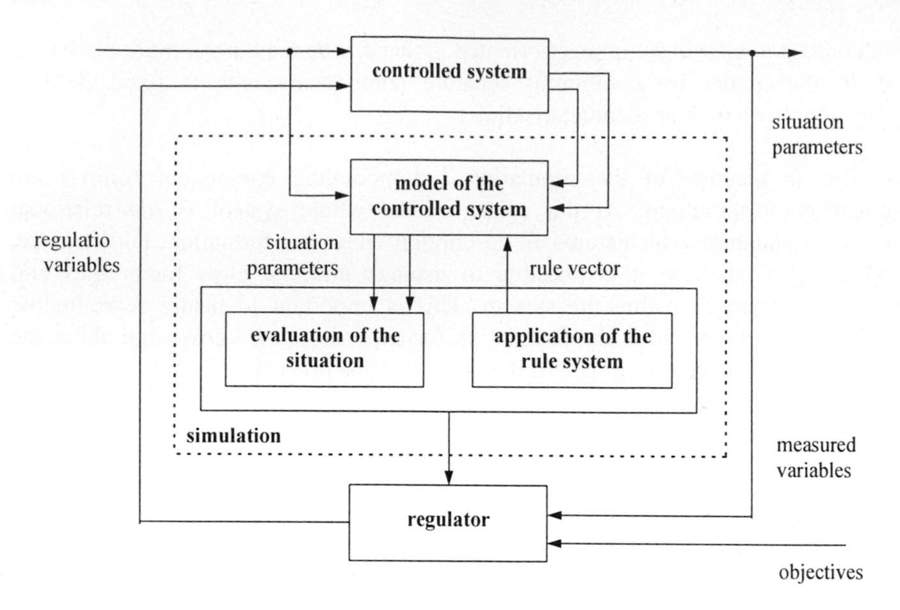

Fig. 2: Representation of the general solution

At the beginning of a simulation experiment, the current state of all production segments are taken over in the simulation model. Algorithms are started within the simulation model through different events (e.g. an order reaches a facility / a facility selects orders for the next processing). These algorithms are used on one hand to prepare for a decision, and on the other hand to make a decision. In general, decisions have to be situation-related. The state of the production segment is evaluated and after that the algorithm schedules - on base of rule systems (see section 4.3.) - the orders. The different weighting of the rules allows to try different control strategies for a production segment. Through the simulation, different control strategies were tested and rated by means of the objective system of the production system. At the end, the best control strategy is submitted to the regulator.

The simulation component is used at the global level of the production control. The necessity of a coordination between the different production segments is caused through the high interdependence, which results from the specific production conditions. To coordinate between the production segments interaction- and coordination-procedures are used.

4.2 Representation of the Multilevel Concept

4.2.1 Structure of the Multilevel Concept in Principal

In the semiconductor industry, the routing of orders takes place on the level of the workstation. At the decision time, the order will be assigned to one of the equipment of the workstation (a workstation consists of equipment, which can do the same processing). On this account the workstation is the lowest decision level, called "local problem solving level". In practice, several workstations were combined to one workarea. Hence, the second decision level results - called "global problem solving level".

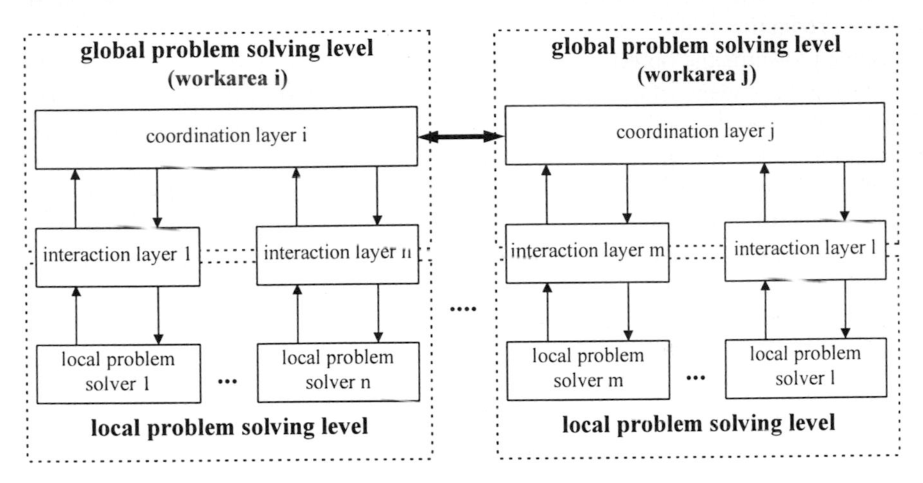

Fig. 3: Architecture of the Solution

4.2.2 Strategy in Principal

The task of the local problem solver is to create a preliminary order sequence under consideration of local objectives. If no definite decision based only upon local information is possible, the interaction layer of the local problem solver tries to find possibilities of improvement by considering further information. One of these possibilities is the consideration of the departure of further orders (Fowler, 1992). These orders will be searched in all of the workareas of the production system. If such orders are found, the interaction layer enables the coordination layer of the global problem solving level. The coordination layer has the tasks (1) to record all inquiries for further orders of its local problem solving levels, (2) to coordinate these inquiries under consideration of all production conditions and (3) to evaluate these inquiries on the basis of a global objective function. (4) Finally, the alternative that fulfills the global objective function best is performed. Thereto the coordination layer sends instructions to the appropriate local problem solvers.

4.3 Description of the Problem Solving Levels

4.3.1 The Local Problem Solving Level

The analysis of the specific production conditions in the semiconductor industry pointed out that the application of priority rules entirely does not suffice for solving the sequencing problem. Therefore, the local problem solving level uses a rule based component, called the RBD-Component (Rule Based Dispatching Component) which consists of hierarchically structured criteria and rule systems. The RBD-Component aggregates three different hierarchies of criteria to a total priority for each charge (Thiel, 1997). Each hierarchy of criteria refers to one or more production conditions. The differentiation of further hierarchies of criteria is oriented towards the objectives to minimise the variation of the due dates and the sequence dependent setup times and to maximise the workload of the charge. The aggregation of the individual parameters in hierarchies of criteria is rulebased. Each hierarchy of criteria indicates which contribution the charge can perform to fulfill the mentioned local objective. Different local strategies can be realised by weighing the hierarchies of criteria differently. The weight of a criterion demonstrates the importance of this criterion and the associated production condition. So it is possible to realise both strategies to minimise tardiness and lateness of orders and strategies to maximise the throughput. A higher flexibility of the solution results from this.

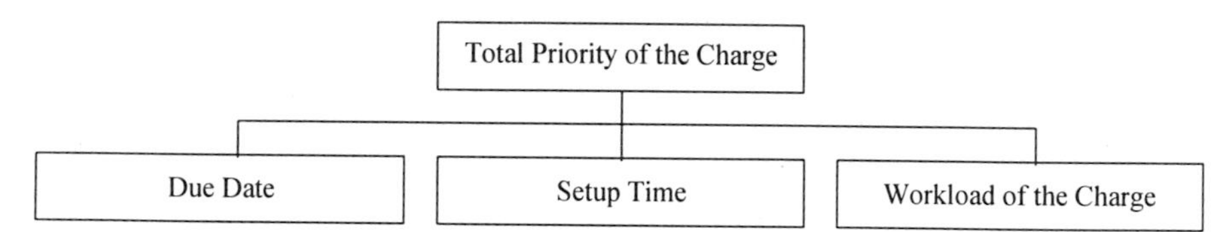

Fig. 4: Aggregation of the Contribution of Hierarchies of Criteria to a Total Priority of the Charge

4.3.2 The Global Problem Solving Level

Objective of the Usage of Interaction- and Coordination-Procedures

Local rules do not consider the interdependencies between production segments. Hence a "good" local sequencing can be found, but a summarisation of "good" local sequences will inevitably not result in the optimal sequencing for the whole production line (Fowler, 1992; Gurnani, 1992; Holthaus, 1993). Therefore, the global problem solving level uses interaction and coordination processes to solve certain manufacturing conditions (for example technological conditional times) and to achieve a better production control with respect to the global objective function. Finally, this is reflected in a better quality of planning.

The range of applications of interaction and coordination procedures derives from the following objectives:

- maximisation of the utilisation of the charge
- maximisation of the capacity utilisation
- minimisation of the amount of setup times
- minimisation of the waiting times
- minimisation of the transgressions of conditional times

Coordination of Different Inquiries at one Production Segment

An essential unsolved problem by applying interaction methods is the decision on realising the arriving inquiries (Holthaus, 1993). When inquiries arrive at a production segment there are two tasks:

1. to evaluate the inquiries by means of local and global criteria
2. to decide whether the inquired tasks should be realised or not.

When a production segment receives such inquiries, its decision component must transfer these inquiries onto the local problem solving level. Based on the local objective criteria, this level determines the change of the benefit to itself the realisation of the inquired task would cause. The global problem solving level of the adjacent production segment compares the given growth of benefit to the benefit of the requesting production segment. Only the inquiry with the highest positive total benefit is confirmed. All other inquiries will be cancelled.

5 Results of Simulation Studies of a Wafer-Fabrication

The presented simulation-based solution was developed within the scope of the research initiative "SMART FABRICATION". "SMART FABRICATION" is a joint project of the German Federal Ministry for Research, Education and Technology (BMBF) and the german semiconductor industry. The aim of the project is to supply innovative production concepts, which will lead on one hand to a marked increase in flexibility and quality, and on the other hand to a decrease in manufacturing costs and a decrease in the length of time of present and future starting-up and running phases in leading-edge production (Thiel, 1997).

The first implemented algorithm has demonstrated its efficiency in comparison with available priority rules or combinations of a priority rule and a setup rule (ManSim, 1998).

criterion	TD	Setup Rule	Algorithm
Processing time [in hours]			
• mean processing time	143	145	133
• min. processing time	93	102	91
• max. processing time	180	182	174
Deviation from due date			
• rate of lateness	17,1%	16,8%	13,2%
• mean lateness [hours]	20	17	13
Capacity utilisation			
• quantity produced	590	595	657
• rate of idle time	30%	29%	32%
• utilisation of charges	68%	68%	71%
• rate of setup time	27%	27%	19%
Queue [in orders]			
• mean length	14,8	15,0	5,7
• max. length	50	46	24

Legend:

TD application of a priority rule which refers to the due dates
Setup Rule combination of a priority rule and a setup rule (ManSim, 1998)
Algorithm application of our solution

Table 1: Results for different methods on a bottleneck

6 Summary

The answer to the dispatching problem in the semiconductor industry requires the integration of all production conditions in one solution. The realisation of the requirements of an equivalent representation of reality creates the assumption of getting valid conclusions from the simulation-results, which relates to the real system. So there is the given basis for a realistic selection of an order schedule.

A model for simulation is developed that supports the order scheduling on a coarse and fine level. Within the scope of the coarse scheduling, a forecast of the due dates of production orders is supported. By way of contrast, the fine scheduling has to determine an efficient order schedule in relation to the actual production situation and the given objectives. By integrating interaction procedures in the solution the dispatcher is enabled to recognise and control interdependencies between the production segments. By integrating specific production conditions and market orientated demands into the model, the deficit of prior production control instruments could be nearly redressed. Experiments in a wafer fabrication have verified the efficiency of the developed algorithm.

References

Beyer, J. (1995): "Konzeption eines simulationsgestützten Informations- und Steuerungsmoduls für die Fertigungssteuerung"; *6. Ilmenauer Wirtschaftsforum;* 26.10.1995; TU Ilmenau

Beyer, J.; Gmilkowsky, P. (1997); "Simulationsgestützte Fertigungssteuerung mit Fuzzy Scheduling und Interaktionsverfahren"; *6. Symposium „Simulation als betriebliche Entscheidungshilfe: Neuere Werkzeuge und Anwendungen aus der Praxis", Proceedings;* 3.3.-5.3.1997; Braunlage

Fowler, J. W.; Phillips, D. T.; Hogg, G. L. (1992): "Real-time of multiproduct bulk-service semiconductor manufacturing processes."; *IEEE Transactions on Semiconductor manufacturing; Proceedings 5;* 158-163

Gmilkowsky, P.; Eckardt, F.; Palleduhn, D. (1997): "Konzeption eines integrierten Systems zur Modellierung, Simulation und Analyse von diskreten Produktionsprozessen"; 10. Ilmenauer Wirtschaftsforum; 6.11.1997; TU Ilmenau

Gurnani, H.; Anupindi, R.; Akella, R. (1992); "Control of Batch Processing Systems in Semiconductor Wafer Fabrication."; *IEEE Transactions on Semiconductor manufacturing; Proceedings 5;* 319-328

Holthaus, O.; Ziegler, H. (1993): "Zur Koordination dezentraler Werkstattsteuerungsregeln"; *Operations Research Proceedings 1993;* 157-163; Berlin

Kahlert, J.; Frank, H. (1993): "Fuzzy-Logik und Fuzzy-Control."; Vieweg Wiesbaden

"ManSim - User Manual". Release 3.10. 1998. TYECIN System Inc.

Müller, J.(Hrsg.). (1993): "Verteilte Künstliche Intelligenz: Methoden und Anwendungen"; BI-Wissenschaftsverlag

Schwinn, J. (1992): "Wissensbasierter CIM-Leitstand" Wiesbaden

Thiel, M.; Schulz, R.; Beyer, J. (1997): "Simulationsbasierte Fertigungssteuerung bei komplexen Prozeßbedingungen am Beispiel der Chipherstellung."; *Produktion als Dienstleistung - 10. Ilmenauer Wirtschaftsforum;* 6.11.1997; TU Ilmenau

Weck, M. (1991): "Simulation in CIM"; Springer-Verlag (CIM-Fachmann)

Process Prototyping of Manufacturing Processes

Bernhard Lenz, Gunther Reinhart, Frank Rick and Hans-Jürgen Trossin
iwb - Institute for Machine Tools and Industrial Management, Technische Universität München, Boltzmannstrasse 15, 85748 Garching near Munich, Germany

Abstract. *In manufacturing technology the term prototyping is usually known in the context of rapid prototyping, the rapid building of physical product models. Usage of this term in the context of process modelling is new and unusual. The concept of process prototyping aims to provide a tool, that enables the user to build up models of manufacturing processes for tests and studies of relevant parameters. From the large number of methods for simulation of manufacturing processes the most suitable have to be selected. Criteria for this selection are shown and explained. As an example, a prototype of the laser welding process, realised in a finite element system, is discussed. Possibilities opened up by this model are evaluated and further improvements are pointed out.*

Keywords. *process prototyping, simulation, manufacturing process, FEM*

1 Introduction

Product development in today's industry is characterised by decreasing development times together with decreasing costs and increasing quality requirements. In view of this, companies try to predict and test product properties as early as possible in the product development cycle. This is achieved by using phyiscal product models, built for example by rapid prototyping techniques, or by means of simulation techniques, e.g. in the context of digital mock-up.

Simultaneously with the products, the related manufacturing processes have to be designed and tested, but in many cases experimental work with the original process is not possible. Either the product model exists only in the computer and there is no physical product model, or experimental work would be too expensive and/or time-consuming. Therefore, a functional environment for product and process development has to provide an experimental area where parameter studies, tests and optimisations of manufacturing processes can be performed.

2 Definition and Concept of Process Prototyping

Process prototyping is a systematical concept for the creation of process models in order to provide information about production processes at early stages of simultaneous process and product development.

In manufacturing technology, there is no exact definition of a prototype by standards or guidelines. The term prototyping is usually known and established in the context of rapid prototyping, that is to say the rapid building of physical product models. Usage of this term in the context of process modelling is new and unusual. Nevertheless, analogies can be determined between the prototype of a product and the prototype of a process.

According to the description of Gebhardt, a prototype of a product corresponds as far as possible with the series product, but differs principally in the type of the production process. In terms of functioning the prototype has to allow the examination of one or more of the product's properties (Gebhardt, 1996). To transfer these basic ideas to a process prototype, not only the literal description but also the purpose of the latter has to be considered.

The main purpose of a process prototype is to increase the maturity of manufacturing processes at early stages of the development process. The abilities of a process prototype do not necessarily depend on the way how it was designed. For this reason, a process prototype can be physical or virtual. To guarantee that all process properties can be examined in the prototype, the level of detail and the interactions which are considered in the model should not be essentially smaller than in the original process.

The following core requirements can be stated:

- The model must be suitable for experiments to be performed.
- Information has to be achieved at the level of physical interactions. Every parameter that describes physical interactions, has to be quantitatively ascertainable.
- The number of physical interactions that are studied should not be substantially smaller than in the original process.
- No expert should be required for preparation and performance of experiments. Results should be provided descriptively and easy to use.

For an application in practice the concept of process prototyping should include serveral superordinate features:

- Provision of qualified methods, which meet the requirements mentioned above.

- A guidance that enables a problem-based choice of a modelling method. The selection of the right modelling technique is of decisive importance for the quality of the process prototype.
- Support in the application of the different methods.
- Proposals of how to integrate this concept in to the business processes of a company.

3 Modelling Methods and Selection of Methods

There is a large number of methods to build up and examine models of manufacturing processes. Within the context of manufacturing technology, the following selection will provide an overview, though without any claim to the completeness (Fig. 1).

Methods can basically be classified into analogue and digital models. Digital means, that the process model exists only in a computer, experimental studies are exclusively computer-based. In contrast, analogue models usually do not need computers and software. In the easiest case only a pencil and a piece of paper are needed to study a process, more complicated problems can be solved by designing a model process.

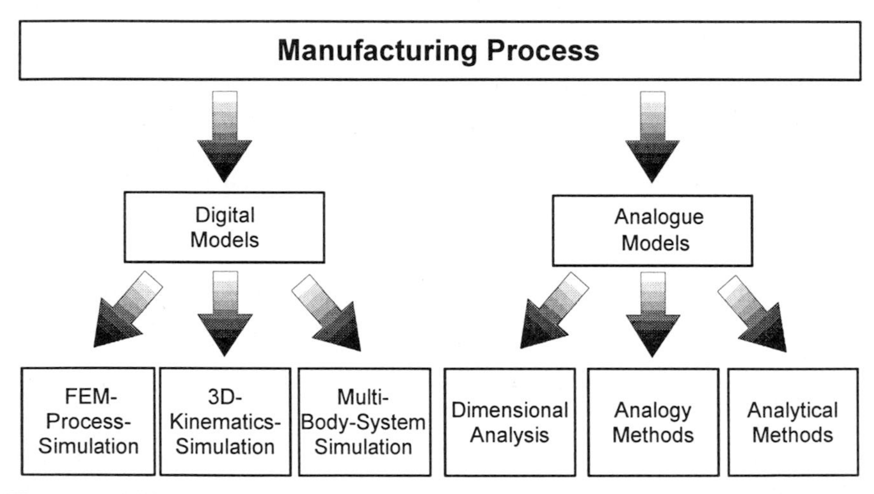

Fig. 1: Modelling techniques used in manufacturing technology

3.1 Analogue Models

Analytical Methods: The process is described by a set of differential equations, which can be solved analytically. Examples of this technique are provided for in turning processes (Milberg, 1971) and for circular sawing (Zäh, 1995).

Analogy Methods: If different processes can be described by differential equations of the same structure, analogies can be used to design a process model. An example Application can be found in Mavroudis (Mavroudis, 1971).

Dimensional Analysis: Two processes are technically equivalent, if they have the same dimensionless representation. Correspondingly, the process describing dimensionless numbers must have the same numerical values in both the model and the original process. The main effort involved in this method is to find a dimensionless description of all relevant process parameters. If this representation is obtained, a scaled model process can be designed. Experimental studies can be performed using the model process. The results are transferred to the real process by scaling equations. This method is state-of-the-art in many engineering sciences (see e.g. Zlokarnik, 1991).

3.2 Digital Models

Multi Body System - Simulation: The Multi Body System - Simulation (MBS) is used for the calculation of the kinetics of systems, which consist of more components. With regard to production processes, MBS can be used to increase the accuracy of machine tools during the process (Reinhart, 1997).

3D- Kinematics - Simulation: Three-dimensional models of the product and the manufacturing plant are examined in a common model, in which all kinematic properties of the objects are additionally implemented. By means of this tool the original sets of motion, e.g. the motion of a robot, can be checked virtually. This method has been used so far for the planning of assembly processes (Rossgoderer, 1995).

Finite Element - Method (FEM): This method is a common tool in product development and partially in process development in different areas of production technology, examples are the design of machine tools (Schneider, 1997), the simulation of sheet metal forming (Tekkaya, 1998) and the simulation of assembly processes (Wisbacher, 1991).

3.3 Selection of Methods

In the definition section of this paper the requirements of a process prototype have been described. Demands on the modelling methods can be derived from these requirements. Obviously one selected method does not necessarily fit every, arbitrarily chosen process. Therefore, the concept of process prototyping should contain a toolbox of methods, that provides the user with the right modelling method for his specific process.

What can the criteria for the selection of a model be?

First of all, process prototyping has to cope with the same laws as any other procedure in production and production planning, as expressed by the classic triangle of costs, quality and time. The more expensive the experiments conducted at the original process are, the more interesting is the use of a process prototype. Furthermore, the method providing highest quality in shortest time has to be selected. Second, as mentioned in the context of the definition of a process prototype, all relevant process parameters and properties must be describable and experimentable in the chosen type of method.

At the Institute for Machine Tools and Industrial Management (iwb) several methods have been examined for process prototyping, regarding the critera mentioned above. Based on those results, only two methods meet these requirements at present, the FEM and the dimensional analysis.

At the iwb the dimensional analysis is used to examine the glass pressing process (Reinhart, 1997). Using this modelling method, experimental investigations of this process are undertaken at room temperature with a composition of $Ca(NO_3)_2$ and KNO_3 as a model liquid instead of soda - lime glass. Generating experimental data with the model liquid, the peak temperature of the process can be scaled from 973K to 375K. However, with an increasing number of relevant process parameters, this method becomes more and more inefficient. For example, at the process of laser welding described in the following section, there are three different states of aggregation of the material, solid, liquid and vapour, which coexist simultaneously. It is very difficult to find a model material, which shows this kind of material behaviour at reduced temperatures.

The Finite - Element Method (FEM) is used at the iwb as a digital method for studies of manufacturing processes. FEM is a tool, which is basically qualified to handle different types of problems. By setting geometric and material properties, boundary and initial conditions as well as the type of solution, the simulation environment is fitted to the specific problem. A process model is generated only in the computer and all experimental data are generated digitally. Parameter studies are obtained by varying initial data as well as boundary conditions and solving the describing differential equations subsequently. At the iwb this method is used to build up a process prototype of laser welding. This will be described in the following section. If a process is very complex and no mathematical model is present, the FEM reaches its limits. The effort for generating a model in a FE - system becomes too great. For example, this method is suitable for the examination of glass pressing processes, because there is no adequate theoretical process model describing the complex material behaviour of glass combined with the problems of heat transfer and fluid flow.

4 Process Prototyping: Finite Element - Analysis of the Laser Welding Process

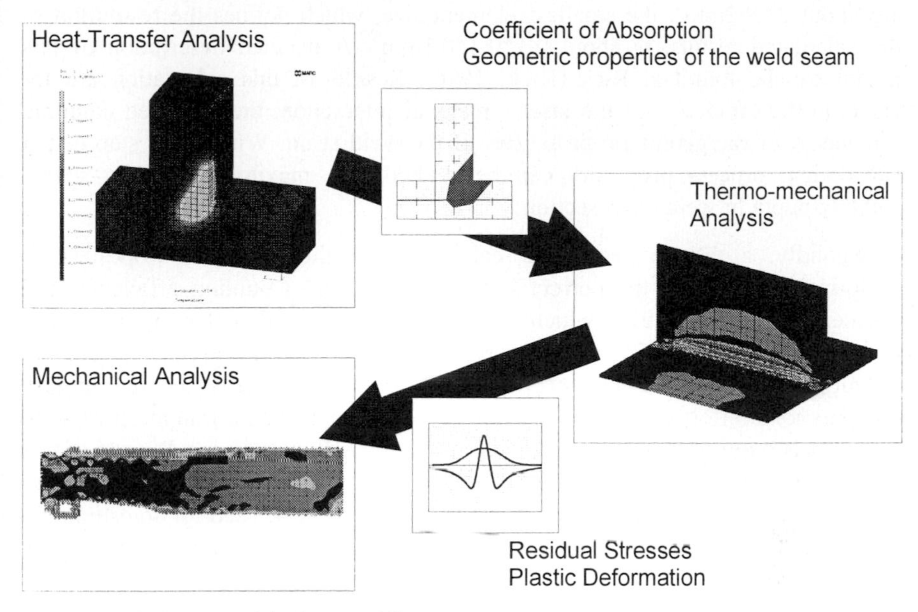

Fig. 2: Analysis steps of the laser welding process

During the last 5 years, the laser has become a standard tool in production technology. However, a high performance laser system for material processing is still a very considerable investment, calling for extended availability and utilisation times of laser systems in production. The potentials of a laser system can only be exploited, if the product and related production process are fitted satisfactorily, which could only be done by experimental studies so far. So, on the one hand, experimental tests are essential, on the other hand economic considerations allow at most only a small number of experimental studies. The laser welding manufacturing process is therefore an appropriate example for process prototyping.

The laser welding process is very complex. Different states of aggregation coexist in a very small area of the work piece. Furthermore, laser material interaction, as well as the material's mechanical reaction to thermal loading, are highly non-linear. To investigate the complete manufacturing process including the resulting product properties, different models are necessary (Fig. 2). Several aspects of the process are examined by different types of models at various levels of detail, but, all parts of the model are based on Finite Element methods. They are realised in the commercial finite element system MARC (MARC, 1998) combined with user-defined subroutines.

First, the formation and motion of the keyhole is calculated in a thermal model of a small part of the workpiece. This model is based on the known models of keyhole formation (see for example (Beck, 1996)). The dimensions of the model are about $2*2*2mm^3$; the smallest element size, which defines the resolution of the calculated results, is about $0.1*0.1*0.1mm^3$. A detailed description of this model can be found in Rick (Rick, 1996). Results of this calculation are for example the efficiency of the laser - material interaction, the so-called coupling efficiency, or the geometric properties of the weld seam. Within this step of the calculation, process properties can be studied, e.g. maximum welding speed, welding depth or weld cross section.

Secondly, a thermo-mechanical calculation of the specimen workpiece is performed, containing the correct type of weld joint. Coupling efficiency and geometry of the weld seam, which were calculated in the first step, are now used as input parameters for the moving heat source. Due to the non-linear thermomechanical behaviour of the material, this step calls for the highest computational efforts. As in the area close to the weld seam a fine mesh of solid elements is required, the model size is limited to small workpieces. Results of this type of calculation are stresses and displacements of the workpiece as a function of time and space. The calculated product properties can be used for an estimation of residual workpiece distortion or for an optimisation of clamping technology (see e.g. (Lenz, 1998)).

In a third step, the resulting properties of a laser-welded product are examined. In the example shown in Fig. 3, a mechanical calculation of an engine bearing under torsional stress was performed. In this example, only the geometry of the weld seam was taken from the thermomechanical calculation. This easy model adequately evaluates critical points of the structure, where failure occurs under cyclic loading. Further efforts now have to be devoted on the question of how to consider the residual stresses and plastic strains, calculated in the second step, as initial conditions in a calculation of fatigue behaviour of complete laser-welded structures. The coupling between the second and the third step of this model therefore has to be improved.

As described in this section, finite element modelling is able to meet criteria for a process prototype if the problem is distributed among different partial models. For the application of such a prototype by a manufacturer, further improvements are needed. Much FEM expertise is required to transfer geometric data from a CAD model to an experimentable process model. Furthermore, interfaces between the different partial models have to be developed in order to obtain a user-friendly simulation environment.

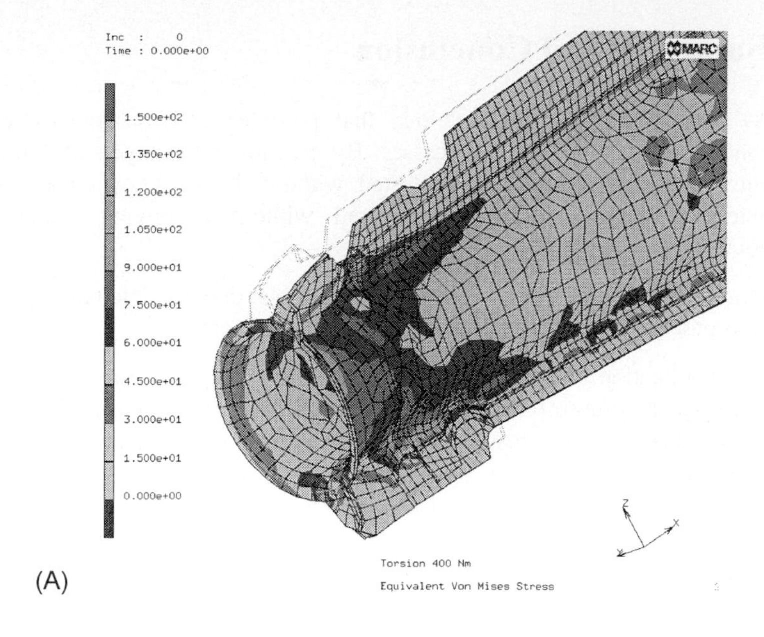

(A)

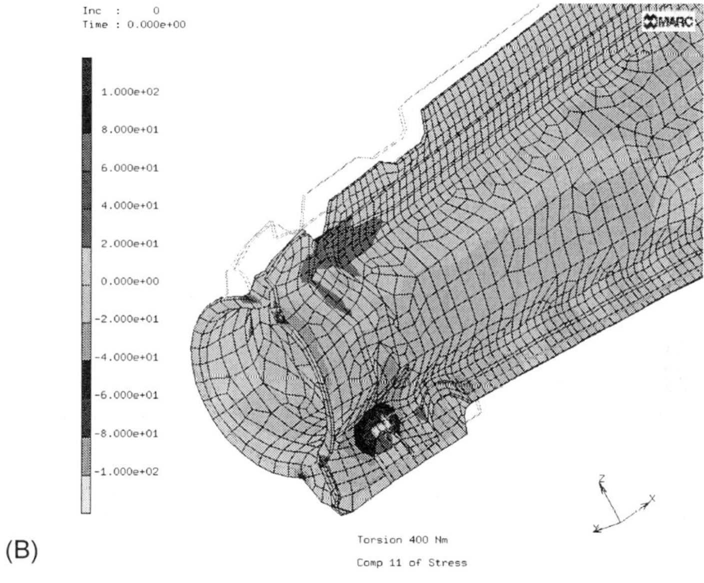

(B)

Fig. 3: Laser welded engine bearing under torsional loading
(A) Von Mises stress (B) Stress in axial direction

5 Summary and Conclusion

Process prototyping is a methodology that provides an environment for the examination of manufacturing processes. By creating a prototype of a process, relevant process properties can be tested without deeper knowledge. Process parameters can be varied and optimised without occupying the original manufacturing systems.

Taking the process of laser welding as an example, one possible realisation of a process prototype was shown based on finite element methods.

For an application of process prototyping in practice, a guideline is necessary, which enables the user to find the most suitable modelling method for his specific process. Furthermore, an answer to the question has to be found as to how a process prototyping system can be integrated in the business process of product development.

Acknowledgement

The authors wish to thank the *Bayerische Forschungsstiftung for funding parts of this work in the project "Laser Assisted Lightweight Construction" (FORLAS II)*, furthermore the *Bundesministerium für Bildung, Wissenschaft, Forschung und Technologie* (BMBF) for support in the context of LASER 2000 and their industrial partners.

References

Beck, M. (1996): A survey of modelling laser deep welding. Proc. of the 6TH European Conference in Laser Treatment of Materials ECLAT '96, Stuttgart. AWT, Wiesbaden 1996

Gebhart, A.(1996): Rapid Prototyping: Werkzeuge für die schnelle Produktentwicklung, Carl Hanser, München 1996

Lenz B. et al (1998): Process Prototyping - Finite Element Analysis of the Laser Welding Process. To be published in Lasers in Engineering 7 (1998)

MARC (1998): MARC K7.2 finite element code, MARC Analysis Research Corporation, Palo Alto, USA

Mavroudis, M. (1971): Modelluntersuchungen über die Ausbildung der Zirkulationszone vor den Blasformen von Hochöfen, Dissertation TU Clausthal, 1971

Milberg, J. (1971): Analytische und experimentelle Untersuchungen zur Stabilitätsgrenze bei der Drehbearbeitung. Dissertation TU Berlin, 1971

Reinhart, G. and Trossin, H.-J. (1998): Process Prototyping - A Low Temperature Model of a Glass Pressing Process. In: Clare, A. G.; Jones, L. E. (ed.): Advances in the Fusion and Processing of Glass II. American Ceramic Society, Westerville, OH, USA, (5) 1998

Reinhart, G. und Weissenberger, M. (1997): Eigenschaftsoptimierung am virtuellen Prototyp - Integrierter Konstruktionsarbeitsplatz für Werkzeugmaschinen. ZWF 92 (1997) 11, pp. 571-574.

Rick, F. et al (1996): Improvment of laser orientated product design using simulation techniques. Proceedings of the 6th European Conference in laser treatment of materials, ECLAT 96, Stuttgart. Wiesbaden, AWT 1996

Rossgoderer, U. and Woenckhaus, C.(1995) : A Concept for Automatical Layout Generation. Proceedings of the 1995 IEEE International Conference on Robotics and Automation, Vol 1, pp 800-805, Nagoya 1995

Schneider, C. and Reinhart, G. (1997): Entwicklung dynamisch beanspruchter Werkzeugmaschinen: In: Streckhart, A (ed) Proceedings of 14TH FEM-Congress, Baden-Baden 1997

Tekkaya, A. (1998): State-of-the-art of Simulation of Sheet Metal Forming: In: Proceedings of the 6th International Conference on Sheet Metal (SHEMET 98), Twente, Twente University Press 1998, pp 53 - 66

Wisbacher, J. (1991): Methoden zur rationellen Automatisierung der Montage von Schnellbefestigungselementen, Berlin, Springer 1991

Zäh, M. (1995): Dynamisches Prozessmodell Kreissägen, Berlin, Springer 1995

Zlokarnik, M (1991): Dimensional Analysis and Scale-up in Chemical Engineering, Berlin, Springer 1991

Advanced Software Improves Decision Quality in Rolling Mill Technology

Holger Müller[1], Thomas Peuker[2] and Günter Wozny[1]

[1] Technical University of Berlin, Institute for Process- and Plant-Design, Sekr. KWT-9, Straße des 17. Juni 135, 10623 Berlin, Germany

[2] Siemens AG, Industrial Projects and Technical Services, ATD MP TM ME, P.O. Box 3240, 91050 Erlangen, Germany

Abstract. *This paper describes a decision-support tool for the design of new mills and the modernisation of older ones, as well as the operational control of foundries. The modular structure and the flexible handling of this tool enables rapid generation, simulation, and evaluation of a number of different plant variants. The tool described here is currently under development. Individual components, such as the flowsheet simulator, have already been developed and are now being used to meet the challenges outlined below.*

Keywords. *steel industry, steel rolling mill, expert system, object-oriented programming, data management*

1 Introduction

The steel industry is going through major changes. Although steel continues to maintain a strong position alongside aluminium, plastics and concrete, there are currently major differences between supply and demand at the local level. Steel production capacities are no longer located where tomorrows customers are expected to be. Over the long term, for example, demand will grow noticeably in Asia (1997: 320 Mio t/a; 2005: 400 Mio t/a (Vondran, 1998)). Thus, new plants will have to be located at or near the centres of future demand. Additionally, the falling price index for steel products (such as rolled steel: 1985=100%, 1997=81%, (Stähler, 1997)) is driving the pace of innovation at an ever faster rate. New steel production capacities are now being made available primarily in the form of electric mini-mills. Considering the fact that a mini-mill requires a capital investment of more than DM 500 million and can produce approximately 2.5 Mio t/a of sheet steel, it becomes clear that the decisions leading to the building of a new plant, the modernisation of an older one, and the optimisation of the operating regimen are of considerable economic significance. To profit from future trends, it has become necessary, more than ever before, to invest in the R&D needed to satisfy market-driven demands for innovation.

Over the last few years, technological operations, such as casting, flat rolling and edging, have been thoroughly optimised. Viewed as a whole, however, there is still significant potential optimisation within steel plants. Not only the sum of individual optimisation measures, but the systematic approach to them, enables decisions to be made that lead to a positive production result. This result involves the optimal plant and operating strategy for a particular product range (with regard to steel type and dimensions) for a particular market segment. A further point of view involves the optimal product spectrum for an existing plant. Throughout these considerations, it is important to keep in mind that there are no ready solutions to all problems, but that individual solutions must be found that are adapted to generalised goals (including the market segment to be targeted, investment willingness) and boundary conditions (cost structure; local situation, including laws; and others) of a plant operator. Doing so effectively requires experience. And, if new approaches are to be followed, as the drive to innovate increasingly requires, then this cannot be done without relying on simulation. Against this background, decisions involving, for example, the increase of throughput, the minimisation of specific energy consumption, the reduction of operating costs (such as roll wear), and/or the growth of profit margins can be made more effectively. This applies both to new plants and to the modernisation of older ones (including the number of stands, the length of the cooling section, the size of the rolls), as well as to the optimisation of the operating regimen (rolling speed, strip thickness).

When considering the process of mini-mills (see Kneppe, 1998), a number of technological challenges become apparent. Mini-mills are very compact and very economical (Fig. 3.3). A caster generates thin slabs (length x width x thickness = 9000 mm x 1600 mm x 250 mm, temperature approx. 1200 °C). A roller hearth furnace then brings the temperature of the slabs to the required level and keeps it there. This is followed by the rolling process in a multi-stand mill, which provides the desired end thickness. The rolling mill can comprise of either a roughing mill and a finishing mill or just a finishing mill. Prior to the coiling of the strip, it is cooled in the cooling section, where the metallurgical properties are stabilised. Guaranteeing the target product characteristics for the finished strip (e.g. thickness = 1 mm, temperature = 890°C at strip speeds greater than 12 m/s) calls for comprehensive technological knowledge and experience.

Decision-making is not a closed process, but comprises of a number of steps: goal identification, process synthesis, evaluation, compromise finding, and decision. Our new ANL-HYBREX tool, for plant design and operation based on a hybrid expert system, enables the generation and evaluation of several promising plants within a reasonable time span. Equipped with convenient options, the tool can evaluate a large number of aspects of plant design, leading to higher decision quality.

2 Flowsheet Simulation in the Steel Industry

In a flowsheet simulation, processes are represented as a network of material flows (Fig. 2.1), with the nodes denoted as unit operations.

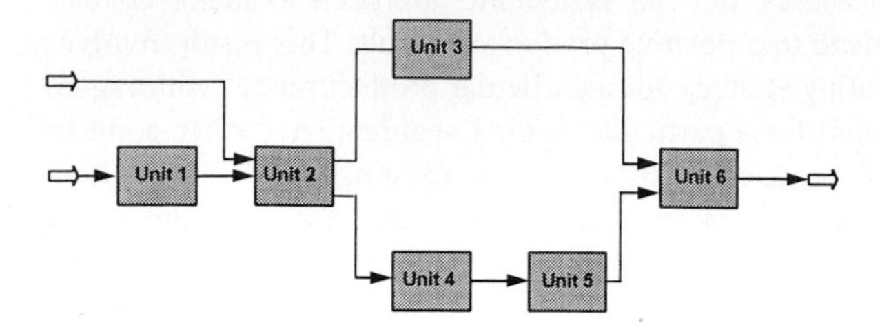

Fig. 2.1: A system as a network of nodes (units) and flows

This way of thinking corresponds to the phase model of production suggested by Polke (Polke, 1994). According to this model, process and state follow one another. A material flow in a particular state serves as the input for a process (= element[1] of the system = unit operation[2]). The process influences the material flow in such a way that a process output generates a material flow with a particular state.

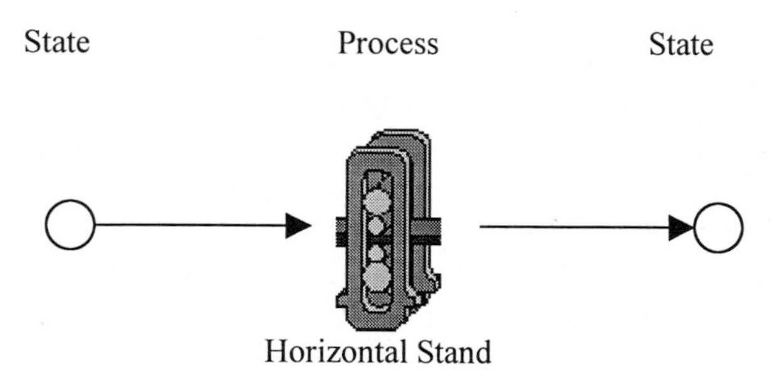

Fig. 2.2: State-Process-State (horizontal stand of a mini-mill)

---

1 Concepts according to system theory

2 Concepts from a technological viewpoint

This procedure is well-suited to the engineer's way of thinking and can be applied to a number of different technical processes, including mini-mills (Fig. 2.2).

The units involved are the caster, the roller hearth furnace, the roller table, the edger, the horizontal stand, the cooling section, and the coiler. The state of the strip (the material to be rolled) is characterised by values associated with geometry, temperature, speed, and material microstructure.

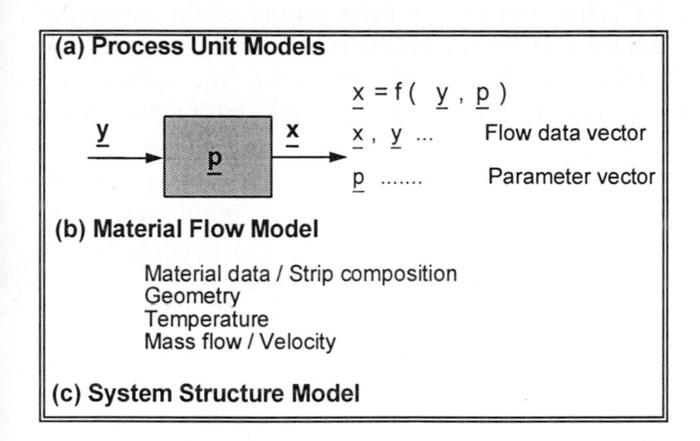

Fig. 2.3: Components of the plant model

This enables the modelling of complex plant structures. The complete plant model is made up of the models of the process units, the model of the material flow, and the model of the system structure. (Fig. 2.3).

A steady state model is sufficient for the analysis of different problem areas. In the first development phase, all important values are viewed independent of time. This also means that strip characteristics (states) are captured over the entire length of the strip. The second development phase uses a discrete model, which takes into account the strip states and unit parameters at defined points in time (quasi-stationary viewpoint).

3 The ANL-HYBREX Tool

3.1 Modularisation and Interfaces

The ANL-HYBREX tool for plant design and operation is developed with the form of modular design (Fig. 3.1). It based on a hybrid expert system. At the core of the ANL-HYBREX tool is a **flowsheet simulator**, which provides a quasi-stationary simulation of the entire process. To keep abreast of the increasing importance of process optimisation, a second important component is the **optimiser**, which works closely together with the simulator. The optimisation procedure can follow several objectives at once, such as maximising throughput while minimising energy consumption, so that more than just one optimisation can be pursued (vector optimisation). The module for **compromise finding** supports the user in the evaluation of individual alternatives. Beside the classic utility analysis, finding compromises involves **strategic knowledge** (operator knowledge and capability), which is available in the form of heuristic rules.

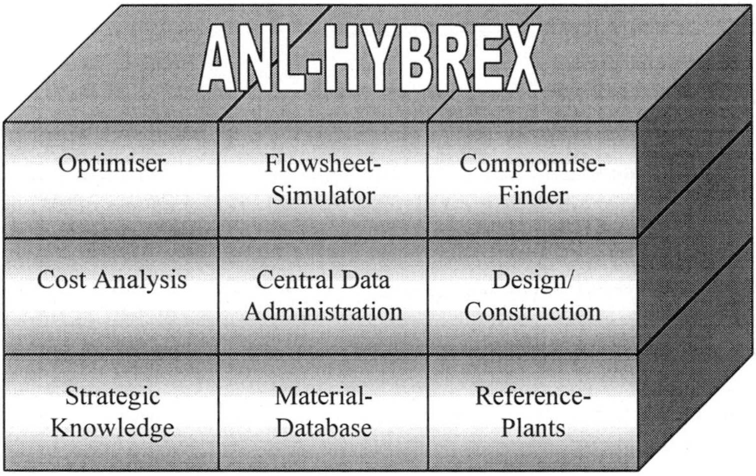

Fig. 3.1: ANL-HYBREX modules

After an optimal mathematical solution has been determined, the plant is **designed and constructed**, with close attention to industry standards (standard product series). This module is also the most important link between theoretical knowledge derived from simulation and practical application. **Reference plants** have already been proven in the field. This makes the information stored in this module on structure, parameterisation, and operating regimen – from the first synthesis step to the configuration of the new plant – of great value. The **cost analysis** module contains prices and business administration models in support of cost-benefit analyses. Information contained in this module helps to determine the economic prospects of the plant and can be input into the compromise finding module as required.

Designed as a relational database, the **central data administration** serves as an interface among all modules and can even be accessed with standard software (Microsoft Access) from outside ANL-HYBREX. All modules communicate with the database via standard access methods (SQL, ODBC, DAO), read all necessary input data out, and write all data from the processing cycle back in. The flowsheet simulator, for example, stores all structure information, parameters of the technological elements, simulation results, and project information and calculation results in the database. In addition to linking the individual modules together, the database also serves as a central pool of data for analysis and evaluation with standard software. The flexible design of the database also permits further modules to be created and integrated into the overall system in the future work. Finally, the subordinate **material database** contains the chemical composition of the steels; parameters describing the flow curve; and physical characteristics, such as density, thermal conductivity, tensile strength, and others.

3.2 ANL-HYBREX/SIM Flowsheet Simulator

Integrated software development (IDE) systems have simplified the creation and implementation of complex programs. Advanced programming technologies, such as OOP, support convenient structuring of very complex technological information. Consistent use of the characteristics of object-oriented programming languages (encapsulating, inheritance, polymorphism) helps to provide a real-life view of processes and systems with their complex interaction. The ANL-HYBREX tool was developed using the **Visual C++** integrated development environment from Microsoft. Visual C++ is a powerful software package providing excellent support in the development of complex program systems with several participating modules and classes. Although the development environment runs exclusively on the Windows 95/98/NT operating system, the computing speed of these computers is only marginally less than that of UNIX-based machines.

The following describes the structure and operation of the ANL-HYBREX/SIM flowsheet simulator.

3.2.1 Structure

The ANL-HYBREX/SIM flowsheet simulator comprises the following components:

- Model generator and simulator
- Class library of the core models
- Class library of the units.

All three components depicted in Fig. 3.2 have been implemented in the C++ programming language. The graphics interface and the class library of the element models are embedded in the MS-Windows operating system, which provides graphic functionality and the visual interface to the user. The class library of the core models, in contrast, is implemented independently of the operating platform in ANSI C++ Standard, which enables universal application of the process models in other simulation environments.

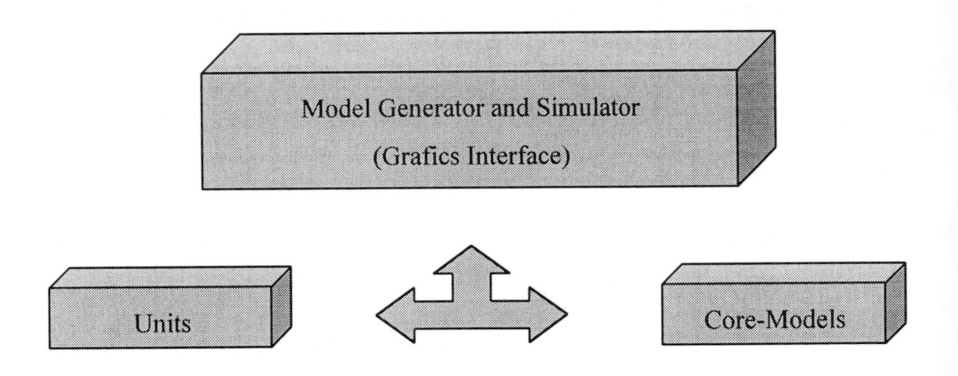

Fig. 3.2: Components of the flowsheet simulator

3.2.2 Model Generator and Simulator

The **model generator** is the direct user interface. With it, a model of the plant, comprising individual units, such as rolling with the flat stand, temperature changes in the roller table and others, can be created and visualised (Fig. 3.3). A distinction is made between technological and non-technological units. Technological units correspond to the real world, meaning the *flat stand* unit, for example, is modelled directly on the real flat stand in the rolling mill. An example of a non-technological unit is the *iteration*, which monitors the output values of a particular unit. If the monitored value deviates from a pre-set target value, then this value is iterated. The entry point in the iteration loop can be any previous unit.

The following units are currently available: flat stand; edging stand; roller table; interstand cooling; caster; furnace; coilbox; coiler; shear; cooling section (technological units); strip editor; and loop editor (non-technological units).

Individual units are combined into a plant (rolling mill) by selecting and placing the units as needed onto the worksheet. Using the mouse, the units are then linked in the layout. A double-click on a particular unit takes the user to an entry template, which displays all parameters and settings for the simulation. While the mill structure is being created on the display, working memory is allocated to the process model of the mill. During the simulation procedure, the **simulator** accesses this internal plant model and calculates the entire plant, block-by-block.

In another mode, the simulator provides automatic calculation of both a strip and the entire annual production. Parameterisation of the units and the strips to be rolled takes place through external files. The templates for the parameter files are generated by the simulator according to the current mill structure and exported in a standard format, such as ASCII. External program tools (e.g. Microsoft Excel) are used to adapt variable parameters, which include not only the strip parameters, but the unit parameters such as the roll diameter or the pass schedule (distribution of thickness reduction over all stands). In this way, any number of variants can be drawn-up without having to manually parameterise individual units.

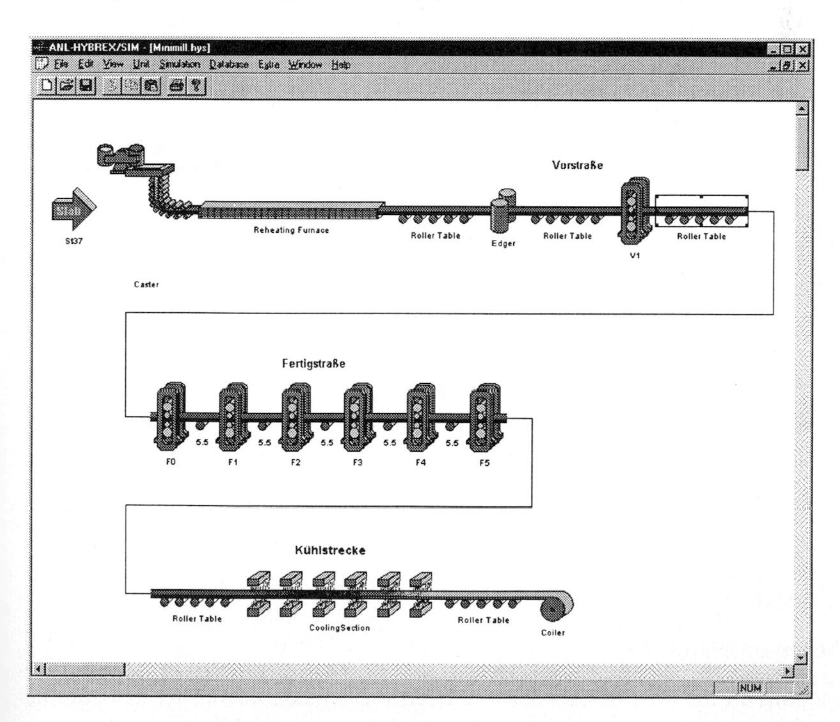

Fig. 3.3: Depiction of a mini-mill flowsheet using ANL-HYBREX/SIM

3.2.3 Class Library of the Core Models

The basic technological operations, such as *rolling in the horizontal stand* or *cooling in the roller table*, include core models that range from simple equations to complex rules for calculation. The same core models are used for individual operations. One example is the operation *rolling in the horizontal stand*, which comprises of the core models for calculating the rolling force, the change in temperature, the material microstructure, and others. The core model for calculation of temperature change is also found in the *rolling in the vertical edger* operation. To avoid multiple implementation, a class library has been developed for the core models (Fig. 3.4).

Individual core models are encapsulated in classes and arranged hierarchically. All classes are implemented in ANSI C++ Standard, making them independent of operating platform. In effect, the class library is a collection of offline process models that can be used not only in ANL-HYBREX, but in other simulation tools as well. Online use of this library in process automation is planned.

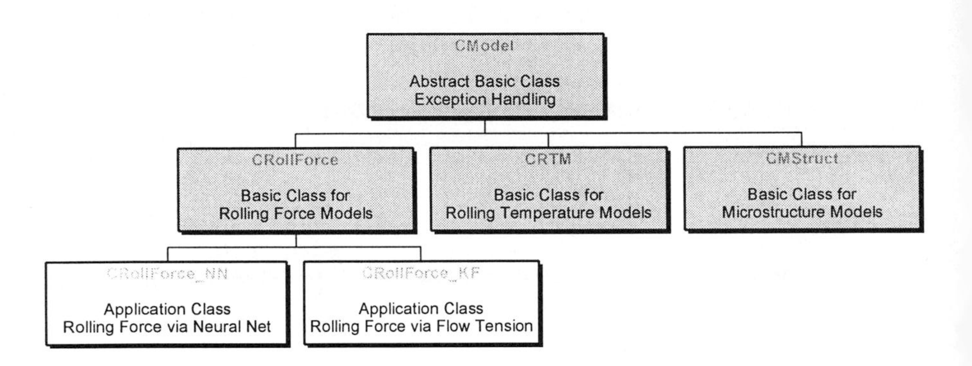

Fig. 3.4: Class library of the core models

The base class defines exception handling, routines that are called up to deal with exceptional situations including:

- Division by zero,
- Physically improbable input parameters,
- Outside the model's domain,
- Convergence problems.

Depending on the type of exception, a decision is made on further programme steps. In any case, every exception is recorded in a log file for analysis, as required, following completion of the simulation.

Further model classes are derived from the base class. The next hierarchical level forms the basis for the applications classes located on the third level. In this second level, member functions define common interfaces for all derivative classes. No programme code is assigned to these classes. They are known as *abstract base classes*, from which nothing concrete can be generated. A uniform programme code is generated instead in the classes of the third hierarchical level and their derivatives. These classes provide unambiguous definition of the desired functionality of the core models. An example: the class *CRollForce* is derived from the *CModel* base class. Defined within *CRollForce* are the parameters that must be transmitted to the calculation function to calculate the rolling force, regardless of whether the calculation is done according to Model A or Model B.

The hierarchical arrangement and encapsulating of the models in classes provides the following advantages:

- Clear interfaces for data exchange – universal application
- Avoidance of redundant code
- Flexible expansion of model base through overwriting of methods
- Uniform error and exception handling.

3.2.4 Class Library of the Units

Similar to the class library of the core models, the units are also collected in a class library (Fig. 3.5). These classes are not independent of the operating platform since the graphics functionality needed to generate the flowsheet – and the mill structure – has been implemented in them.

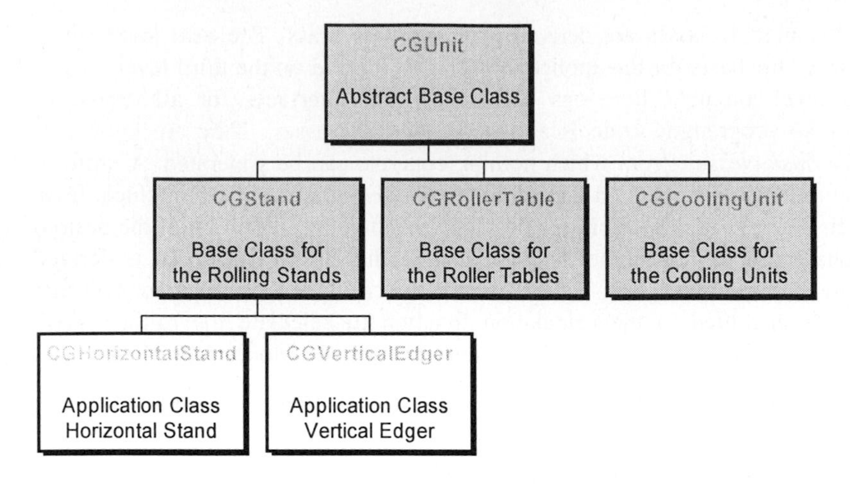

Fig 3.5: Class library of the units

The individual classes represent technological (e.g. flat stand, roller table) and non-technological (e.g. iteration unit) elements. All elements are derived from a common class, for which the basic functionality has been defined. This includes the drawing of the elements on the display: the acceptance/handover of current values for the strip to be rolled and methods for the loading and storage of visual information (position, size, arrangement, symbols). Added to these are the handling routines for particular user reactions (mouse click, shift, zoom, etc.). Derived from this base class are element-specific base classes for each unit type (e.g. stand, roller table), for which element-specific characteristics and methods have been defined, such as input templates for parameterising the units, functions for loading and storage of unit parameters, and database interfaces. The base classes are abstract, meaning that nothing concrete can be generated from them. Classes derived from the base classes include the characteristics and methods of the special technological and non-technological units. Within them, the member variables are defined, which ensure the functionality of the core models.

4 Practical Applications

The following shows how the ANL-HYBREX tool can be used to solve real problems.

For an existing European hot-rolling mill, a study was conducted to pinpoint current throughput capacities and how they could be increased. The product range of the plant involves typical representatives of 25 product classes (with regard to finished strip width and thickness). For these products, the rolling process was simulated and the throughput of the plant determined. Because the desired throughput was not achieved, further studies have been conducted.

For product No. 8 (entry temperature in the first stand = 970°C, finished strip temperature = 890°C, finished strip thickness = 1.5 mm, portion of annual production by mass = 2.338%), the following relationships were recorded through simulation and plotted as graphs (**Fig. 4.1**):

Finished strip temp.: T = *f*(rolling speed)

Motor load in F4: load = $P/P_N * 100 = f$(rolling speed)

Throughput : m = *f*(rolling speed)

(The rolling speed at all stands is characterised by the strip entry speed at the first stand (F0). Observations were made at the F4 stand (bottleneck stand) because it shows the greatest motor load (power P to rated power P_N).)

These studies allow the following valuable conclusions to be drawn:

1. The throughput has a maximum. It does not increase continuously with increasing rolling speed. (If the motor is driven at more than 100% load, the thermal overload requires that an additional rolling pause t_{Pause} = 15 s + $t_{add.}$ be adhered to, which lowers the effective throughput.)
2. Through the required finished strip temperature of 890°C, the working point is given at v_in_F0 = 0.86 m/s. This is "to the right of the optimum," meaning that the overload of the motor leads to a reduction of throughput (m=357 t/h).
3. There are several ways to increase throughput :
   a) Were larger motors (limit case: infinitely large) to be used, the throughput would grow continuously with the rolling speed. At the given working point, a throughput of approx. 700 t/h could thus be achieved.
   b) Through reduction of the rolling speed (v_in_F0 -> 0,60 m/s), the throughput could be increased to a maximum of 500 t/h. One way to do so would be by increasing the entry temperature of the strip into the first stand, in order to reach the required finished strip temperature of 890°C at a lower rolling speed.
   c) By changing the pass schedule toward the reduction of the load on the F4 stand, the motor load on F4 could be increased, reducing the necessary pause time and enhancing throughput.

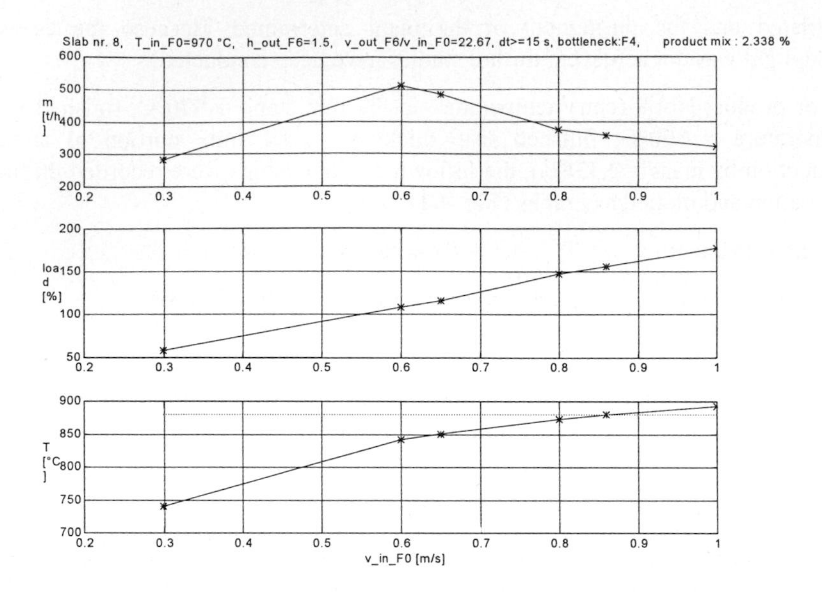

Fig. 4.1: Throughput as a function of rolling speed

The effects outlined here occur with some of the products. By following the suggestions made under (3a), for the entire product spectrum, however, throughput can be increased by approximately 10%. An estimate of the growth in profits (not including the investment needed to strengthen the motors) has shown an annual increase in the (DMark) double-digit million range is definitely possible. In contrast, the cost of replacement of stronger motors is lower.

References

Vondran, R. (1998): "Aussichten für den Weltstahlmarkt bis zum Jahr 2000." In : *Stahl und Eisen* 118 (1998), Nr. 4, S. 33-39

Stähler, K. (1997): "Stahl '97 – Ein Werkstoff für Fortschritt durch Innovation." In : *Stahl und Eisen* 117 (1997), Nr. 12, S. 40-44

Kneppe, G.; Rosenthal, D. (1998): "Warmbandproduktion: Herausforderung für das neue Jahrhundert." In : *Stahl und Eisen* 118 (1998), Nr. 7, S. 61-68

Polke, M. (1994): *Prozeßleittechnik*, Oldenbourg, 1994, 2. erw. Aufl.

Lohe, B.; Futterer, E. (1995): "Stationäre Flowsheet-Simulation" in Schuler, H. [Hrsg.]: *Prozeßsimulation* VCH, Weinheim, 1995

Coad, P.; Yourdon, E. (1991): *Object-oriented design*. (Prentice Hall), New Jersey, 1991

Process Optimisation in the Packaging Industry of the Energy Industry: Optimisation of Weight Limits

Jochen Biedermann, Ludwig Cromme and Stephan Würll

Chair for Numerical and Applied Mathematics, Brandenburg Technical University Cottbus, Universitätsplatz 3/4, D-03044 Cottbus, Germany
namall@math.tu-cottbus.de

Abstract. *The production of brown coal briquette bundles in a completely automated production plant imposes strict quality requirements on the pressing of the briquettes. Thus the packaging of bundles which consist of several layers of briquettes requires that each layer has approximately the same length. As the pressing of briquettes with exactly predefined measurements cannot be mastered technically, a part of the production must be picked out with a suitable quality surveillance procedure. The quality surveillance procedure that was used in production could be mathematically described exactly and was subject to an in-depth analysis. It could be verified that, besides the question for the optimal choice of single parameters, it is essential for this particular application that the performance of a quality surveillance procedure does not depend on the dynamic behaviour of the factors of production. The output of the factory could be increased considerably by eliminating these problems and optimising the production plant. Besides the mathematical modelling, this result was based on the optimisation and simulation by a computer-based model.*

Keywords. *optimization, process control, production simulation, brown coal plant*

1 Introduction

Real world problems in the field of production planning and process control often lead to dynamic and non-deterministic optimisation problems. Computer simulation as part of computer integrated manufacturing (CIM) is a powerful tool to improve the efficiency of the industrial production. To improve the output rate of a complex packing process (fully automated production) for brown coal briquette bundles at a factory in the East German brown coal industry we designed a simulation system combining statistics, visualisation and optimisation routines using state-of-the-art object-orientated software design techniques. With the optimisation part we found optimal settings of the numerous production parameters such as weight limits, briquette and bundle transport velocities, sensor positions, and delays etc. for different production conditions. Due to our simulation studies it was possible to realise a strong increase in the production output of the plant. The whole system serves as a basis for an on-line process control. Furthermore, the system allows the simulation and investigation of changes and extensions to the plant at minimal costs and without disturbing the production process.

1.1 The Factory

In a complex production process – at a briquette factory of one of the three German brown coal mining companies – briquette bundles of three different sizes and target weights (10 kg and 25 kg) are automatically packed and palletised.

The briquettes created by six electrically driven brown coal presses have a fixed width and height. Their thickness is approximately normally distributed with an electronically controllable mean value (adjustable target thickness) and their density varies in time with changing coal quality (grain sizes and water content).

Packing machines stack up to 5 layers of briquettes to form a bundle (see Fig 1.1-1.2). Any layers with an unacceptable length are rejected immediately. After the packing process the created bundles have to pass special calibrated scales and all bundles below the target weight are picked out.

After creating a new briquette bundle the packing machines have to push the bundle on a collecting line which transports the bundles to the scales (see Fig. 1.3).

Therefore the packing machines have to wait for sufficient gaps in the bundle stream on the collecting line. An optical sensor for each packing machine recognises such a gap. While waiting for a gap, a packing machine cannot continue working and (because the briquette presses cannot be stopped) during this time period all incoming briquettes have to be thrown away. For this reason each packing machine has got a short queue for a maximum of 3 bundles at its exit.

One problem is to minimise the amount of unacceptable layers whereas another one is to optimise the bundle stream on the collecting line and subsequent conveyors so that there are almost always sufficient gaps for all packing machines. In this article we will analyse only the first problem due to lack of space.

1.2 Minimising the Amount of Unacceptable Layers

A briquette bundle consists of a fixed number of briquette layers (3 or 5) each made up of a fixed number of briquettes (6 or 9). An example for the 25-kg briquette bundle is presented in Fig. 1.1.

The aim of the packing process is the production of briquette layers with an acceptable length (the packing machines cannot handle layers which are too short or too long) and of sufficient weight (so that, at minimum the target weight of the bundle can be reached). In order to minimise the probability of producing underweight bundles the following quality-control-algorithm (see Fig. 1.1–1.2) was used:

1. $i := 1$, $g^1_{\text{targ}} := \gamma_1$
2. a layer l is grasped by the claw arm
3. if the length of layer l is not within $[\delta_{\min}, \delta_{\max}]$ reject it and go to 2.
4. if the weight of layer l is below g^i_{targ} reject it and go to 2.
5. if $i = 5$ STOP, a new bundle has been packed
6. $i := i + 1$, $g^i_{\text{targ}} := \gamma_i - \sum_{k=1}^{i-1} g_k$, go to 2.

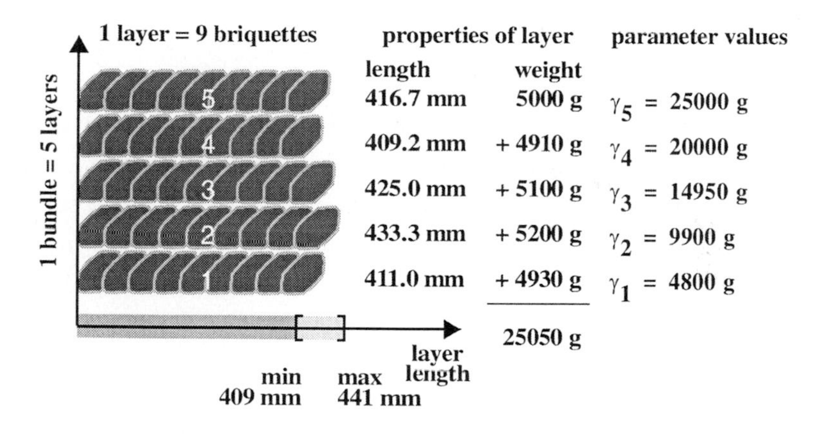

Fig. 1.1. Example of a packed briquette bundle

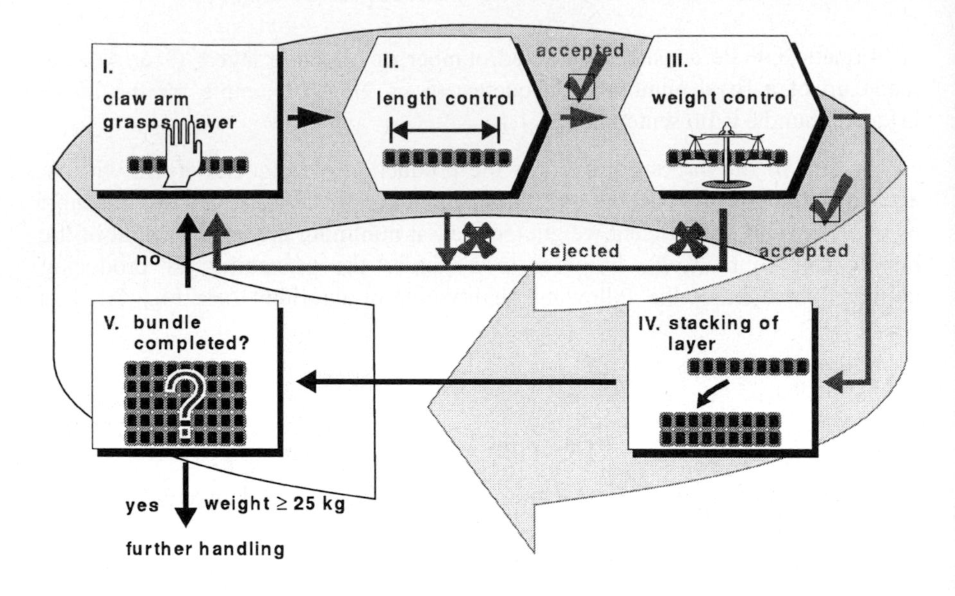

Fig. 1.2. Schematic representation of the quality control

1.3 Problem Definition

As already indicated the problem consists of determining optimal weight limits $\gamma_1,\ldots,\gamma_4$ to a predefined normal distribution $D \sim N(\mu_D, \sigma_D^2)$ of the layer length, accompanying specific briquette mass ρ, average length of layers μ_D, minimal acceptable layer length $\delta_{\min}$ and maximal layer length $\delta_{\max} > \delta_{\min}$ subject to weight limits $\gamma_0 = 0g$, $\gamma_5 = 25000\text{g}$ in the sense of the already mentioned requirements.

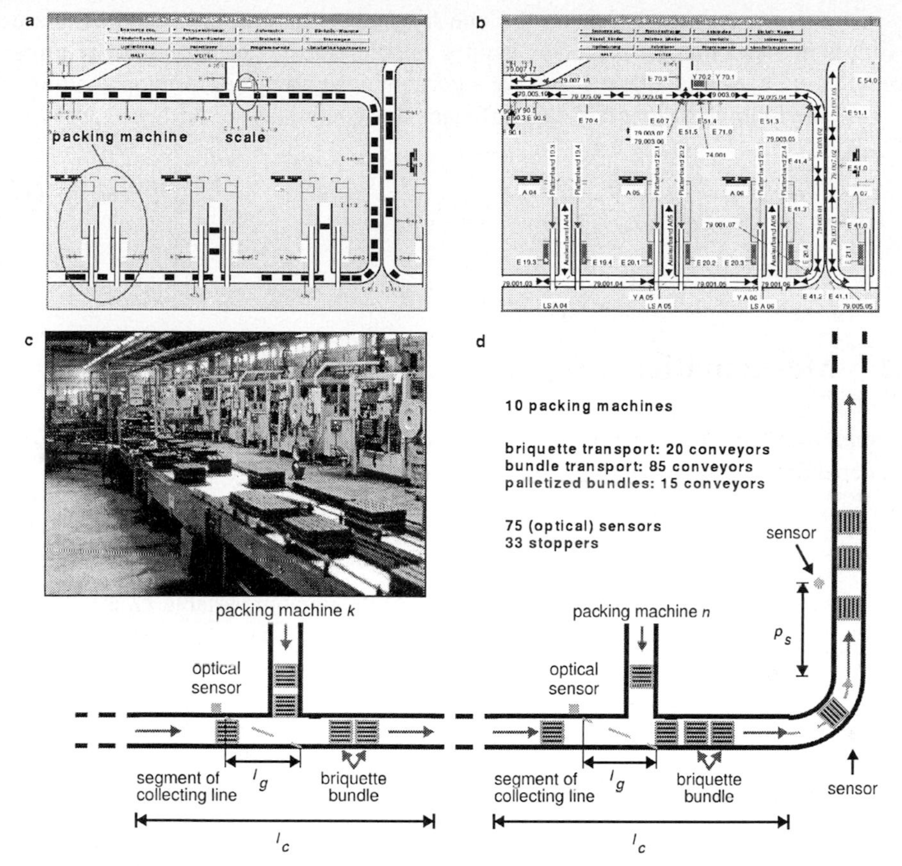

Fig. 1.3. (a) User interface (GUI) of simulation environment, (b) GUI with notation of the segments – illustrating the complexity of the plant, (c) automated line for 25 kg-bundles at Briquette Factory "Mitte", (d) schematic representation of a part of the conveyor system

For the problem definition to make sense, it must be possible to produce a bundle in a finite number of steps. This will surely not be possible, even if those layers that possess the maximum attainable weight, would not suffice to reach the target weight. Due to this consideration we get the restriction

$$\rho > \frac{\gamma_5}{5\delta_{\max}},$$

since the above task is not solvable at all. The answer to the problem of making the optimal parameter choice is a triviality if $\rho \geq \gamma_5/(5\delta_{\min})$, the optimal parameter values are then given by $\gamma_i = i\delta_{\min}\rho$.

In order to guarantee that on the one hand no redundant side conditions are formulated and that on the other hand a bundle can be arranged from a finite number of layers, i.e. that the packing algorithm always reaches the stop condition $i = 5$, the following necessary and sufficient side conditions are further required:

$$\left.\begin{array}{l}\delta_{\min}\rho \leq \gamma_{i+1} - \gamma_i \\ \gamma_{i+1} - \gamma_i < \delta_{\max}\rho\end{array}\right\} i = 0,1,\ldots,4 \qquad (1)$$

2 Mathematical Model

Under the assumption that the layer lengths D and corresponding weights G are independently and identically normal distributed, i. e.

$$D \sim N(\mu_D, \sigma_D^2),\ G = \rho \cdot D,\ G \sim N(\mu_G, \sigma_G^2),$$

in which $\mu_G = \mu_D \rho$, $\sigma_G = \sigma_D \rho$, the density of the random variable G is

$$\varphi(x) = \frac{1}{\sigma_G \sqrt{2\pi}} \cdot e^{-\frac{1}{2}\left(\frac{x-\mu_G}{\sigma_G}\right)^2},$$

and the accompanying distribution function is defined by the following expression:

$$\Phi(x) = \int_{-\infty}^{x} \varphi(u)\,du = \frac{1}{2}\left(1 + \operatorname{erf}\left(\frac{x - \mu_G}{\sigma_G \sqrt{2}}\right)\right)$$

$\Phi(x)$ is the probability that a layer gripped by chance has at most the weight x. The acceptance probability that a gripped layer will be accepted, i.e. it has at least the weight g_{targ}^{i} and that its length lies within the tolerance interval $[\delta_{\min}, \delta_{\max}]$, is:

$$\begin{aligned} p(g_{\text{targ}}^{i}) &= P(D \in [\delta_{\min}, \delta_{\max}] \wedge G \geq g_{\text{targ}}^{i}) \\ &= P(G \in [\max\{\delta_{\min}\rho, g_{\text{targ}}^{i}\}, \delta_{\max}\rho]) \\ &= \begin{cases} \Phi(\delta_{\max}\rho) - \Phi(\max\{\delta_{\min}\rho, g_{\text{targ}}^{i}\}) & \text{if} \quad g_{\text{targ}}^{i} < \delta_{\max}\rho \\ 0 & \text{otherwise} \end{cases} \end{aligned}$$

Since the least weight g^i_{targ} currently demanded depends on the current value of the limit γ_i and the weights $g_1,\ldots,g_{i-1}$ of the already accepted layers, g^i_{targ} is given by the expression

$$g^i_{\text{targ}} = \gamma_i - \sum_{k=1}^{i-1} g_k \, .$$

Let $A^i = A^i(\gamma_1,\ldots,\gamma_i) \subseteq \Re^i$ denote the set of acceptable tuples of layer weights $(g_1,\ldots,g_i)^T \in \Re^i$, i.e.

$$A^1 = [\max\{\delta_{\min}\rho, \gamma_1\}, \delta_{\max}\rho] = [\gamma_1, \delta_{\max}\rho]$$

$$A^i = \left\{ (g_1,\ldots,g_i)^T \middle| (g_1,\ldots,g_{i-1})^T \in A^{i-1}, g_i \in \left[\max\left\{ \delta_{\min}\rho, \gamma_i - \sum_{k=1}^{i-1} g_k \right\}, \delta_{\max}\rho \right] \right\}$$

$$i = 2,3,4,5$$

Due to the side conditions the A^i are always non-empty and convex.

By X_i we mark the random variable which describes how many layers have to be tested to accept a layer taken by chance at the ith location of the bundle. G^i_{targ} (probability density function $\varphi^i_{\text{targ}} : \Re^i \to [0,1]$) is the random variable whose realisation is g^i_{targ}. The probability distribution of the discreet random variable X_i, for given value g^i_{targ}, is then given by the following expression.

$$P\left(X_i = j \middle| G^i_{\text{targ}} = g^i_{\text{targ}}\right) = \left[1 - p\left(g^i_{\text{targ}}\right)\right]^{j-1} p\left(g^i_{\text{targ}}\right), \; j \in \aleph$$

As

$$E\left(X_i \middle| G^i_{\text{targ}} = g^i_{\text{targ}}\right) = \sum_{j=1}^{\infty} k \cdot P\left(X_i = j \middle| G^i_{\text{targ}} = g^i_{\text{targ}}\right) = \frac{1}{p\left(g^i_{\text{targ}}\right)}$$

holds, the expectation value is given by the integrals

$$E(X_1) = \frac{1}{\Phi(\delta_{\max}\rho) - \Phi(\gamma_1)}$$

$$E(X_{i+1}) = \frac{\int_{A^i} \varphi^i_{\text{targ}}(g_1,\ldots,g_i) \cdot \frac{1}{p\left(g^i_{\text{targ}}\right)} dA^i}{\int_{A^i} \varphi^i_{\text{targ}}(g_1,\ldots,g_i)\, dA^i}, i = 1,2,3,4 \quad . \tag{2}$$

To go into details, the expectation values are:

$$E(X_2) = E(X_1) \cdot \int_{\gamma_1}^{\delta_{\max}\rho} \frac{\varphi(g_1)}{p(\gamma_2 - g_1)} dg_1 = \frac{\int_{\gamma_1}^{\delta_{\max}\rho} \frac{\varphi(g_1)}{p(\gamma_2 - g_1)} dg_1}{\int_{\gamma_1}^{\delta_{\max}\rho} \varphi(g_1)\, dg_1} \tag{3}$$

$$E(X_3) = \frac{\int_{\gamma_1}^{\delta_{\max}\rho} \int_{\max\{\delta_{\min}\rho, \gamma_2 - g_1\}}^{\delta_{\max}\rho} \frac{\varphi(g_1)\varphi(g_2)}{p(\gamma_3 - g_1 - g_2)} dg_2 dg_1}{\int_{\gamma_1}^{\delta_{\max}\rho} \int_{\max\{\delta_{\min}\rho, \gamma_2 - g_1\}}^{\delta_{\max}\rho} \varphi(g_1)\varphi(g_2)\, dg_2 dg_1} \tag{4}$$

$$E(X_4) = \frac{\int_{\gamma_1}^{\delta_{\max}\rho} \int_{\max\{\delta_{\min}\rho, \gamma_2 - g_1\}}^{\delta_{\max}\rho} \int_{\max\{\delta_{\min}\rho, \gamma_3 - g_1 - g_2\}}^{\delta_{\max}\rho} \frac{\varphi(g_1)\varphi(g_2)\varphi(g_3)}{p(\gamma_4 - g_1 - g_2 - g_3)} dg_3 dg_2 dg_1}{\int_{\gamma_1}^{\delta_{\max}\rho} \int_{\max\{\delta_{\min}\rho, \gamma_2 - g_1\}}^{\delta_{\max}\rho} \int_{\max\{\delta_{\min}\rho, \gamma_3 - g_1 - g_2\}}^{\delta_{\max}\rho} \varphi(g_1)\varphi(g_2)\varphi(g_3)\, dg_3 dg_2 dg_1} \tag{5}$$

$$E(X_5) = \frac{\int_{\gamma_1}^{\delta_{\max}\rho} \cdots \int_{\max\{\delta_{\min}\rho, \gamma_4 - g_1 - g_2 - g_3\}}^{\delta_{\max}\rho} \frac{\varphi(g_1)\varphi(g_2)\varphi(g_3)\varphi(g_4)}{p(\gamma_5 - g_1 - g_2 - g_3 - g_4)} dg_4 \ldots dg_1}{\int_{\gamma_1}^{\delta_{\max}\rho} \cdots \int_{\max\{\delta_{\min}\rho, \gamma_4 - g_1 - g_2 - g_3\}}^{\delta_{\max}\rho} \varphi(g_1)\varphi(g_2)\varphi(g_3)\varphi(g_4)\, dg_4 \ldots dg_1} \tag{6}$$

The meaning of the side conditions $\gamma_{i+1} - \gamma_i < \delta_{\max}\rho$ can now be read from the integral expressions directly. The estimate $g_{\text{targ}}^{i} = \gamma_i - \sum_{k=1}^{i-1} g_k \leq \gamma_i - \gamma_{i-1} < \delta_{\max}\rho$ shows that the lower integration limits ($\max\{\delta_{\min}\rho, g_{\text{targ}}^{i}\}$) can never coincide with the upper limits. In addition, the probability $p(g_{\text{targ}}^{i})$ is always different from zero.

For the case $\gamma_i = \gamma_{i-1} + \delta_{\min}\rho$, $i = 1, \ldots, 5$ the weight limit is not relevant since in this case every layer whose length lies within the limits $[\delta_{\min}, \delta_{\max}]$ can be accepted. The expectation value simplifies in this special case to $E(X_i) = \dfrac{1}{p(\delta_{\max}\rho)}$.

Our aim is to minimise the expected number of layers needed for the composition of a bundle. This results in the following nonlinear optimisation problem dependant on the parameters $\gamma_1,\ldots,\gamma_4$:

$$
\begin{aligned}
&Z(\gamma_1,\gamma_2,\gamma_3,\gamma_4)=\sum_{i=1}^{5}E(X_i)\rightarrow \min \\
&\text{on}\quad \mathrm{V}=\{(\gamma_1,\gamma_2,\gamma_3,\gamma_4)|\delta_{\min}\,\rho\le\gamma_{i+1}-\gamma_i\le\delta_{\max}\,\rho, i=0,1,\ldots,4\} \\
&\text{s.t.}\quad \rho>\frac{\gamma_5}{5\delta_{\max}}
\end{aligned}
\tag{7}
$$

3 Properties of the Optimisation Problem

Numerical optimisation methods make different assumptions about the objective function and are, as a rule, more efficient if far-reaching requirements on the objective function (e.g. continuity, differentiability) and side conditions (e.g. closure) are fulfilled. We therefore derive some properties of the objective function in the sequel.

The objective function Z is continuous on V since $p(\cdot)$ is positive on V. Furthermore, Z is positive on V and it can even be shown that $E(X_i)>1$, since the domain of integration is non-empty, the functions $\varphi(\cdot)$ and $p(\cdot)$ are positive on the domain of integration and $p(\cdot)<1$ on V ($i=2,3,4,5$). To investigate the differentiability of the function Z, with respect to the variables $\gamma_1,\ldots,\gamma_4$, turns out to be much more difficult.

For the partial derivative of the inner integral of the denominator of $E(X_3)$ with respect to γ_2 holds

$$
\frac{\partial}{\partial\gamma_2}\int_{\max\{\delta_{\min}\rho,\gamma_2-g_1\}}^{\delta_{\max}\rho}\varphi(g_1)\varphi(g_2)\,dg_2
=\begin{cases}-\varphi(g_1)\varphi(\gamma_2-g_1) & \text{if}\quad \gamma_2>g_1+\delta_{\min}\rho\\ 0 & \text{if}\quad \gamma_2<g_1+\delta_{\min}\rho\end{cases}
$$

This follows by applying Leibniz' rule for the derivation of an integral with variable integration limits (cf. Fichtenholz, 1972) (or see the Appendix). The integral *does not have* a continuous derivative with respect to γ_2 and it seems reasonable to suspect that Z is not differentiable.

However, a detailed examination shows that Z is (totally) differentiable in contrast to this expectation. This is a result of the fact that the jump discontinuity in the derivation of the integral with respect to γ_2 (see equation (4)) is smoothed by the following integration with respect to g_1 (!!). Using Lemma 5 we get (Cromme, 1998):

Lemma 1 The objective function of problem (7) is smooth and (totally) differentiable with respect to the parameters γ_i, $i = 1,\ldots,4$. The set of all feasible points is depictable as the intersection of a finite number of partly open, partly closed half-spaces.

4 Numerical Evaluation of the Objective Function

To evaluate the objective function (7) a multidimensional numerical integration method is required. This can in principle be achieved by recursive application of the following one-dimensional quadrature formula.

$$\begin{gathered} \int_{a_1}^{b_1}\int_{a_2(x_1)}^{b_2(x_1)} \cdots \int_{a_n(x_1,\ldots,x_{n-1})}^{b_n(x_1,\ldots,x_{n-1})} f(x_1,\ldots,x_n)dx_n \ldots dx_1 = \int_{a_1}^{b_1} g(x_1)\,dx_1 \\ g(x_1) = \int_{a_2(x_1)}^{b_2(x_1)} \cdots \int_{a_n(x_1,\ldots,x_{n-1})}^{b_n(x_1,\ldots,x_{n-1})} f(x_1,\ldots,x_n)\,dx_n \ldots dx_2 \end{gathered} \tag{8}$$

However, since the numerical effort grows fast for increasing n, even for smooth functions and adaptive knot choice, this method suffers from inefficiency. Beyond this fault, known error bounds are not transferable to the recursive application case for one-dimensional quadrature methods directly. As the function $g(x_1)$ in formula (8) itself is given as an approximate value and cannot be determined with machine precision the application of formula (8) does not have to be recommended. In practice, therefore, one also realises an unfavourable convergence behaviour and concerns oneself with the exact quadrature of a linear or square approximation of $f(x_1,\ldots,x_n)$, if no high precision is demanded.

There also exist algorithms for multiple integrals analogously to the development of adaptive quadrature methods for one-dimensional integration. But here the following difficulties arise:

- As a rule, multidimensional numerical integration methods are formulated only for special integration areas e.g. for rectangular areas. IMSL® offers only the possibility to calculate integrals of the form $\int_{a_1}^{b_1}\int_{a_2(x)}^{b_2(x)} f(x,y)\,dy\,dx$ or

$\int_{a_1}^{b_1}\ldots\int_{a_n}^{b_n} f(x_1,\ldots,x_n)\,dx_n\ldots dx_1$ (Werner, 1992).

- Adaptive multidimensional integration methods which allow a reliable error estimation are offered only in a few, mostly commercial, numerical libraries (e.g. IMSL®).
- The value of the objective function on the given integration area cannot be determined directly with any of the currently available numerical libraries.

For these reasons two different approaches were pursued and the results were compared with each other: Romberg's one-dimensional quadrature formula used recursively, and algorithms from the class library Cubpack++ (Cools, 1994) were used because of their easy applicability to areas with almost arbitrary boundaries.

4.1 Romberg's Quadrature Formula

When evaluating the objective function Z (see equation (7)) by use of Romberg's one-dimensional quadrature method for $f \in C^{2k+2}[a,b]$, $k \geq 0$, the following estimate can be used to calculate an upper bound of the error of the numerical integration, by means of Euler-Maclaurin's summation formula (Werner, 1992):

$$\left|\int_a^b f(x)dx - T^{(k)}(f;a,b)\right| \leq 3(b-a)^{2k+3}\frac{|B_{2k+2}|}{(2k+2)!}\max_{\xi\in[a,b]}\left|f^{(2k+2)}(\xi)\right|$$

in which the term B_{2k+2} is a Bernoulli number, defined as coefficient of the power series expansion $\frac{x}{e^x-1} = \sum_{k=0}^{\infty} B_k \frac{x^k}{k!}$, $|x| < 2\pi$.

Unfortunately, this result is not valid for the case of the multidimensional (recursive) integration.

4.2 Adaptive Quadrature / Cubature by Means of DQNC79 and Cubpack++

As an alternative to Romberg's quadrature method, the integral (3) was evaluated by applying routine DQNC79 of the SLATEC numerical library. (More precisely: The FORTRAN code of DQNC79 was translated by f2c into C.) This is an adaptive quadrature method on the base of one 7 points Newton-Côtes formula.

Thereby the quadrature error and the number of the function evaluations could be checked precisely.

The C++ class library Cubpack++ provides the user with routines for automatic cubature. With the object-orientated implementation of Cubpack++ this library is not static but can be enlarged, in principle, arbitrarily. Those areas which are not included in Cubpack++ can be simply added by inheritance. For these areas the user must carry out a transformation of the integral on an area for which a cubature formula exists. In practice there might be only a few integration areas which cannot be described and integrated with the provided regions. As far as it is known to us, this is worked on to enlarge the functionality of Cubpack++ on general n-dimensional integration.

For these reasons the integrals (4)-(6) were evaluated by a combination of Cubpack++ and DQNC79.

5 Solving the Optimisation Task

5.1 Is the Problem Formulated Suitably?

Independently of the choice of a certain optimisation algorithm till now the question has remained open, of whether a local minimum can be located at the border ∂V of V. If an optimisation algorithm creates a sequence $\vec{\gamma}^n = (\gamma_1^n, \gamma_2^n, \gamma_3^n, \gamma_4^n)_{n \in \mathbb{N}}$ of parameter values converging towards a point $\lim_{n \to \infty} \vec{\gamma}^n = \vec{\gamma}^* \in \partial V$ at the border of V, the expectation value $E(X_i)$ is not defined in the limit $\vec{\gamma}^*$, if $\lim_{n \to \infty}(\gamma_i^n - \gamma_{i-1}^n) = \delta_{\max}\rho$, $\gamma_0^n := 0$ holds. In this case $p(\gamma_i^* - g_{\text{targ}}^i) = 0$ appears in the denominator of equation (2). Therefore we want to prove now that no optimum of (7) can be a boundary element of the set $\overline{V} - V$.

To judge purely heuristically whether the expectation value $E(X_i)$ remains finite in the limit value $\lim_{n \to \infty}(\gamma_i^n - \gamma_{i-1}^n) = \delta_{\max}\rho$, or grows beyond all limits, is unsatisfactory. Therefore, we now want to examine exactly if the expectation values grow beyond all limits at the partly open boundary of V.

In the following we want to use an elementary lemma whose proof follows from the mean value theorem of the differential calculus directly.

Lemma 2 If f is differentiable on $[a,b]$ and $\left|\frac{d}{dx}f(x)\right| \leq L$ is valid for all $x \in [a,b]$, then $|f(x_1) - f(x_2)| \leq L|x_1 - x_2|$ for arbitrary $x_1, x_2 \in [a,b]$.

We can prove now that the objective function of (7) grows at the partly open boundary to infinity.

Proposition 3 Let $\vec{\gamma}^n = (\gamma_1^n, \gamma_2^n, \gamma_3^n, \gamma_4^n)_{n \in \mathbb{N}}$ be a sequence of feasible parameter values. If $\lim_{n\to\infty} \vec{\gamma}^n = \vec{\gamma}^* \in \partial V$ and there exists at least one index i that fulfills the condition $\lim_{n\to\infty}(\gamma_i^n - \gamma_{i-1}^n) = \delta_{\max}\rho$, $\gamma_0^n := 0$, then the sequence of the values of the objective function grows beyond all limits, i.e. $\lim_{n\to\infty} Z(\gamma_1^n, \gamma_2^n, \gamma_3^n, \gamma_4^n) = \infty$.

Proof: We want to look at the cases $i = 1$, $i = 2$ etc. in succession.

1: $i = 1$: From $\lim_{n\to\infty} \gamma_1^n = \delta_{\max}\rho$ follows

$$\lim_{n\to\infty} E(X_1) = \lim_{n\to\infty} \frac{1}{\Phi(\delta_{\max}\rho) - \Phi(\gamma_1^n)} = \frac{1}{\Phi(\delta_{\max}\rho) - \lim_{n\to\infty}\Phi(\gamma_1^n)} = \infty$$

Thus, the expectation value $E(X_1)$ is unrestricted in the limit and therefore the objective function is, too.

1. $i = 2$: Without loss of generality we may assume $\lim_{n\to\infty} \gamma_1^n \neq \delta_{\max}\rho$, because otherwise we could argue with point 1. Be $\lim_{n\to\infty} \gamma_1^n = \gamma_1^* < \delta_{\max}\rho$ in the following. The function $\Phi(x)$ is Lipschitz continuous with respect to the constant $L = 1/(\sigma_G\sqrt{2\pi})$, i.e. $|\Phi(x_1) - \Phi(x_2)| \leq L|x_1 - x_2|$. With the auxiliary variable $M = \int_{\delta_{\min}\rho}^{\delta_{\max}\rho} \varphi(g)\,dg$, $m = \min\{\varphi(g) | g \in [\delta_{\min}\rho, \delta_{\max}\rho]\}$ we get:

$$
\begin{aligned}
E(X_2) &= \frac{\displaystyle\int_{\gamma_1^n}^{\delta_{MAX}\rho} \frac{\varphi(g_1)}{p(\gamma_2^n - g_1)} dg_1}{\displaystyle\int_{\gamma_1^n}^{\delta_{MAX}\rho} \varphi(g_1) dg_1} \geq \frac{m}{M} \int_{\gamma_1^n}^{\delta_{MAX}\rho} \frac{1}{p(\gamma_2^n - g_1)} dg_1 \\
&= \frac{m}{M} \int_{\gamma_1^n}^{\delta_{MAX}\rho} \frac{1}{\Phi(\delta_{MAX}\rho) - \Phi(\max\{\delta_{MIN}\rho, \gamma_2^n - g_1\})} dg_1 \\
&\geq \frac{m}{M} \int_{\gamma_1^n}^{\delta_{MAX}\rho} \frac{\sigma_G \sqrt{2\pi}}{\delta_{MAX}\rho - \max\{\delta_{MIN}\rho, \gamma_2^n - g_1\}} dg_1 \\
&= \frac{m\sigma_G\sqrt{2\pi}}{M} \left(\ln\left(\delta_{MAX}\rho - (\gamma_2^n - g_1)\right)\Big|_{\gamma_1^n}^{\min\{\delta_{MAX}\rho, \gamma_2^n - \delta_{MIN}\rho\}} + \frac{\delta_{MAX}\rho - \min\{\delta_{MAX}\rho, \gamma_2^n - \delta_{MIN}\rho\}}{\delta_{MAX}\rho - \delta_{MIN}\rho} \right) \\
&= \begin{cases} \dfrac{m\sigma_G\sqrt{2\pi}}{M}\left(\ln\left(2\delta_{MAX}\rho - \gamma_2^n\right) - \ln\left(\delta_{MAX}\rho - (\gamma_2^n - \gamma_1^n)\right)\right), & \delta_{MAX}\rho \leq \gamma_2^n - \delta_{MIN}\rho \\ \dfrac{m\sigma_G\sqrt{2\pi}}{M}\left(\ln(\delta_{MAX}\rho - \delta_{MIN}\rho) - \ln\left(\delta_{MAX}\rho - (\gamma_2^n - \gamma_1^n)\right) + \dfrac{\delta_{MAX}\rho + \delta_{MIN}\rho - \gamma_2^n}{\delta_{MAX}\rho - \delta_{MIN}\rho}\right), & \delta_{MAX}\rho > \gamma_2^n - \delta_{MIN}\rho \end{cases}
\end{aligned} \tag{9}
$$

Because of the assumption $\lim_{n\to\infty}(\gamma_2^n - \gamma_1^n) = \delta_{\max}\rho$ for all $\varepsilon > 0$ there exists one index $n_0 = n_0(\varepsilon)$, so that $\delta_{\max}\rho - (\gamma_2^n - \gamma_1^n) \leq \varepsilon$ for all $n \geq n_0$. Substituting this into (9) gives us:

$$
\begin{aligned}
E(X_2) \geq & \frac{m\sigma_G\sqrt{2\pi}}{M}\left(\ln\left(\delta_{\max}\rho - \left(\gamma_2^n - \min\{\delta_{\max}\rho, \gamma_2^n - \delta_{\min}\rho\}\right)\right) - \ln\varepsilon\right) \\
& + \frac{m\sigma_G\sqrt{2\pi}}{M} \cdot \frac{\delta_{\max}\rho - \min\{\delta_{\max}\rho, \gamma_2^n - \delta_{\min}\rho\}}{\delta_{\max}\rho - \delta_{\min}\rho} =: e(\varepsilon)
\end{aligned}
$$

Since $\varepsilon > 0$ can be chosen arbitrarily small one gets the divergent minorant $\lim_{\substack{\gamma_2^n - \gamma_1^n \to \delta_{ma}\rho \\ n\to\infty}} E(X_2) \geq \lim_{\substack{\varepsilon\to 0^+ \\ n \geq n(\varepsilon)}} e(\varepsilon) = \infty$.

1. $i = 3$: Without loss of generality one can suppose $\lim_{n\to\infty}\gamma_1^n \neq \delta_{\max}\rho$ and $\lim_{n\to\infty}(\gamma_2^n - \gamma_1^n) \neq \delta_{\max}\rho$. The construction of one divergent minorant is carried out like in the case $i = 2$. The complexity of the expressions grows, though. The integration area of $E(X_3)$ looks as follows:

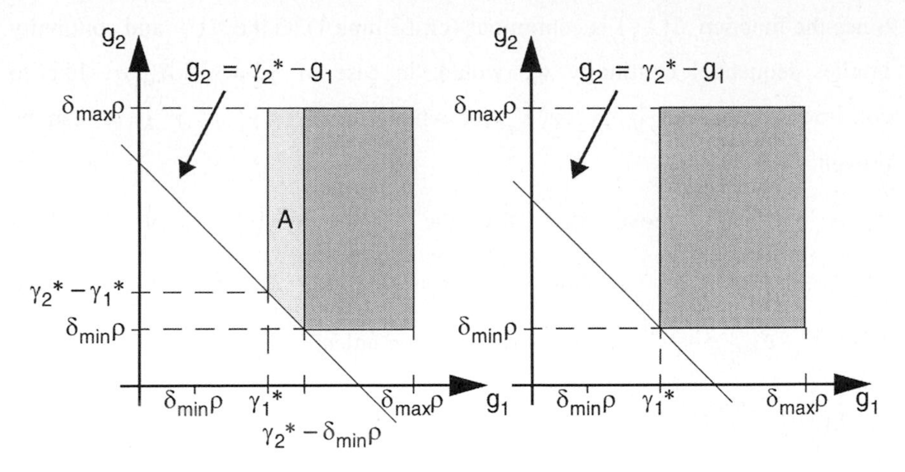

Fig. 5.1. Integration area of $E(X_3)$ for the cases $\gamma_2^* > \gamma_1^* + \delta_{\min}\rho$ (on the left) and $\gamma_2^* = \gamma_1^* + \delta_{\min}\rho$ (on the right).

In the case of $\gamma_2^* > \gamma_1^* + \delta_{\min}\rho$ we can calculate a lower estimate on $E(X_3)$ since the integrands are positive. Doing this we want to restrict the integration to the area marked by A.

$$\begin{aligned}
E(X_3) &= \frac{\displaystyle\int_{\gamma_1^n}^{\delta_{\max}\rho} \int_{\max\{\delta_{\min}\rho,\gamma_2^n - g_1\}}^{\delta_{\max}\rho} \frac{\varphi(g_1)\varphi(g_2)}{p(\gamma_3^n - g_1 - g_2)} dg_2 dg_1}{\displaystyle\int_{\gamma_1^n}^{\delta_{\max}\rho} \int_{\max\{\delta_{\min}\rho,\gamma_2^n - g_1\}}^{\delta_{\max}\rho} \varphi(g_1)\varphi(g_2)\, dg_2 dg_1} \\
&\geq \frac{m^2}{M^2} \int_{\gamma_1^n}^{\delta_{\max}\rho} \int_{\max\{\delta_{\min}\rho,\gamma_2^n - g_1\}}^{\delta_{\max}\rho} \frac{1}{p(\gamma_3^n - g_1 - g_2)} dg_2 dg_1 \\
&\geq \frac{m^2}{M^2} \int_{\gamma_1^n}^{\delta_{\max}\rho} \int_{\max\{\delta_{\min}\rho,\gamma_2^n - g_1\}}^{\delta_{\max}\rho} \frac{\sigma_G\sqrt{2\pi}}{\delta_{\max}\rho - (\gamma_3^n - g_1 - g_2)} dg_2 dg_1 \\
&= \frac{m^2\sigma_G\sqrt{2\pi}}{M^2} \left[\left(2\delta_{\max}\rho - (\gamma_3^n - g_1)\right) \cdot \ln\left(2\delta_{\max}\rho - (\gamma_3^n - g_1)\right) - g_1 \Big|_{\gamma_1^n}^{\gamma_2^n - \delta_{\min}\rho} \right] \\
&\quad - \frac{m^2\sigma_G\sqrt{2\pi}}{M^2} \ln\left(\delta_{\max}\rho - (\gamma_3^n - \gamma_2^n)\right) \cdot \left((\gamma_2^n - \delta_{\min}\rho) - \gamma_1^n \right)
\end{aligned} \tag{10}$$

It is easy to prove that last expression in equation (10) grows beyond all limits as $n \to \infty$.

Since the function $E(X_3)$ is continuous (cf. Lemma 1) on the set V and continuity implies sequential continuity we would, in case of $\gamma_2^* = \gamma_1^* + \delta_{\min}\rho$, like to construct a sequence $(\gamma_1^n, \gamma_2^n, \gamma_3^n)_{n\in\mathbb{N}}$ for which $\lim_{n\to\infty} E(X_3, \gamma_1^n, \gamma_2^n, \gamma_3^n) = \infty$ can be proven.

It is easy to prove that for the sequence $(\gamma_1^n, \gamma_2^n, \gamma_3^n)_{n\in\mathbb{N}}$, where $\lim_{n\to\infty} \gamma_1^n = \gamma_1^* < \delta_{MAX}\rho$ is chosen arbitrarily, $\gamma_2^n := \gamma_1^n + \delta_{MIN}\rho + \frac{1}{n+k}$ and $\gamma_3^n := \gamma_2^n + \delta_{MAX}\rho - \mathrm{e}^{-n^2}$ the following equation holds:

$$E(X_3) \geq \frac{m^2\sigma_G\sqrt{2\pi}}{M^2}\left[\left(2\delta_{\max}\rho - (\gamma_3^n - g_1)\right)\cdot \ln\left(2\delta_{\max}\rho - (\gamma_3^n - g_1)\right) - g_1\Big|_{\gamma_1^n}^{\gamma_2^n - \delta_{\min}\rho}\right]$$
$$- \frac{m^2\sigma_G\sqrt{2\pi}}{M^2}\ln\left(\mathrm{e}^{-n^2}\right)\cdot\frac{1}{n+k}$$

which implies $\lim_{n\to\infty} E(X_3) = \infty$.

We do not want to mention explicitly the representation of the cases $i = 4,5$ at this point since the methodology was described adequately and the terms tend to be quite long.

We have thus proven that all accumulation points of a descent sequence fulfill all restrictions in (7).

5.2 Results of the Optimisation

For the initial values the old (to be improved) parameter values of the quality surveillance procedure of the packing machines

$$(\tilde{\gamma}_1, \tilde{\gamma}_2, \tilde{\gamma}_3, \tilde{\gamma}_4) = (4800\text{g}, 9900\text{g}, 14950\text{g}, 20000\text{g}) \tag{11}$$

are used, if $(\tilde{\gamma}_1, \tilde{\gamma}_2, \tilde{\gamma}_3, \tilde{\gamma}_4) \in V$. Otherwise one chooses another feasible initial value with the following scheme:

$$(\tilde{\gamma}_1, \tilde{\gamma}_2, \tilde{\gamma}_3, \tilde{\gamma}_4) = (5000\text{g} - 4w, 10000\text{g} - 2w, 15000\text{g} - w, 20000\text{g})$$
$$w \in [0\text{g}, 50\text{g}]$$

The evaluation of the objective function is very time consuming because of the necessarily numerical integrations. That is why there are methods preferred which need just a few function evaluations.

A typical representative of such procedures is the method of Hooke and Jeeves (Hooke, 1961). In Fig. 5.2 you can find the values of the objective function in the optimum, which are given by the method to the different parameter combinations. To the expense of the received solution the procedure needs between 250 and 400 function evaluations depending on the parameter combination.

As a further procedure the downhill-simplex-method of Nelder and Mead (Nelder, 1965) is tested. With this method one partly gets, but can not improve, the solutions found by Hooke and Jeeves' method. The method usually requires more than 600 function evaluations to reach the optimum. It is well known that the downhill-simplex-method is not very efficient in the number of function evaluations (see also Press, 1992). Therefore, a listing of the results at this point has not been produced.

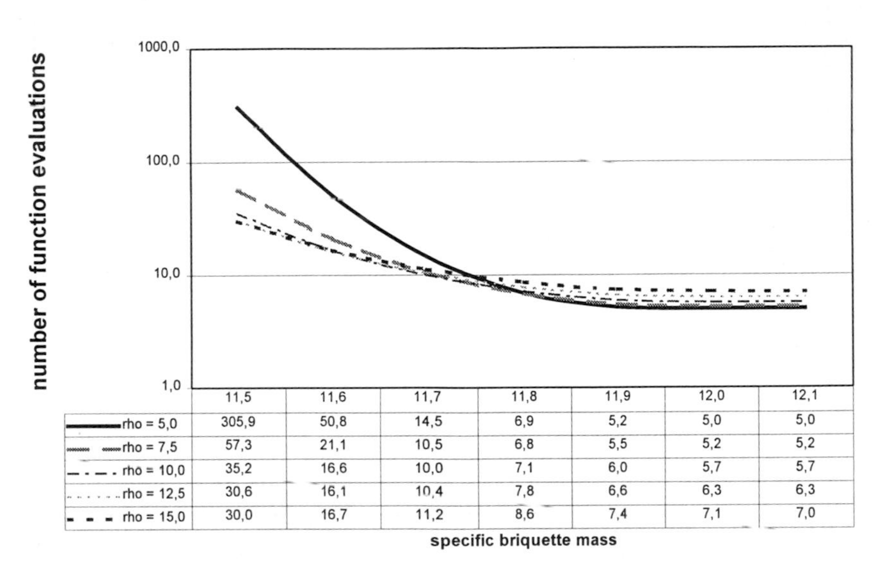

	11,5	11,6	11,7	11,8	11,9	12,0	12,1
rho = 5,0	305,9	50,8	14,5	6,9	5,2	5,0	5,0
rho = 7,5	57,3	21,1	10,5	6,8	5,5	5,2	5,2
rho = 10,0	35,2	16,6	10,0	7,1	6,0	5,7	5,7
rho = 12,5	30,6	16,1	10,4	7,8	6,6	6,3	6,3
rho = 15,0	30,0	16,7	11,2	8,6	7,4	7,1	7,0

Fig. 5.2. Scored minimums for varying specific briquette mass and different values of the scatter of the layer length.

There is an alternative method of conjugated directions of Powell (see Acton, 1990, pp. 464-467). It produces the same results as the ones of Hooke and Jeeves and is even a little better in the number of function evaluations.

Also, with Simulated Annealing the results of Hooke and Jeeves' method could neither be reached nor improved. The number of function evaluations rises as expected up to 3000, with that the computing time also rises by approximately a factor of 10.

If one exploits the differentiability of the objective function (see section 3), approximates the gradient numerically and then uses a gradient method, it only takes 2 to 15 gradient steps (depending on the initial value), i.e. between 50 and 560 function evaluations, to reach the optimum (see Fig. 5.3). The received results are not as good as the ones that the method of Hooke and Jeeves delivered, as the numerical differentiation amplifies the propagation of errors of the numerical integration – but we have not yet analysed this in depth. To weigh up the advantages and disadvantages of the use of differentiability of the objective function for that very special problem one can conclude that it is disadvantageous to use a numerical approximation of the gradient. Therefore, alternative procedures are still tested.

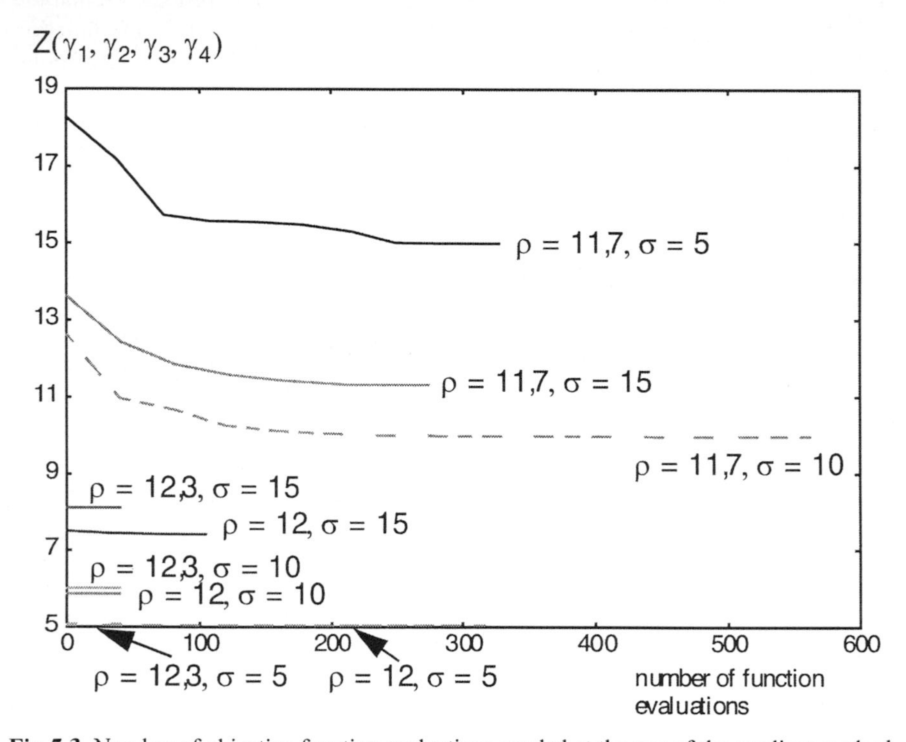

Fig 5.3. Number of objective function evaluations needed at the use of the gradient method together with numerical differentiation.

Finally, one can realise that the optimal values of the weight limits $(\gamma_1, \gamma_2, \gamma_3, \gamma_4)$ are highly dependent on the parameters σ_D and ρ which vary during the ongoing operation of the plant (see Table 1 in the Appendix).

5.3 Interpretation

The relatively even form of the curves in Fig. 5.2 leads one to suppose that, by the weighing of single layers, a high briquette bundle production rate can be reached.

Unfortunately, this is only correct if the optimal values $(\gamma_1^*, \gamma_2^*, \gamma_3^*, \gamma_4^*)$ are exactly matched to the given specific briquette mass ρ. So, in the production process this cannot be used, since the value of ρ is subject to natural variations, usually the value varies within hours. The main problem is that even small changes of the optimal parameter setting $(\gamma_1^*, \gamma_2^*, \gamma_3^*, \gamma_4^*)$ lead to significant changes of the value $Z(\gamma_1, \gamma_2, \gamma_3, \gamma_4)$ of the objective function. This means that a minor misadjustment of the weight limits may result in seriously production losses.

On the other hand the length limits δ_{MIN} and δ_{MAX} as well as the mean value μ_D are usually chosen constantly for longer periods of time or can be well controlled.

Therefore the idea of developing a computer-aided system which determines optimal parameter settings and modifies them on-line depending on the actual values of the influence values σ_D and ρ was rejected.

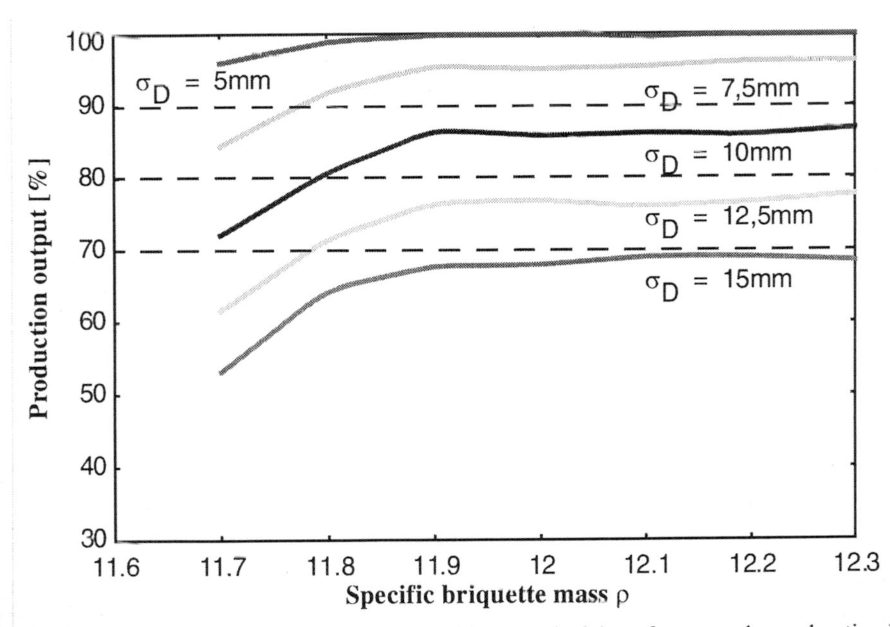

Fig. 5.4. Maximum production output at suitable control of the reference value and optimal weight limits, depending on specific briquette mass ρ. Lower and upper length limit were chosen as $\delta_{mn} = 409\text{mm}$, $\delta_{\max} = 441\text{mm}$.

In addition to this it has been examined how the values of the objective function are influenced if the mean layer length μ_D, as additional variable is considered, since μ_D can be supervised by a well performing 3-point briquette thickness control at the briquette presses. The maximum attainable packing quota can be seen in Fig. 5.4. There it is obvious that the primary relevant quantity, for achieving high packing quotas, is the suitable choice of a mean layer length that can be nearly optimally adjusted by means of the briquette thickness control. However, this again can be arbitrarily deteriorated by a non-optimal choice of the weight limits.

Therefore, it was obvious to examine the results that can be achieved if there is no weight control of single layers and to adjust the mean briquette thickness optimally instead. The attainable packing quotas can be seen in Fig. 5.5. Under the specific conditions of this particular application the strategy "Bundling without weight control" proves to be competitive to the quality surveillance procedure based on a check of a single layer weights for values $\rho > 11.9\, g/\mathrm{mm}$. We no longer have the disadvantage of being dependent on several highly sensitive parameters.

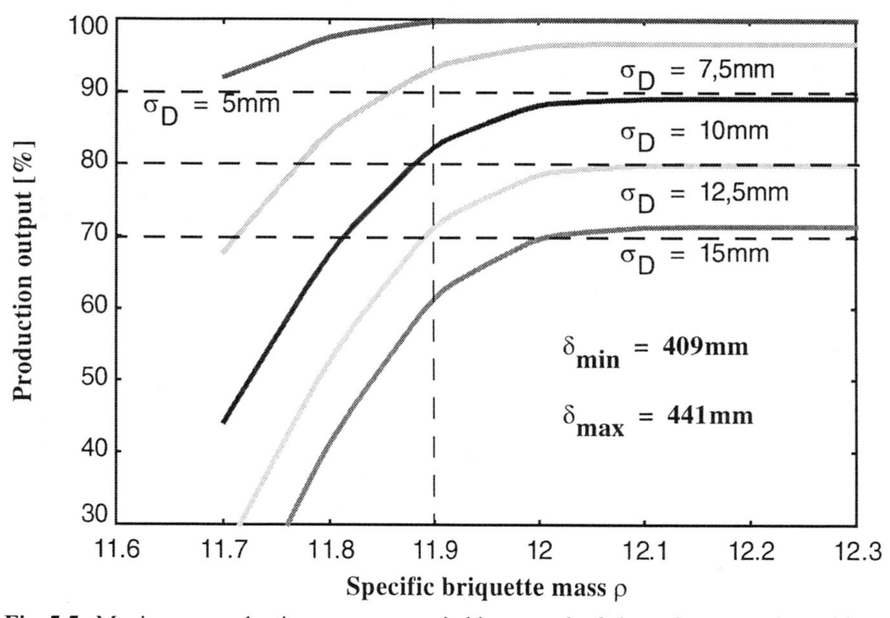

Fig 5.5. Maximum production output at suitable control of the reference value without weight control, depending on specific briquette mass ρ. Lower and upper length limit were chosen as $\delta_{min} = 409\mathrm{mm}$, $\delta_{\max} = 441\mathrm{mm}$.

This may even lead to the recommendation to run the production without weighing each layer if the parameters $\gamma_1,\ldots,\gamma_4$ cannot be held optimal at all times.

These calculated optimal values of the reference value setting, presented in Fig. 5.6, have the feature of being almost independent of the scatter of the value σ_D of a certain production unit. In addition, to guarantee a nearly optimal production output, it is sufficient to adjust the reference value at the centre of the interval, limited by $\delta_{\min}$ and $\delta_{\max}$, as long as the specific briquette mass does not fall below 12 g/mm.

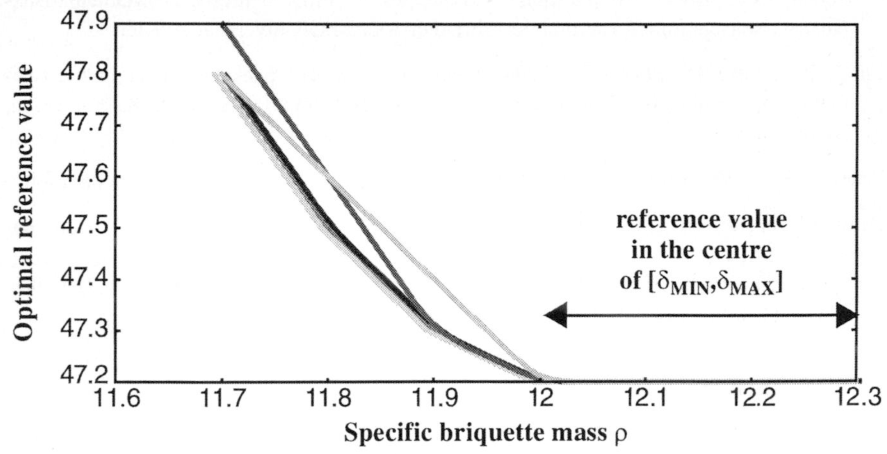

Fig. 5.6. Optimal setting of the reference value in dependence of the specific briquette mass ρ. Lower and upper length limit were chosen as $\delta_{mm} = 409\text{mm}$, $\delta_{\max} = 441\text{mm}$.

6 Conclusion

An exact mathematical analysis shows that the quality surveillance procedure may be very helpful in increasing the output of the production line. The optimal parameter settings though are sensitive to changes of the side conditions. Therefore sensitivity must be checked and may lead to the recommendation to run the production line without layer weight control.

References

Acton, F.S. (1990): Numerical Methods that Work, The Mathematical Association, Washington

Biedermann, J.; Cromme, L., Rhede, L., Weber, K., Würll, S. (1997): Simulation and optimization of a production process in brown coal industry. In: Proceedings of the 15th IMACS World Congress on Scientific Computation, Modelling and Applied Mathematics, volume 5: Systems Engineering, Wissenschaft & Technik Verlag, Berlin, pp. 707–712

Biedermann, J.; Cromme, L., Weber, K, Würll, S., Rehn, F., Rhede, L. (1997): Modellierung von Verpackungsprozessen für Braunkohlebriketts. Braunkohle / Surface Mining, 49(4):369–377

Bornholdt, S. (1997): The Statistical Mechanics Approach to Genetic Algorithm: Annealing Schedules from Population Dynamics. PhD thesis, Mathematisch-Naturwissenschaftliche Fakultät der Christian-Albrechts-Universität zu Kiel

Cools, R.; Laurie, D.; Pluym, L. (1994): Cubpack++: A C++ package for automatic two-dimensional cubature, Report TW 220, Dept. Of Computer Science, K.U. Leuven, Netherlands

Cromme, L. (1998): Eine erweiterte Leibnizsche Regel und Anwendungen. In preparation

De la Maza, M.; Tidor, B. (1993): An analysis of selection procedures with particular attention paid to proportional and boltzmann selection." In: Forrest, S. (ed.) Proceedings of the Fifth International Conference on Genetic Algorithms (ICGA '93), Morgan Kaufmann Publishers, pp. 124–131

Fichtenholz, G.M. (1972): Differential- und Integralrechnung II, VEB Deutscher Verlag der Wissenschaften, Berlin

Hooke, R.; Jeeves, T.A. (1961): Direct Search Solution of Numerical and Statistical Problems, Journal of the ACM 8, pp. 212-229

IMSL®, Fortran and C application development tools from Visual Numerics (1994), Manual "Fortran subroutines for mathematical applications", vol 1, pp. 615-620

Nelder, J.A.; Mead, R. (1965): A simplex method for function minimization, Computer Journal 7, pp. 308-313

Press, W.H. (1992): Numerical Recipes in C, 2nd Edition, Cambridge University Press, Cambridge

Prügel-Bennett, A.; Shapiro, J.L. (1994): Analysis of genetic algorithms using statistical mechanics. Physical Review Letters, 72(9), pp. 1305–1309

Prügel-Bennett, A.; Shapiro, J.L. (1997): The dynamics of a genetic algorithm for simple random ising systems. Physica D, 104, 75–114

Werner, J. (1992): Numerische Mathematik I, Vieweg Verlag, Braunschweig

Appendix

Used Mathematical Propositions

Theorem 4 (Leibniz theorem, cf. Fichtenholz 1972) Let $f : D \to \Re$, $f = f(g,\gamma)$ be a continuous function, defined on the rectangle $D = [a,b] \times [c,d]$

with partial derivatives with respect to the variables g and γ. If these partial derivatives are continuous and if the curves $\alpha:[c,d]\to\Re$ and $\beta:[c,d]\to\Re$ fulfill the relations $\alpha([c,d])\subseteq[a,b]$ and $\beta([c,d])\subseteq[a,b]$ on D then, if α and β are continuously differentiable on D the following holds:

$$\frac{\partial}{\partial\gamma}\int_{\alpha(\gamma)}^{\beta(\gamma)} f(g,\gamma)dg=\int_{\alpha(\gamma)}^{\beta(\gamma)}\frac{\partial}{\partial\gamma}f(g,\gamma)dg+\beta'(\gamma)f(\beta(\gamma),\gamma)-\alpha'(\gamma)f(\alpha(\gamma),\gamma)$$

Lemma 5 (Extension of Leibniz' theorem) The function $f:\Re^2\to\Re$, $f=f(g,\gamma)$, be continuous with respect to the variables g and γ and have for $\gamma-g\neq k$, $k\in\Re$, the continuous partial derivative $\frac{\partial f}{\partial\gamma}$. A jump discontinuity may be on the line $\gamma-g=k$. Then the partial derivative of $\int_a^b f(g,\gamma)dg$ with respect to γ exists. The derivative $\frac{\partial f}{\partial\gamma}$ is a continuous function and the following holds:

$$\frac{\partial}{\partial\gamma}\int_{\alpha}^{\beta} f(g,\gamma)\,dg=\int_{\alpha}^{\beta}\frac{\partial}{\partial\gamma}f(g,\gamma)\,dg\,.$$

$[\delta_{\text{MIN}}, \delta_{\text{MAX}}] = [409\text{mm}, 441\text{mm}]$, $\mu_D = 423\text{mm}$								
σ_D	ρ	$(\gamma_1^*, \gamma_2^*, \gamma_3^*, \gamma_4^*)$	$E(X_1$	$E(X_2$	$E(X_3)$	$E(X_4$	$E(X_5$	$Z(\gamma_1^*, \gamma_2^*, \gamma_3^*, \gamma_4^*)$
5	11.5	(4987, 9988, 14990, 19994)	59.982	59.425	60.143	61.479	64.845	305.874
	11.6	(4981, 9981, 14985, 19991)	9.924	9.757	9.899	10.208	11.005	50.793 (-178.448)[1]
	11.7	(4970, 9968, 14973, 19984)	2.759	2.732	2.804	2.942	3.270	14.506 (-9.835)
	11.8	(4941, 9938, 14948, 19968)	1.245	1.288	1.344	1.422	1.579	6.879 (-0.697)
	11.9	(4867, 9877, 14904, 19945)	1.003	1.013	1.027	1.048	1.090	5.182 (-0.076)
	12.0	(4908, 9816, 14848, 19916)	1.003	1.003	1.003	1.003	1.005	5.016 (-0.005)
	12.1	(4949, 9898, 14847, 19881)	1.003	1.003	1.003	1.003	1.003	5.014 (0.000)
7.5	11.5	(4975, 9976, 14981, 19988)	10.935	10.919	11.160	11.583	12.709	57.305
	11.6	(4963, 9963, 14969, 19981)	4.007	3.974	4.081	4.279	4.755	21.096 (-17.891)
	11.7	(4945, 9941, 14951, 19969)	1.958	1.963	2.035	2.158	2.428	10.541 (-2.584)
	11.8	(4909, 9903, 14920, 19951)	1.224	1.270	1.327	1.406	1.562	6.788 (-0.401)
	11.9	(4867, 9849, 14878, 19928)	1.041	1.062	1.085	1.117	1.182	5.488 (-0.112)
	12.0	(4908, 9816, 14834, 19904)	1.041	1.041	1.041	1.044	1.053	5.221 (-0.024)
	12.1	(4949, 9898, 14847, 19876)	1.041	1.041	1.041	1.041	1.041	5.204 (-0.001)
10	11.5	(4967, 9968, 14973, 19983)	6.618	6.651	6.834	7.146	7.993	35.242
	11.6	(4949, 9948, 14957, 19973)	3.110	3.113	3.215	3.391	3.821	16.649 (-7.693)
	11.7	(4925, 9920, 14933, 19958)	1.829	1.842	1.926	2.053	2.326	9.976 (-1.529)
	11.8	(4883, 9875, 14897, 19938)	1.276	1.331	1.396	1.481	1.650	7.134 (-0.352)
	11.9	(4867, 9826, 14857, 19915)	1.132	1.153	1.181	1.220	1.297	5.983 (-0.139)
	12.0	(4908, 9816, 14818, 19893)	1.132	1.132	1.133	1.139	1.154	5.690 (-0.041)
	12.1	(4949, 9898, 14847, 19871)	1.132	1.132	1.132	1.132	1.133	5.661 (-0.002)
12.5	11.5	(4961, 9962, 14969, 19980)	5.695	5.743	5.916	6.208	6.994	30.556
	11.6	(4940, 9939, 14949, 19968)	2.978	2.996	3.099	3.274	3.710	16.056 (-5.432)
	11.7	(4911, 9905, 14920, 19951)	1.890	1.904	2.008	2.146	2.439	10.387 (-1.258)
	11.8	(4866, 9856, 14882, 19928)	1.376	1.453	1.522	1.617	1.805	7.773 (-0.359)
	11.9	(4867, 9812, 14843, 19906)	1.260	1.280	1.312	1.354	1.444	6.649 (-0.164)
	12.0	(4908, 9816, 14810, 19886)	1.260	1.260	1.261	1.268	1.288	6.338 (-0.053)
15	11.5	(4958, 9959, 14966, 19979)	5.571	5.629	5.807	6.103	6.894	30.004
	11.6	(4933, 9933, 14944, 19965)	3.079	3.107	3.214	3.398	3.858	16.656 (-4.714)
	11.7	(4901, 9895, 14912, 19946)	2.031	2.049	2.167	2.317	2.638	11.201 (-1.192)
	11.8	(4856, 9842, 14871, 19922)	1.507	1.603	1.683	1.789	1.999	8.581 (-0.387)
	11.9	(4867, 9803, 14834, 19900)	1.409	1.428	1.464	1.511	1.612	7.424 (-0.189)

Table 1: Results of the optimisation with the procedure of Hooke and Jeeves.

[1] The difference between the value of the objective function $Z(4800\text{g}, 9900\text{g}, 14950\text{g}, 20000\text{g})$ for the initial (default) parameter values and the optimal parameter setting is indicated in brackets provided that the default parameter setting was feasible.

Process Model for the Abrasion Mechanism during Micro Abrasive Blasting (MAB) of Brittle Materials

Martin Bothen and Lothar Kiesewetter

Chair of Microsystems Production Technologies, Brandenburg Technical University of Cottbus, Universitätsplatz 3/4, D-03044 Cottbus, Germany,

Abstract. *In micro abrasive blasting (MAB), a fine abrasive powder is accelerated in an air stream. When the powder hits the target with a velocity near the velocity of sound, material is abraded from the target surface. In this paper it is shown for the first time that, by comparing experiment and theory, the abrasion mechanism during micro abrasive blasting can be described mathematically. By investigating the effective mechanism for blasting of brittle materials and the concentration distribution of the blast medium in the jet of compressed air a process model for micro abrasive blasting can be established. With only few experimentally measured values it is possible to apply the procedure for computer simulation of profile forming structuring techniques.*

Keywords. *micro abrasive blasting, abrasive powder, nozzle, air jet, process, model, abrasion mechanism, brittle materials, fracture toughness, microcracks, profile geometry*

1 Introduction

The material abrading effect of MAB is based on the energy transmission of the potential pressure energy contained in the compressed air to the blast medium grains in the jet in the form of kinetic energy, Fig. 1. Depending on velocity and grain type the kinetic energy of the emitted blast medium this will result in hammering, consolidating, polishing, grinding, purifying, roughening, lapping, deburring, drilling or separating when it hits the surface

The application of MAB, using blast media of the finest grain size and directed open jet guiding, is seen as a new possibility of abrasive processing for the creation of deep structures in substrate materials of microsystem technology. Nozzles with openings as small as 250 μm are used. As abrasives, hard blast media aluminium oxide, silicon carbide and glass with grain diameters from 10 μm to 80 μm are used.

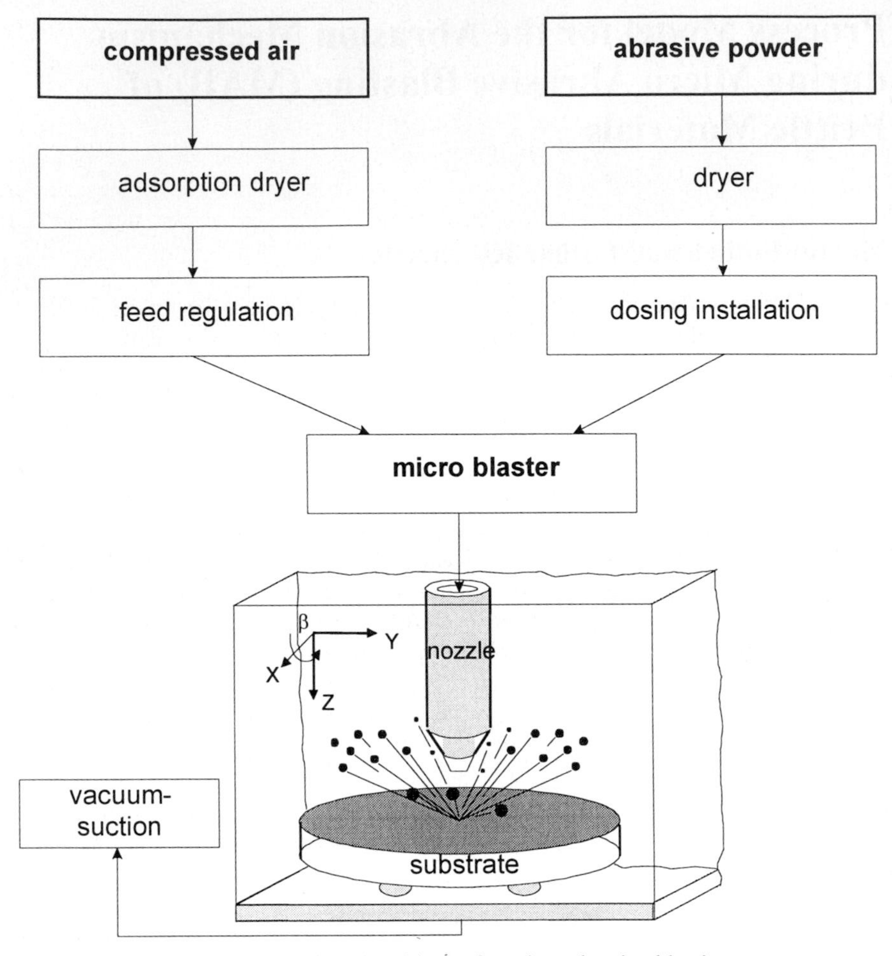

Fig. 1: Functional principle of the micro blaster for micro abrasive blasting.

2 Abrasion Mechanism during Blasting of Brittle Materials

The most commonly used materials in the components, such as silicon, quartz, glass and various ceramics, are brittle at room temperature. Brittle materials are not able to degrade defect spots through dislocation. When the maximum load, which depends on the biggest crystal defect present its position in the crystal and the strength of the material is exceeded, a failure occurs. The strength is described through the material parameter K_{1c}, which can be derived from failure mechanics.

In MAB, the abrasion mechanism is based on the knocking off and knocking out of material through the impact of the small jet grains. Through the impulse of the hitting grain this energy is added and the material is fragmented. The abrasion rate depends on the hardness, the form, the kinetic energy and the number of particles hitting the substrate in the specified time. Every single impulse can lead to several microcracks. When a sufficient number of microcracks per time unit are caused in the material, so that the individual microcracks meet and cross, the smallest material fractions come off the substrate material.

Through material abrasion, the surface of the material increases. The correlation between the energy W_f, which is applied for the creation of the new surface, and the specific surface energy γ_f, according to the theory of Griffith for brittle materials (Griffith, 1921) is:

$$\gamma_f = \frac{W_f}{A_M} = \frac{K_{1c}^2}{E'} \qquad \text{with } W_f = \eta_S \cdot W_{kin} \tag{1}$$

γ_f = specific surface energy; $[\gamma_f]$ = N/m
W_f = energy for the creation of the new material surface; $[W_f]$ = Nm
A_M = material surface created through fractures; $[A_M] = m^2$
K_{1c} = fracture toughness; $[K_{IC}] = \sqrt{N^2/m^3}$
E' = E-modulus of the substrate material; $[E'] = N/m^2$
W_{kin} = kinetic energy of the hitting blast medium grains; $[W_{kin}]$ = Nm
η_S = energy efficiency for the creation of the new material surface; $[\eta_S] = 1$

Altogether, the new material surfaces that were created through the many individual microcracks result, in material abrasion. The material abrasion volume, V_M, is proportional to the newly created material surface:

$$V_M = \eta_M \cdot A_M = \eta_M \cdot \frac{W_f}{K_{1C}^2} E' \tag{2}$$

η_M = proportionality factor between the newly created material surface and the abraded volume; $[\eta_M]$ = m

3 Process Model for the Abrasion Mechanism during MAB

The geometry of the blasted structure can be calculated mathematically when the concentration distribution of the jet particles per time unit within the jet, the kinetic energy of the jet particles during the impact on the substrate surface, the material properties of the substrate material and jet grain as well as the blasting

time are known. For the process model to be developed for the abrasion mechanism during MAB, the following simplifying assumptions are made:
- The number of particles leaving the nozzle per time unit is constant.
- The flow propagation is constant over time.

3.1 Velocity of the Blast Medium Grains in the Compressed Air Jet

The blast medium is accelerated by means of compressed air. When cylindrical nozzles are used, the air flows from the nozzle at the velocity of sound and carries the ambient air along. The friction at the boundary layer jet-ambient air, on the one hand, leads to to a retarded acceleration of the air flow and, on the other hand, the ambient air is sucked which leads to an expansion of the jet, Fig. 2.

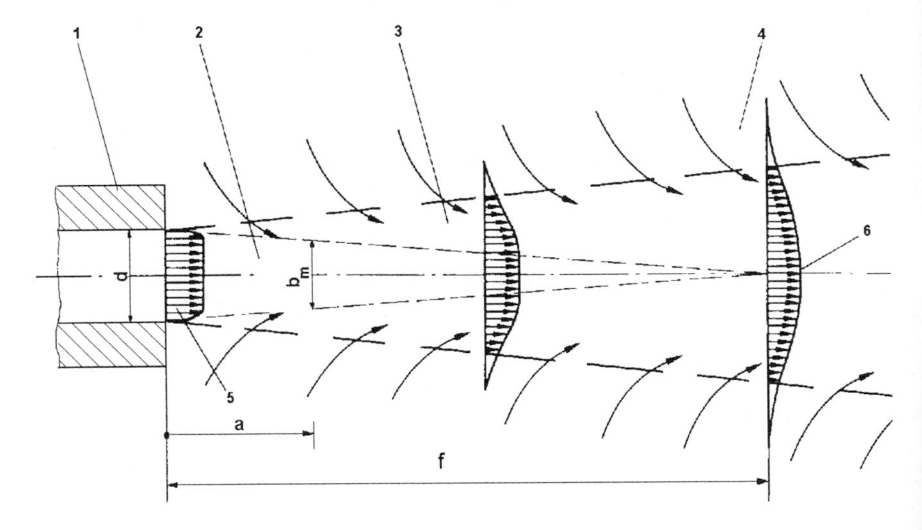

Fig. 2: Velocity distribution in the open jet: 1 nozzle, 2 core area, 3 mixing area, 4 ambient air, 5 velocity of the air flowing out, 6 velocity distribution in the open jet; $b_m(a)$ = jet middle area; a = distance to the nozzle, d = nozzle diameter, f = length of the core area.

The dashed line in Fig. 2 demonstrates the jet expansion. It connects positions in which the flow velocity in the same flow profile has fallen to 10% of v_{max}.

The velocity profile in the jet changes from an approximately rectangular form, position 5 in Fig. 2, to a distribution according to a mathematical exponential function, position 6. In the transition area the jet middle area b_m decreases with the distance a to the nozzle.

Due to the inertia of the grains there is a velocity deficit at the outlet of the nozzle between the jet particles and the air. The grains leave the nozzle at approx. 120m/s while the compressed air flows out at the velocity of sound. Thus also after leaving the nozzle the blast medium grains are further accelerated by the air flowing out until the air and the blast medium grains have reached the same velocity.

3.2 Concentration Distribution of the Jet Middle Grains in the Compressed Air Jet

For the concentration distribution, C(x,y,z,t), of the jet particles per time unit in the open jet, the diffusion equation is chosen as the approach. The diffusion equation describes the mixing of the jet middle grains with the air. A great air-molecular friction, as well as an increased jet pressure, results in an expansion of the jet. (Liu, 1992)

Using the Laplace transformation yields a normal distribution function as a solution (Sugaku, 1986). With a circular jet nozzle a diffusion in x and y-direction takes place. The equation for the concentration distribution of the jet middle grains in the open jet for this nozzle form is:

$$C(x,y,z) = C_0 \cdot \frac{\exp\left(-\left(\frac{x}{\sqrt{2}\cdot\lambda_x(z)}\right)^2\right)}{\sqrt{2\pi}\cdot\lambda_x(z)} \cdot \frac{\exp\left(-\left(\frac{y}{\sqrt{2}\cdot\lambda_y(z)}\right)^2\right)}{\sqrt{2\pi}\cdot\lambda_y(z)} \cdot dx \cdot dy \tag{3}$$

C(x,y,z) = number of grains penetrating the place S(x,y,z) at the time Δt; [C] = Number/s

C_0 = number of grains leaving the nozzle at the time Δt; $[C_0]$ = number/s

$\lambda_x(z)$ = jet expansion coefficient in x-direction; $[\lambda] = m$

$\lambda_y(z)$ = jet expansion coefficient in y-direction; $[\lambda] = m$

When the diffusion is the same in x and y-direction, equation (3) with $r^2 = x^2 + y^2$ and $\lambda_x = \lambda_y = \lambda$ can be simplified:

$$C(r,z) = C_0 \cdot \frac{\exp\left(-\left(\frac{r}{\sqrt{2}\cdot\lambda(z)}\right)^2\right)}{2\pi\cdot\lambda(z)^2} \cdot r \cdot dr \cdot d\varphi \tag{4}$$

When the circular nozzle is moved at a feed velocity v_v in the direction of the y-axis, the area interspersed by the jet is a function of the time t, Fig. 3.

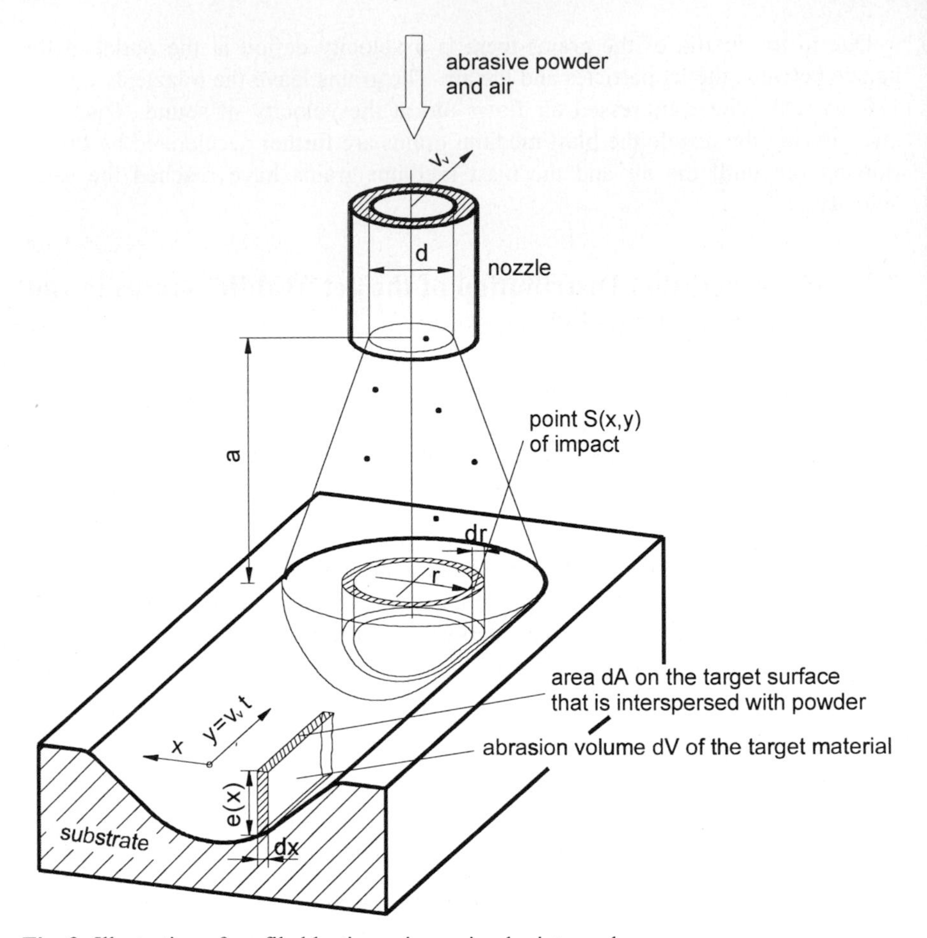

Fig. 3: Illustration of profile blasting using a circular jet nozzle.

The number of grains penetrating the place S(x,y,z) is thus:

$$C(x,y,z)=\frac{C_0}{v_v}\cdot\frac{1}{\sqrt{2\pi}\cdot\lambda(z)}\cdot\exp\left(-\left(\frac{x}{\sqrt{2}\cdot\lambda(z)}\right)^2\right)\cdot dx\cdot dv_v \tag{5}$$

v_v = lateral feed velocity of the nozzle in y-direction; $[v_v]$ = m/s

3.3 Calculation of the Abrasion Profile

The abrasion volume at the place S(x,y,z=a) caused by the impact of the grains, can be calculated when the energy of the individual grain is known. On condition that all grains have the same mass and hit the substrate surface with the same velocity, the abrasion e(x,y) at the place S(x,y,z=a) is proportional to the number of the grains hitting. The abrasion volume per time unit is calculated using equation (2) as follows:

$$V_M = \int e(x,y)\cdot dx\cdot dy = \eta_M \cdot \eta_S \cdot \frac{E'}{K_{1C}^2}\cdot W_{kin,1}\cdot C_0 \tag{6}$$

When a circular nozzle is used which moves at a feed velocity v_v in the direction of the y-axis, the following abrasion depth is calculated at the place S(x,y):

$$e(x,y) = \eta_M \cdot \eta_S \cdot \frac{E'}{K_{1C}^2}\cdot W_{kin,1}\cdot \frac{C_0}{v_v}\cdot \frac{1}{\sqrt{2\pi}\cdot\lambda(z)}\cdot \exp\left(-\left(\frac{x}{\sqrt{2}\cdot\lambda(z)}\right)^2\right) \tag{7}$$

This equation can be simplified when v_v = constant and e(x,y) = e(x):

$$e(x) = e_{100}\cdot \exp\left(-\left(\frac{x}{\sqrt{2}\cdot\lambda(z)}\right)^2\right) \tag{8}$$

$$e_{100} = e(x=0) = \eta_M \cdot \eta_S \cdot \frac{E'}{K_{1C}^2}\cdot W_{kin,1}\cdot \frac{C_0}{v_v}\cdot \frac{1}{\sqrt{2\pi}\cdot\lambda(z)} \tag{9}$$

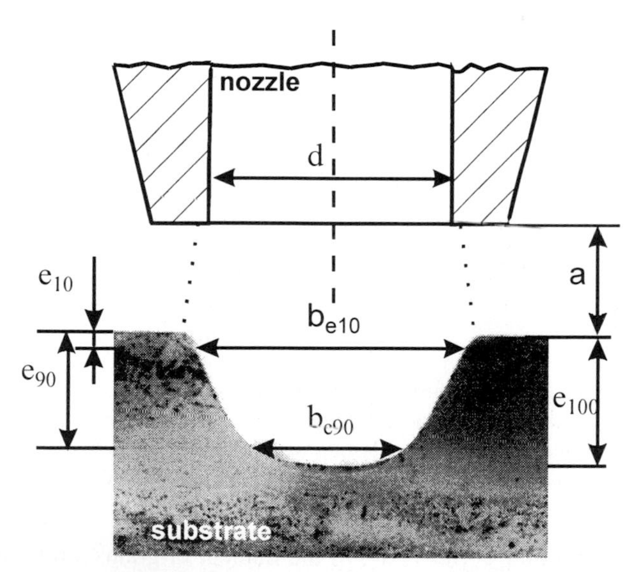

Fig. 4: Profile of the treated substrate area: d = nozzle diameter; a = nozzle distance, e_{10} = abrasion depth at 10% of the maximum depth; b_{e10} = profile width at the abrasion depth e_{10}, b_{e90} = profile width at the abrasion depth e_{90}.

Thus, with the experimental determination of the abrasion depth e_{100} as a function of the nozzle distance a during vertical blasting of a substrate area the profile form can be determined.

Fig. 4 shows a typical profile of a blasted substrate surface when the nozzle jet hits the substrate surface vertically. The abrasion depth e(x) at the place x is correlated regarding its parameters with the velocity distribution according to Fig. 2 and the concentration distribution of the jet middle grains in the jet.

3.4 Influence of the Nozzle Distance on Abrasion Volume, Maximum Abrasion Depth and Profile Width

The abrasion volume per time unit, as a function of the nozzle distance, a is illustrated in Fig. 5. During blasting, the distance of the substrate surface from the nozzle increases due to material abrasion. When the nozzle distance a is the distance of the nozzle to the original surface, for abrasion depths $e(x) \ll a$ the simplification $z \cong a =$ constant can be made.

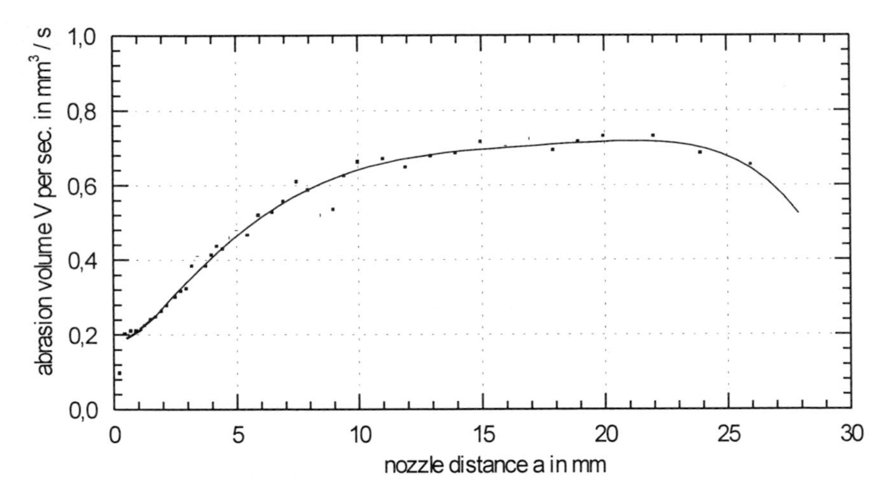

Fig. 5: Influence of the nozzle distance a to the abrasion volume V per second blasting time in silicon. Test parameters: abrasive powder Al_2O_3, medium grain size 25 μm, nozzle diameter d = 0.8 mm, blasting angle $\beta = 90°$, air pressure $p_ü = 6$ bar, blast medium passage 6.5 g/min, air passage 40 l/min, nozzle feed $v_v = 1.7$ mm/min.

Due to the law of the conservation of the number of grains hitting which effect material abrasion equals the number of grains leaving the nozzle at the same time Δt. The abrasion volume according to equation (6) is thus only dependent on the kinetic energy of the grains hitting. As shown in Fig. 5, the kinetic energy of the jet grains is maximal at distances which corresponds to 20 times the nozzle diameter.

Fig. 6 presents the abrasion depth e_{100}, dependant on the nozzle distance a. The abrasion depth, e_{100} also increases with increasing nozzle distance a, but only up to approx. ten times the nozzle diameter because due to the diffusion in the jet the number of blast medium grains effecting abrasion e_{100} decreases with the distance a.

Diffusion causes an expansion of the blasted structures. In Fig. 7 the jet expansion's dependence on the nozzle distance is illustrated. The limit value e_{10} means that the abrasion depth at this position amounts to 10% of the maximum abrasion depth e_{100}. The position of e_{10} is proportional to the nozzle distance a and the nozzle diameter d.

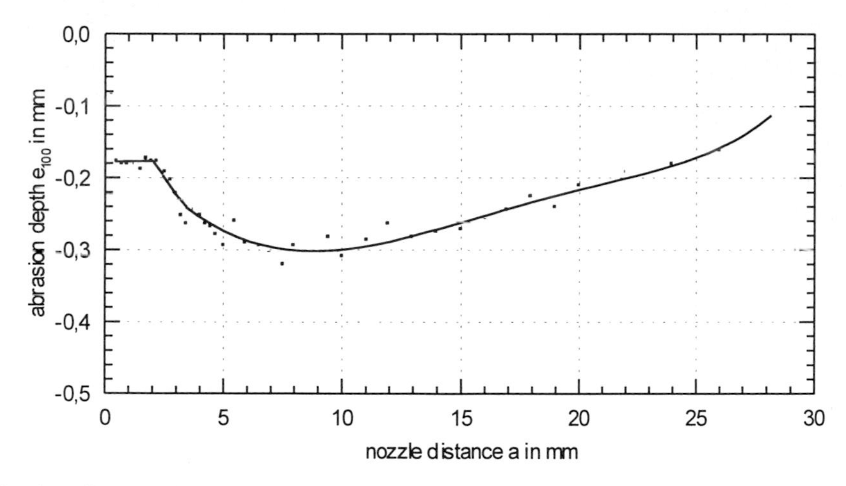

Fig. 6: Influence of the nozzle distance a on the abrasion depth e_{100}. Test parameters as in Fig. 5.

3.5 Influence of the Blast Time on the Profile Geometry

The effect of the MAB depends on the impact angle of the powder the substrate surface. However, during blasting the geometry of the substrate surface and thus the impact angle of the jet grains change. For the iterative calculation of the blasted structure the total blast time t is divided into several time intervals Δt: $t = i \cdot \Delta t$. In order to simplify the process model, the change of the trajectory of the blast medium grains of maximally 10° (Fig. 2) caused by the nozzle expansion is neglected so that in the first time interval Δt_1 all grains hit the substrate surface vertically.

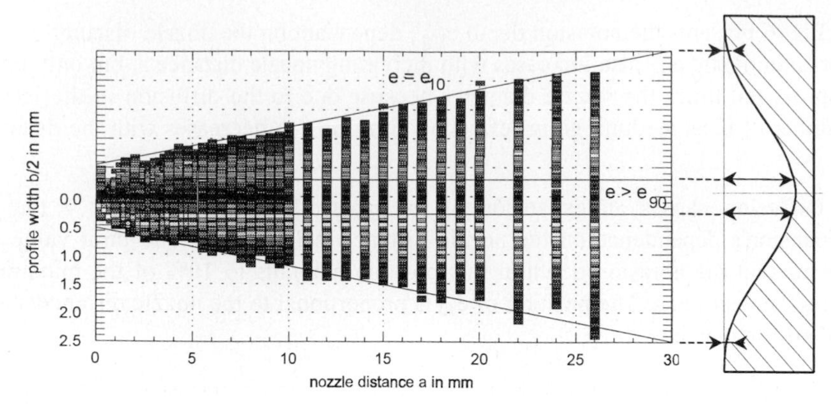

Fig. 7: Influence of the nozzle distance a on the profile width as well as the form. With increasing nozzle distance a the profile width b_{e10}, see Fig.4, increases by 2x4°. Test parameters as in Fig. 5.

The powder quantity that hits the substrate surface in the first time interval, Δt, effects the material abrasion depth e_1 (x,t_1), according to equation (8):

$$e_1(x,t_1) = \frac{e_{100}}{i} \cdot \exp\left(-\left(\frac{x}{\sqrt{2}\cdot\lambda(z)}\right)^2\right) \tag{10}$$

The further calculation of the abrasion requires the superposition of the radius-dependent density distribution of the grains in the jet with the changed impact angle of the grains which is caused by the abrasion that has already taken place. The profile angle, α_P , at the impact place of the grain is calculated using the formula $\alpha_P = \arctan(\Delta e(x)/\Delta x) = \arctan(e'(x))$, Fig. 8.

For the second time interval, the abrasion depth at the position x is:

$$e_2(x,t_2) = \frac{e_{100}}{i} \cdot \exp\left(-\left(\frac{x}{\sqrt{2}\cdot\lambda(z)}\right)^2\right) \cdot \cos(\alpha_{P,1}) \tag{11}$$

For the j-th time interval the abrasion quantity is calculated analogously:

$$e_j(x,t_j) = \frac{e_{100}}{i} \cdot \exp\left(-\left(\frac{x}{\sqrt{2}\cdot\lambda(z)}\right)^2\right) \cdot \cos(\alpha_{P,j-1}) \tag{12}$$

The total profil a depth e(x) in the location x, at the time t=i Δt is:

$$e(x) = \sum_{j=1}^{i} \frac{e_{100}}{i} \cdot \exp\left(-\left(\frac{x}{\sqrt{2} \cdot \lambda(z)} \right)^2 \right) \cdot \cos\left(\alpha_{P,j-1}\right) \tag{13}$$

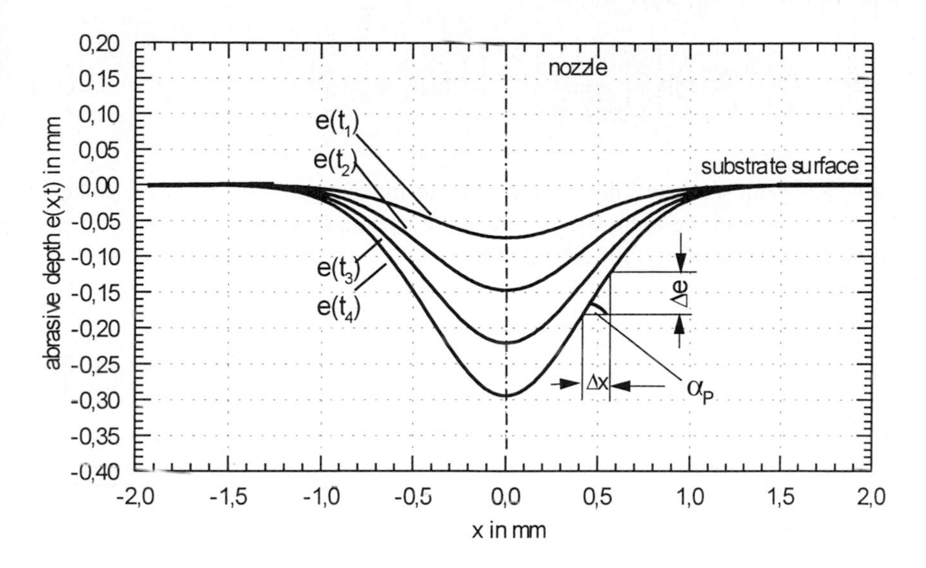

Fig. 8: Calculated abrasion depth e(x) as a function of the time t.

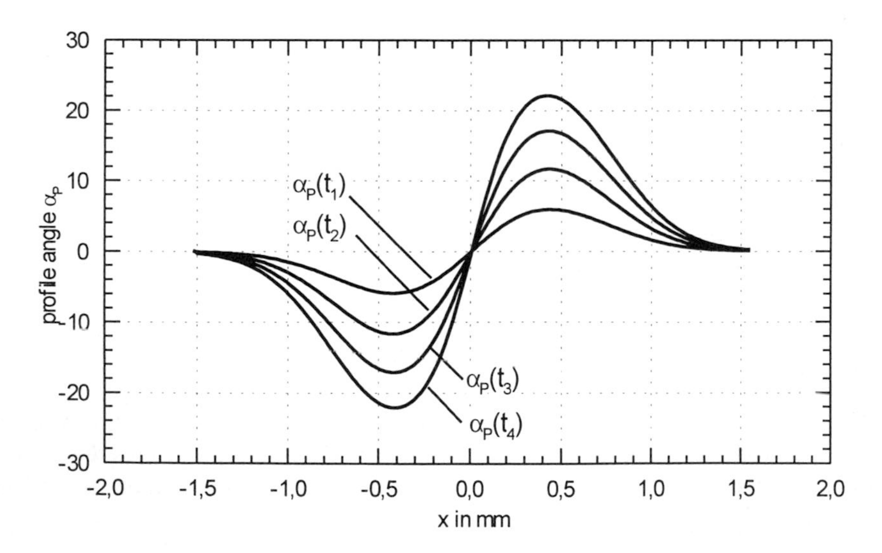

Fig. 9: Calculated profile angle α of the profile illustrated in Fig. 8 as a function of the time t.

3.6 Influence of the Jet Angle on the Abrasion Profile

When the nozzle does not blast vertically but at an angle of β < 90° on the substrate surface, equation (13) is:

$$e(x) = \sum_{j=1}^{i} \frac{e_{100}}{i} \cdot \exp\left(-\left(\frac{x}{\sqrt{2} \cdot \lambda(z)} \right)^2 \right) \cdot \sin\left(\beta - \alpha_{P,j-1}\right) \tag{14}$$

With increasing abrasion, the position of the maximum abrasion depth changes (Fig. 10). Through the oblique position of the nozzle, it is even possible to form a profile flank angle of α > 90° on one side.

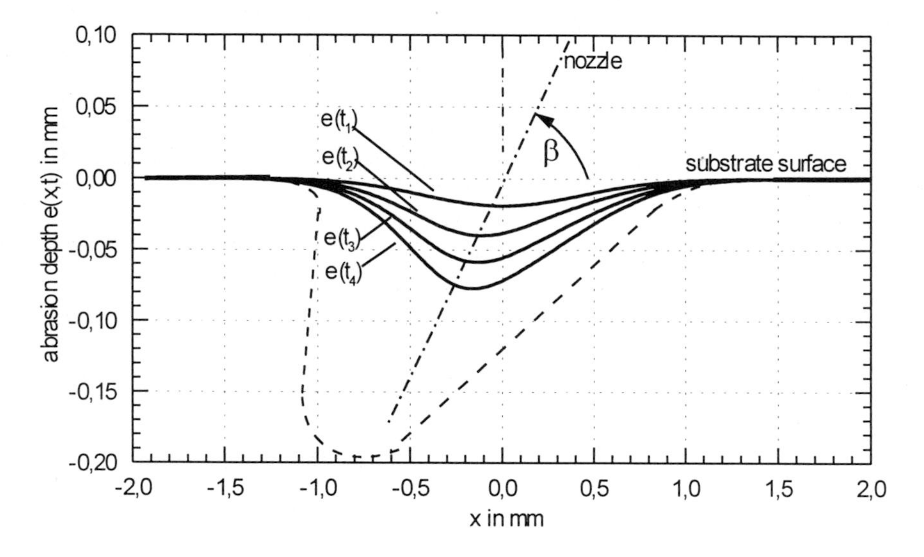

Fig. 10: Calculated abrasion depth e(x) as a function of the time t at a jet angle of β = 15°.

3.7 Influence of the Nozzle Diameter on the Profile Geometry

The formula for the concentration distribution of the grains in the jet assumes a circular nozzle opening. For a nozzle distance larger than the 15 times the nozzle diameter, the nozzle opening can be treated in a good approximation as a point source.

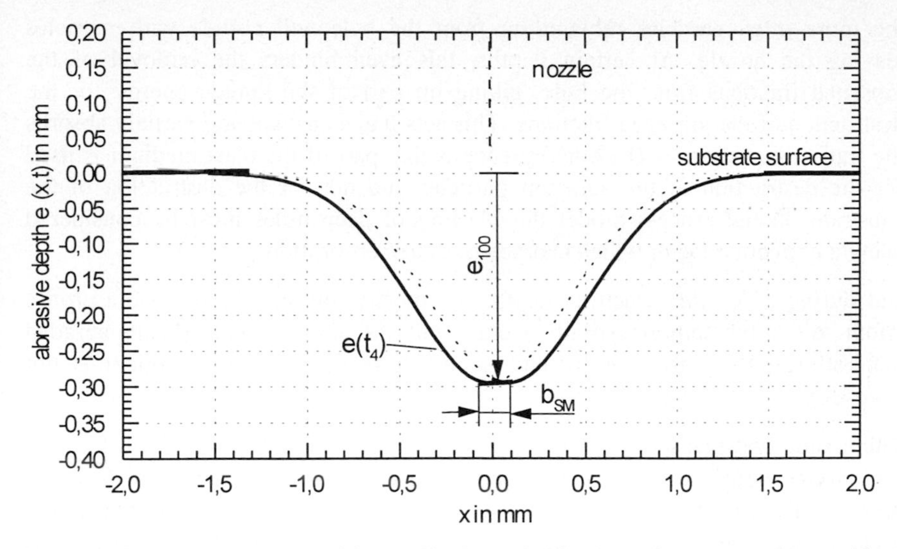

Fig. 11: Profile displacement of the profile shown in Fig. 8 at t_4 in correspondence with an expansion of the jet middle area b_{SM}; the ratio of nozzle distance to nozzle diameter is a/d=8.

For a practically usable nozzle distance, which is smaller than this critical nozzle distance, the calculated profile curve around the jet middle area b_{SM} is expanded centrally, Fig. 11:

$$\text{für } |x| \geq \frac{b_{SM}}{2}: \; e(x) = e_{100}(z) \cdot \frac{\exp\left(-\left(\frac{|x| - \frac{b_{SM}}{2}}{\sqrt{2} \cdot \lambda(z)}\right)^2\right)}{2\pi \cdot \lambda(z)^2} \tag{15}$$

$$\text{für } |x| < \frac{b_{SM}}{2}: e(x) = e_{100}(z) \cdot \frac{1}{2\pi \cdot \lambda(z)^2} \tag{16}$$

The jet middle area, b_{SM}, is dependent on the nozzle diameter, the nozzle distance a and the jet pressure.

4 Limitations of the Process Model

The assumption that there are no collisions of the grains in the jet, finds its limitations for the blasting of deep holes. The deeper the jet penetrates the hole, the more often particles rebounding from the hole will collide with particles leaving the nozzle. At certain depths, this even hinders the removal of the loosened fractions from the hole, taking up part of the kinetic energy for the destruction of the loosened fractions. This acts like a cushion and partialy absorbs the transmitted energy. The consequence is that part of the blast medium is used for the destruction of the abrasion particles and not for the destruction of the substrate. In the process model the blasting of deep holes must be considered using a correction factor which characterizes this absorption.

But during MAB the structural depths are mostly low in relation to the profile width so that the compressed air together with the blast medium and the abraded material can flow off laterally and a correction of the process model is not necessary.

The assumption that the jet particles are slowed down through an air cushion on the surface before they hit the silicon plate could not be confirmed. Blasting narrow rails resulted in the same abrasion as the blasting of plane areas. Due to the great mass of the blast medium and the low density of the air the compressed air cushion on the substrate surface hardly influences the blasting performance.

References

Griffith, A.A.(1921): *The phenomena of rupture and flow in solids.* Phil. Trans. Roy. Soc. London, 1921. A221: p. 163-198.

Liu, X. (1992) *Beitrag zur Simulation des pneumatischen und elektrostatischen Lackierprozesses*. Produktionstechnik Berlin, ed. G. Spur. Vol. 103. 1992; München, Berlin: Carl Hanser Verlag.

Sugaku, I. (1986): *Encyclopedic dictionary of mathematics*. Vol. 2. Edition. 1986.

Model Based Approaches for Closed-Loop Quality Control in Injection Moulding

Jonannes Wortberg and Mahmud Al-Haj Mustafa

University of Essen, Institute for Design Theory and Plastics Machinery, Schützenbahn 70, 45127 Essen

Abstract. *For illustrating the injection moulding process a partial neural network structure is developed which takes into account the relationships between machine settings, process-sequence variables and quality parameters. The model is actualised continuously by an adaptation mechanism. It predicts the quality of the product which is defined individually by a quality vector and corresponding specification limits. During production the predicted quality vector is compared with the calculated intervention limits for each quality parameter. If a quality value lies out of its intervention interval the actual quality deviation is calculated and passed on to the controller. The relationship between changes of certain machine settings and its qualitative effects on the quality values must be accurately known for controller design. This relationship can be ascertained through an experimental plan or empirical experience of the machine operators. The design of experiments technique is applied in the first examination of moulds. The resulting data contains the influences of varied machine settings on the defined quality values. The collected data, changes of manipulated values and corresponding quality value changes, can be fixed in an expert system. This expert system can be extended with empirical experience know-how of the process which the machine operators have. Alternatively both kinds of knowledge could be integrated in a fuzzy module which is able to calculate the necessary manipulations of the machine settings. First experiments show that this hybrid design is able to compensate disturbances on quality parameters as batch variations to a certain degree.*

Keywords. *Quality Control, Neural Networks, Fuzzy Control, Quality Management, Injection Moulding, Hybrid Models, MIMO-Control*

1 Introduction

Increasing customer's requirements, strong foreign competitors and an intensified environmental legislation force the plastics processors to increase productivity with a simultaneous orientation to the main thoughts of Total Quality

Management (TQM). New methods and approaches are in demand which qualify the companies to realise these goals. For this purpose a set of quality management methods is available, which cover all phases of the production process, from raw materials to the finished product.

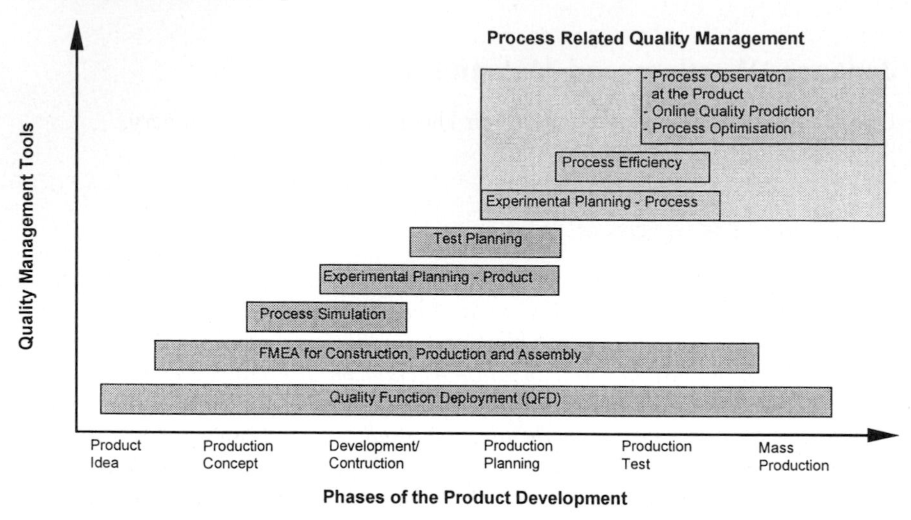

Fig. 1.: Process-Related Quality Management in the Product Development

Process-related quality management methods represent the highest stage of the QM measures taken during the manufacture of the product (Fig. 1). Its objectives are process monitoring, on-line quality prediction and, based on these, process optimisation and quality control. The rigorous application of these methods must lead to higher process reliability and a reduction in process costs (Bourdon, 1997). The methods require models that describe the complex relationships between process parameters and product properties. An important prerequisite is the acquisition and analysis of all relevant process and product parameters.

2 Characteristics of the Plastics Injection Moulding Process

The plastics injection moulding process is characterised by many complex interactions between raw material, process and product characteristics. Changes in the raw material characteristics, different compositions of virgin material and regrind, as well as changes in regrind particle size and geometry and in regrind temperature, lead to problems in processing that frequently result in scrap production and in changed machine settings. Normally these influences are compounded by effects caused by the machine.

Finally, effects like wear of the machine, oil quality or environmental influences are of importance in the case of quality deviations. Wear of the plasticating screw leads to a worse output behaviour, a longer melt residence time in the screw and to melt inhomogeneities. All these influences are reflected in quality characteristics of the product.

3 Possibilities of Modelling

The modelling of plastics processing processes in closed mathematical or physical models has not yet been carried out with sufficient accuracy. We are not yet able to formulate the complex multidimensional flow, deformation and heat-exchange processes that occur throughout the process in such a way as to allow precise derivations of product quality (the sum of all the relevant quality features). In addition, the model quality depends greatly on the necessary simplifications, assumptions and material functions, and on the boundary and initial conditions. It is impossible to assess the effect on quality of the scattering of influencing variables that is always found in running processes as a result of the raw material, ambient conditions and the machine itself.

With the implementation of CAD-systems and the advanced development of flow pattern simulation programs data on FEM-basis is now available. These programs allow an estimation of the flow processes in the mould as well as the calculation of the quality relevant process variables such as the pressure profile in front of the screw and of certain quality parameters like warpage and shrinkage. In the future experiments with selective parameter variations could be carried out (Bourdon, 1997).

The obtained results are not precise enough to forecast the product quality with sufficient accuracy but in the future they could be combined with heuristic model building techniques to so-called hybrid models.

A more exact description of the entire process can be obtained from heuristic modelling processes, which represent the relationships between machine setting variables, process variables and quality variables by means of the data records obtained during or preceding production. Statistical processes, such as multiple regression, and, more recently, neural networks, have proved successful.

3.1 Modelling in the Running Process

For such a complex process as the injection moulding process an intelligent strategy for data acquisition and for a subsequent adaptation of the heuristic model must be implemented. The plastics injection moulding process is exposed to permanent scattering and disruptive disturbances that result from different sources. To include these real production stages the process data must be determined entirely from the "natural" scattering behaviour of the process during normal production (including the corrections to manipulated variables that become necessary during the observation period). The necessary parameterisation of the degrees of freedom in the models is carried out automatically, in the background, by evaluating the scatterings ascertained on the running process. Numerous applications are known (Häußler 1994; IKV 1995; Potente 1997; Schmidberger 1996; Wortberg 1997; Hanning 1994). The advantages are an uninterrupted implementation of the process models in the running production and the acquisition of the "actual" process.

3.2 Statistical Experimental Design (SED)

Often, the statistical experimental design (SED) is used for determining the data records from production (Bourdon 1996; Bourdon 1997; Michaeli 1996; Vaculic 1996). Selected machine parameters are varied to achieve a survey of the qualitative and quantitative influence on certain quality parameters.

SED permits a good quantitative description but there are critical disadvantages. On one hand the production must be interrupted for the realisation of the experiments. On the other hand not all machine parameters can be included because of the high expenditures. Also, this kind of data acquisition cannot include the "real" process behaviour.

Nevertheless, the SED proved successful in the first examination of moulds. It provides an initial meaningful machine setting for finding stable operating points (Bourdon, 1997). Additionally, the effects of changed machine settings on the quality parameters can be identified without disturbances. This information is necessary for implementation of quality control.

3.3 Model Adaptation

Changes in the process sequence as a result of changes in the machine settings by the operator or the occurrence of disturbance factors are expressed as changed machine setting, process-sequence and quality parameters. Because the data base is continually being expanded with information about the current process state, it must be regularly checked whether the model used still has sufficient predictive quality. If deviations in the predicted results or changes in the process control are ascertained, it will be necessary to adapt the model, taking into account all the current process states (Fig. 2). Only an automatically initiated model adaptation permits a good long-term quantitative description of product quality and provides the prerequisites for optimisation and closed-loop control concepts (Häußler 1994; IKV 1995; Wortberg 1997).

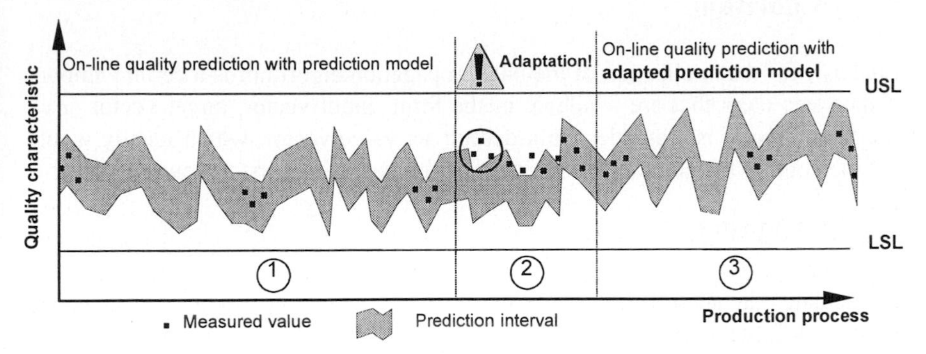

Fig. 2.: Principal of the adaptation mechanism

4 Neural Networks

Neural networks belong to the heuristic model building techniques. Artificial neural networks are structures of simple processing elements with weighted and directed connections between the elements. In most cases, the processing elements are arranged in layers and perform only one non-linear transmission function. The neural network works as an input/output system: it obtains an input vector and produces a corresponding output vector $\hat{y}$. The interrelations between x and y are established through a supervised learning procedure with exemplary data set pairs (x_p,y_p). Training of an artificial neural network means the adaptation of the connection weights, so that the interrelations contained in the training examples perform as well as possible.

For a description of the complex relationships in plastics processing, which are characterised by non-linearities and reciprocal effects, it is advisable to use neural networks. In addition, attributive as well as continuous quality parameters can be modelled. Attributive characteristics are parameters that cannot be measured and that only can be estimated subjectively. In practice, MLP-networks have proven successful.

4.1 Training the Network with the Backpropagation Error Algorithm

The basis for the application of the backpropagation algorithm is a certain amount of data sets pairs that are available in the form: input vector, target vector (x,y). The target vector is equivalent to a default set value vector, which ideally would be given out by the neural network in a feedforward pass. First the output vector

$$\hat{y} = NN(x|w)$$

is calculated in accordance to the actual input values x and to the so called network weights $w=(w_1,\ldots,w_G)$. Thereafter, the calculated output is compared with the desired values of the corresponding data set. The cause for the deviation is propagated back through the network to the input value, whereby the weights are changed, thus the term "backpropagation error". This procedure is applied gradually for all learning examples x,y. The whole cycle can be repeated until a certain error limit is accomplished. The single-error of a neuron in the output layer can be calculated:

$$E(w) = \left(y_i - NN(x_i|w)\right)^2$$

The minimisation of the whole network error requires the modification of the weights w_{ij}, as the only independent variables in the network. The rule for changing the weights corresponds to performing steepest descent on a surface in weight space whose height at any point in weight space is equal to the error measure.

In practice, the data set is divided into a learning and a test data set. The neural network is trained with the learning data set which contains 70% of the original data set. With increased training periods the aberration in the network is minimised. In addition, the aberration in the network for the test data set is calculated every cycle. The test data set contains data pairs which are unknown for the neural network. That means the network must generalise the learned relationships to illustrate these unknown data patterns correctly. The training for the network is finished when the aberration for the test data set has reached a minimum.

4.2 Part Network Structure for the Illustration of the Plastics Injection Moulding Process

To appreciate the complexity of the injection moulding process, a partial network structure must be developed. This structure must take into account the relationships between the machine settings and the process and between the process and product quality (Fig. 3). The first network contains the interrelations between the machine setting and the process-sequence variables.

$$\hat{x} = NN_1(u|w)$$

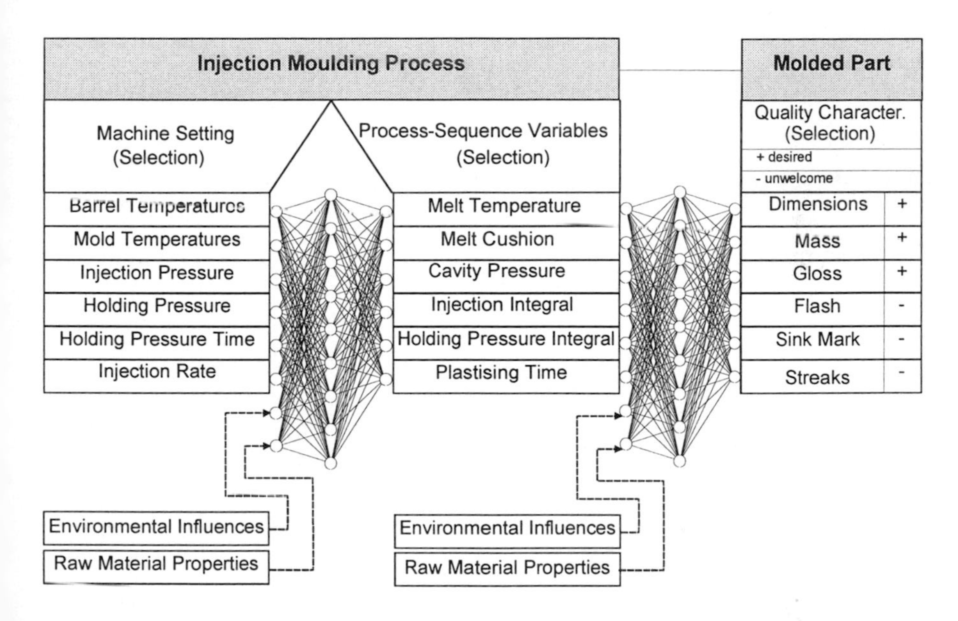

Fig. 3.: Illustration of the Injection Moulding Process in a Neural Network Part Structure

In this case u is the manipulated value vector. This network produces a mean value of the resulting process-sequence variables if the machine settings are constant. That means the output of this network does not contain the process scattering and the influence of the disturbing variables. In order to include the usually existing scattering of the process-sequence variables, the output values of the first network are combined suitably with the actual scattering process variables. These combined values are the input into the 'Quality Expert Network' which contains the relationship between process-sequence and quality variables.

$$\hat{y} = NN_2(\hat{x}, x|w)$$

$$\hat{y} = NN_2(NN_1(u|w), x|w)$$

Thus the changes in machine settings as well as the scatterings of process-sequence and quality parameters of the actual operating point can be illustrated. In addition, the dynamic behaviour of the machine can be accounted for.

This process model represents the basis for closed-loop quality control, which is able to detect disturbing influences and to predict quality deviations (Häußler, 1994).

5 Closed-Loop Quality Control

5.1 First Basic Approaches in Injection Moulding

The wishes for reproducible production conditions have led to the development of partially and fully closed-loop controlled injection moulding machines which have replaced the conventional controlled injection moulding machines. The decline of the price of sensors and electronic modules was a main reason for this development.

Further approaches for closed-loop quality control try to keep a certain process variable constant by tracking a certain manipulated variable. The prerequisite for this approach is the fact that this chosen process variable must describe all quality variables of the injection moulded article. In reality however, the quality specification of an injection moulded part is influenced by a set of process and boundary conditions. The relationship is to a certain extent non-linear and influenced by interactions.

5.2 Advanced Quality Concept

In addition, the idea of quality must not be exclusively understood as a product specification. Factors such as productivity or energy use that represent a dimension of process suitability must be handled as a controlled variable similar to the product specification. For example, it might be the wish to reduce the energy consumption during certain hours.

The closed-loop quality control is linked with the quality optimisation of a process. Quality optimisation tends to reduce the process scattering and to stabilise and economically optimise operating points. The goal of the closed-loop quality control is to conserve optimised operating conditions by eliminating all disturbing influences. The closed-loop control of non-optimised operating points is always possible but not economically reasonable.

5.3 Model Based Approaches

The disadvantages of the first approaches have led to model based closed-loop quality control concepts. In (Vaculic, 1996) a concept with a state controller is presented where the controlled system is described by a statistical process model. In addition, certain characteristic values are generated to better include the existing process conditions in front of the screw and in the cavity. The data for the model was collected by statistical experimental plans. For the closed-loop control of the machine settings standard control methods such as the least squares algorithm or jump experiments are utilised. The assumed time invariance between process-sequence and quality parameters is the basis of this approach.

A further approach (Schmidberger, 1996) uses neural networks for the illustration of the interrelations between process-sequence and quality parameters. For the closed-loop control a second model is integrated which describes the relationship between machine settings and process variables. If the prescribed specification limits are exceeded, the control algorithm is activated.

In (Potente, 1997) it is shown that melt inhomogeneities could be compensated with model based concepts. The mass scatterings caused by batch variations could be reduced with the control algorithm.

5.4 MIMO Closed-Loop Concepts

Frequently predictive controllers in combination with adaptation strategies are applied for MIMO (Multi Input Multi Output)-systems (Beerhold 1996; Fritz 1996; Korn 1996; Martinetz 1996). Neural Networks are used particularly for non-linear systems to describe the control system. The basic idea of predictive control is the calculation of the optimal manipulated variable with the model of the control system. The calculation of the optimum can be carried out with classical methods or with genetic algorithms (Offergeld, 1995). This approach premises a time invariant representation between the manipulated variables and the quality parameters which does not exist in reality.

It seems to be easier to interpret the control system as a linear multi value model and to calculate the manipulated variable changes through partial models (Unbehauen, 1995).

5.5 Requirements of the Injection Moulding Process

The control of the injection moulding process requires a MIMO-control system, whereby the process controlled system is non-linear and time variant. In addition the system is not "totally variable" (Unbehauen, 1995) because the output - in this case the quality variables - cannot be completely observed. An online measurement of the quality variables is in principal possible, but too costly for many parameters. Also, certain parameters, can not be measured online, and certain parameters such as dimensions, accomplish their final values after different periods.

The process model identifies the quality values whereby, a difference between the predicted quality values $\hat{y}(k)$ and the real quality values $y(k)$ still remains.

Another difficulty is that the illustration of the process controlled system could not be inverted. That means that the direct calculation of the manipulated values out of the process model is not possible.

Two principles have to be considered for the controller design: The controller should either have good guide behaviour or good disturbance behaviour. A good guide behaviour would allow the machine operator to influence selectively certain quality parameters and to change them quantitatively. This aspect is important and interesting in the first examination and setting-up of moulds.

Good disturbance behaviour is a prerequisite for mass production. Normally the production is characterised by an unchanged operating point. The negative influences of disturbances on product quality must be eliminated by the closed-loop controller.

5.6 Hybrid Model Approach for Controller Design

The following approach for implementing a closed-loop quality control system bases on a hybrid design. The controller is designed for optimal disturbance behaviour.

Generally the process control system can be formulated for the time periods $k=Ti$, $i=0,1...$, whereby the time periods are equal to the actual part counter, as follows.

$$y(k+1) = f\left(x(k), u(k), z(k)\right)$$

The model of the process control system is described in the preceding chapters. It is a part neural network structure which has an additional adaptative mechanism.

This model produces a prediction of all quality variables in the form of a quality vector.

$$\hat{y} = \left[\hat{y}_1, \hat{y}_2, \ldots, \hat{y}_m\right]^T.$$

Continuous quality values are defined via a specification interval:

$$y_d = \left[(y_{d1USL}, y_{d1LSL}), (y_{d2USL}, y_{d2LSL}), \ldots, (y_{dmUSL}, y_{dmLSL})\right]^T$$

Attributive values have the following form:

$$y_{di} = \begin{cases} 1 \text{ if feature exists} \\ 0 \text{ if feature does not exist} \end{cases}$$

All values are customer dependent.

The specification limits must be converted into intervention limits which consider the error of the measuring instrument as well as the prediction interval of the model.

$$y_{EG} = \left[(y_{d1UIL}, y_{d1LIL}), (y_{d2UIL}, y_{d2LIL}), \ldots, (y_{dmUIL}, y_{dmLIL})\right]^T$$

Now the predicted quality vector $\hat{y}$ is compared with the calculated intervention limits y_{UIL} and y_{LIL}. If a quality value lies out of its intervention interval, the actual quality deviation is calculated and passed on to the controller. For a stable controller design it is advisable to change the manipulated value until the quality deviation has occurred repeatedly.

The control deviation is passed on to the controller which calculates the manipulated valuc vector:

$$u = \left[u_1, u_2, \ldots, u_n\right]^T$$

with

$$u(k+1) = u(k) + p \cdot \dot{u}(k) \text{ with p = constant}$$

The relationship between changes of the manipulated values and its qualitative effects on the quality values must be accurately known for controller design. This relationship can be ascertained through an experimental plan or in the future through results of a flow pattern simulation and, or instead with, empirical experience of the machine operators. Usually Jump experiments during production are not possible.

5.6.1 Controller Design with Data from Experimental Plans

As previously described, the design of the experimental technique is applied first in the examination of moulds. The resulting data contains the influences of varied machine settings on the defined quality values. The limitation to a few machine setting values is disadvantageous. The collected data, changes of manipulated values and corresponding quality value changes, is fixed in a matrix which includes for every experimental point the qualitative changes of the quality values $\dot{y}(k)$ dependent on the changes of the manipulated value $\dot{u}(k)$:

$$\dot{y}_i(k) \approx \frac{y_i(k) - y_i(k+1)}{p} \text{ and } \dot{u}_i(k) \approx \frac{u_i(k) - u_i(k-1)}{p}$$

The Matrix M is illustrated as follows:

$$M = \left[\begin{array}{ccccc|ccccc} \dot{y}_1(0) & \cdot & \cdot & \cdot & \dot{y}_m(0) & \dot{u}_1(0) & \cdot & \cdot & \cdot & \dot{u}_n(0) \\ \cdot & \cdot & & & \cdot & \cdot & \cdot & & & \cdot \\ \cdot & & \cdot & & \cdot & \cdot & & \cdot & & \cdot \\ \cdot & & & \cdot & \cdot & \cdot & & & \cdot & \cdot \\ \dot{y}_1(K) & \cdot & \cdot & \cdot & \dot{y}_m(K) & \dot{u}_1(K) & \cdot & \cdot & \cdot & \dot{u}_n(K) \end{array}\right]$$

with K=Amount of all experimental points

After the calculation of the control deviation several, suggestions of manipulated value changes are ascertained with the matrix M and standard optimisation tools. Thereby, a certain manipulated variable intervention is picked out of the matrix via a form of pattern comparison. The picked out intervention should compensate the control deviation (Fig. 4).

Four aspects are important by implementing this strategy. The illustration between machine settings and quality values is linearised. Additionally, only the machine settings which are considered in the experimental plan are manipulated. Moreover, the ascertained relations can not be extrapolated arbitrarily. Finally, an adverse oscillating behaviour might occur because the suggested changes of the machine settings are only qualitative linearised manipulations.

The possibility of individual configuration and expansion of the matrix is an advantage over a predictive controller which uses polyoptimisation for the calculation of the optimal intervention. For example, the matrix can be extended with empirical knowledge of the process which the machine operators have. Additionly, individual parameters can be extracted out of the quality vector and be controlled separately.

5.6.2 Controller Design using an Expert System

The mentioned disadvantages of the described approach have led to a new approach which tries to incorporate simulation results with empirical knowledge of the machine operators in an expert system. The latter system would calculate the necessary manipulations of the machine settings. With flow pattern simulation, the influences of changes in machine settings on certain process and quality parameters can be investigated.

Empirical experience knowledge of the machine operators exists in form of measures which have been taken in the past to solve certain problems. This

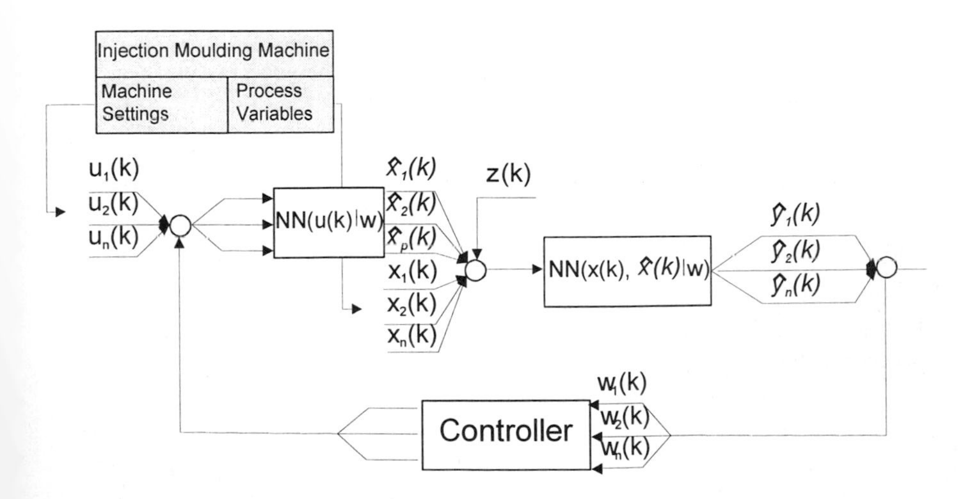

Fig. 4.: Diagrammatic Illustration of the Closed-Loop Quality Control Strategy

includes strategies for production optimisation as well as the necessary manipulations for compensation of quality deviations.

Both kinds of knowledge could be integrated in a fuzzy module which, in the future, will be able to calculate the necessary manipulations of the machine settings. The advantage of such a module lies in the integration of different knowledge, in particular the knowledge of the machine operators which exists usually in spoken form. Additionally non-linear relationships could be included.

5.7 Experimental Results

The first approach is being tested presently in the laboratories of the Institute for Design Theory and Plastics Machinery with a fully hydraulically injection moulding machine which has a central computer interface. An external PC is connected and collects data through the interface. The engaged mould has cavity pressure sensors which are linked to the control of the injection moulding machine. The PC collects the data, builds up the model and starts the quality prediction. In a second step, the PC calculates the necessary interventions in the machine settings.

In a laboratory experiment, a batch variation and the breakdown of the feed zone cooling have been simulated. The mass and the outer diameter of the injection moulded part are predicted. A need to manually readjust the settings was avoided.

The part neural network structure has been trained with the collected data. Fig. 5 shows that the process model could detect the actual quality deviations correctly. This is confirmed by the calculated correlation coefficient of 98% for the outer diameter.

For the configuration of the controller, a data set was recorded beforehand with the help of an experimental plan in order to ascertain the necessary influences of machine setting variations on the quality characteristics.

After the training and the calculation of the matrix M, the quality prediction is started. The external PC collects the data online which are written directly as inputs for the neural network. The predicted quality values are compared immediately with the defined intervention limits.

For testing the closed-loop control, the disturbances have been generated again. If the intervention limits are exceeded, the controller starts to calculate the necessary changes in machine setting.

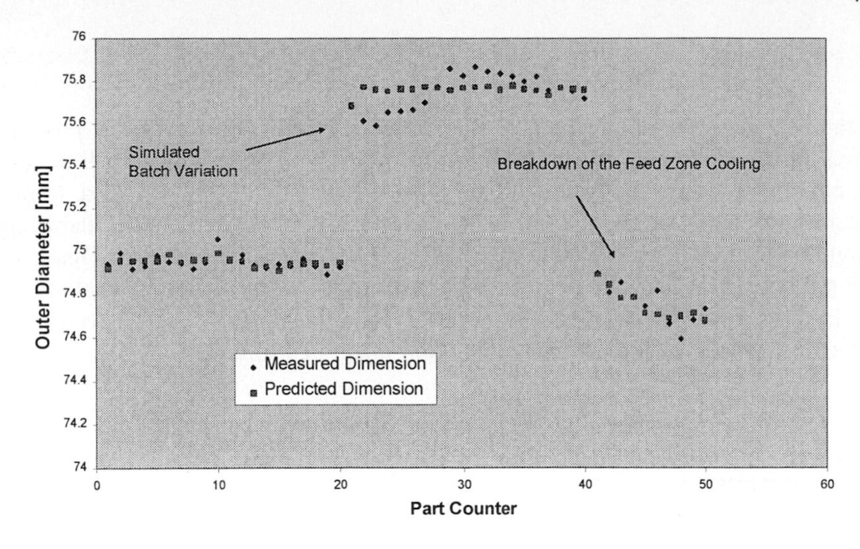

Fig. 5.: Predicted and Measured Dimension at Simulated Process Disturbances

As Fig. 6 shows, the influences of a batch variation on the outer diameter could be compensated to a certain degree. But optical defects occurred in the injection moulded parts because they have not been included in the calculation of the model as a quality parameter.

These first results are the basis for further experiments that will be carried out in the future.

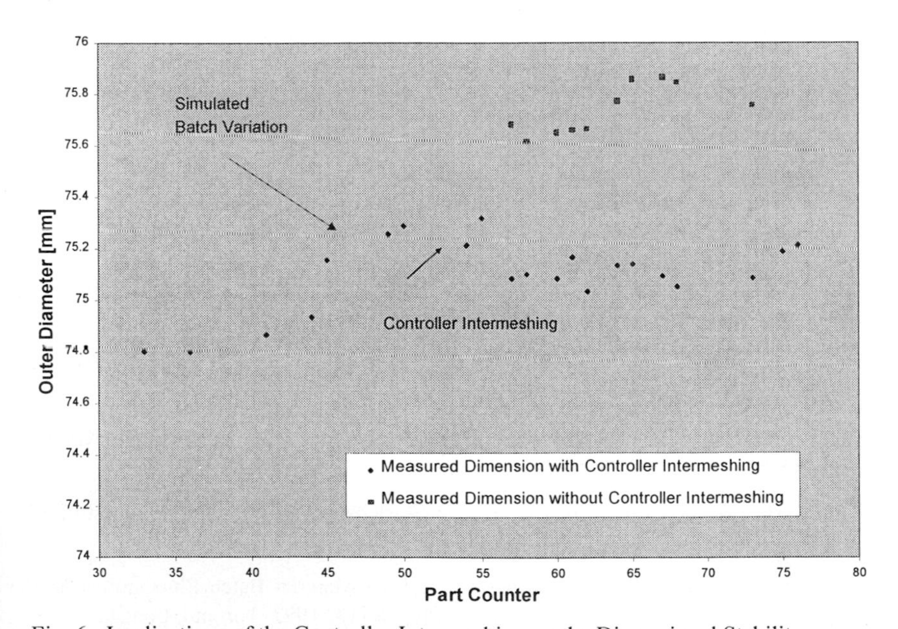

Fig. 6.: Implications of the Controller Intermeshing on the Dimensional Stability

6 Outlook

In the near future, an improvement of modelling can be expected, since process data acquisition is being continually improved. More efficient sensors will permit the effects of disturbance factors on the process behaviour to be identified. Many experts are convinced that, in the next ten years, methods will be available to provide self-monitoring and self-optimising processes (VDMA, 1996). This task will not only be a challenge to plastics machine manufacturers, but especially also to plastics processors, since the information processing will go well beyond the scope of the machine and requires production-integrated solutions.

References

Beerhold, J. R. (1996): Stabile adaptive Regelung nichtlinearer Mehrgrößen-Systeme mit neuronalen RBF-Netzen am Beispiel von Mehrgelenkrobotern; Automatisierungstechnik 44; 12/1996; p. 577-583

Bourdon, R. (1996): Qualitätssicherung entlang der Prozeßkette, Qualität und Zuverlässigkeit 41; 12/1996; p. 1402-1407.

Bourdon, R. (1997): Robuste Maschineneinstellung und Verbesserung der Prozeßbedingungen; contribution to the seminar „Prozeßnahe Qualitätssicherung"; 18./19.3.1997 in Würzburg

Fritz, H. (1996): Modellgestützte neuronale Geschwindungsregelung von Kraftfahrzeugen; Automatisierungstechnik 44; 5/1996; p. 252-257

Häußler, J. (1994): Eine Qualitätssicherungsstrategie für die Kunststoffverarbeitung auf der Basis künstlicher Neuronaler Netzwerke; Ph. D. thesis at the university of Essen; 1994

IKV (Hrsg.) (1995): „Qualitätssicherung beim Spritzgießen - Neue Ansätze"; VDI-Verlag; Düsseldorf 1995

Korn, U., Driescher, A. (1996): Ein Multimodellansatz für prädikative Regelungen mit neuronalen Netzen, Automatisierungstechnik 44; 1/1996, p. 10-20

Martinetz, Th., Gramckow, O., Protzel, P., Sörgel, G. (1996): Neuronale Netze zur Steuerung von Walzstraßen; Automatisierungstechnische Praxis 38; 10/1996; p. 28-42

Michaeli, W., Bluhm, R., Vaculik, R., Wybitul, K. (1994): Formteilfehler sicher erkennen; Kunststoffe 84; 8/1994; p. 979-982

Michaeli, W., Franke, M., Schmidt, G. (1996): Neuronale Netze zur Qualitätssicherung; Plastics-Spezial; 6/1996; p. 34-37

Offergeld, H., Lochner, J. (1995): Darwinsche Betriebspunktoptimierung; Kunststoffe 85; 1/1995; p. 71-75

Potente, H., Ujma, A. (1997): The Influence of Raw Material Batch Fluctuations on the Injection Moulding Process and Product; SPE-ANTEC 1997; Toronto, Canada

Schmidberger, E., Fischer, G. (1996): Einsatz Neuronaler Netze zur Prozeß- und Qualitätsüberwachung beim Kunststoffspritzgießen; contribution to the IPA-technology forum „Moderne Informationstechnologien in der Prozeßüberwachung"; 4.12.96 in Stuttgart

Unbehauen, H. (1995): Regelungstechnik III; Vieweg Verlag; Wiesbaden 1995

Vaculic, R. (1996): Regelung der Formteilqualität beim Spritzgießen auf der Basis statistischer Prozeßmodelle; Ph. D. thesis at the university of Aachen; 1996

VDMA (Hrsg.) (1996): Zukunftstechnologien der Kunststoffindustrie - DELPHI-STUDIE; VDMA 1996

Wortberg, J., Walter, A., Al-Haj Mustafa, M. (1997): Process-Related Quality Management - Possibilities, Limitations and Applications of Process Modelling in Plastics Processing; Plast Europe 97; 11/1997, p. 1664 - 1668

Wortberg, J. (1996): Qualitätssicherung in der Kunststoffverarbeitung; Carl Hanser Verlag; München, Wien 1996

Hanning, D. (1994): Continuous Process Control - Qualitätssicherung im Kunststoffverarbeitungsprozeß auf der Basis statistischer Prozeßmodelle; Ph. D. thesis at the university of Paderborn; 1994

Pharmaceutical Process Modelling

S.A MacGregor[1], L.B. Newnes[1], J.N. Staniforth[2], R.C.L. Lamming[3], M.J. Tobyn[2], Li Ming[1], G.R. Kay[2], M.D. Horrill[2] and D.W. Hajee[3]

[1] Department of Mechanical Engineering
[2] Pharmaceutical Technology Group, Department of Pharmacy and Pharmacolgy
[3] School of Management
University of Bath, Bath, BA2 7AY, UK

Abstract. *The use of Computational Fluid Dynamics in the design of a pharmaceutical processor is described. A simple computational model of the process was developed using a single phase flow. The results of this preliminary study showed good agreement with experimental tests. The model was further developed to consider the effect of solid particles on the processes occurring within a single vessel pharmaceutical processor. Having obtained converged solutions the model was used to optimise the design of the processor. The design optimisation was done on the basis of comparisons with conventional pharmaceutical processing equipment. Optimisation of the process parameters are important as they are known to affect the functionality of the final product.*

Keywords. *Pharmaceutical processors, powder processing computational fluid dynamics, two phase flow.*

1 Introduction

In the preparation of powders for tabletting, the most commonly used techniques are those of wet granulation and direct compression. Conventional wet granulation techniques may involve as many as seven individual sub-processes, depicted in figure 1. These processes are often carried out in separate vessels, thus making the overall process inefficient and with reduced quality due to the need for transportation of materials, holding the powders *in situ* and long changeover times between batches.

In an attempt to rationalise the process by reducing the number of processing vessels used, there are examples of single pot processors, which combine some but not all of the individual sub-processes (Robin, 1994; MacGregor, 1996; Newnes, 1996).

The current study focuses on the design of a single vessel in which all seven sub-processes are carried out. Computational Fluid Dynamics (CFD), has been used extensivley in the design phase of the processor. The use of CFD is now commonplace and there are many examples of its application (MacGregor, 1988; Biffin, 1984; Gupta, 1984).

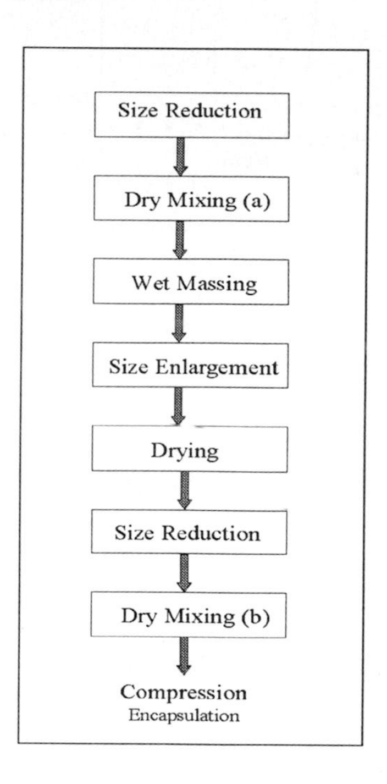

Fig. 1: Sub-processes for Wet Granulation

2 Physical Model

A schematic diagram of the single vessel processor is shown in figure 2. The vessel consists of a cylindrical section which has a cone at the base. The purpose of the cone is to fill the central region to avoid the settling out of powder in regions where the velocity is low. There are two sets of air inlets to the vessel, one which provides air for the fluidising of the powder, this is of low velocity but high volume flow rate. The second supplies the ring of high pressure nozzles around the periphery of the vessel. These high velocity jets provide energy for the various unit sub-processes. The outlet from the vessel passes through a cyclone dust separator where any powder, which escapes from the processing section is collected and returned to the main vessel through a jet pump.

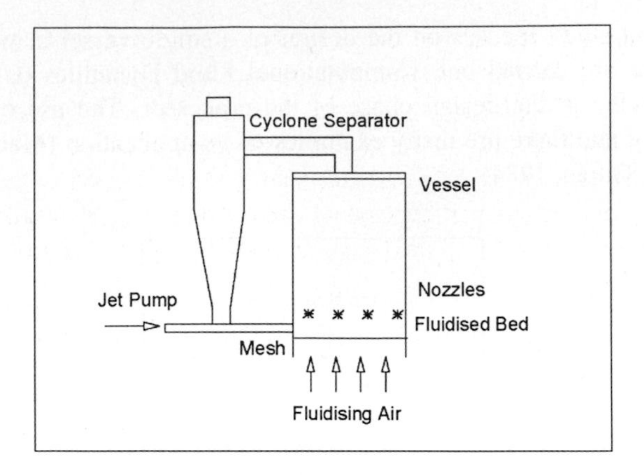

Fig. 2: Schematic diagram of the single vessel procesor

3 Computational Fluid Dynamics

The CFD model was constructed using the code STAR_CD and consisted of a three dimensional axi-symmetrical sector of the vessel. The computational grid consisted of 4000 cells and is shown in figure 3. The grid has been refined in regions where the velocity gradients are anticipated to be large.

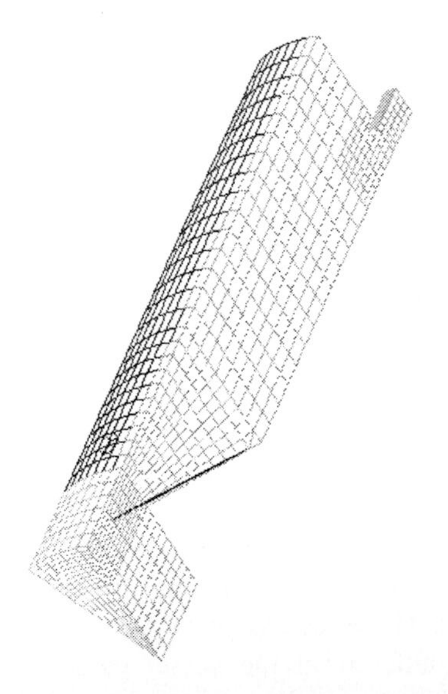

Fig. 3: CFD model of single vessel processor

The software offered a number of options in the use of turbulence models, however it was found that the best convergence and most consistant results were achieved using a *K*-ε model. The model was run for both single and two phase flow cases. In the latter, the solids loaded was varied between 10-40% by volume. The bottom two layers of the computational grid are loaded with particles. This investigation focused on particles which had a diameter of 50μm. This particle size was representative of typical powders used in pharmaceutical industry for tablet production.

A number of cases were studied in which the inlet conditions were varied, these are summarised in Table 1.

Case	Fluidising Velocity	Nozzle Velocity
3.1.1.1	0.2ms^{-1}	2.00ms^{-1}
B	0.5ms^{-1}	2.00ms^{-1}
C	0.2ms^{-1}	3.68ms^{-1}
D	0.3ms^{-1}	3.00ms^{-1}

Table 1: Computational Fluid Dynamic Inlet Conditions

The values for the fluidising and nozzle velocities were selected based on preliminary experimental testing.

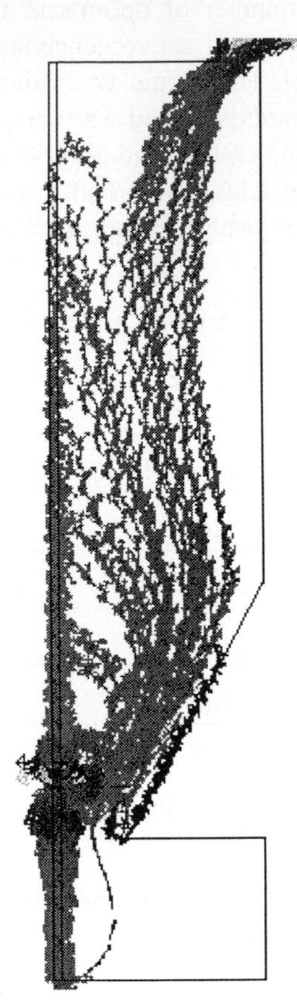

Fig. 4: Computational results for single vessel processor

4 Results

In all cases the flow fields were found to be dominated by large regions of recirculating flow. In examples in which the fluidising velocity has been increased above 0.2 ms^{-1} (cases B and D, table 1) it was found that there was a rapid transportation of the powder from the base of the vessel to the exit in the top wall. Figure 4 shows case C, it can be seen that the flow in the main vessel is dominated by a large recirculation zone. This case was found to be the optimum in terms of powder processing potential. The flow is such that particles will be recirculated many times before they escape from the vessel. This is advantageous in the mixing of the various powders in tablet production.

In the case of size reduction it is important that there are as many particle interactions as possible to ensure that the necessary reduction is achieved. Figure 5 shows typical size reduction results for granular lactose for various inlet pressures to the high pressure nozzles. Due to the low volume flow rate from these nozzles the inlet velocities do not have a significant effect on the general flow through the vessel. These nozzles provide the energy to achieve size reduction. Figure 5 shows that, initially with an inlet pressure of 10 psi the size reduction process is relatively slow. As the presssure is increased the processing time is reduced. Increasing the pressure beyond 20psi results in very little reduction in the time to achieve the required reductions. At this point the flow through the high pressure nozzles is choked and therefore the increases in pressure do not result in a significant increase in energy supplied to the process. This is supported by the results of the CFD studies which suggest that only the fluidising air supply will have any real affects on the overall structure of the flow in the main vessel.

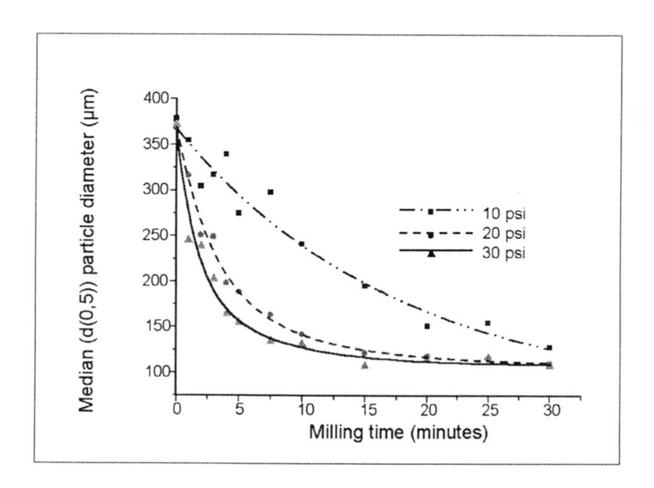

Fig. 5: Size Reduction Curves for Granular Lactose

5 Conclusion

Results of a computational study of a single vessel pharmaceutical processor have been presented. Two phase flow predictions show that by optimising the position of the inlet nozzles and the inlet flow rates the performance can be enhanced in terms of both mixing and size reduction. The performance is comparable with conventional pharmaceutical processors.

References

Biffin, M. (1984): Improved cyclone dust seperators for hot gass clean up. PhD Thesis, University of Wales.

Gupta, A K et al. (1984): Swirl flows. Abacus Press.

MacGregor, S A et al. (1988): Coal fired cyclone combustors. Coal combustion: science and technologyof industrial and utility applications, Hemisphere Publishing, New York, pp657-664.

MacGregor, S. A. et al. (1996): A preliminary study of a single vessel pharmaeutical processor. Proc Inst Mech Eng Part E: Journal of process mechanical engineering, 210, pp121-124.

Newnes, L. B. et al. (1996): Flexible pharmaceutical powder production, IJCIM, 9, 3, pp227-233.

Robin, P. et al. (1994): Rationale for selection of a single-pot manufacturing process using microwave/vacuum drying. Pharmaceutical Technology 18, 28.

Part 6 Modelling and Simulation of Water Systems

Modelling and Simulation of Ground Water Processes

Peter-Wolfgang Gräber

Dresden University of Technology, Institute of Waste Management and Contaminated Sites, Pratzschwitzer Str. 15, D-01796 Pirna, Germany
graeber@rcs.urz.tu-dresden.de

Abstract. *To control and monitor pumping wells, drainage systems in mines and excavations pits, deposits and industrial or agriculture contaminants, it is necessary to know the processes in the soil and groundwater zone. There are physical, chemical and/or biological processes. The physical processes are described mathematically by a system of partial differential equations (PDEs). The groundwater flow described by the heat-conduction equation and the transport of substances by the convection-diffusion equation. For a few of the simplified groundwater relationships one can use simple equations. For the decision support of the water experts and the simulation of a few important processes in the soil and groundwater zone a Computer Aided Engineering (CAE)-Groundwater system is developed at the Dresden University of Technology.*

Keywords. *groundwater, modelling, simulation, computer aided engineering*

1 Introduction

The processes in the soil and groundwater zone are composed of physical, chemical and biological steps. An important aid in research for the geohydrologist is found in the simulation and visualisation of these processes. The main research areas in groundwater management especially in controlling and monitoring, are of

- groundwater in water plants

  The groundwater in water plants has a big significance for the drinking water production. In Germany, about 80% of the drinking water is supplied from underground sources. This can be direct groundwater or river infiltration water. Therefore, it is needed to simulate the groundwater processes, in order to know the possibilities of quantity and quality of water filtration.

- excavation pits and open-cast mining drainage systems

In the Region of Middle Germany, near the city of Leipzig, in Lausitz, near the city of Cottbus and in Rhineland, near the town of Colleen, exist many large open-cast mining pits. Each region has an area of 100 km by 100 km. In those areas conflicts of interest exist between different users, especially between the mining corporations, the water plant enterprises and agriculture farmers.

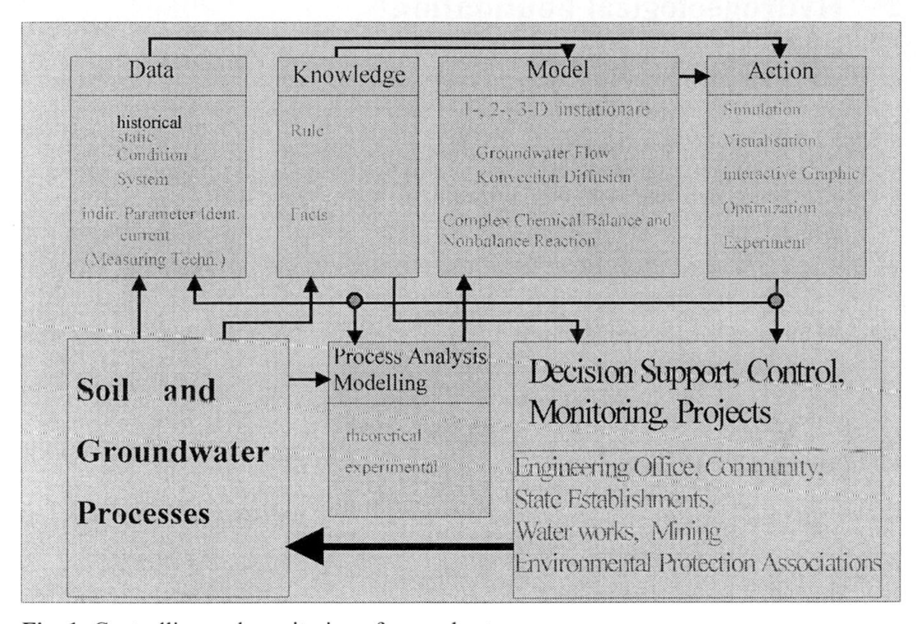

Fig. 1: Controlling and monitoring of groundwater processes

as well as in the redevelopment of

- landfills, contaminated sites and industrially or agriculturally contaminated areas

  The main problems for the groundwater is, when the landfills come into contact with the groundwater. The rain falls on the deposits and washes out different materials, which are unhealthy for humans. The task is to prevent these materials from reaching the groundwater and working their way from the landfills toward the pumping wells of the water plants. Such processes often are bad for health. On the basis of the very big time retardation, the impact appears only now, though the causes of the underground pollution took place 30 or 100 years ago. Such causes of migration can be e.g. chemicals enterprises, military areas, gasworks and gas stations. A large risk comes from the disposal sites placed in the open-cast mining pits. After the increase of groundwater level, the pollutants are washed out and stream into the groundwater flow.

The research, the monitoring and the controlling of such complex systems is possible only with well-founded models and simulation technologies by means of extensive scenario analysis.

2 Hydrogeological Foundation

To describe the dynamic flow and the coupled energy- and mass transport processes as well as the energy and mass conversion in the groundwater zone the following equations are used:

- the dynamic basis equation of the mass flow $\vec{v} = k \operatorname{grad} h$
- the mass balance equation $\operatorname{div} \vec{v} = S \frac{\delta h}{\delta t} - w$
- the boundary conditions for the mass flow

For each phase (air, water and soil) and for each migrant in the soil the following equation must be used:

- the dynamics base equation
  - transport by dispersion $\vec{g}_1 = \vec{D} \operatorname{grad} P$
  - transport by convection $\vec{g}_1 = \vec{v} P$
- the balance equation $\operatorname{div} \vec{g} = (n_0 + \alpha) \frac{\delta P}{\delta t} - w_g$
- the boundary conditions for convection and dispersion.

Explanation:

$\vec{v}$	Speed of groundwater flow	$\vec{D}$	dispersion
k	Conductivity coefficient	P	quality potential
h	Water level	α	Sorption coefficient
S	Storage coefficient	w, w_g	Source or drain intensity
$\vec{g}_1, \vec{g}_2, \vec{g}_3, \vec{g}$	Specific quality flow		

In addition to these equations, chemical reactions (mass conversion) and biological processes might be considered in sink/source terms

The mathematical model consists of a system of ordinary and partial differential equations and algebraic equations, whose coefficients are usually a function of space, time and potential. Therefore, the system is strongly non-linear and the parameters are location and time variables. The process is composed of a higher complexity, a negative conditioning, a greater range of time constants as well as a greater uncertainty of input parameters.

The combination of the equation's results in two non-linear partial differential equations of second order.:

- the conduction equation for the groundwater flow (elliptical PDE)

$$\operatorname{div}(k \operatorname{grad} h) = S\frac{\delta h}{\delta t} - w \tag{1}$$

- the convection-diffusion-equation for the transport of substances (hyperbolic PDE)

$$\operatorname{div}\left(\vec{D} \operatorname{grad} P - \vec{v}\right) = (n_0 + \alpha)\frac{\delta P}{\delta t} - w_g \tag{2}$$

The coupling of the mass and quality flow can be accomplished, using the properties of the water quality (e.g. temperature, mass concentration, kinematical viscosity and density) and the properties of the groundwater flow (gradient of water level, storage change, internal sink and sources).

Often, instead of this complex kind of the groundwater model simplified equations are used, where one or the other process is neglected. One gets a fundamental simplification about the specific view of the process, e.g. the model of the flow, transport processes or the chemical reactions. Essential simplifications can be achieved by the reduction of the dimension of space coordinates to the one space coordinate and/or the time dependence. Any number of examples should show this.

2.1 Pumping Wells

Under the condition: a simplified flow and the use of cylindrical coordinates and an integral transformation over the altitude z, the so-called GIRINSKIJ-Potential Φ, the PDE is:

stationary flow:

$$\frac{d^2\Phi}{dr^2} + \frac{1}{r}\frac{d\Phi}{dr} + \frac{w}{k} = 0 \tag{3}$$

(Leaky aquifer):

$$\frac{d^2 Z}{dr^2} + \frac{1}{r}\frac{dZ}{dr} - \frac{Z}{B^2} = 0 \tag{4}$$

nonstationary flow: $$\frac{\partial^2 Z}{\partial r^2}+\frac{1}{r}\frac{\partial Z}{\partial r}-\frac{Z}{B^2}=a\frac{\partial Z}{\partial t} \qquad (5)$$

THEIS and others have solved these equations by analytical methods. These equations have a big importance for the projects with local characters e.g. for the calculation of the groundwater level by pumping wells in excavation pits. Moreover, based on these equations it is possible to identify the parameters of the groundwater, the coefficient of the permeability and of the storage.

2.2 Plane Filtration

The plane geofiltration equation represents a fundamental equation as a model for the groundwater flow in the saturated zone besides the THEIS-equation. Using the DUPUIT-acceptance, which characterises a simplified aquifer, an integral transformation for the description of profile permeability comes to the transmissibility T. The plane filtration equation is:

$$\operatorname{div}\left(T_{(x,y)}\,\operatorname{grad} z_R\right)=S\frac{\partial z_R}{\partial t}-w_N \qquad (6)$$

With a multilayer aquifer system, this equation can be used for each layer. The hydraulic coupling between the layers can be taken into consideration by a vertical flow.

This equation forms the basis for most of the hydrogeological big area models. In this case, also for the open cast mining areas in Germany, the Middle Germany district, the Lausitz area and the Rhine district.

2.3 One-Dimensional Transport of Substances

The modelling of one-dimensional transport processes of substances plays a big part. This is because these models are solved by analytical methods and these processes form the base of the calibration of the models. Such processes take place e.g. at column experiments in the lab, tracer experiments in the field.

Examples of one-dimensional equations are:

Transport of heat by rainfalls in the unsaturated soil through water (index W) and through air (index L)

$$\frac{\partial}{\partial z}\left(k_W\frac{\partial}{\partial z}\left(\frac{p_W}{\rho_W}+z\right)\right)=\frac{\partial n_W}{\partial t}-w_W \qquad (7)$$

$$\frac{\partial}{\partial z}\left(k_L\frac{\partial}{\partial z}\left(\frac{p_L}{\rho_L}+z\right)\right)=\frac{\partial n_L}{\partial t}-w_L \qquad (8)$$

One-dimensional transport by:

Convection:
$$\varepsilon \frac{\partial C}{\partial t} = -q \frac{\partial C}{\partial x} \tag{9}$$

Dispersion:
$$\frac{\partial^2 C}{\partial z^2} = a \frac{\partial C}{\partial t} \tag{10}$$

Dispersion and convection:
$$MD_1 \frac{\partial^2 C}{\partial x^2} - q \frac{\partial C}{\partial x} = \varepsilon \frac{\partial C}{\partial t} + \lambda C - w \tag{11}$$

Three cases are distinguished, in which λ and/or w are equal or unequal to zero.

2.4 Multiphase Flow

At the modelling of the multiphase flow, several phases in the porous media, the soil or the aquifer, are to be taken into consideration simultaneously. In (Schäfer, 1996) the relationships for a three phase system are set up. Corresponding the equation (2) follows at the neglect of the part of dispersion

$$\operatorname{div}\left((\rho_\alpha \vec{v}_\alpha\right) + \frac{\partial\left(\phi \rho_\alpha S_\alpha\right)}{\partial t} = \rho_\alpha \tag{12}$$

In this case, α represents a general fluid phase. In this three-phase-system the water phase (α=w), NAPL (n) and air (a) are to be taken into consideration. Here, NAPL is the abbreviation of Non-Aqueous-Phase-Liquid, e.g. oil and the derivatives.

Under the neglect of the impulse law between the fluid phases, the DARCY-law can be extended to the multiphase system.

$$\vec{v}_\alpha = -\frac{k \cdot k_{r\alpha}}{\mu_\alpha}\left(\operatorname{grad} p_\alpha + \rho_\alpha \cdot \vec{g}\right) \tag{13}$$

The interaction between the different phases are described by additional equations.

$S_w + S_n + S_a = 1$ (the pore room filled up by the sum of the three phases)

$p_n - p_w = P_{Cnw(Sw,\, Sa)}$ (Relationship between Capillary pressure - Saturation)

$p_a - p_n = P_{Can(Sw,\, Sa)}$

$k_{r\alpha} = k_{r\alpha(Sw,\, Sa)}$ (Relative Permeability-Saturation relationship)

In many practical applications this nonlinear equation system can be reduced to one movement phase for water and for NAPL under the assumption that the air is infinitely mobile.

In the one-dimensional process of the displacement of oil by water, the analytical solution by BUCKLEY and LEVERETT (Buckley, 1942) is used. This solution describes the nonstationary processes. By the description of the relative permeability-saturation curve the COREY-function can be stated.

$$k_{rw} = S^{*4} \tag{14}$$

$$k_{rn} = \left(1 - S^*\right)^2 \cdot \left(1 - S^{*2}\right) \tag{15}$$

with: $S^* = \dfrac{(S_w - S_{wr})}{(1 - S_{wr} - S_{nr})}$ $\qquad S_{wr} = S_{nr} = 0{,}2$

For the two-dimensional case and the three-phase-system, air/NAPL/water, the following first signs are investigated:

For the capillary pressure-saturation relationship the method of PARKER (Parker, 1987) is used for the estimation of parameters.

$$P_{Cnw} = p_n - p_w = \frac{1}{\alpha_{vG} \cdot \beta_{nw}} \left(S_e^{\frac{n_{vG}}{(1-n_{vG})}} - 1 \right)^{\frac{1}{n_{vG}}} \tag{16}$$

$$P_{Can} = p_a - p_n = \frac{1}{\alpha_{vG} \cdot \beta_{an}} \left(\left(\frac{S_n + S_w - S_{wr}}{1 - S_{wr}} \right)^{\frac{n_{vG}}{(1-n_{vg})}} - 1 \right)^{\frac{1}{n_{vG}}} \tag{17}$$

with: $S_e = \dfrac{S_w - S_{wr}}{1 - S_{wr}}$; $\qquad \beta_{nw} = \dfrac{\sigma_{aw}}{\sigma_{nw}}$; $\qquad \beta_{an} = \dfrac{\sigma_{aw}}{\sigma_{an}}$

For the Relative Permeability-Saturation relationship of the non wetting fluid (NAPL) a model of STONE (Aziz; 1979) is used.

$$k_{rn} = \frac{S_n (1 - S_{wr}) k_{rnw} \cdot k_{ran}}{(1 - S_w)(S_n + S_w - S_{wr})} \tag{18}$$

in which k_{rnw} and k_{ran} represent the Relative Permeability-Saturation relationship of the NAPL-Phase in the two-phases-system, (water/NAPL) and (air/NAPL). The parameters S_{nr} and k_{rncw}, which are found in the original kind of the STONE-model, cover the values „0“ and „1“. For the water phase, the relationship of PARKER (Parker, 1987) is used.

3 Simulation System CAE-Groundwater

The simulation is connected very closely with the modelling. For the different kinds of the PDE, various methods and technologies can be used for the simulation. There it is possible to put in continuous methods, e.g. analytical methods or integral transformations, and discrete methods, e.g. finite difference, finite elements methods. In the following, some methods are shown.

The CAE-Groundwater System developed at the Institute of Waste Management and Contaminated Sites of the Dresden University of Technology should support the expert, e.g. hydrogeologist, civil engineer, water engineer, at the control and monitoring of the groundwater processes.

With the aid of tools from the CAE-Groundwater System the data input can be realised, as well as the data pre-processing, the calculation and simulation, representation of the results effectively.

3.1 The Data Base

The data base is divided into three parts.

- Information
- Processing
- Model

There are the information data, which are collected by the explorer or the observer. Generally they are structured into the point, line or time data. The information data are collected from different water engineering offices. The CAE-System must make available programmes, which convert such different data in a uniform structure. Here, the interface to the Geographic Information Systems (GIS) is to be made available too. Especially in the state of Saxony, exists an Environment Information System (EIS) with special parts, the Subject Information Systems (SIS). Such SIS's are used e.g. for the soil, groundwater, geology.

3.2 The Simulation

For the simulation of the processes in the soil and groundwater zone, a large number of computer programmes exist. Generally the methods of the solution of the PDEs are divided in to different classes.

By the task:

- solution of the groundwater flow problems
- solution of the transport of substances
- solution of the inverse problems, the parameter identification

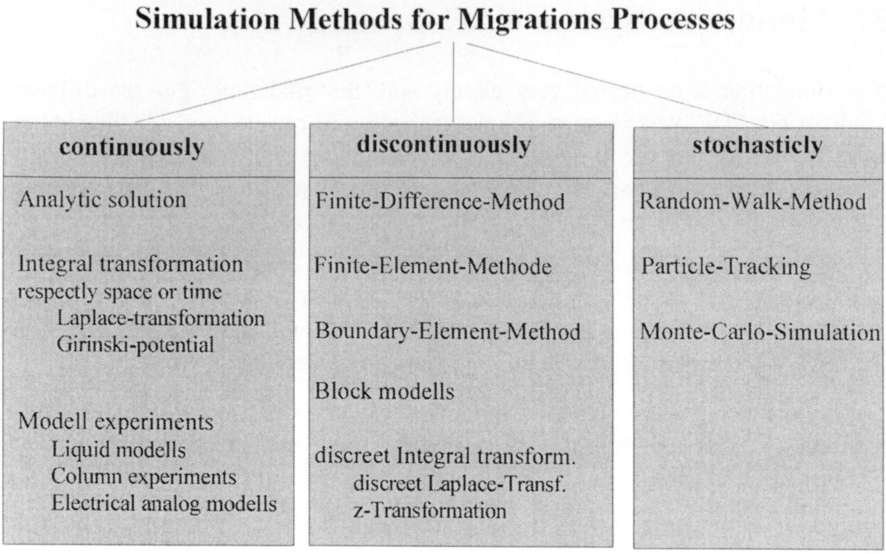

Simulation Methods for Migrations Processes

continuously	discontinuously	stochasticly
Analytic solution	Finite-Difference-Method	Random-Walk-Method
Integral transformation respectly space or time Laplace-transformation Girinski-potential	Finite-Element-Methode	Particle-Tracking
	Boundary-Element-Method	Monte-Carlo-Simulation
	Block modells	
Modell experiments Liquid modells Column experiments Electrical analog modells	discreet Integral transform. discreet Laplace-Transf. z-Transformation	

Fig 2: Simulation methods

By the dimension (space and time):

- 1-, 2- and 3-D
- stationary and nonstationary

By the water content:

- saturated
- unsaturated zone

As the mathematical solution methods are needed, the insertion of:

- analytical solution for the simple groundwater flow relations
- numerical solutions, (FDM, FVM, FEM) for the more complex relations
- statistical numerical solution (Random-Walk, Particle-Tracking) for the simulation of spreading pollution.

All computation results are put into special files. Such results are

- groundwater level
- volume quantity flowing through the boundary conditions
- concentration of substances
- identified parameters

3.3 The Data Representation

For the representation of input data and the results of the simulation, several tools can be used. It is possible to use the graphic subroutines in the CAE-Groundwater System to display the isolines, the stream lines or the 1-D representations.

An export interface to many external graphic routines, e.g. SURFER, GRAPHER, ArcInfo is possible, too.

4 Conclusion

The modelling and simulation of processes in the soil and groundwater zone is one of the prerequisites for controlling and monitoring of such applications such as pumping wells in water plants or drainage systems in mines or excavation pits. The physical processes, the groundwater flow and the transport of substances are described mathematically by a system of PDEs. For the simulation a CAE-Groundwater System was developed at the Dresden University of Technology.

References

Arnold, W. (1991): Beitrag zur Modellierung und digitalen Simulation von Mehrphasen-/Mehrmigranten-Prozessen im Untergrund, Diplomarbeit, Techn. Univ. Dresden, Fak. Bau-, Wasser- und Forstwesen, 1991

Aziz, K.; Settari, A. (1979): Petroleum Reservoirs Simulation, Elseviers Applied Science, New York, 1979

Buckley, S. E.; Leverett, M. C. (1942): Mechanism of Fluid Displacements in Sands. Transaction of the AIME, 146: pp 107 - 116, 1942

Busch, K.-F.; Luckner, L.; Tiemer, K. (1993): Geohydraulik, Gebrüder Bornträger, Berlin, Stuttgart. 1993

Gottschalk, Th.. (1994): Analyse und Testung von Software im Rahmen der Altlastenbehandlung, Diplomarbeit, Techn. Univ. Dresden Fak. Forst- Geo- und Hydrowissenschaften, 1994

Gräber, P.-W. (1990): Entwicklung von CAD-Elementen zur Beeinflussung nichttechnischer Prozesse im Boden- und Grundwasserbereich, Wissensch. Zeitschrift der Techn. Univ. Dresden, Dresden, 39(1990), H. 5, S. 181-187

Gräber, P.-W. (1991): Beitrag zur Entwicklung von geräte- und programmtechnischen Komponenten für die Steuerung und Überwachung nichttechnischer Systeme am Beispiel des Boden- und Grundwasserbereiches; Habilitationsschrift Techn. Univ. Dresden, Fakultät für Elektrotechnik/Elektronik, 1991, 175 S.

Gräber, P.-W.; Gutt, B.; Kemmesies; O.; Arnold, W. (1991): Einheitliche Window- und Schnittstellengestaltung für das Pre- und Postprozessing innerhalb der CAE-Software für Boden- und Grundwasserprozesse; Dresden: Techn. Univ. Dresden, Proc. 2. Dresdner Informatiktage, 1991, S. 83 - 91

Kinzelbach, W.; Rausch, R. (1995): Grundwassermodellierung, Gebrüder Bornträger, Berlin, Stuttgart. 1995

Kolrep, H. (1994): Entwicklung eines Windows-gestützten CAE-System „Grundwasser“ , Diplomarbeit, Techn. Univ. Dresden, Fak. für Informatik, 1994

Parker, J. C.; Lenhard, R. J.; Kuppusamy, T. (1987): A parametric model for constitutive properties governing multiphase flow in porous media, Water Res. Res., 23(4), pp 618 - 624, 1987

Schäfer, G.; Helmig, R.; Thiez, P. L. (1996): Vergleich numerischer Mehrphasen-strömungsmodelle zur Simulation von NAPL-Migrationsvorgänge in porösen Medien, Fortschr. in der Simulationstechnik, Bd. 10, Vieweg Verlag, Braunschweig / Wiesbaden 1996, S. 355 - 360

Modelling Approaches for Water Flow and Solute Transport through a Porous Medium

Willibald Loiskandl

Universität für Bodenkultur Wien, Inst. f. Hydraulik und landeskulturelle Wasserwirtschaft, Muthgasse 18, 1190 Vienna, Austria

Abstract. *This paper aims to contribute to a better understanding of simulation applications. To concretise the topic, modelling approaches of the spreading of a contamination plume in groundwater, the radial symmetric flow towards an auger hole and an activated carbon filter simulation are presented. Common to all examples is the flow property - the porous medium.*

Keywords. *Flow through porous medium, solute transport, water flow modelling, simulation tools*

1 Introduction

Solving engineering problems is performed in three steps (Kreyszig, 1988): *Modelling*, translating given physical or other information and data into a mathematical form – a mathematical model (e.g. differential equations), *Solving*, applying mathematical methods (e.g. numerical procedures) and *Interpreting*, understanding the meaning and the implications of the mathematical solution for the original problem. These steps together are incorporated into Simulation. The term "Simulation" may be defined in various ways. In this work a distinction is made between simulation tools (step 1 and step 2 defining mathematical solving methods) and simulation models (step 2 applying mathematical methods and step 3). In spite of the fast development of simulation tools, one should be aware that the governing equations have remained unchanged for many cases. The importance of the knowledge of the physical and mathematical background has to be pointed out very clear. Especially by using commercial software products, a detailed mathematical description is a necessity. To some extent the transparency of a simulation tool can be considered as a quality criteria. The confidence in a simulation model depends on the validity of the problem considered and the plausibility of the results.

The success of a simulation is dependent on the quality and quantity of the observed or measured data and also the initial and boundary conditions. Combined with the required data, a model may serve as a design or forecasting tool. In the presented work, the flow through a porous media, modelling and simulation may contribute to a better understanding of the physical processes. In general, the evaluation of different scenarios is carried out with a reasonable effort. By performing sensitivity analysis the impacts and uncertainties of the involved parameter are estimated.

2 Examples on Practical Applications

The modelling approaches are concretised by means of real world applications. Common to all examples is the description of the movement of water through a porous media by variations of the general flow equation (Verruit, 1970).

$$\frac{\partial}{\partial x}\left(k_{xx}\frac{\partial \phi}{\partial x}\right)+\frac{\partial}{\partial y}\left(k_{yy}\frac{\partial \phi}{\partial y}\right)+\frac{\partial}{\partial z}\left(k_{zz}\frac{\partial \phi}{\partial z}\right)+W=\mu_s\frac{\partial \phi}{\partial t} \tag{1}$$

The first porous flow example is a case study of a regional groundwater flow problem, utilising standard software. The goal of the study was the evaluation of the propagation of a contamination plume. For radial symmetric flow problems, like the flow towards an auger hole or the flow in an activated carbon filter, the development of a simulation tool was considered necessary. The developed simulation models were compared with laboratory experiments. The physical experiments were performed with sand-box models (cylindrical segments).

2.1 Contamination Plume

For groundwater modelling, a large number of two- and three-dimensional flow and transport models are available. Hence, the most crucial question is how to select the best product for a given purpose. Besides the technical requirements, financial and other user dependent criteria may also play a role. Nevertheless in many cases an adaptation of the software is necessary, meaning that one has to go back to step 1. This was the case for the propagation of a contamination plume in a shallow river basin along side the Danube, hence the reason for presenting this example. For the given task the software MOC (Konikow, 1978) (the name stems from method of characteristics) for nearly horizontal 2-D and single phase flow was found to be appropriate.

The original software was written for confined groundwater and steady state boundary conditions. Inside the flow domain, transient conditions are allowed due to pumping. To calculate unconfined groundwater and to consider varying boundary conditions adjustments have to be made. For an easier data handling all read statements are omitted and MOC is embedded as a subroutine into a main program (Kammerer, 1995).

The mathematical background of MOC is summarised in Table 1. For the flow module the method of finite differences (MFD) is used. Equation 2 is the 2-D difference form of Equation 1 for nearly horizontal flow and Equations 3 and 4 describe the velocity vector by a centred and forward difference scheme respectively. The solution of the transport and dispersion Equation 4 is based on the method of characteristics. Equations 6,7,8 are the three characteristics, where F represents the source/sink term. Equation 8 is solved again by a finite difference scheme.

A calculation starts by putting an equal number of particles (4, 9 or 16) in each segment of the flow domain, representing the concentration of the segment. Each particle is moved according to the calculated flow velocities corresponding to a characteristic.

$$T_{xx[i-1/2,j]}\left(\frac{\phi_{i-1,j}^{n}-\phi_{i,j}^{n}}{\Delta x^{2}}\right)+T_{xx[i+1/2,j]}\left(\frac{\phi_{i+1,j}^{n}-\phi_{i,j}^{n}}{\Delta x^{2}}\right)+T_{yy[i,j-1/2]}\left(\frac{\phi_{i,j-1}^{n}-\phi_{i,j}^{n}}{\Delta y^{2}}\right) \quad (2)$$

$$+T_{yy[i,j+1/2]}\left(\frac{\phi_{i,j+1}^{n}-\phi_{i,j}^{n}}{\Delta y^{2}}\right)=\mu\left(\frac{\phi_{i,j}^{n}-\phi_{i,j}^{n-1}}{\Delta t}\right)+\frac{q_{w(i,j)}}{\Delta x\Delta y}-\frac{k_{z}}{b}\left(H_{s(i,j)}-\phi_{i,j}^{n}\right)$$

$$v_{x(i,j)}=\frac{k_{xx(i,j)}}{n_{e}}\cdot\frac{\left(\phi_{i-1,j}^{n}-\phi_{i+1,j}^{n}\right)}{2\Delta x} \quad (3)$$

$$v_{x(i+1/2,j)}=\frac{k_{xx(i+1/2,j)}}{n_{e}}\cdot\frac{\left(\phi_{i,j}^{n}-\phi_{i+1,j}^{n}\right)}{\Delta x} \quad (4)$$

$$\frac{\partial c}{\partial t}=\frac{1}{b}\frac{\partial}{\partial x_{i}}\left(bD_{ij}\frac{\partial c}{\partial x_{j}}\right)-v_{x}\frac{\partial c}{\partial x}-v_{y}\frac{\partial c}{\partial y}+F \quad (5)$$

$$\frac{dx}{dt}=v_{x} \quad (6) \qquad \frac{dy}{dt}=v_{y} \quad (7) \qquad \frac{d(c)}{dt}=\frac{1}{b}\frac{\partial}{\partial x_{i}}\left(bD_{ij}\frac{\partial c}{\partial x_{j}}\right)+F \quad (8)$$

$$F=\frac{c\left(\mu\frac{\partial\phi}{\partial t}+W-n_{e}\frac{\partial b}{\partial t}\right)-c'W}{n_{e}b}$$

Table 1: Flow and transport equations for MOC

If sufficient information is available, the simulation of the groundwater flow provides satisfying results in most cases. This holds for the simulated groundwater basin with respect to flow calculations. The Danube to the south and an impermeable rock formation parallel to the stream enclose the river basin on two opposite sides. Groundwater observation wells were used for defining the boundary conditions upstream and downstream. Measurements of groundwater heads and of other hydrological inputs are accessible through the Austrian central hydrological office. Measurements of the concentrations in the basin could not be provided with the same accuracy as the flow parameters. The partly missing values were found through calibration by inverse modelling on a trial-and-error basis. The simulation task was performed in two steps. First, the model was calibrated with groundwater heads for one month and then, secondly validated for the calculation period of six years (Fig. 1).

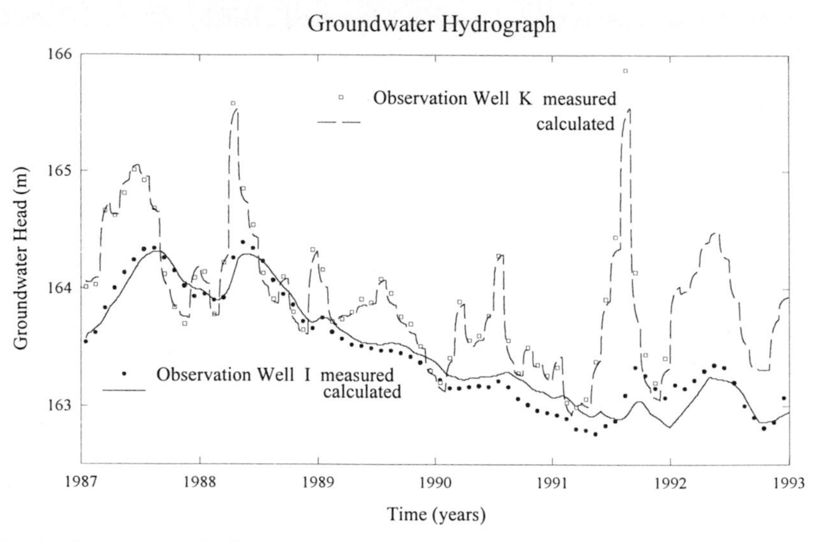

Fig. 1: Groundwater hydrographs at two observation wells in the river basin. Well K is located near the Danube and well I at a remote location.

The hydrograph dependence on the distance from the river boundary is clearly visible. The good agreement between measured and calculated groundwater heads provide a sound base for the subsequent simulation of convective contamination transport. At two locations measurements of HOX concentrations were available for the calibration (Fig. 2). The measurements started after the proposed input of the contaminant. The actual amount of HOX added to the groundwater was unknown. By modelling an estimate of HOX input was inversely calculated. The results in Fig. 2 prove the reliability of the simulation. Discharge well 1 is located close to the contaminant input site and observation well M is situated about 3,5 km downstream.

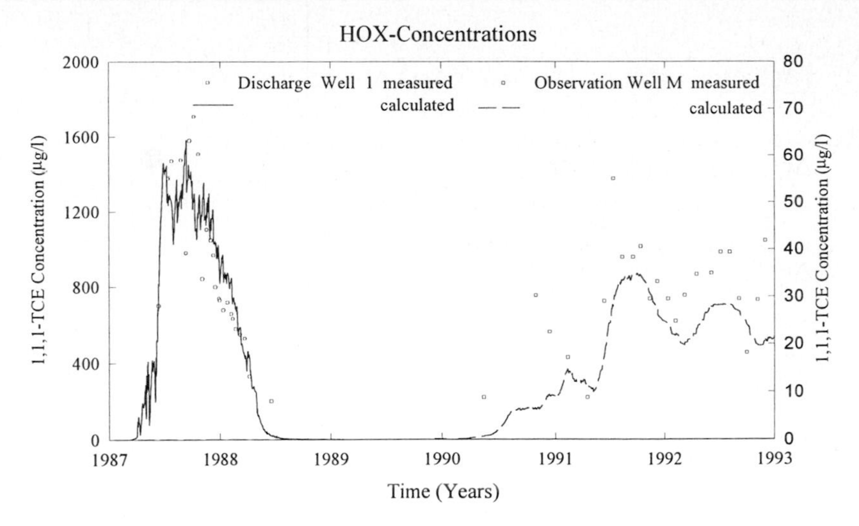

Fig. 2: Measured and calculated contamination concentrations at two observation points.

The model output provides monthly plots of groundwater head distributions and of the propagation of the contamination plume (Kastanek, 1994).

2.2 Radial Symmetric Flow Towards an Auger Hole

For the flow simulation of seepage water to an auger hole, a steady state model was expanded including the movement of a phreatic water table. The transient flow model (BORINS using finite differences) could be compared with a finite element solution for steady state conditions (Loiskandl, 1998). The differences were mostly smaller than 1%, so it can be concluded that the numerical method is of minor importance for the calculation result.

Inside the flow domain Laplace's equation a variation of Equation 1 is also valid. For the numerical procedure the discrete form is:

$$A\phi_{i,j} + B\phi_{i-1,j} + C\phi_{i+1,j} + D\phi_{i,j+1} + E\phi_{i,j-1} = 0 \tag{9}$$

and the coefficients are derived according to the finite difference scheme from the distances (Δr, Δz) of a centre point to the neighbouring points (Table 2, Fig. 3).

$A = -\left(\frac{1}{\Delta r_{i-1}\,\Delta r_M} + \frac{1}{\Delta r_i \Delta r_M} + \frac{1}{\Delta z_j\,\Delta z_M} + \frac{1}{\Delta z_{j-1}\,\Delta z_M}\right)$ $B = \frac{1}{\Delta r_{i-1}\,\Delta r_M} - \frac{1}{r_i\left(\Delta r_i + \Delta r_{i-1}\right)}$ $C = \frac{1}{\Delta r_i\,\Delta r_M} + \frac{1}{r_i\left(\Delta r_i + \Delta r_{i-1}\right)}$ $D = \frac{1}{\Delta z_j\,\Delta z_M}$ $E = \frac{1}{\Delta z_{j-1}\,\Delta z_M}$ $\Delta r_M = \frac{\Delta r_i + \Delta r_{i-1}}{2}$; $\Delta z_M = \frac{\Delta z_j + \Delta z_{j-1}}{2}$	j+1, Δ z_j, Δ r_{i-1}, Δ r_i, i-1, i,j, i+, Δ z_{j-1}, j-1 Fig. 3: Indices and incre-Ments in z, r i = 1- n number of horizontal increments j = 1-m number of vertical increments

Table 2: Coefficients for finite difference scheme

The movement of the phreatic surface in time and space is found by means of a Leapfrog scheme (Rushton, 1979).

$$H_i^{n+1} - H_i^{n-1} = \frac{2\,k\,\Delta t}{n_e}\left[\left(1 - \frac{\partial\phi}{\partial z}\right)\left(\frac{H_{i+1}^n - H_{i-1}^n}{\Delta\,r_i + \Delta\,r_{i-1}}\right)^2 - \frac{\partial\phi}{\partial z}\right] \tag{10}$$

where the term $\partial\phi/\partial z$ is approximated with:

$$\frac{\partial\phi}{\partial z} \cong p'(z = z_0) = \phi_0\,\frac{2\Delta z_1 - \Delta z_2}{\Delta z_1\left(\Delta z_1 + \Delta z_2\right)} + \phi_1\left(\frac{1}{\Delta z_1} + \frac{1}{\Delta z_2}\right) - \phi_2\,\frac{\Delta z_1}{\Delta z_2\left(\Delta z_1 + \Delta z_2\right)} \tag{11}$$

The developed model facilitates the easier interpretation of field measurements for the estimation of the saturated hydraulic conductivity and the impact of the draw down on the water table. The model also enables smoothing of oscillations of the measurements. The result is not so much dependant on the beginning of the flow process as it is for the usual steady state calculation.

To prove the transient calculation the water table in the auger hole is kept constant. The step-by-step solution approaches an analytical solution (Fig. 4).

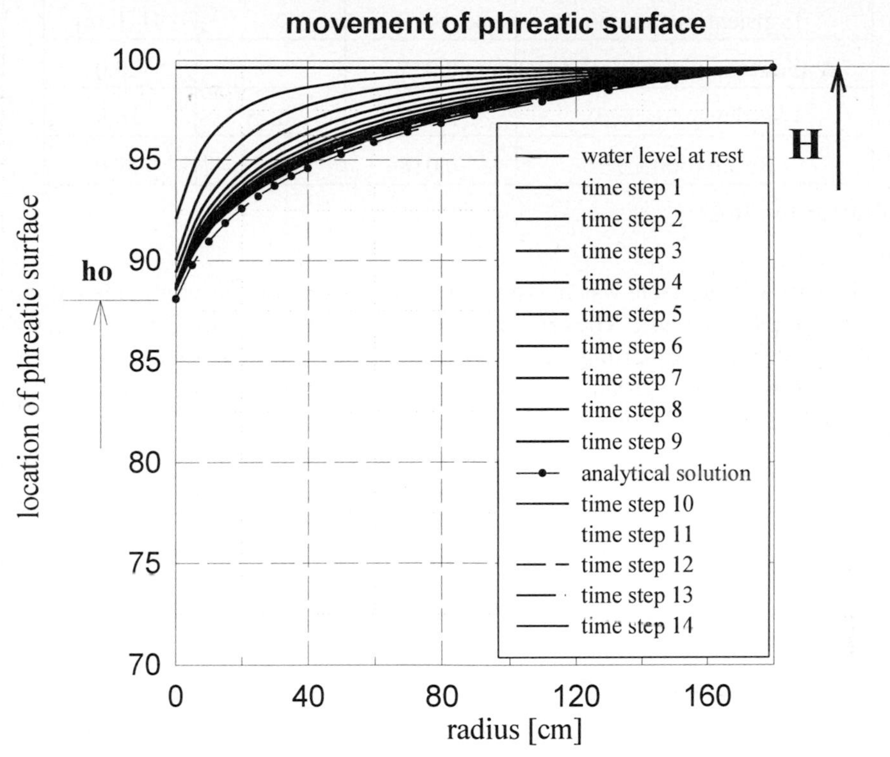

Fig. 4: Movement of water table towards a steady state condition, h_o is lowered at t=0 and then kept constant

By means of the auger hole methods (Ernst, 1950) the rising water table velocity dh/dz after the sudden lowering and the geometry represented by a shape factor C are related.

$$k = C\frac{dh}{dz} \tag{12}$$

To test the computed results laboratory experiments, with a sector model (length 1,8 m, aquifer depth 1,0 m, auger hole 0,1 m, sector 15°) providing well known physical conditions, were performed at the institute for Rural Water Management. As an example test 10-22 is calculated and compared with the measurements (Tab. 3).

Transient calculation	k_f [cm/s]	S_y	H_B [cm]
Without draw down (solid)	0,027	-	20,0
1st estimate (dashed)	0,020	0,25	16,9
Result (dotted)	0,018	0,085	17,0

Table 3: Test 10-22

The rising water table was measured with a pressure transducer mounted at the bottom of the auger hole. The calculation was started by neglecting the draw down with the steady state starting values. To improve the match a storage coefficient was chosen for the transient calculation (1st estimate). By further iterations on a trial and error basis the final result was achieved (Fig. 5).

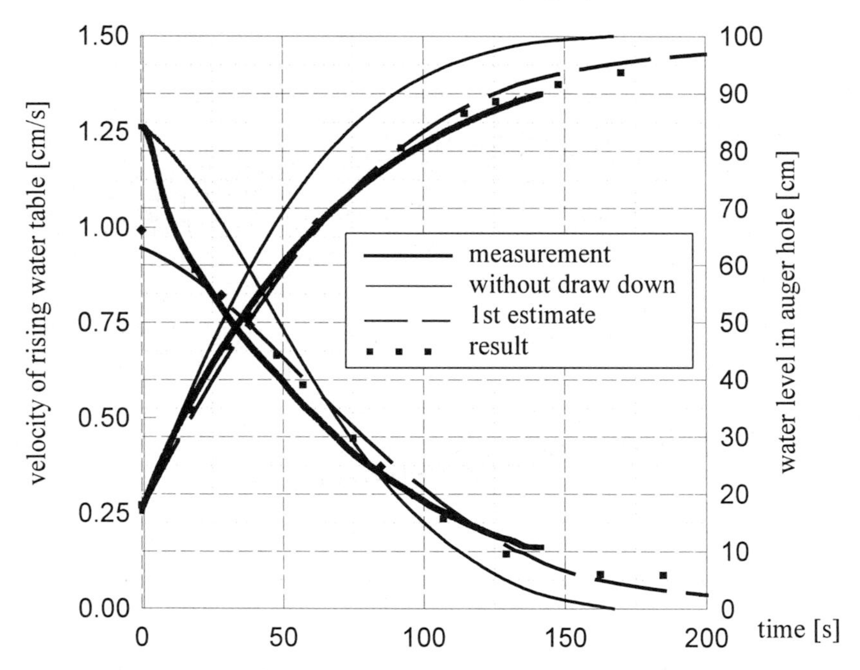

Fig. 5: Iteration procedure for permeability evaluation from measured rise of water level or velocity

2.3 Simulation of an Activated Carbon Filter

The evaluation of the flow pattern in an activated carbon adsorber presented the basis for an activated carbon adsorption model. Many parts of the program (subroutines) for the hydraulic calculations could be used from the simulation software (BORINS) for the auger hole method, but the draw down of the water table was determined directly for the steady state conditions of the activated carbon filter operation. The results of the numerical flow model (FILTER) were compared as before with physical model tests (Table 4) and other simulation software SEEPW and ASM.

Test NR.	1.3	1.1	2.1	3.1	4.1
Hydraulic head (inflow) [m]	0,29	0,38	0,50	0,625	1,303
Flow rate (measured) [l/s]	0,66	1,33	2,00	2,66	5,92

Table 4: Flow rates and hydraulic heads at the inflow of laboratory test with a physical sector model of 15 °

Measurements and calculations are in a good agreement (Fig. 6).

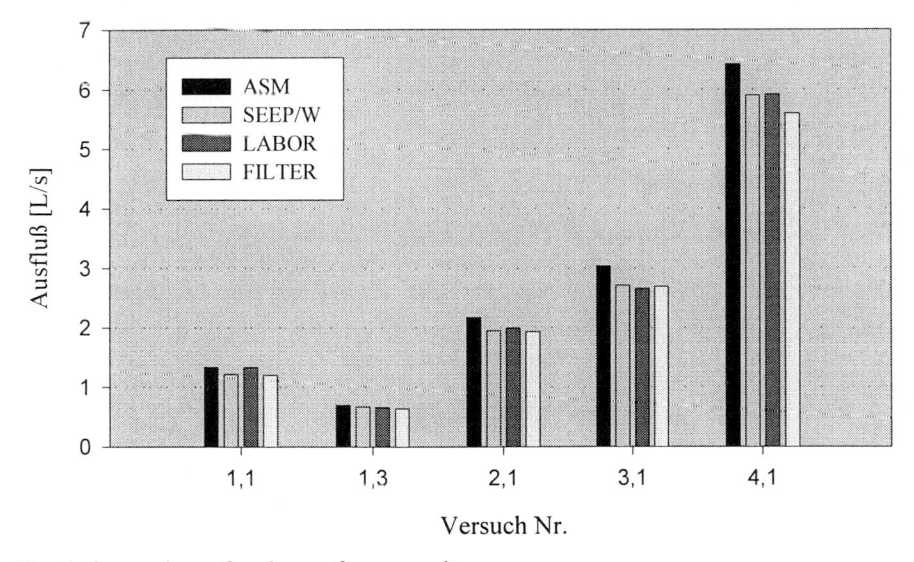

Fig. 6: Comparison of various software results

The next step was the combination of the flow model with an adsorption model (Schachner, 1998) to obtain an activated carbon adsorber filter model. The new KSVA procedure (Weingartner, 1993) ((**K**)continuos **S**elective **V**ariable flow velocity **A**dsorber) implies hydraulic conditions that vary with time and space (Loiskandl, 1994). For usual operation with constant flow rates, the components of the velocity vector $v_r = v_r(r,z)$ and $v_z = v_z(r,z)$ are quasi-steady state, this means that the flow pattern is calculated only once for each sector. The concentration is a function of time and space, representing the movement of the breakthrough curve $c_i = c_i(r,z,t)$. The diffusion transport equation reads as follows:

$$n_e \frac{\partial c(\vec{r},t)}{\partial t} + \nabla \cdot \left[\vec{v} \cdot c(\vec{r},t) \right] + \rho_f \frac{\partial \bar{q}_{i,tot}(c_1(\vec{r},t),\ldots,c_N(\vec{r},t))}{\partial t} = 0 \quad (13)$$

$\bar{q}_{i,tot}(r,z,t)$ is the mean loading of a representative adsorber particle. By integrating over the flow domain the total loading of the filter is evaluated.

$$q_{i,tot} = \int_V \bar{q}_{i,tot}(r,z,t)dV \quad (14)$$

The local concentration change is described by two mechanisms, the transport through a film layer and a diffusion process into the carbon particle (Table 5). For the first, the film concentration gradient (c_i-c_i*) determines the mass transfer $n_{L,i}$ of chemicals towards the particle surface. In general, the mass transfer coefficient $\beta_{L,i}$ depends on the local flow velocity. The transport inside the particle is described by a diffusion equation. The diffusion parameter Ds,i is assumed to be independent of concentration and reactions between the chemicals due to low concentration.

Film transport	Diffusion process
$n_{L,i} = \beta_{L,i}(c_i - c_i^*)$ (15)	$\frac{\partial q_i}{\partial t} = D_{s,i}\left(\frac{\partial^2 q_i}{\partial r_p^2} + \frac{2}{r_p}\frac{\partial q_i}{\partial r_p}\right)$ (16)
Boundary condition: $\frac{\partial q_i}{\partial r_p}(r_p = 0, t) = 0$	$\frac{\rho_f D_{s,i}}{1 - n_e}\frac{\partial q_i}{\partial r_p}(r_p = d_p/2, t) = \beta_{l,i}(c_i - c_i^*)$
Initial condition: $q_i(r_p, t = 0) = 0$	

Table 5: Transport processes for adsorption model

The link between solution and mass adsorbed is provided by the ideal adsorbed solution theory (IAST (Radke, 1972)).

$$c_i^* = \frac{q_i^*}{\sum_{j=1}^{N} q_j^*}\left[\frac{\sum_{j=1}^{N} q_j^*/n_j}{K_i/n_i}\right]^{1/n_i} \qquad q_i^* \equiv q_i(r,z,r_p = d_p/2,t). \quad (17)$$

The mean loading of the representative carbon particle follows by integrating over the volume of the sphere.

$$\bar{q}_{i,tot}(r,z,t) = \frac{24}{d_p^3} \int_0^{d_p/2} q(r,z,r_c,t)\, r_c^2\, dr_c \quad , \tag{18}$$

The use of a higher order programming language enabled the inclusion of the flow model as a subroutine in the simulation tool HAM (Schäfer, 1996) (hydrodynamic adsorber model). The adsorption processes for competing substances in the activated carbon filter could be modelled successfully utilising the IAST (ideal adsorption solution theory).

For LINEAR FILTER, LUCA and KSVA filter operation methods the operation time was compared. LINEAR FILTER operation stands for a single layer carbon bed, whereas the LUCA and the KSVA procedures are based on charging shallow beds one after another. LUCA operates discontinuously and with a uniform velocity throughout the filter (piston flow). The KSVA operation enables a continuous charging of the layers with a variable flow velocity.

A filter column (diameter 40 cm) and a length of 1m was calculated for the different procedures. Two substances methylene blue and phenol were used. The outflow concentration limit to switch to a new layer was set to 20% of the inflow methylene blue concentration.

	LINEAR FILTER		LUCA	KSVA	LUCA -linear	KSVA -linear	KSVA -LUCA
l_0 [cm]	$t_{20\%}$ [h]	Sector	$t_{20\%}$[h]	$t_{20\%}$[h]	Difference [h]		
10	84	1	84	57	0	-27	-27
20	248	2	258	234	10	-14	-24
30	446	3	472	459	26	13	-13
40	664	4	711	712	47	48	1
50	893	5	964	979	71	86	15
60	1127	6	1224	1254	97	127	30
70	1364	7	1488	1532	124	168	44
80	1601	8	1753	1812	152	211	59
90	1839	9	2020	2092	181	253	72
100	2076	10	2286	2373	210	297	87

Table 5: Comparison of operation time for linear Filter, Luca and KSVA procedures

For the Luca procedure, the charging of layers one after other is immediately obvious. For the KSVA procedure is the effective filter length for the first sector is due to the outflow at the side wall shorter than 10 cm hence the activated carbon volume is less as well. For a larger number of sectors the KSVA procedure may increase the efficiency. For the KSVA procedure the concentration and the loading are given as an example (Fig. 7).

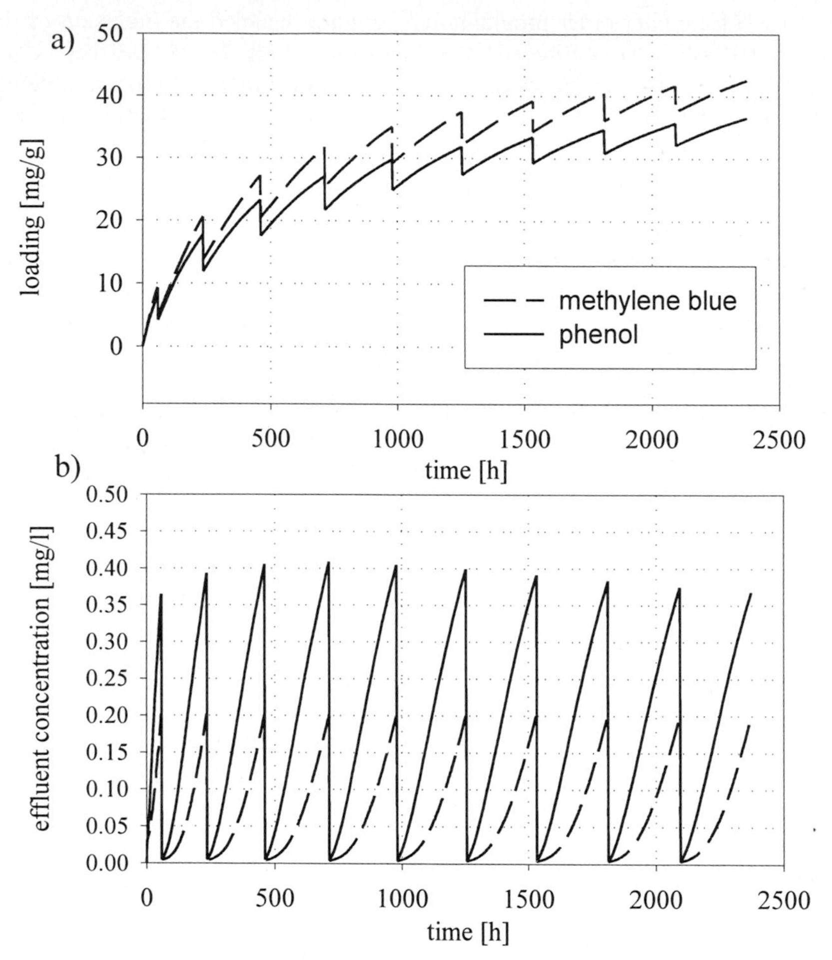

Fig. 7: Loading a) and outflow concentration b) of the KSVA procedure

The advantages of a sectored operation and the additional increase of efficiency by applying a variable flow velocity for a selective removal of organic traces are shown. Provisions are made to use different adsorption and desorption isotherms.

3 Conclusions

The use of simulation tools as a means for forecasting and for parameter estimation was demonstrated and the governing modelling approaches were presented. The importance of the physical and mathematical background was pointed out very distinctly. The comparison of different software products showed that the boundary conditions and input data play the key role rather than the numerical methods. By working with the method of finite differences for most calculations a personal computer (Pentium) was sufficient. With the help of physical models verifications of the developed user tools were performed. As a side effect, the limitations of the physical models are demonstrated. The exception is the groundwater model, which was transferred to a main frame computer after the developing phase for the final calculation.

The examples presented in this paper showed the antagonism between pure application and new development of a simulation tool. Simulation can not replace measurements (field data, laboratory experiments), but may extend the interpretation substantially. Measurements and simulations are always together in a symbiosis. Not to be underestimated, is the value of the creation of a physical/ mathematical model. By working with the governing equations a deeper insight into the simulated processes, the necessary assumptions and limitations are achieved automatically. The application of simulation is always bound to be a learning process, or in other words, learning by doing.

References

ERNST, L. F. (1950): A New Formula for the Calculation of the Permeability Factor with the Auger Hole Method, Agricultural Experiment Station, TNO Groningen

KAMMERER, G.; LOISKANDL, W. and ZELEZO, G. (1995): Numerical Simulation of Pollution in a Groundwater Basin in Austria, Computational Mechanics Publications,VI:37-47, Southampton-Boston

KASTANEK, F.; LOISKANDL, W.; ZELEZO, G.T.; KAMMERER G. und KUBU, G. (1994): Gutachten - Korneuburger Bucht, Institut für Hydraulik und landeskulturelle Wasserwirtschaft, Universität für Bodenkultur Wien, unveröffentlicht

KONIKOW, L.F. and BREDEHOEFT, J.D. (1978): Techniques of Water Resources Investigations of the United States Geological Survey, Chapter 2, Computer Model of two-dimensional Solute Transport and Dispersion in Ground Water, United States Government Printing Office, Washington

KREYSZIG, E. (1988): Advanced Engineering Mathematics, sixth edition, John Wiley & Sons, New York

LOISKANDL, W. (1998): Simulationsanwendungen bei der Strömung durch ein poröses Medium, Wiener Mitteilungen Abwasser - Wasser - Gewässer, Band 149

LOISKANDL, W.; RASSINGER, M.; SCHÄFER, E. and WEINGARTNER, A. (1995): Simulation of the adsorption and the flow pattern in an activated carbon adsorber, Proceedings of Eurosim ´95: 909-914, Elsevier, Amsterdam-Oxford-New York-Tokyo

RADKE, C.J. and PRAUSNITZ, J.M. (1972): Thermodynamics of Multi-Solute Adsorption from Dilute Liquid Solutions Amer.Inst.Chem.Eng. J. 18: 761-768

RUSHTON, K.R. and REDSHAW, C. (1979): Seepage and Groundwater Flow, Wiley and Sons, Chichester, New York, Brisbane, Toronto

SCHACHNER, H.; RASSINGER, M.; LOISKANDL, W.; SCHÄFER, E. und WEINGARTNER, A. (1998): Mathematische Beschreibung des Simulationsmodells HAM (Hydrodynamic Adsorption Model), Die Bodenkultur 48(4):217-227

SCHÄFER, E. and LOISKANDL, W. (1996): Application of Simulations in the Field of Water Resources and Water Treatment, Modeling and Simulation, Proceedings ESM-Budapest: 1053-1055

VERRUIJT, A. (1970): Theory of Groundwater Flow, Macmillan and Co. Ltd., London and Basingstoke

WEINGARTNER, A. (1994): The KSVA-Procedure – A New Procedure for the Selective Removal of Organic Traces, Water Supply, 14:145-158

Nomenclature

b	saturated thickness of aquifer
c_i^*	concentration at adsorbent surface
c_i	concentration of substance i in fluid
c	concentration of dissolved chemical
$c`$	c of dissolved chemical in source
C	shape factor
d_p	particle diameter
$D_{s,i}$	diffusion coefficient
Dij	axial dispersion
h	water level in auger hole
H	phreatic surface elevation
H_s	hydraulic head in source bed, stream
i,j	space indices
K_i	Freundlich coefficient
k_{xx}, k_{yy}, k_{zz}	hydraulic conductivities
N	Number of substances
n	time index
n_i, n_j	Freundlich exponent
n_e	porosity
$\dot{n}_{l,i}$	mass flow liquid phase
q_i	loading of substance
q_i^*	loading of substance at particle surface
$q_{i,tot}$	total loading
$\overline{q}_{i,tot}$	mean loading of particle
qw	volumetric rate of withdrawal
r, r_p	polar co-ordinates
t	time
T_{xx}, T_{yy}	transmissivities
v_x, v_y, v_r, v_z	components of flow velocity
$\vec{v}$	velocity vector
V	volume
W	source/sink term
x, y, z	cartesian co-ordinates
ϕ	hydraulic head
$\beta_{l,i}$	mass transfer coefficient liquid phase
ρ_f	density of the adsorbent
μ	storage coefficient
$\Delta x, \Delta y, \Delta z, \Delta r$	space increment
Δt	time increment

Space-Time-Modelling of Ascending Salty Sewage Water in the Werra-Potash District *(Middle Germany)*

Raimund Rödel

Institut of Geography at the-University of Greifswald, Jahnstr. 16,
17487 Greifswald, Germany
roedel@rz.uni-greifswald.de

Abstract. *Since 1925 salty sewage water of the potash-industry (Werra-region, Middle Germany) has been deposited in the geological setup. A quasi-closed system was assumed. It was found out that a part of the salt water ascend to the surface again. Time series-analysis allow a precise analysis of the outcoming salt water. To separate trend and periodical shares the time series were filtered numerically. The Fourier-transformation allows an investigation of the periodic shares. After the evaluation of the data it seems possible that two bodies of water united into one single body of waste water. The spreading of salty water in the underground was modelled by the concept of cellular machines.*

Keywords. *time series-analysis, groundwater modelling, numerical filtering, periodogram analysis, cellular automata*

1 Introduction

In considering the arrow of time, the study of geosystems has become an important aspect to understand fluctuations such as accelerations and retardations. Fluctuations mean the periodic shares, accelerations and retardations mean the behaviour of trends. According to the findings of system theory these phenomena have a great importance in the modelling changes of composition balances and energy in past, present and future. (AURADA, 1992a)

In the Werra-potash district in the middle of Germany the formation of Zechstein ranges from the Werra-salinar with the potash horizons of Thuringia and Hesse, to the Leine-salinar. Leine-salinar falls into a stratum of halite and karstified limestone on the top. The limestone is known as sheet dolomite. Layers of clay- and siltstone are situated above and below the sheet dolomite. Therefore it is assumed as a quasi-closed system. Because of its high cavity volume, sheet dolomite is suitable for the deposition of the salt-containing waste water from the production of potash fertiliser. The permeability coefficiency of this layer is comparable with a medium-grained sand. It is assumed, that heavy salt water flows off into the depth because of a synclinal structure underground. (DVWK, 1993)

In figure one you can see the surface of the sheet dolomite in the Werra-potash district. A trough is visible, which is known as the Eiterfeldian trough. A line shows the assumed front of the flowing off of waste brine. This line is based on ground water measurements and the underlying assumption, that heavy salt water substitutes formation water. The river Werra with the potash fertiliser factories and the disposal wells are also visible. There are disposal well regions around Widdershausen, Eichhorst and Kleinensee. Important flood measuring points are situated in Unterrohn, Tiefenort, Widdershausen and Gerstungen. (DVWK, 1993)

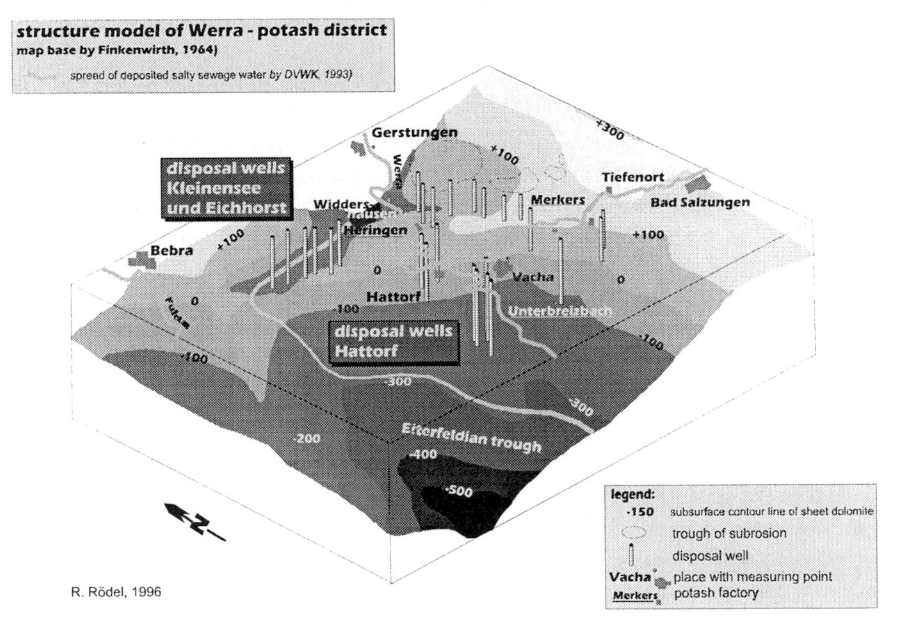

Fig. 1

Shortly after the feeding in the waste brine into the karstified dolomite in 1925, they found out that the formation water in the sheet dolomite and a part of the disposal salt water had ascended to the surface again. Favoured paths are the geological zones of weakness along faults and basaltic dikes. The amount of this ascending water is not be regulated and not exactly known.

The quantity of diffuse ascending sewage water has been balanced by AURADA. (AURADA, 1992B)

The following picture (Fig. 2) shows the underlying idea. The diffuse ascending of salt-containing waste water be comes obvious in the differences of chloride and hardness freight-measurements between two flood measuring points. The flow chart shows all measuring points in the Werra-potash district and the equation used to define their freights. The most important zones of diffuse ascending sewage water are between Unterrohn and Tiefenort as well as Widdershausen and Gerstungen. A common consent of the flood measuring point in Unterrohn was defined to show the natural salty freights, also known as geogenic background.

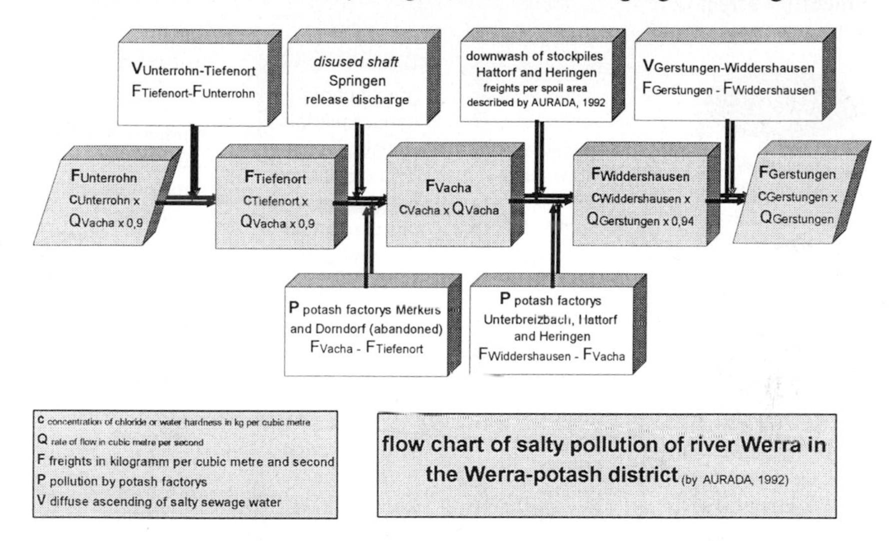

Fig. 2

2 Methods / Results

Referring to what was said in the beginning, stochastic is a better tool to describe the natural processes with the keys of numeric models. The study of AURADA (AURADA, 1992B) is based on annual meaning freight values up to 1976. Long detailed time series with daily values of thirty years have now allowed a precise analysis of outcoming salt water that reaches the river Werra. In this case, the changes and development of ascending sewage water should be investigated.

At first the time series were filtered by Gaussian high- and low pass filters. The method is designed to divide time series into a share of short time fluctuations and a share of trend. Statistical methods such as modelling of autoregressive processes presuppose that the trend is disposed from the original time series. (SCHÖNWIESE, 1992)

Specially chosen filter weights enables the division of time series into short time fluctuations as well as into the share of monthly trend. Taking all this into consideration, it is possible to model autoregressive processes and to compare the annual trend with annual results of deposited waste brine.

We leave the question of trend influenced share in the time series aside and look at the share of periodic fluctuations. This is the high-pass filtered part. Estimating the autoregressive processes it is possible to obtain knowledge about the memory effect.

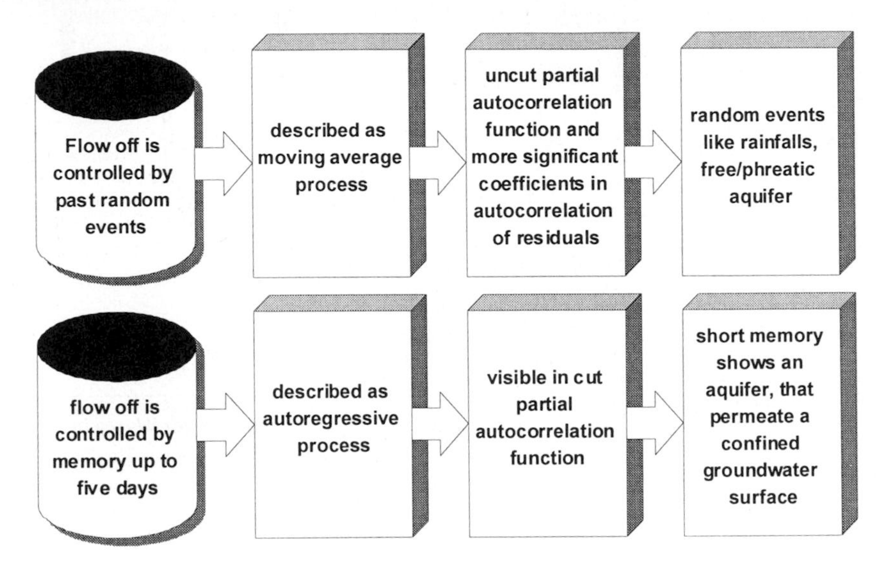

Fig. 3

The partial autocorrelation of chloride and hardness freights of geogenic background as well as hardness freights of assumed ascending water may be considered to describe autoregressive processes in the eighth ordinal. The autocorrelation function of residues is cut later than lag ten. A moving average process with ten or more ordinals lies behind the autoregressive process. Therefore a long memory effect is assumed in there freights.

In this case, the question is how the stochastic processes describe the controlling of ground water discharge. A common consent is that autoregressive processes show a to have short time memory. Moving average processes are controlled by past random events. There are two ways of controlling aquifer discharge. On the one hand, the discharge is controlled by past random events like rainfalls in a karstified water bearing stratum. This process is known as moving average process. It is assumed, that the freights of geogenic background and the freights of hardness in Unterrohn, which were described as moving average rocess, came out of a phreatic aquifer. It is possible to regard it as the aquifer of formation water in sheet dolomite.

On the other hand these is the ascending of salt-containing water, which is visible in the time series of chloride freights between Unterrohn and Tiefenort. There, groundwater discharge is controlled by short time memory. The underlying process is an autoregressive one. It is possible to draw the conclusion, that the layer of heavy salt water lies below the formation water in sheet dolomite. Only with time has the salt water permeated the confined groundwater surface. The short time memory of their discharge can be explained by this process.

In summary, we can say that the analysis of periodic shares supports the thesis of ascending waste brine. Another investigation into periodic shares is Fourier-analysis. The one gets a power spectrum. If one has a long time series, a moving power spectrum shows fluctuations of amplitude. (SCHÖNWIESE, 1992)

Figure four shows the moving power spectrum of trend-eliminated chloride freights. Fluctuations of amplitude in the time series between 1976 and 1984 are seen here. Look at the y-axis. The chart indicates an amplitude of over thirty kilogramms per second.

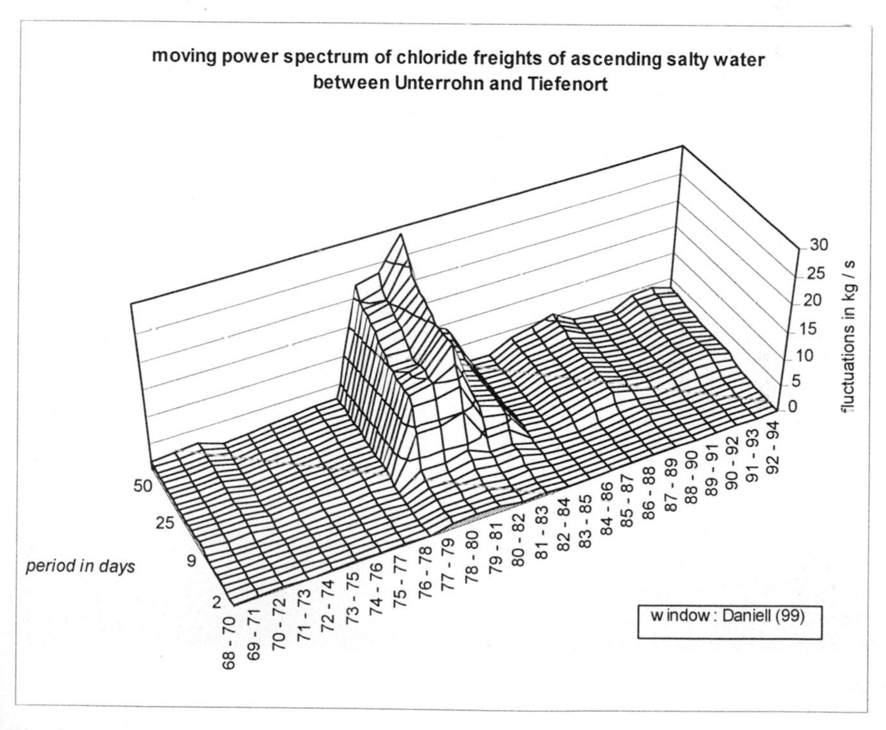

Fig. 4

It is interesting to know, that chloride and hardness freights reached peaks within this time. A thaw-salt down-wash from streets was assumed, because some of these freight peaks are at the same time, as high water in spring. One knows that thaw salt is magnesiumchloride.

I must admit that difficulties can arise by using the model of thaw-salt wash-down. Provided that the freight peaks result from the wash-down of magnesiumchloride, the question arises why there were such peaks during summer high waters, for instance in 1980 and 1981.

One has to take into account that measuring results in the flood containment area may be inaccurate in times of high water. Yet there must be another explanation for the freight peaks, because they only appeared between 1976 and 1984.

The next chart (Fig. 5) shows a comparison of the yearly trend of ascending chloride water with the rate of yearly deposited waste brine. The black line shows the results of yearly trend of ascending chloride water between Unterrohn and Tiefenort. Bars demonstrate the yearly means of disposal chloride waste brine of the three disposal wells.

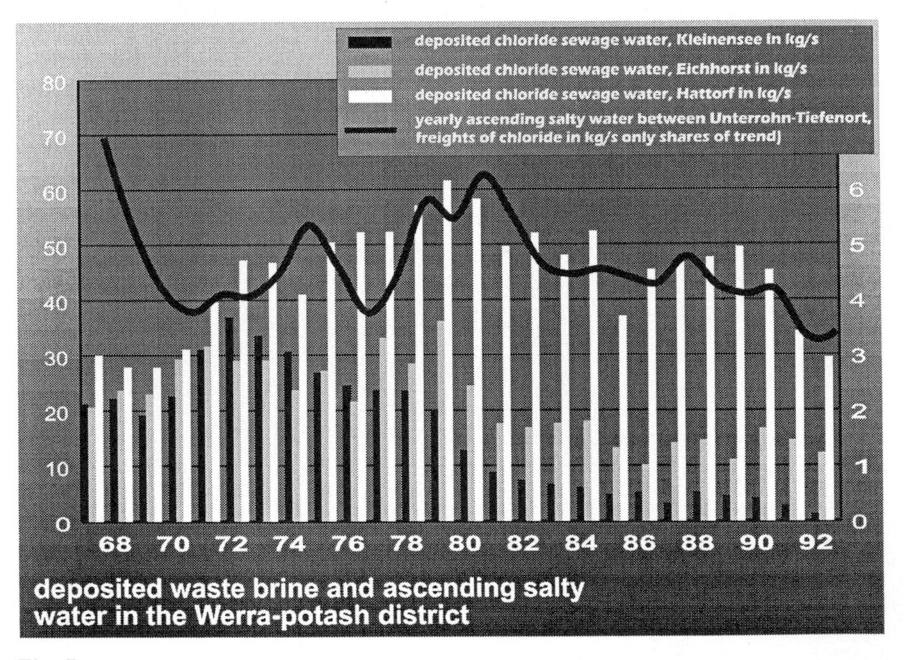

Fig. 5

Thuringia disposed its waste brine into the sheet dolomite only up to 1968. The end of the waste brine deposition carried out by Thuringia is visible on the chart up to 1971 by the descending black line.

Since 1979, a coherence has been detected between yearly values of ascending salt water in the region of Tiefenort and the rate of deposited waste brine. In statistical modelling this coherence may be considered as a multiple linear regression. In that assumption a coefficiency of determination of 0,75 was calculated. That means that 75 per cent of the freight level line is determined by the rate of deposited waste brine. Only the values of the disposal wells of Eichhorst are significant at 95% confidence-level.

In contrast, a coherence between the yearly chloride freights of ascending salt water in the region of Widdershausen and Gerstungen and the rates of deposited waste brine up to 1979 is observed. In the assumption of also multiple linear coherence a coefficient was found from a determination of 0,62.

Linear regression may be considered as not to be the best from this point of view. But let us come back to what was said at the beginning. Taking all that into consideration, there is a possibility to explain the freight peaks of chloride between 1976 and 1980.

What has happened? In common consent, the heavy sewage water that has been deposited in the underground, flows off into the centre of Eiterfeldian trough. But, if as the saying goes, the exception proves the rule and therefore this is not possible, as there is confined formation water in the geological set-up.

In my opinion the disposed waste brine does not substitute the confined formation water in sheet dolomite, in that the volume of disposal site is much smaller than geologists had assumed. The coherence between the rate of disposal waste brine and ascending salt water in the region Widdershausen-Gerstungen assumes that salty springs were controlled by the rate of disposal in Widdershausen up until 1979. Because of the continual depositing of waste brine by potash factories in Hesse the disposal site nearly overflowed in 1980.

Geologists (DVWK, 1993) assumed that there were two bodies of water in the setup because there is a swell between the disposal wells in Thuringia and Hesse where the salt water was pumped underground. After the evaluation of the data it seems possible that the two bodies of water united into a single body of water around 1980. Disposal waste brine did flow from the higher disposal wells near Eichhorst over the swell into the region between Unterrohn and Tiefenort. So it is possible to explain the significant coherence between disposal wells in Eichhorst and the freights of ascending salt water in Tiefenort up to 1980.

Summarising we can say that the assertion of a disposal site for waste brine from potash fertiliser production (DVWK, 1993) in the Werra-potash district cannot bye proved. The fact is that the ascent of waste brine is indisputable. The argument, that the heavy sewage water flows off into the depth of Eiterfeldian trough is not convincing.

3 Concluding Remarks

In order to understand the spatial processes better, the spreading of salty water in sheet dolomite was modelled by the concept of cellular machines. The model validates the results of time series analysis and simulates the movement of wasted water.

0,05	**0,2**	**0,05**
0,05	**0,4**	**0**
0,2	**0,05**	**0**

Fig. 6: The used Moore-raster

The spreading of the known yearly amount of deposited waste brine was simulated with the help of the above represented Moore-raster (Fig. 6). FINKENWIRTH AND FRITSCHE IN (DVWK, 1993) notice a spreading velocity between the control wells Friedewald and Weisenborn (distance 2.7 km) of nearly 450 m per year. This velocity itself reflects nearly by distance of raster of 500 x 500 m. In this raster field the movement was simulated. Programming can be done on an EXCEL table page.

40 per cent of the salty water remains in a raster cell after one simulation step, as one can see. Another part moves correspondingly Fig. 6 into the neighbour cells.

A weak salt load spreads relatively quick over the entire area, as the simulation results prove. Amounts of salt of more than 10 t per raster cell remain much longer around the disposal wells and spread slowly, like the definition of the Moore-cells, underground. The non-linear spreading of the main front of the waste brine body is simulated well with the help of the model.

The arriving front of salty sewage water between Gerstungen and Widdershausen in 1940 is a reference point of the model-plausibility. A diffuse ascending of salty water was also reported at this time (HOPPE, 1962) .

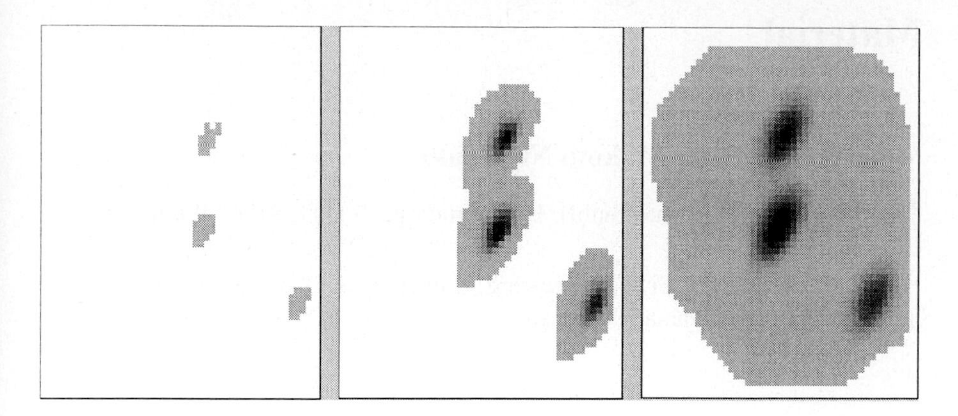

Fig. 7: two-dimensional spreading model of underground-deposite salty water in the Werra-potash district. Model works with Moore-neighbours (cell-distance 500x500 m = spreading in yearly steps). Dark areas past time-steps of 2 years, 12 years and 35 years are the most salt-loaded reaches.

References

AURADA, K.-D. (1992a): Prolegomenon einer Ökologistik Angewandter Geographie. In: Petermanns Geographische Mitteilungen 136 (2+3/1992), p. 81-91.

AURADA, K.-D. (1992b): Auswirkungen der Salzabwässereinleitungen und -deponien im Südharz- und Werra-Kalirevier... . In: Neue Bergbautechnik 22 (6/1992), p. 203-212.

DVWK (1993): Salz in Werra und Weser. DVWK, Bonn, 1993.

HOPPE, W. (1962): Grundlagen, Auswirkungen und Aussichten der Kaliabwasser-versenkung im Werra-Kaligebiet. In: Geologie 11 (1962/9), p. 1059-1086.

SCHÖNWIESE, C.-D. (1992): Praktische Statistik für Meteorologen und Geowissenschaftler. Berlin, Stuttgart, 1992.

RÖDEL, R. (1997): Zeitreihenanalyse von aufsteigendem Salzabwasser im Werra-Kalirevier. In: Wasser & Boden 12/1997, p. 27-30.

Mass- and Energy Migration in Low Pervious Material

Wido Arnold[1] **and Makoto Nishigaki**[2]

[1] igi Niedermeyer Institute GmbH, Hohentrüdinger Str. 11, 91747 Westheim, Germany.
[2] GEL, Department of Civil Engineering, University of Okayama, 3-1-1 Tsushima-naka, Okayama 700, Japan.

Abstract. *A theoretical treatment of miscible phase concept and its three-dimensional realisation will be presented. As a practical background, the simulation of water-, air- and vapour migration around waste repository in the subsurface, under special consideration of the impact of an additional thermal gradient servers as a second driving force. One approach is to handle hazardous gases that may be the result of the burial of such materials beneath thick layers of compacted clay, where the clay's small pores may act as a diffusive barrier that retards the release of harmful gases to the atmosphere. The mass migration is considered to take place in the water phase, air phase as well as in the vapour phase. The concept of condensation/evaporation is used to interconnect the vapour and water phase. Gaseous flux through the subsurface low pervious material cannot be described exactly by Fick's law. Applying Hirschfelder's approach for dilute gases the diffusion operator was divided into three components, the binary diffusion, the Knudsen diffusion, and the thermal diffusion, which regards the impact of the thermal gradient. In contrast to the mass migration, heat migration is considered in four phases among which migration through the solid matrix plays a dominant role. Both of the migration equations were interconnected by the terms of relative permeability, partial pressure, viscosity and moisture contents.*

Keywords. *numeric simulation, multiphase flow, coupled modells, mass migration, energy migration, fem*

1 Introduction

Multiphase flow in porous media is a complex process encountered in many fields of engineering interest, such as oil recovery from reservoir rocks, aquifer pollution by liquid wastes and soil reconstitution, and agricultural irrigation.

The aim of this paper is focused on problems of theoretical treatment of the two-phase flow, as well as to its practical realisation. One area of such interest is the question of how to use geological formations as depositories for nuclear waste. Due to the long-term character of such facilities, after a certain period a thermal gradient can be observed. The flow conditions around the deposit become partially unsaturated.

A system in the subsurface is described by the simultaneous existence of phases. According to Fredlund and Rahardjo (Fredlund, 1993) a phase is defined to have different properties from the contiguous material and to have a bounding surface. Under unsaturated conditions two mobile phases can be observed, water and air. In addition to this, water may pass the contractile water-air skin and appear as a third vapour phase.

The modelling of geothermal reservoirs must address the coupled movement of heat and groundwater. Mercer and Faust (Faust, 1979) introduced the theoretical approach, the numerical treatment and the practical realisation of a geohydrothermal quasi three-dimensional field problem. Well known and wide-spread are the Finite Element based program SUTRA (Voss, 1984) and TOUGH (Pruess, 1987) an Integrated Finite Difference simulator for multi dimensional coupled fluid-heat flow of multiphase multicomponent fluid mixtures in porous and fractured media. Flow occurs under pressure, viscous, and gravity forces according Darcy's law, with interference between the phases, defined by user specified functions. Difficulties encountered in the analysis of coupled migration phenomena can be summarised as follows. The inclusion of thermal effects introduces additional parameters and coefficients that are more difficult to measure or to determine.

Coupled problems are refereed to as highly non-linear, interactions between dependent variables that cause numerical problems. Therefore, powerful numerical methods must be applied for solution of the governing equations

Moreover, in cases, if the ratio of permeability is extremely small, i.e. compacted clay or Bentonite, effects of the so-called *Knudsen*-diffusion as well as the Klinkenberg effect becomes considerable.

2 Governing Model

Modelling of multiphase-migration involves a number of distinct steps, proceeding from a conceptual to a mathematical to a numerical model.

2.1 Heat Transfer

Energy migrates through the three mobile phases as well as through the soil matrix. After adding up all migration paths, the energy-migration equation can be formulated as follows:

$$\begin{aligned} \frac{\partial T\langle \rho C\rangle}{\partial t} = & +\left(c_l^m q_l^m \rho_l + c_v^m q_v^m \rho + c_a^m q_a^m \rho\right)\nabla T \\ & +\nabla(\Lambda \nabla T) \\ & -F_V^V \\ & +Q^V \end{aligned} \tag{1}$$

The left side of Eq. 1 refers to the summation of sensible heat through all three phase. The first term on the right side describes the energy convection part, the second term the conduction part. The term F_V^V represents the energy losses due to the exchange processes between the fluid and the vapour-phase and the last part represents the energy sinks and sources by an external source. The advective heat flux $\langle \rho c\rangle$ is defined by Eq 2:

$$\begin{aligned} \langle \rho C\rangle = & +(1-\Phi)\rho_s c_s \\ & +\Theta \rho_l c_l \\ & +(\Phi-\Theta)(\rho_v c_v + \rho_a c_a) \end{aligned} \tag{2}$$

2.2 Mass-Migration

According to Peaceman (Peaceman, 1977) the motion force of the liquid phase can be formulated as follows:

$$q_l^m = \frac{K_{ij} K_l}{\mu_l} \nabla(p_l + \rho_l g z) \tag{3}$$

Hirschfelder (Hirschfelder, 1954) showed that the momentum equation of a dilute gas is identical to its components in some cases if the gas velocity is replaced by the mass average velocity of the mixture.

$$v = \frac{m_1 n_1 v_1 + m_2 n_2 v_2}{m_1 n_1 + m_2 n_2} \tag{4}$$

Introducing in such a way the impact of the *Knudsen* diffusion (second term), the Binary diffusion (third term) and the Thermo-diffusion (fourth term), for the vapour phase the mass equation can be written as:

$$q_v^m = \frac{K_{ij}K_v}{\mu_v}\nabla(p_v + \rho_v gz)$$

$$-(\Phi - \Theta)\rho_v \frac{D_{Ka}D_{Kv}}{\rho_g D_B + \rho_v D_{Ka} + \rho_a D_{Kv}}\nabla(p_g + \rho gz)$$

$$-(\Phi - \Theta)\rho_v \frac{\frac{\rho_g}{\rho_v} D_B D_{Kv}}{\rho_g D_B + \rho_v D_{Ka} + \rho_a D_{Kv}}\nabla(p_v + \rho_g gz)$$

$$-(\Phi - \Theta)\rho_g \frac{R_a D_{Ka} D_T}{\rho_g D_B + \rho_v D_{Ka} + \rho_a D_{Kv}}\nabla T \qquad (5)$$

and analog for the air-phase:

$$q_a^m = \frac{K_{ij}K_g}{\mu_a}\nabla(p_a + \rho_a gz)$$

$$-(\Phi - \Theta)\rho_a \frac{D_{Ka}D_{Kv}}{\rho_g D_B + \rho_v D_{Ka} + \rho_a D_{Kv}}\nabla(p_g + \rho_g gz)$$

$$-(\Phi - \Theta)\rho_a \frac{\frac{\rho_g}{\rho_a} D_B D_{Kv}}{\rho_g D_B + \rho_v D_{Ka} + \rho_a D_{Kv}}\nabla(p_a + \rho_g gz)$$

$$-(\Phi - \Theta)\rho_g \frac{R_a D_{Ka} D_T}{\rho_g D_B + \rho_v D_{Ka} + \rho_a D_{Kv}}\nabla T \qquad (6)$$

Bear (Bear, 1990) gave the adequate conservation equation for each phase the following description. Without regarding external sources and sinks, and under the assumption that the porosity of the solid matrix is not a function of the state variables, the equation can be written as:

$$\frac{\partial \Theta_l^V \rho_l}{\partial t} = -\nabla(\rho_l q_l^M) - f_{l \to v}^l \qquad (7)$$

$$\frac{\partial \Theta_g^V \rho_v}{\partial t} = -\nabla(\rho_v q_v^M) - f_{v \to l}^l \qquad (8)$$

$$\frac{\partial \Theta_g^V \rho_a}{\partial t} = -\nabla(\rho_a q_a^M) \qquad (9)$$

where Q_x^v refers to the volumetric moisture content of phase x, q_x^m is the mass weighted discharge of phase x, ρ_x stands for the density of phase x, $f_{v \to l}^v$ is the volumetric rate of vapourisation and $f_{v \to l}^{v}$ is the volumetric rate of condensation, t refers to the time and l,v,a are indexes of the liquid, vapour and air phase.

2.3 Assumptions and Simplifications

In developing the partial differential equations governing the simultaneous two-phase flow, the following assumptions were made in order to simplify the model. The non-wetting phases are assumed to be an ideal gas and the wetting phase as to be a *Boussineq* fluid, the phases are continuous, and the existence of a chemical and thermodynamic phase equilibrium. Parts of a phase, which is dissolved in another one, are not regarded. Moreover, elastic-, plastic- or thermo-elastic soil behaviours are not considered.

3 Parameter and Interactions

3.1 Diffusions

Effects of gaseous diffusion in soil has traditionally been simulated by application of a air-diffusion coefficient which could be suitably modified according to soil conditions, especially the pore tortuosity. This approach is based on the assumption that pore-wall interactions are neglected and gaseous migration in soils is only influenced by the collisions with other molecules, the so-called bulk-migration. With decreasing the pore size or decreasing the pressure, the frequency of pore-wall interaction increases drastically. This effect is called Knudsen-transport or Knudsen-diffusion.

The effects of Knudsen-diffusion have been known since the beginning of the century (Knudsen, 1909) and have already been discussed theoretically concerning the flow of gas through porous media (Scheidegger, 1970). Nevertheless, they have received only little attention in geo-hydraulic models. According to Kennard (Kennard, 1938) and Present (1958) for tiny cylindrical capillary pores the coefficient of *Knudsen*-diffusion can be obtained by the following formula:

$$D_K = \frac{2}{3} r \left[\frac{8RT}{\pi} \right] \tag{10}$$

Eldridge and Brown (Eldridge, 1976) showed that Eq. 11 is not only limited to cylindrical pore's shapes, but rather the ratio between the pore's minimal axis and the maximal axis show a greater influence to the coefficient of *Knudsen*-Diffusion. A ratio of 10 yields a value of D_k that is around 75% of that predicted, based on a circular pore with an equivalent area. In most cases, the aspect ratio is less than 3.5 which led to an approximate error of 10% of D_k.

Introducing the impact of tortuosity τ Eq. (8) can be formulated for the two gaseous phases as:

$$D_{Kphase} = \frac{2}{3}\frac{r_{Pore}}{\tau}\sqrt{\frac{8R_{Phase}\,T}{\pi}} \tag{11}$$

According to Bird (Bird, 1960) the binary diffusion D_B was obtained using the following formula:

$$D_B = 4.4*10^{-5}\frac{T^{2.334}}{(P_V + P_A)\tau} \tag{12}$$

The thermo-diffusion coefficient D_T is defined as follow:

$$D_T = \frac{\lambda}{[(1-\Phi)\rho_S c_S + \Theta\rho_S c_S + (\Phi-\Theta)(\rho_V c_V + \rho_A c_A)]} \tag{13}$$

3.2 Densities

The density of pure water increases within a temperature between 0^0C and 3.98^0C, density of water reaches its maximum within and above this temperature the density again decreases. Considering the compressibility and thermal expansively of pure water the density can be computed by using the following equation:

$$\rho_l = p_{l0}[1 - \beta_T(T - T_0) + \beta_P(P - P_0)] \tag{14}$$

where p_{l0} is the reference density under the reference condition P_0 and T_0, β_T refers to the thermal expansively and β_P the water compressibility. Under soil mechanical considerations and assuming that no air is dissolved in water the temperature has a stronger impact than pressure. Within a temperature limit between 0^0C and 320^0C the equation (17) is used to specify the temperature-density relation.

$$\rho_l = 999.222 - 8.6637*10^{-2}T - 2.9747*10^{-3}T^2 \tag{15}$$

Dry air and vapour are defined as to behave as an ideal gas. According to Dalton's law, the density value for the both phase can be defined as follow:

$$\rho_x = \frac{\varpi_x}{R_x T}p_v \tag{16}$$

3.3 Viscosities

Fluids resist a changing of their forms or the actions of shearing, the resisting forces are defined by:

$$F = \mu A \frac{d\vec{v}}{d\vec{n}} \tag{17}$$

This resistance, according to Newton's law is characterised by two parameters called the dynamic viscosity, μ, and the kinematic viscosity, ν. The kinematic viscosity is given as:

$$\nu = \frac{\mu}{\rho} \tag{18}$$

The dynamic viscosity is a function of pressure, temperature and dissolved components. However, for most of the civil engineering application the influence of pressure can be neglected. The variation of water-viscosity with temperature can be obtained from a steam-table or alternatively calculated by the formula that was proposed by Gray (Gray, 1972):

$$\mu_L = 2.414 * 10^{-5} 10^{\left(\frac{247.8}{T-140}\right)} \tag{19}$$

To get the viscosity for the vapour-phase, the following formula (Gray, 1972) was employed:

$$\mu_V = 8.04 * 10^{-6} + 4.2 * 10^{-8} (T - 273.15) \tag{20}$$

In areas, where only air or is present, the air viscosity is calculated using:

$$\mu_A = 1.71 * 10^{-5} + 4.386 * 10\ T - 1.056 * 10^{-11} T^2 \tag{21}$$

In the case of dilute gases Falta et al. (Falta, 1992) recommended the usage of Wilkes semi-empirical method of multicomponent gases to calculate the gas viscosity μ_g. The first notation was given by Bird (Bird, 1960):

$$\mu_g = \frac{\mu_g^c}{1 + \Phi_{c,aw} \left(\frac{\chi_g^{aw}}{\chi_g^c} \right)} + \frac{\mu_g^{aw}}{1 + \Phi_{aw,c} \left(\frac{\chi_g^c}{\chi_g^{aw}} \right)} \tag{22}$$

where μ_g^{aw} refers the air-water viscosity which can be obtained from the modified kinetic gas-theory (Hirschfelder, 1954), χ_g^{aw} represents the sum of the vapour and air mole fraction and χ_g^c is the chemical vapour mole fraction. The interfraction parameters $\Phi_{c,aw}$ and $\Phi_{aw,c}$ are defined as follow:

$$\Theta_{c,aw} = \frac{\left(1+\sqrt{\frac{\mu_g^c}{\mu_g^{aw}}}\sqrt{\frac{M_{wt}^{aw}}{M_{wt}^c}}\right)^2}{2*\sqrt{2*\sqrt{1+\frac{M_{wt}^c}{M_{wt}^{aw}}}}} \tag{23}$$

and

$$\Theta_{aw,c} = \frac{\left(1+\sqrt{\frac{\mu_g^{aw}}{\mu_g^c}}\sqrt{\frac{M_{wt}^{aw}}{M_{wt}^c}}\right)^2}{2*\sqrt{2*\sqrt{1+\frac{M_{wt}^{aw}}{M_{wt}^c}}}} \tag{24}$$

According to Reid (Reid, 1987) the vapour viscosity, μ_g^c, is computed from the corresponding stated method:

$$\mu_g^c = \frac{0.606\frac{T}{T_{crit}}F_p^0}{0.1706\left[\frac{T_{crit}}{\left(M_{wt}^c\right)^3\left(P_{crit}\right)^4}\right]^{\frac{1}{6}}} \tag{25}$$

with:

$$F_p^0 = 1 \qquad if\, 0 \leq \eta_{dr} < 0.022$$
$$F_p^0 = 1 + 30.55\left(0.292 - Z_{crit}\right)^{1.72} \qquad if\, 0.022 \leq \eta_{dr} < 0.075 \quad \text{and}$$
$$F_p^0 = 1 + 30.55\left(0.292 - Z_{crit}\right)^{1.72}\left|0.96 + 0.1\left(T_r - 0.7\right)\right| \qquad if\, 0.075 \leq \eta_{dr}$$

$$\eta_{dr} = 52.46\eta_d^2\frac{P_{crit}}{T_{crit}^2} \tag{26}$$

where Z_{crit} refers the critical compressiblity factor. According to Falta et al. (Falta, 1992) comparisons with experimental data had shown that for an air-water composition and at temperatures up to 150 ^{0}C the accuracy of Wilkes semiempirical approach lies within a range of 4%.

3.4 Pressure-Saturation Relation

Accurate descriptions of soil hydraulic properties are essential to simulate the migration of water and air in the unsaturated zone. Several models for estimating these relationships exists. Two of the most common used functions are those proposed by Brooks and Corey (Brooks, 1966) and van Genuchten (1980). These models are popular because they are simple and provide a wide applicability.

$$k_l = k_0 \sqrt{\frac{S_r}{S_s}} \left[\frac{1-\left(1-S_r^{1/m}\right)^m}{1-\left(1-S_s^{1/m}\right)^m} \right]^2 \tag{27}$$

where

$$S_r = \frac{(\Theta_r - \Theta_s)}{(\Phi - \Theta_s)} \tag{28}$$

The gaseous permeability is obtained using the simple relation (Eq. 29), which works, for most of the cases, sufficiently correctly.

$$k_g = 1 - k_l \tag{29}$$

3.5 Heat-Related Parameters

The ratio of vapourisation can be obtained using Eq. 30:

$$F_V^V = c(\Theta - \Theta_r)(p_{vs} - p_v) \tag{30}$$

By calibration, the numerical model has to vary this value in order to enforce local vapour/liquid equilibrium in the defined domain.

$$c = 10^5 \sqrt{k_g} \tag{31}$$

Following Hardley (Hardley, 1985), in the present model the factor c is estimated employing Eq. 32.

$$L = 2662 - 0.94571(\overline{T} - 273) - 0.00090094(\overline{T} - 273)^2 \tag{32}$$

Eq. 32, taken from a steam table, is applied to calculate the value of the latent heat of vapourisation.

4 Solution

4.1 Analytical Solutions

A large number of analytical models have been presented to simulate part of complex multiphase flow systems under various assumptions. Analytical solutions are subject to more restrictive assumptions, but are generally easy to evaluate and provide exact answers, for which round-off and computational errors are negligible. The complicated framework of the governing equation can be analytically solved by applying certain simplifications. For example, if the entire domain is dominated by a single phase, the system equations can be simplified to a parabolic differential equation for the fluid pressure. On the other hand, if the fluids are incompressible, the flow-direction is one-dimensional dominate and the capillary pressure tends to zero, then the system of equations can be reduced to a nonlinear hyperbolic equation, the Buckley-Leverett equation. Buckley and Leverett (1942) solved a one-dimensional flow problem, without gravity and neglecting capillarity and phase-compressibility. Using the Buckley-Leverett approach, immiscible flow can be treated as a sharp interface problem.

When more complex flow problems have to be examined, which considers capillary effects between the individual phases by solving the governing phase flow equations in combination with constitutive relations for saturation and relative permeability, analytical models are inadequate.

In order to verify the applicability of the proposed model the de-coupled migration paths were verified analytically using the solution.

$$\frac{\partial T}{\partial t} = D_x \frac{\partial^2 T}{\partial t^2} + K_X \frac{\partial T}{\partial t} \tag{33}$$

The simplified heat-transfer equation (Eq. 33) can be replaced by the following analytical formulation (Eq. 34) and can be applied for the initial condition defined by $T_{t=0} = 0$ and the boundary conditions $T_{x=0} = 1,\ T_{x=\infty} = 0.$

$$T(x,t) = \frac{1}{2}\left[erfc\left(\frac{x - K_x t}{4 D_x t}\right) + exp\left(\frac{K_x x}{D_x}\right) erfc\left(\frac{x + K_x t}{4 D_x t}\right)\right] \tag{34}$$

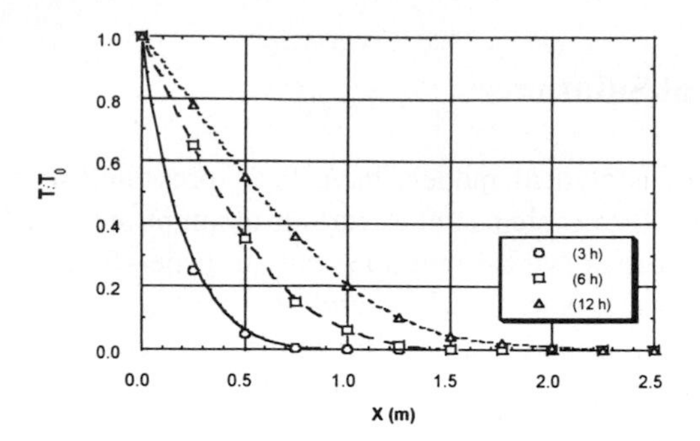

Fig. 1 Comparison of energy migration with analytical solution

4.2 Numerical Realisation

Especially near and around constructions, in cases of zones of large anisotropy and local inhomogenities as well as by processes with more than one driving force, which vectors are not parallel to each others, the simulation should include all three dimensions.

The spatial discretisation was carried out using the *Galerkin*-approach of the Finite Element Method (FEM). FEM gives more opportunities to treat porous media with large anisotropy, like fractured rocks that is dominant in the baserock of Japan. Linear, quadratic and cubic shape functions for triangle, rectangle tetrahedon and hexahedron elements

The advantage of using iterative solvers consists in a reduced demand of hardware-resources especially. They consist of a sequence of similar operations, which work by repeatedly improving an approximated solution until a predefined accuracy. Since the access to the global matrix is carried out via a matrix-vector product, or its transposed pedant, the iterative solvers can be formulated in the form of a template.

Normally, an FEM-stiffness is diagonally dominated, but in order to improve the convergence behaviour the matrix-condition is increased by using a pre-conditioner. Controlled by the euclidean norm, a scaling of the matrix improves the runtime-behavior and the result-quality. GMRES with a LU-preconditioner showed, for the presented test case, the best runtime performance, but for most the problems, CGS was more stable and sufficiently fast.

5 Verification

Two test sets were used in order to test and to verify the several program features. For the first, infiltration experiment results are used, conducted in homogenous soils. The second case is restricted to specify the quantitative temperature to the two-phase flow regime.

5.1 Isothermal Flow

The often referred to experimental results of a pounded infiltration in vertical column published by Touma et al. (Touma, 1984) were used. The column was filled with an Air-dried sand graded from 0.02 to 1 mm, with approximately 50% weight less than 0.3mm, having a bulk-density of 1.67 g/cm^3 and a mean porosity of 0.37 cm^3/cm^3. The water head is set to 2.3 cm above the soil surface and water moves at the lower boundary.

Fig. 2 and Fig 3 illustrated the unsaturated relations of the experimental conditions.

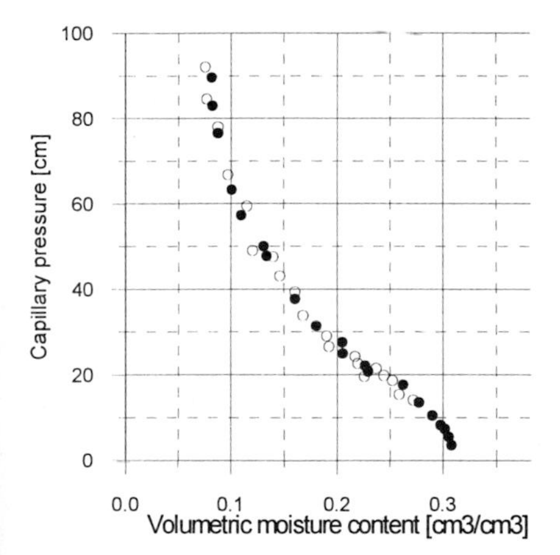

Fig. 2 Capillary pressure vs. volumetric moisture contents

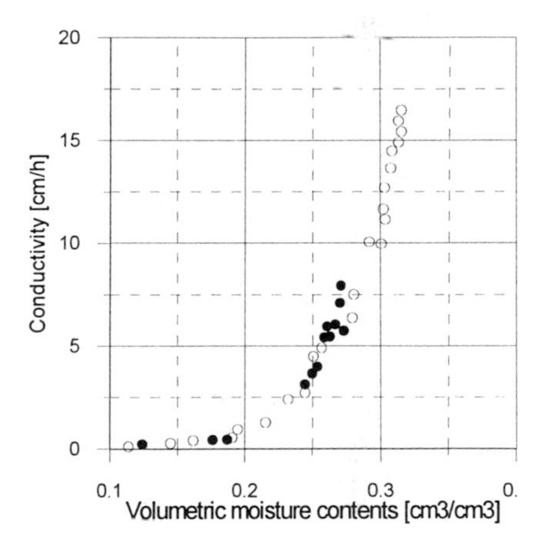

Fig. 3 Conductivity vs. volumetric moisture contents

By modelling water movement in soil, the air is generally considered as to escape freely. In case of a shallow water table, impermeable layers or lateral obstructions air may not escape and which will lower the infiltration rate. Touma et al. (Touma, 1984) specified these air-effects by carrying out the infiltration experiment in two ways, with regarding air-effects, marked by the open points (entrapped air cannot freely escape), and without air (filled points) effects.

Fig. 5 shows the moisture profiles obtained by seven time steps, after 6, 12, 24, 36, 48, 60 and 72 minutes (from top to bottom).

Due to escaping air bubbles, Touma et al. (Touma, 1984) observed that the natural saturation of 0.312 was reduced to 0.272 during the experiment. This effects can be observed in Fig 4. After correcting the unsaturated condition in order to take into account this phenomenon, the proposed model was found to be in good agreement with the actual results.

The three-dimensional numerical model consists of 3420 nodes, 5 elements in x- and y- direction and 94 elements in z-direction. It appears to be sufficient to apply eight-node brick elements with a linear shape function and to carry out the Gauss-Legendre integration using two Gaussian points in each direction. In order to solve the pressure-stiffness matrix with a range of 10260 and the temperature-stiffness matrix it took approximately 30 hours on a Sparc20 workstation.

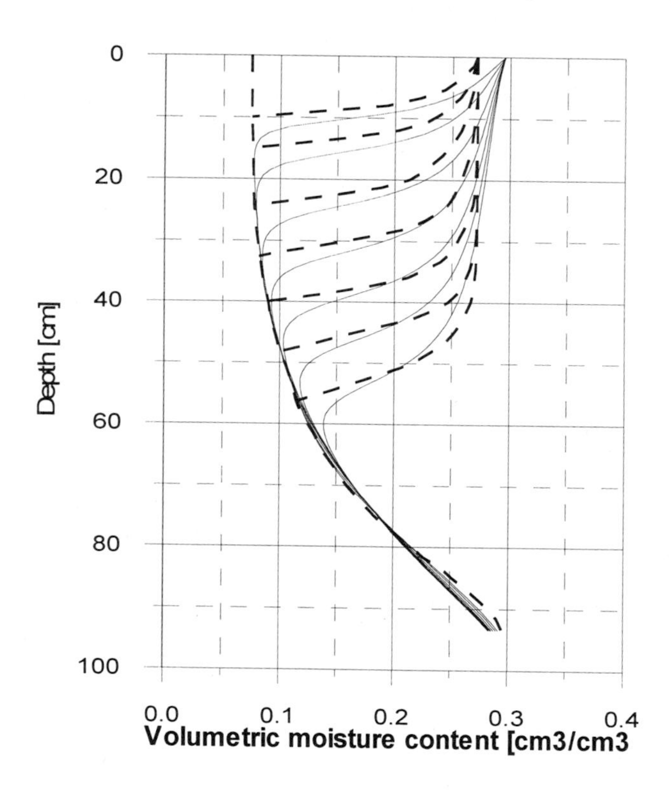

Fig. 4 Comparison of experimental results (dashed line) with numerical results.

5.2 Non-Isothermal Flow

In order to verify the qualitative impact of non-isothermal flow a similar infiltration experiment as the previous one was defined. A tank is taken, filled with dry loose material, with initial moisture content of 0.08 and an intrinsic permeability K of 0.00001 $^{cm}/_{s}$.

The infiltration only takes place towards the above boundary, and pressure heat was set to 20 cm above the upper boundary.

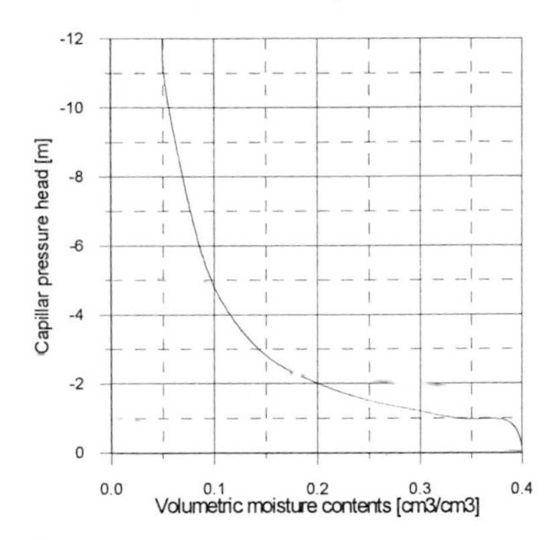

Fig. 5 Capillary pressure vs. volumetric moisture contents relation.

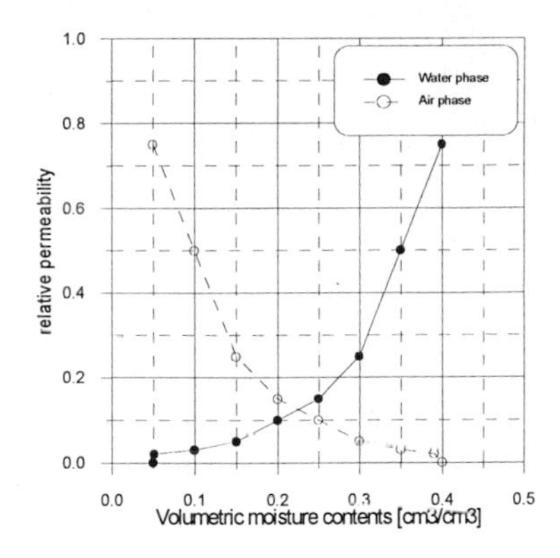

Fig. 6 Relative permeabilities vs. volumetric moisture contents

The unsaturated condition were defined according to common sand, capillary pressure vs. volumetric moisture content relation (Fig. 5) and relative permeability vs. volumetric moisture content relation (Fig. 6). The settings for thermal conductivity parameters and the heat capacities are provided in Table 1.

Table 1 Heat-migration related parameters for case II

	Thermal Conductivity $\left[\frac{W}{m\,K}\right]$	Heat Capacity $\left[\frac{J}{kg\,^0C}\right]$
Soil	1.74	840
Water	0.65	4200
Vapour	0.03	2000
Air	0.03	1012

It is seen from the figures that the Two-Phase flow is a temperature dependent phenomenon. Although the programme can simulate the temperature dependent effect on the two-phase flow behaviour, the dependency is not noticed as expected.

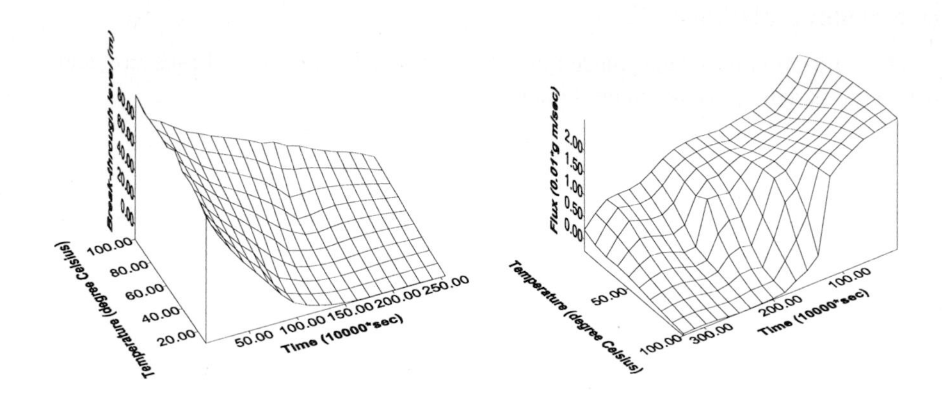

Fig. 7 Break-through curves vs. temperature (moisture content tracked)

Fig. 8 Temperature-depend Flux-vs. time

As seen in Fig. 7 the temperature scale is graduated in 10 °C, the break-through depth of moisture content, 0.2, belonging to the predefined moisture content of 0.08 is time-dependent and the temperature of the medium as well.

In Fig. 8, the flux rate is tried to be correlated with the temperature and the elapsed time of flow process. It is clear from the figure that the flux rate decreases with time, the maximum at the beginning when the soil column is dry and the consistent decrease of flux rate as the column gradually becomes wet. At higher temperature, the rate of decrease is high compared to the process at lower temperature.

The proposed model is verified with the test results from the reference in isothermal and non-isothermal conditions and are found to be in good agreement with the actual results.

Notation

D_B	Coefficient for binary diffusion	p_l	Partial pressure (liquid)
D_T	Coefficient for thermal diffusion (air)	p_v	Partial pressure (vapor)
D_{Kv}	Knudsen – diffusion coefficient (vapor)	p_a	Partial pressure (air)
D_{Ka}	Knudsen – diffusion coefficient	p_p	Partial pressure (gas)
$f^v_{l \to v}$	Volumetric rate of vaporation	R_v	Ideal gas constant (vapor)
$f^v_{v \to l}$	Volumetric rate of condensation	R_a	Ideal gas constant (air)
j_l	Mass flux vector (liquid)	r_{pore}	pore radius
j_v	Mass flux vector (vapor)	t	time
j_a	Mass flux vector (air)	Θ_l	Volumetric moisture content
K_{ij}	Intrinsic permeability	Θ_g	$(\Phi - \Theta_l)$
k_l	Permeablility (liquid)	τ	Tortuosity soil matrix
k_g	Permeablility (gas)	Φ	Porosity
		ρ_l	Density (liquid)
		ρ_v	Density (vapor)
		ρ_a	Density (air)
		ρ_g	Density (air)

References

Bear,J.; Bachmat, Y. (1990): Introduction to Modeling of Transport phenomena in Porous Media; Kluwer Acadamic Publisher

Bird, R. Stewar; W.E.; Lightfoot, E.N. (1960): Transport Phenomena; John Wiley & Sons

Brooks, R.H.; Corey, E.N. (1966): Properties of porius media affecting fluid flows; J. Irrig. Drain. Div. Am. Soc. Civil Eng.; Vol. 92; pp. 61-68

Clifford, S.M.; Hittel, D.: The stability of ground ice in the equatorial region of Mars; J. Geophys; Res. 88:2456-2474

Eldridge, B.D.; Brown, L.F. (1976): The effects of cross sectional pore shape on Knudsen diffusioin in porous meterial; Am. Inst. Chem. Eng; J. 22; 942:944

Falta, R. W.; Pruess, K.; Javandel, I.; Witherspoon, P. A. (1992): Numerical Modeling of Steam Injection for the Removal of Nonaqueous Phase Liquids From the Subsurface; 1. Numerical Formulation; Water Res. Res., Vol. 28; No. 2; Pp. 433-449

Faust, C.R.; Mercer, J.W. (1979): Geothermal reservoir Simulation; 1. Mathematical models for liquid and vapor dominated hydrothermal systems; Water Res. Res.; Vol. 15; No. 1; pp. 23-30

Fredlund,D.G.; Rahardjo, H. (1993): Soil Mechanics for Unsaturated Soils; John Wiley and Sons, Inc.

Gray, D.E. (1972): American Institute of Physics Handbook; Third Ed.; McGraw-Hill; New York; NY

Hardley, G.R. (1985): Thermal conductivity of Ratially Saturated Materials; Sandia National Laboratories; Alubuquerque; New Mexico

Hirschfelder, J.O. (1954): Molecular theory of Gases and Liquids; John Wiley and Sons; Inc.; New York

Kennard, E. H. (1938): Kinetic theory of gases; Mc-Graw-Hill; New York

Knudsen, M. (1909): Die Gesetze der Molekularströmung und der inneren Reibungsströmung der Gase durch Röhren; Ann. Physik 73; 6296-6300

Meiri, D. (1981): Two-Phase Flow Simulation of Air storage in an Aquifer; Water Res. Res.; Vol. 17; No. 2; pp. 1360-1366

Peacemen, D.W. (1977): Fundamental of numerical reservoir simulation; Elsevier Scientific Publishing Company

Pruess, K. (1987): TOUGH User's Guide; Lawrence Berkley Laboratory; report LBL-20700

Reid, R.C., Prausnitz, J.M., Poling, B.E (1987): The Properties of Gases and Liquids; 4th ed.; McGraw-Hill

Rothfield, L.B. (1963): Gaseous counterdiffusion in catalyst pellets; Am. Inst. Chem. Eng; J. 9; 19-24

Scheidegger, A.E. (1970): The physics of flow through porous media; Macmillan; New York

Touma, J.; Vachaud,G.; Parlange, J.-Y. (1984): Air and Water Flow in a sealed, pounded vertical soil Column: Experiment and Model; Soil Sci.; Vol. 137; No. 3

Voss, C.I. (1990): Sutra (Version V-069-2D); Program description and program notes; U.S; Geological Survey

Modelling Philosophy in the Structural Analysis of Dams

Dimitar Kisliakov

University of Architecture, Civil Engineering and Geodesy (UACEG),
1 Chr. Smirnenski Blvd., Sofia 1421, Bulgaria

Abstract. *The paper will dicuss some statements about the structural analysis of dams published in other papers. The main groups of problems concerning the adequate modelling of the dam structure, analysed in more detail are: development of new methods and possibilities for their application (informatics, theory of probabilities, fuzzy set theory); the system approach in the structural analysis of dams; physical and mathematical modelling of the investigated structure: effort required, role of the active Codes; hierarchical levels in the structural modelling of dams; problem solving with insufficient input information (floods and earthquakes); validation, verification and calibration of the structural model; uncertainty modelling in the structural analysis of dams.*

Keywords. *dam engineering, structural analysis, model development*

1 Introduction

Dam Engineering is a special field of Structural Engineering, where environmental-technical complexes of great importance for the life in large regions are designed, constructed and operated. The basic principles of the structural analysis philosophy in this field must be very precisely formulated and argued because of the special physical phenomena involved in the construction and operation of the dam structures, and of the risks connected with their possible failures.

The reason for this paper are the works (Lombardi, 1995; Lombardi, 1996) which deal with the basic principles of Structural Engineering and Applied Science. These works have already been critically discussed in detail in (Kisliakov, 1998). The aim of the present paper is to bring further the analysis of some fundamental problems in the field of Dam Engineering formulated in (Lombardi, 1995; Lombardi, 1996) as well as to discuss the related principles and approaches in the modelling phase of the structural analysis process.

2 Changing Approaches in Scientific Thinking and Development of New Methods

2.1 The Modern Computational Methods

Simultaneously, with the extremely intensive development of the numerical methods for engineering analysis, of their software applications and with the increase of the performance of the hardware platforms, more and more often the question is set about whether the structural engineer is in fact now able to design a dam better by means of these modern computational tools. Definitely yes, in our opinion. Nowadays there are more perfect tools and possibilities for a more perfect design of dams. But one should clearly distinguish these things. These possibilities are available and they are present to a great extent due to the development of the computational technologies (numerical methods, hard- and software), but they are only possibilities. This development has enabled, also, the creation of tools (for example the FEM-implementations) for much more perfect modelling (in the sense of adequacy to the physical reality) of the environmental-technical system (ETS "dam" and its connections / interactions with the surrounding world. The structural engineer nowadays knows more about the behaviour of such an ETS. Although whether any design team would succeed to actually achieve more during the structural analysis process depends on the qualification of its members and on their skills / possibilities to implement it, since all modern analysis approaches require corresponding special education and implementation equipment. All methods are based on some assumptions and also have their limits of application. Every professional is clearly aware of this point in any case considered. It should be emphasised that the analysis methods in structural engineering are being continuously improved and further developed with the aim of more adequate representation of the physical reality - especially recently when computational intensity is no longer an qualitative problem.

One more issue should be emphasised when the use of informatics in Dam Engineering is considered - the possibility for creation of large data bases. Since the knowledge about the actual behaviour of different types of dams, subjected to some types of loading (strong earthquakes for example), is quite limited. The creation of, and the access to, such data bases (via the Internet?) would be of great importance for both the design and operation of dams.

2.2 Probabilities and Chaos

As far as probabilities, chaos and fuzzy sets are considered in (Lombardi, 1995; Lombardi, 1996), completely different scientific domains have been confused there. The mathematical probability is related to the random events, i.e. such ones which might happen or not happen under the same conditions. "The same conditions" means: as far as these are under the control of the experimenter (Crandall, 1963; Kisliakov, 1981). The random quantities are characterised by the distribution function and by the probability density function. The probabilistic evaluations are connected with the law of large numbers. They may be applied to phenomena whose essential features are already well known, in order to account for the less known but not so important features. There is no room for them when quite unknown phenomena are considered. It may not be stated in this connection that they are "only a modest mantle hastily thrown over reality to hide our ignorance of it" (Lombardi, 1996). The statistical description is an averaging, so hence the basic characteristic of every random quantity is the *mathematical expectation*. When processes in the time domain are considered, for example the seismic loading on a structure, the *theory of random processes* should be applied. It should be emphasised, that the probabilistic approach may be applied for both linear and non-linear problems. However, randomness is one of the essential components of uncertainty in general; this issue will be commented on in more detail later.

The theory of deterministic chaos is a completely different thing (Schuster, 1984). Here, unpredictability of the process appears *in an absolutely determined system*, which *does not contain any random parameters*. Moreover, only in non-linear systems! The dynamic chaos cannot be mentioned in connection with the structural analysis of dams because it is carried out for static loads. When seismic excitation is considered, it is a random process, and thus deterministic chaos cannot take place in this case either. Finally, the non-linearities in the dam-foundation-reservoir system are of such a nature that hardly anyone would be able to construct the corresponding non-linear differential equations and their phase space.

The situation with the probabilistic methods is quite different. They are necessary for the statistical analysis of the hydrological factors and when the parameters of the design floods and earthquakes have to be specified. These parameters cannot be determined exactly, but certain probabilities can be assigned to their values on the basis of the data already available. Thus, the procedures of seismic or flood hazard assessment results in the determining of the probability that some key parameter value of the flood or of the earthquake has been reached (of exceeded). The determining of these probabilities is one of the important implementations of the probabilistic approach, which is a powerful tool for the modelling of natural hazards.

In general, the theory of probabilities should be mentioned as inevitable in constructing of fault and event trees. This process is a very important part of the probabilistic risk analysis (PRA) methodology for important engineering facilities (dams) (Nielsen, 1994; Lafitte, 1993) and actually constitutes the first and the last steps of the PRA procedure which deals with the parameters of the impact and with the consequences. The implementation of probabilistic methods in the seismic risk analysis of dams is also discussed in (Kisliakov, 1997).

When applications of the probabilistic approach are considered, one more question often discussed should be mentioned here - the problem of the safety factor. It is well known that this factor has a probabilistic nature. It corresponds to some (sufficiently large) probability of non-failure. Moreover, in the statistical theory of dimensioning or, as it is called, the theory of reliability, failure is considered as a result of either exceeding some dangerous level, or of damage accumulation, or of fatigue. When the first problem is considered, Rice's formula is used for the expected exceedance number, $\mathbf{v}_a^+$, over some defined level **a** of the stationary random process for a time unit:

$$v_a^+ = \int_0^\infty \dot{x} p(a, \dot{x}) d\dot{x} . \tag{1}$$

For a narrow-band random process with a Gaussian distribution and zero mean value we get (Crandall, 1963):

$$v_0^+ = \frac{1}{2\pi} \frac{\sigma_{\dot{x}}}{\sigma_x} , \tag{2}$$

where σ is the standard deviation (RMS value).

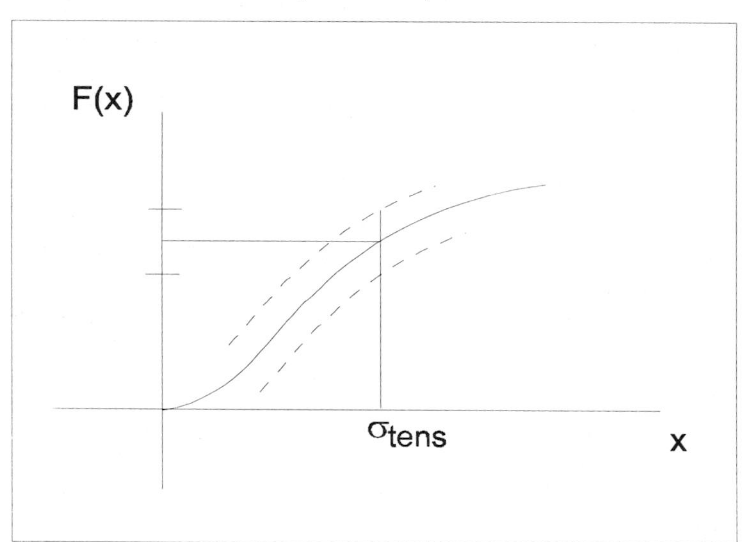

Fig. 1: Distribution function for the exceedance of some dangerous level

In the case of a dam, the load (and hence the stress and strain) time histories, although random, are not stationary. That is why $\mathbf{v}_a^+$ has to be calculated not after

(1) but after another, more complicated formula. A huge statistical database would be needed for the use of formulas of such kind. It is much simpler to consider the exceedance of some dangerous level as a random event (rather than a process). Furthermore, the distribution function of this event can be sought, which is of the type displayed in Figure 1. Here **x** is a stress, and $F(x)=P(X<x)$ is the probability of non-failure, **X** being the expected stress value in the material ($\sigma_{max,tens}$). For this value that the probability of a non-failure is to be obtained, which should be considered as sufficient (with the corresponding safety margin) by the expert decision, and this constitutes the dimensioning process according to the theory of reliability. Figure 10 in (Lombardi, 1995; Lombardi, 1996) looks very similar to Figure 1, but it has different sense there, and is given without any analytical background. Moreover, in our opinion the safety factor for tension for concrete in the numerical example presented in (Lombardi, 1995; Lombardi, 1996) must be higher than for compression, and not vice versa.

2.3 Fuzziness and Uncertainty Modelling

Special attention should be paid to the question of the fuzzy set theory (Ragot, 1993; Kaufmann, 1977). Unlike the classical set theory, here an element **x** may not only belong or not belong to the set **A**; there is a characteristic function $\mu_A(x)$, $0 \leq \mu_A \leq 1$, which states that **x** belongs to **A** to some extent. "Fuzzy Set Theory (FST) is a theory about vagueness and uncertainty; it enables ill-defined concepts to be used for ill-defined situations!" (Ragot, 1993). Although the FST and the theory of probabilities have much in common, they differ substantially in their axioms (Ragot, 1993). FST may prove to be useful when witness descriptions of a natural event are collected. In such cases, a numerical characteristic could hardly be assigned to such descriptions; on the other hand, a great amount of such descriptions (set!) can lead to valuable conclusions about the event.

FST may also prove to be useful when the extent of adequacy of some structural model to the real structure has to be evaluated. If in this model the non-exact hypotheses of linearity and elasticity have been included deliberately, we could further comment to what extent the corresponding equations are true. Thus, one comes to the idea of the fuzzy relations theory (Kisliakov, 1996).

Since fuzziness is regarded as one of the essential components of uncertainty, a special attention should be paid to this relation as well as to the more general problem of uncertainty modelling (the following remarks on this issue are based on (Davis, 1996)). Uncertainty is an essential feature of every modelling process as well as of decision making since the latter one is also based on models. All models are abstractions from the reality. The process of abstraction for developing a model is always based on assumptions and associated uncertainties. Every model has a structure and parameters, and there are uncertainties in both. These uncertainties need to be recognised and properly managed.

Fuzziness is just one of the three uncertainty components. Both remaining ones are suggested to be incompleteness and randomness (Davis, 1996). Since a model is not the reality by definition, it is incomplete by definition, too. Moreover, incompleteness exists also because there is always something we do not know. Randomness is the lack of specific pattern in the occurrence of the investigated phenomenon.

Although there are already some well elaborated implementations of uncertainty modelling procedures, even for Civil Engineering structures (Sanchez-Silva, 1996), the possible applications of this approach in Dam Engineering seem to be, in our opinion, one of the modern challenges in this field because of the great complexity of the system to be analysed. In this connection, the problem of uncertainty modelling in specifying the material and loading parameters (especially those of floods and earthquakes) and the parameters of the geologic conditions for a given dam site is open and requires intensive research. However, we dare state that, in our opinion, even with the recently developed more powerful approaches to uncertainty modelling such as, for example, the Interval Probability Theory (IPT) (Cui, 1990), the construction of a so-called "open world" model of a dam-reservoir-foundation system could hardly be possible in the near future due to the uniqueness and complexity of such systems. Nowadays, the problems with the description of earthquake and flood hazards are solved by means of special probabilistic procedures developed to deal with the lack of information specific for these events and with their random nature. It should be mentioned that since theory of evidence and FST are special cases of IPT (Davis, 1996), future application of IPT for more realistic flood and seismic hazard assessment and for uncertainty modelling in Dam Engineering at all would be a large field for research activities.

3 The Modelling Process in Structural Analysis of Dams

The modelling process is an inevitable part of the feasibility study, design, construction and operation of each dam-reservoir system. The development of a model is an abstract and approximate representation (based on some assumptions) of the physical object considered. The aim of the model is to represent the real behaviour (performance) of the real object in its real environment. On the basis of such representation a prognosis can be made and decision making can be carried out about the parameters of this behaviour under different conditions. This prognosis has to be as adequate as possible to the physical reality. This adequacy is, in general, the most important criterion for the quality of the engineering model.

It is beyond the scope of this paper to discuss the general principles and methods of Modelling in Engineering. In the following, only some particular problems in the field of structural modelling in Dam Engineering will be discussed.

In general, the modelling process for a dam structure follows the sequence presented in (Van Oortmerssen, 1991) for the development of a computational fluid dynamics code. The specific features of a dam structural model result from the great complexity of the dam structure (in the sense of physical phenomena presented) and from its intensive interaction with the environment. So, it is often more reasonable to talk about an environmental-technical dam-reservoir-foundation system rather than about the dam only. Moreover, each dam is a unique set of topographical, geological, climatic and structural parameters. Hence, each case study requires a strongly individual approach.

3.1 Structural Modelling of Dams - Aims and Problems

The main phases of the modelling process in dam structural analysis are as follows:

- Development of a conceptual physical model of the real structure under consideration. This step is based on appropriate assumptions and leads to simplification of the reality in the sense of interacting physical objects and phenomena presented. Appropriate boundary conditions for the model have also to be defined.
- Formulation of continuous mathematical model. At this stage, the mathematical description of the conceptual physical model is formulated, as well as for its boundary conditions.
- Discretization of the continuous mathematical model. This step enables further numerical solution of the mathematical equations by means of computer or of any approximate procedure in the most simple case.
- Implementation of the numerical solution of the discretized equations in a computer code.
- Calibration of the computational model. This is a very important step at which proper values have to be assigned to the parameters of the developed model. The further use of the model strongly depends on the properly performed calibration. When an existing dam is modelled, the problem of identification of the real structural parameters becomes of key importance for the implementation of the model and for the following calculations. In this case, the process of calibration of the model consists of assignment to the structural parameters of values for which the calculated structural response is identical (in terms of key parameter values) with the measured one for a known loading.

This procedure is often implemented when for example the dynamic characteristics of a dam have to be obtained by means of full-scale dynamic testing.

- Actual computation.
- Analysis of the results and decision making.

The above shortly presented the general flow of the modelling process of dam structural analysis which has been well traced and elaborated in detail during many years of developments in Dam Engineering. Besides, most of these general steps are common for many types of engineering facilities. Some engineers, especially when deeply involved in specific expert calculations, do not even recognize these modelling steps, moreover, when using modern highly sophisticated software. We share the opinion that the structural engineer should be clearly aware of his current position in the global modelling process in every time instant in order to minimize the risk of errors.

The aim of the obtained computational model is to enable simulation of all possible foreseen modes of structural behaviour for the possible impacts on the structure and its interactions with the environment at all stages of construction and operation, as well as optimization of the type of the structure and its parameters for example in the case of a newly designed dam. Thus, the structural model can serve not only as a tool for prediction of the parameters of structural behaviour under possible loading conditions but also as a powerful tool for decision making and operational management.

The large dams are usually constructed over several years. This leads to the change of the structural parameters with the time, and this phenomenon has also to be accounted for in the modelling process. The special features of the civil construction process with respect to the possibilities for the modelling of its design are in details analysed in (Platt, 1994). The application of these techniques in the field of Dam Engineering is still an open problem.

3.2 Verification and Validation

Special attention should be paid to the verification and validation processes for the developed model.

Verification of a computational model consists of a check that the computer program is a correct representation of the mathematical model used. It is usually not a duty of the structural engineer to perform verification of the computer program he uses, moreover if the latter is a well established one. But if he has coded by himself any special calculation procedure he has to be very careful with its verification.

Validation is the confirmation that the developed computational model is an adequate representation of the physical reality (in the sense of enough good approximation). Hence, validation is a broader activity which includes verification and comparisons of the results with real measured values and experimental benchmark results. Nowadays, significant experience has been already collected with some types of problems in Dam Engineering, and there are many modelling assumptions which have proved to be a good approximation of the reality. Most of the professional computer programs available have been also validated with numerous benchmark examples and experimental results.

More details concerning verification and validation of a computer code are presented in (Van Oortmerssen, 1991).

3.3 Hierarchical and Sophistication Levels

Each step of the general modelling process can be analysed at different hierarchical levels. The lower the hierarchical level is, the clearer the structure of the problem to be solved. The main principles of hierarchical organization of information have been defined in (Sanchez-Silva, 1996). The hierarchical description and the mathematics of IPT have been used to manage features of fuzziness and uncertainty in the numerical example solved there of a simple frame structure. An application of hierarchical levels in seismic risk assessment of concrete gravity dams has been discussed in more details in (Kisliakov, 1997).

On the other hand, different levels of sophistication for each stage of the modelling process are also one of its essential features (Kisliakov, 1997). The diving into a deeper level of sophistication in the structural modelling, i.e. the "refinement" of the model obtained in the sense of increasing rigorousity, should be performed through continuous iterations simultaneously with the identification of the dangerous scenarios for the structure considered. The identification of potential deficiencies in the structural performance and the construction of damage / failure scenarios, respectively, goes on together with the increasing sophistication of the analysis. A typical implementation of such progressive methodology in the seismic analysis of gravity dams is in details presented in (Ghrib, 1997).

3.4 The System Approach

An interesting question arises about the role of the system approach in the structural analysis of dams. In our opinion, this approach has not the aim to hide the cause-result relations in dam analysis (Lombardi, 1995). On the contrary, the components of the dam-reservoir-foundation (DRF) system (when regarded as a system) at each hierarchical level, as well as their interrelations are functions of time and some physical nature. The system analysis of the DRF system can help better understanding the cause-result relationships within it. Such analysis has been successfully performed in (Malakhanov, 1991). As a result, classification of all main problems connected with the feasibility, design, construction and operation has been obtained. Furthermore, classification has been developed with the quality characteristics for any hydraulic structure with relation to the different branches of human life on the different possible hierarchical levels. As a result, important proposals have been made for improvement of the design criteria for hydraulic structures (dams in particular). This approach becomes even more useful when the environmental impact / interaction of a DRF system is analysed. In our opinion, the recently developed Interacting Object Process Modelling (IOPM) approach (Blockley, 1995 ; Sanchez-Silva, 1996) actually represents the further development of the system approach.

3.5 Effort Required and the Role of Codes

A key question for the solution of every modelling problem is the one about the effort required in connection with the necessary sophistication levels of the solution. On the one hand, every dam has to meet the safety requirements of the currently active national codes with respect to the design, construction and operation phases. On the other hand, the development of the technical knowledge is always faster than the upgrade of the codes. Besides, the application of a very sophisticated solution is usually very expensive. However, the development and the revisions of the design criteria is also related to the development of the construction technologies. Thus, it is a duty of the design team to ensure both fulfilment of the code requirements and a state-of-the-art design solution for a minimum of costs. This formulation makes the problem very difficult. Of course, every model is an approximation of the reality. The question about the effort needed in the modelling phase is a typical example for optimization problem. In this sense, the code requirements represent some limit for the "minimum of adequacy" which has to be provided by the engineer for not allowing any compromises with the structural and / or operational reliability of the dam.

The safety requirements themselves are closely related to the definitions of "risk" and "residual risk", as well as to the problem of the so called "acceptable" levels of risk for a given dam. This group of problems remains beyond the scope of this paper. Moreover, it is still open (Risk-Based Dam Safety Evaluations, 1998), and its solution (if possible at all) would require a very broad professional and social discussion on all possible national and international levels.

4 Closure

The only aim of this analysis was to put onto a correct base the investigations leading to the construction of complex facilities of great importance, as well as to make the expert decisions easier. The main conclusions which can be drawn from the performed analysis are, in our opinion, as follows:

- the qualities and the possibilities for application of new methods and theories in the field of Structural Engineering (Dam Engineering) should be estimated very carefully with the correct reference to their theoretical base;
- the further development of all methods for modelling of particular physical phenomena in the structural performance of dams is indeed necessary for more adequate representation of the physical reality and thus for better understanding of the behaviour of dams. For example such as: constitutive models for the mass concrete under dynamic loading, behaviour of zoned embankments under seismic loading, earthquake-induced dam-reservoir interaction etc;
- the implementation of recently developed methods for modelling of uncertainty and "open world" problems in the field of Dam Engineering needs intensive research activities.

References

Blockley, D.I. (1995): “Computers in Engineering Risk and Hazard Management”, Archives in Computational Methods in Engineering, Vol.2, 2, 1995

Crandall S. H. & Mark, W.D. (1963): "Random Vibration in Mechanical Systems", Academic Press, 1963

Cui, W. & Blockley, D.I. (1990): “Interval Probability Theory for Evidential Support”, Int. Journal of Intelligent Systems, 1, 1990

Davis, J.P. & Blockley, D.I. (1996): “On Modelling Uncertainty”, Hydroinformatics’96, Müller (ed.), Balkema, Rotterdam, 1996

Ghrib, F.; Leger, P. ; Tinawi, R.; Lupien, R.; Veilleux, M. (1997): "Seismic Safety Evaluation of Gravity Dams", The Int. J. on Hydropower & Dams, Issue two, 1997

Kaufmann, A. (1977): "Introduction a la theorie des sous-ensembles flous, in Fuzzy Set Theory", Masson, Paris, 1977

Kisliakov, D (1997): "Seismic Risk Analysis of Concrete Gravity Dams - Problems and Solutions", Proc. Int. Conf. Hydropower'97, Trondheim, Norway, 30.June-2.July 1997, Broch, Lysne, Flatabö & Helland-Hensen (Eds.), Balkema, Rotterdam, 1997

Kisliakov, D. (1998) "Once Again on the Limits of Structural Analysis of Dams", National Sc. Conf. with Int. Participation "Water Resources - Use and Protection", 23-25. Sept. 1998, Institute of Water Problems, BAS, Sofia, Bulgaria, 1998

Kisliakov, S. (1981): “Stokhasticheski metodi v prilozhnata mekhanika” (in Bulgarian), Tekhnika, Sofia, 1981

Kisliakov, S. (1996): "Chaotic Oscillations in a Stochastic Environment", Chaos, Solitons and Fractals, Vol.7, No.10, 1996

Lafitte, R. (1993): "Probabilistic Risk Analysis of Large Dams: Its Value and Limits", Int. Water Power & Dam Construction, Vol.45, No.3, 1993

Lombardi, G. (1995): "Des Limites de l'Analyse dans l'Etude des Barrages", In: Proceedings of the International Conference : Research and Development in the Field of Dams, Crans-Montana, Switzerland, 7-9. Sept. 1995, Swiss National Committee on Large Dams, 1995

Lombardi, G. (1996): “On the limits of structural analysis of dams”, The Int. Journal on Hydropower & Dams, Issue Five, 1996

Malakhanov, V.V. (1991): “Sovershenstvovanie proektirovaniya gidrosooruzheniy” (in Russian), Sovremennie problemi gidrotekhniki, A.N. Rasskazov, G.A.Vorobyev (eds.), Moskva, 1991

Nielsen, N.M.; Vick, S.G. & Hartford, D.N.D. (1994): "Risk Analysis in British Columbia" Int. Water Power & Dam Construction, Vol.45, No.8, 1994

Platt, D.G. & Blockley, D.I. (1994): “Process Modelling in Civil Engineering Design”, Design Studies, Vol.15, No.3, 1994

Ragot, J. & Lamotte, L. (1993): "Fuzzy Logic Control", Int. J. Systems Sci., 24, 1825-1848, 1993

Sanchez-Silva, M.; Blockley, D.I. & Taylor, C.A. (1996): “Uncertainty Modelling of Earthquake Hazards”, Microcomputers in Civil Engineering, Vol.11, 1996

Schuster, H. G. (1984): "Deterministic Chaos", Physik-Verlag, Weinheim, 1984

Van Oortmerssen, G. (1991): “Assessment of the Validity of Computer Models for the Prediction of the Dynamics of Floating Structures”, In: W.G. Price, P. Temarel & A.J. Keane (eds.), Dynamics of Marine Vehicles and Structures in Waves, Elsevier, London, 1991

“Risk-Based Dam Safety Evaluations” (1998), Workshop Report, The Int. Journal on Hydropower & Dams, Issue One, Two, 1998

Time Series Analysis of Water Quality Data

Albrecht Gnauck

Brandenburg Technical University of Cottbus, P.O.B. 10 13 44, D-03013 Cottbus, Germany

Abstract. *Interrelations between variables of freshwater quality were investigated by time series analysis methods. Especially wavelet transformations are used to identify the dynamic ecosystem behaviour. The basic idea is to analyse ecological processes according to time and scale. The original ecological signal can be represented in terms of wavelet expansions while data operations can be performed by the corresponding wavelet coefficients. Data taken from investigations of sediments of hypertrophic riverine lakes showed different interrelations between water temperature and soluble reactive phosphorus of pore water under aerobic and anaerobic conditions. Each time if the temperature of the upper sediment layer is higher than the temperature of the pelagic water the phosphate content of the overlying water increases and vice versa. Therefore, one topic of wavelet analysis was the qualitative identification of the water temperature-phosphate-sediment subsystem.*

Keywords. time series analysis, water quality, wavelet analysis, phosphate remobilisation, sediment interaction

1 Introduction

Time series analysis methods have become standard methods for evaluation of results obtanied from measurements and observations in ecology for a long time (Patten, 1972; Thomann, 1972; Powell, 1995). Mostly, the time and space behaviour of ecological processes is analysed by Fourier analysis, correlation functions and spectral analysis (Straškraba, 1985). While they are based on linear system theory (Parzen, 1970; Box, 1994) practically useful results may also be obtained from linearisations of nonlinear systems (Dobesch, 1970). For the purpose of analysing time series different transformations were introduced. The most well known representative is the Fourier Transformation (FT). Applying FT on stationary signals some informations of the frequency content of the signal will be obtained. Because this condition is not satisfied in most cases adaptations and extensions must be introduced, such as Short-Time Fourier Transformation (STFT). But a point by point localisation of the frequency information could not be reached. The possibility to perform local analysis can be achieved with wavelets. The areas of analysis reaches from detecting discontinuities, breakdown points, self-similarities and long-term evolution up to compressing or filtering signals and proving on causality (Mallat, 1998; Hubbard, 1997; Misiti, 1997). By Wavelet Transformation (WT) the windowing technique, such as STFT, was continued. As a result time-scale representations of the signals can be given.

Since about 25 years ago, eutrophication of freshwater ecosystems has been recognised as an essential problem of water quality management. One problem is the formulation of nutrient budgets with changing loading rates. Under anaerobic conditions, mostly during summer time, the phosphorus release from sediment causes a secondary loading of the water bodies. During such periods of anaerobiosis, soluble reactive phosphorus fraction can change substantially (Bengtson, 1975; Canale, 1976; Ryding, 1977). In contrast to pelagic waters, inorganic and organic components of phosphorus and nitrogen show much higher values in sediments. Direct interrelations exist between sediment and water quality influenced by physical, chemical, and biological driving forces. dependant on these internal and external ecosystem conditions, the sediment layer works either as nutrient trap or as nutrient source. Phosphate remobilisation from sediment can be considered as a result of contradictory processes of matter changes (Lijklema, 1980). The depth of sediment layers changes between 1 mm and 15 cm (Hayes, 1955; Holden, 1961; Lijklema, 1983; Schettler, 1993; Furrer, 1993). One goal of the modelling framework consists in the quantification of nutrient remobilisation from sediments. As a first step, time series analysis methods are used to identify interrelations between essential variables of pelagic water and sediments. Time series are sequences of observations taken sequentially in time (Box, 1994). Intrinsic features of time series are that adjacent observations are dependant. Therefore, time series analysis methods are concerned with data analysis techniques. One of the important applications of these methods is the determination of a transfer function of a system subject to a given series of input data. In this paper, a qualitative modelling framework of phosphate remobilisation, influenced by water temperature is presented. In contrast to the classical procedures modern wavelet analysis is used.

2 Experimental Area and Data Material

The River Havel belongs to the greatest tributaries on the right hand side of the River Elbe. The source region lies in the south of the Mecklenburg Lake Area. The length of the River Havel is 325 km. The very small elevation difference between source of River Havel (63 m above sea level) and the mouth into the River Elbe (22 m above sea level) characterises a low land river. Fig. 1 shows a generalised representation of the experimental area.

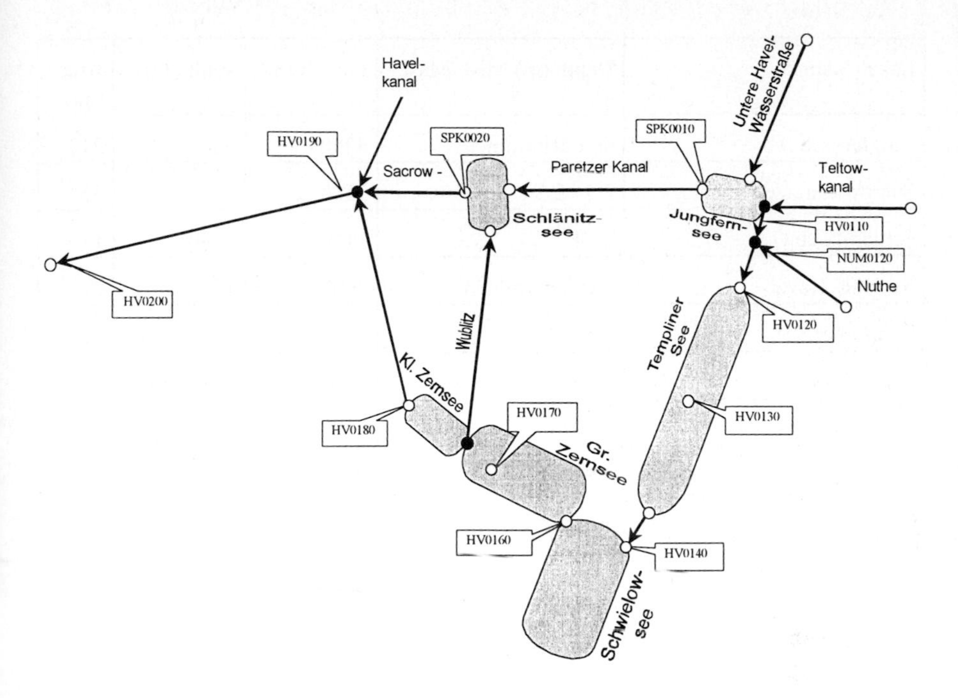

Fig. 1: Generalised graph of the experimental area

In this figure the shallow lake system of the River Havel is presented. Measuring points (blank points) and nodes of the hydrologic system (dark points) characterise the different parts of the river system. The numbers of the measuring points in the course of the river Havel are given in bubbles as indicated. Numerous hydraulic works and banked-up water levels influence the water levels and flows along the course of the river. In contrast to rivers in hilly mountain regions the slope of the River Havel is extremely low. For low-flow situations a slope of the water level of 2 cm/km is observed. The watershed is characterised by high evaporation rates. Only 25% of precipitation contribute to flow. Mostly, the active sediment layer of the River Havel is given by 2 cm to 6 cm. Table 1 contains some characteristics of the riverine lakes under investigation.

Lake/Channel	Depth (m) Med./Max.	Length (m)	Width (m)	Area (ha)
Jungfernsee	non determined	4250	500	212
Sacrow-Paretzer-Kanal	2.5 / 2.5	7500	50	35
Schlänitzsee	2 / 4	1100	1000	110
Ketziner Havel	non determined	3000	125	37
Templiner See	4.5 / 17	5500	1300	715
Schwielowsee	3 / 9	5200	1800	890
Werdersche Havel	non determined	4250	1000	425
Großer und Kleiner Zernsee	3.5 / 8	5000	1250	320

Table 1: Hydromorphological characteristics of riverine lakes of River Havel

In Table 2 a general overview on the characteristics of the river stretches and the main usage of the investigated area is presented.

Measuring point	Code	Characterisation
Potsdam-Humboldt Bridge	Hv0110	canalised river, influence of the region of Berlin
Potsdam-Kiewitt	Hv0120	canalised river, influence of the region of Potsdam
Templiner See	Hv0130	riverine lake, recreation area, tourism
Caputh	Hv0140	riverine lake, recreation area, tourism
Baumgartenbrück	Hv0160	riverine lake, recreation area, tourism
Werder	Hv0170	riverine lake, special agricultural area, tourism
Alt Töplitz/Phöben	Hv0180	riverine lake, agricultural area
Ketzin	Hv0190	canalised river, agricultural area
Brandenburg	Hv0200	canalised river, urbanised area

Table 2: Measuring points and areal characteristics of the River Havel

Fig. 2 shows the overall effect of reducing external loads. For the year 1995 the basic phosphate concentration is much lower than in 1990. The nutrient concentrations of effluents of sewage water treatment plants are diminished rigorously according to the outlines of German environmental laws (Klose 1995; Kalbe 1997). Furthermore, a reduced usage of fertilisers and changes in the land usage of the agricultural areas have diminished the amount of nutrients from diffuse pollution sources.

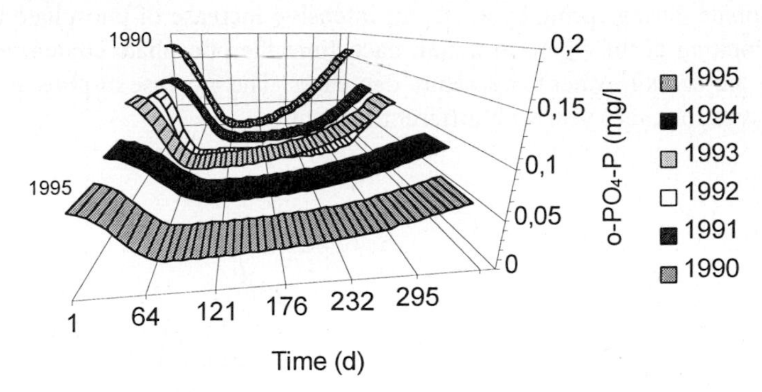

Fig. 2: Orthophosphate balances of the Lower River Havel

Along the course of the River Havel riverine lakes mainly cause changes of water quality levels. Different hydraulic conditions between canalised and riverine stretches influence the intensity and kinetics of the chemical and biological reactions. In Fig. 3 a comparison of orthophosphate concentrations of different sub-watersheds at Potsdam-Humboldt Bridge (canalised river stretch) and at Ketzin (riverine lake region) is shown. The amplitudes are much higher at measuring point Ketzin, which can be explained by a higher secondary load caused by remobilisation of phosphate from sediments (Schettler, 1993; Klose, 1995). For time series analysis samples are taken into consideration from 9 measuring points along the course of the river. Important intercorrelations found during experimental work were confirmed by an extended multivariate statistical analysis. Fig. 3 shows phosphate time series of different sub-watersheds.

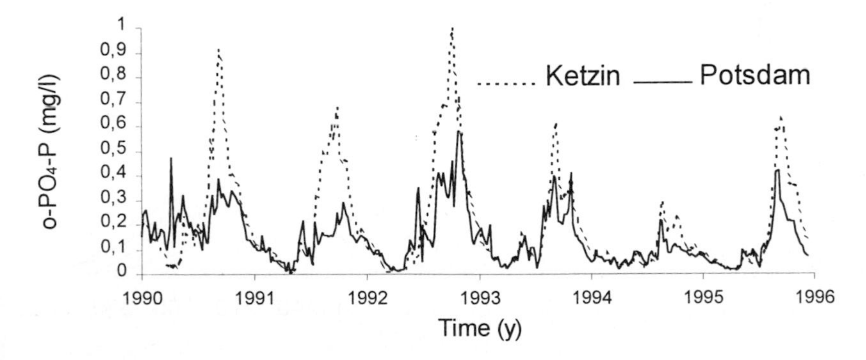

Fig 3: Phosphate concentrations for two different sub-watersheds

Special interest was given to the analysis of interrelations between water temperature and phosphate-phosphorus. For some measuring points the same time behaviour of water temperature and phosphate could be observed. Fig. 4 shows a typical overall effect between the variables considered. After the diminishing of phosphate during spring by algae, an intensive increase of phosphate takes place. But looking at the figure in detail, each time the phosphate content of the water body increases if water temperature decreases. The increase of phosphate could be observed for every year with different intensities.

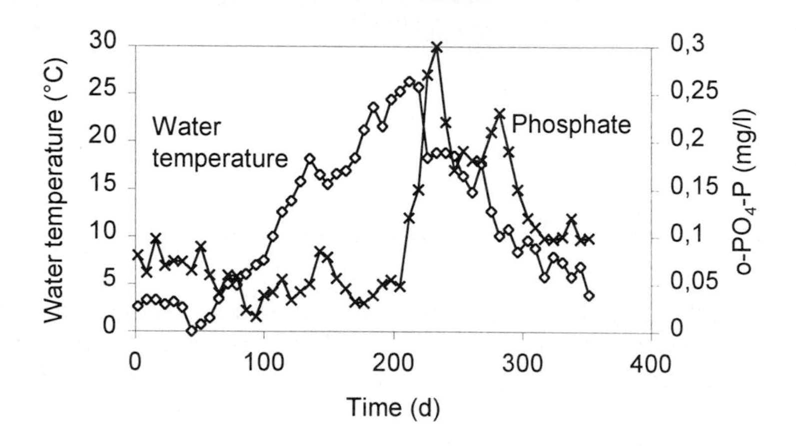

Fig. 4: Dynamic correlation between water temperature and phosphate-phosphorus

In Fig. 5 the time curves of the phosphate concentrations of pelagic water are shown. After the diminishing orthophosphate phosphorus concentrations caused by algae during the spring period, an enormous increase of phosphate concentrations in the late summer and autumn period could be observed.

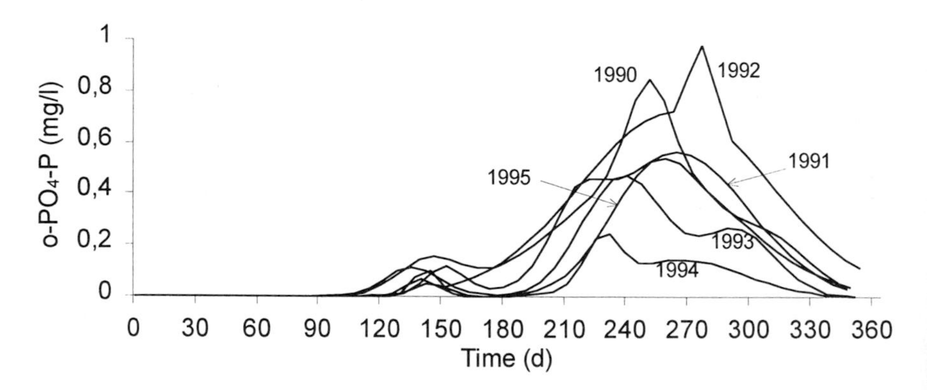

Fig. 5: Time-dependent behaviour of phosphate in shallow lakes of River Havel

3 Wavelet Analysis of Data

Wavelets are functions that satisfy certain mathematical requirements. They are used in representing data or other functions (Daubechies, 1988; Chui, 1992; Kaiser, 1994; Mallat, 1997). The fundamental idea of wavelet analysis is to identify signals according to time and scale. Temporal analysis is performed with a contracted, high frequency version of the prototype wavelet, while frequency analysis is performed with a dilated, low frequency version of the same wavelet. Because the original ecological signal can be represented in terms of a wavelet extension, data operations can be performed using just the corresponding wavelet coefficients.

3.1 Wavelet Transformations of Signals

Using the Continuous Wavelet Transformation (CWT), a lot of wavelet coefficients will be obtained (Mallat, 1997). The Discrete Wavelet Transformation (DWT) is comparable to Fast Fourier Transformation (FFT). The resulting coefficients are subsets of the set of coefficients obtained by CWT but just as accurate. A various kind of mother-wavelets and wavelet analysis methods are available. The choice of the best wavelet is mostly done manually. One has to check the desired information which should be obtained, the sort of signal and noise that has to be analysed and the algorithm of analysis (continuous, discrete, wavelet package).

3.2 Wavelet Decomposition of Signals

The general purpose of signal decomposition is to get the maximum amount of information stored in the time and space behaviour of the system components. In the case of riverine lakes, a wavelet of the Daubechies family was selected. In Fig. 6 the multi-level decomposition of the water temperature and is shown at level 5 with a of db3.

There is different information at each coefficient level. So one can remove one or more levels or analyse a level separately. In Fig. 6a the results for water temperature time series are given.

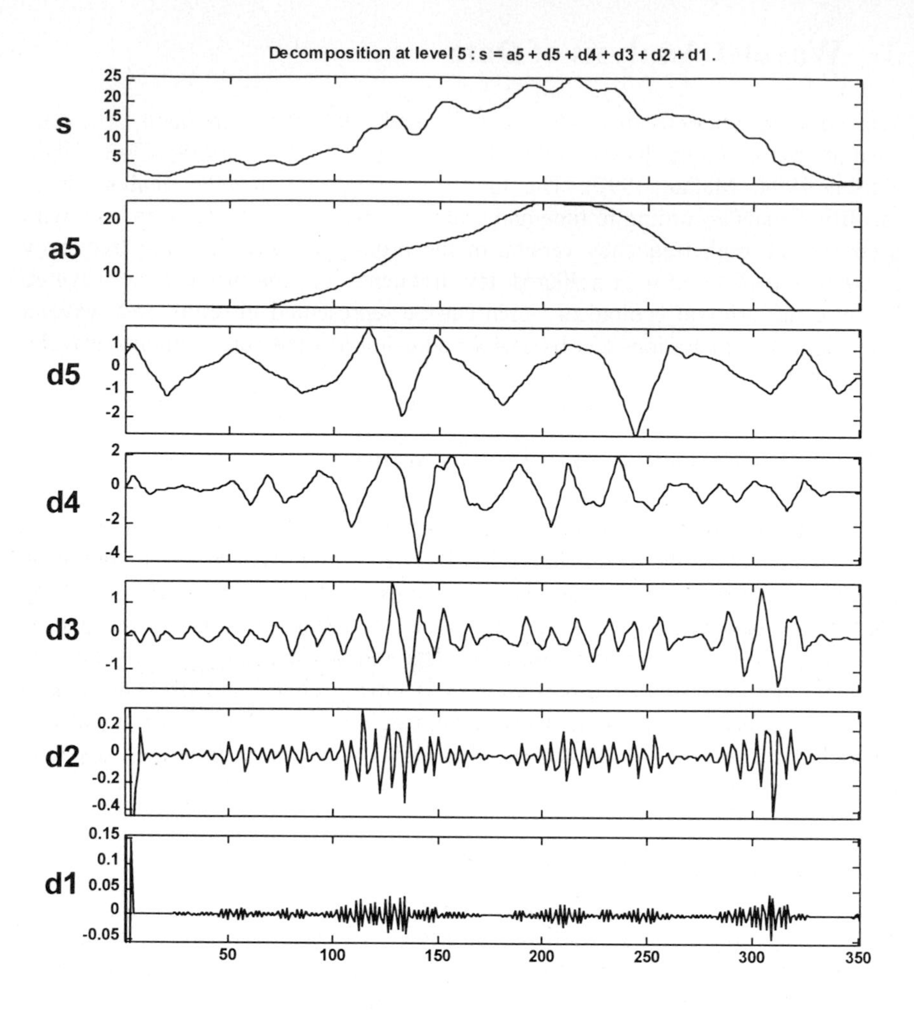

Fig. 6a: Wavelet decomposition of water temperature time series

The coefficient a5 represents the low-frequency content. It is called approximation while the coefficients d1 - d5, showing the high-frequency content, are called details. In the case of the time series of the River Havel, the coefficient a5 represents the seasonal component. Both coefficients d5 and d4 select the shorter time fluctuations of water temperature and orthophosphate. From coefficient d3 one can see artefacts of the approximation of raw data.

Fig. 6b contains the results of wavelet decomposition of the orthophosphate time series. They will be interpreted analogously. Before wavelet decomposition was executed the data were transformed equidistantly using cubic splines. Between longer unmeasured periods (at day 250 for orthophosphate and from day 135 to day 310 for water temperature) d3 detects such discontinuities.

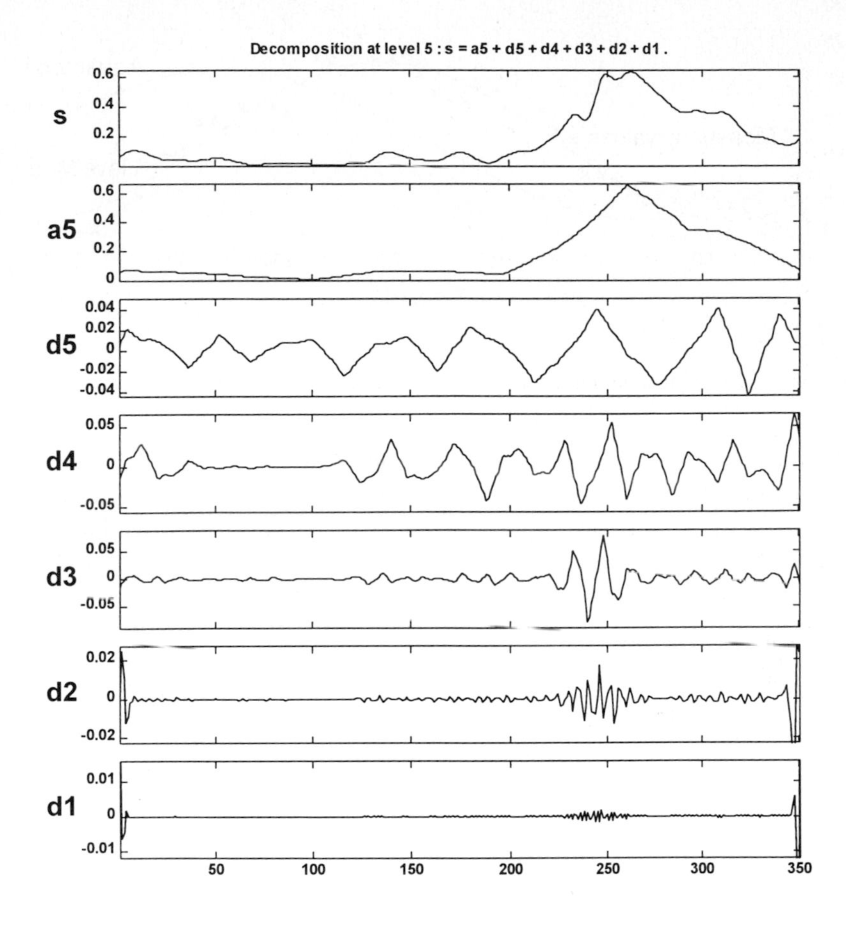

Fig. 6b: Wavelet decomposition of orthophosphate-phosphorus time series

3.3 Event-Based Ecomposition

The time-varying behaviour of water temperature was modeled by sinusoidal and by Gaussian functions. In Fig. 7 a comparison is made between the sinusoidal models of AQUAMOD 1 (Straškraba, 1985) and of HavelMod (Gnauck, 1998). The maxima of the water temperature curves show a difference of about 75 days. The warming up of shallow riverine lakes causes this time lag, which occurs faster than in deep reservoirs. On the other hand, the water temperature of shallow water bodies is decreased by cooling effects due to diminished air temperatures much more earlier than in reservoirs.

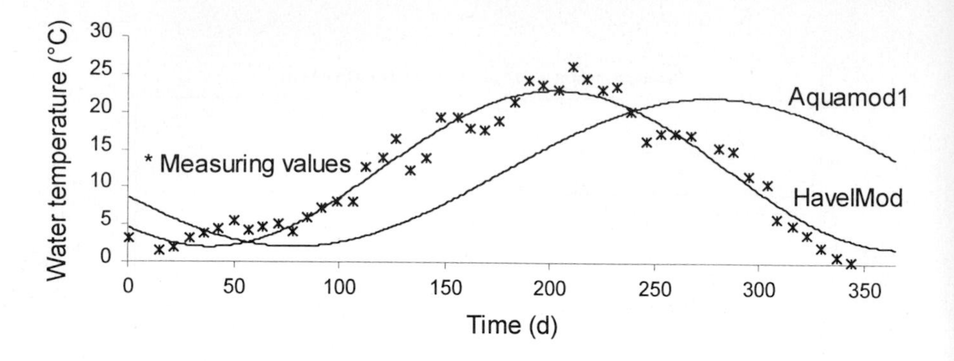

Fig. 7: Approximation of water temperature curves by sinusoidal functions

Fig. 8 shows the same time series fitted by different Gaussian functions. The approximation carried out using with four Gaussian functions (bell shaped) varying in height, width and position and one asymmetric Gaussian function varying in height, width, position and tailing. The parameters were fitted using a quasi-Newton least square algorithm.

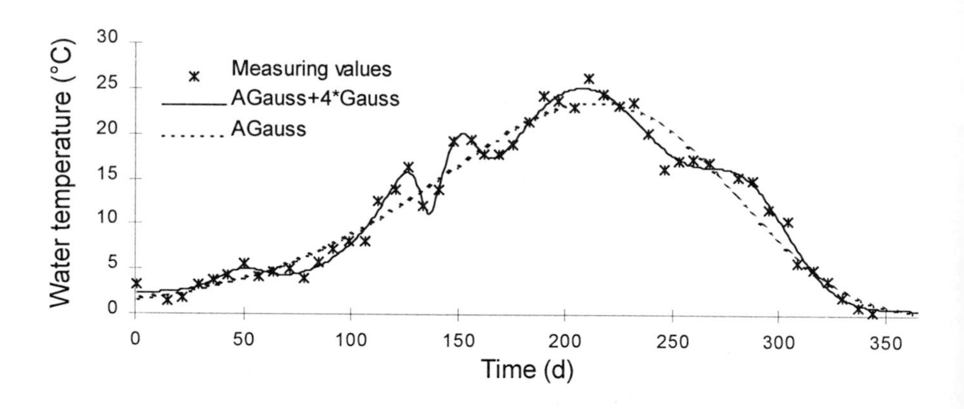

Fig. 8: Approximation of water temperature curve by Gaussian functions

Compared with the sinusoidal models the Gaussian models of water temperature show a better approximation. The same procedure was used to decompose the time series of orthophosphated phosphorus by bell shaped Gaussian functions. A distinction could be identified between the basic content of phosphate-phosphorus of the pelagic water body and the remobilised portion of phosphate-phosphorus. For some measuring points, the same results could be obtained where the remobilised portion grows up to 500% of the basic phosphorus content (Fig. 9).

After some small remobilisation phases in the first half of a year the second half is characterised by two events with very high remobilisation rates.

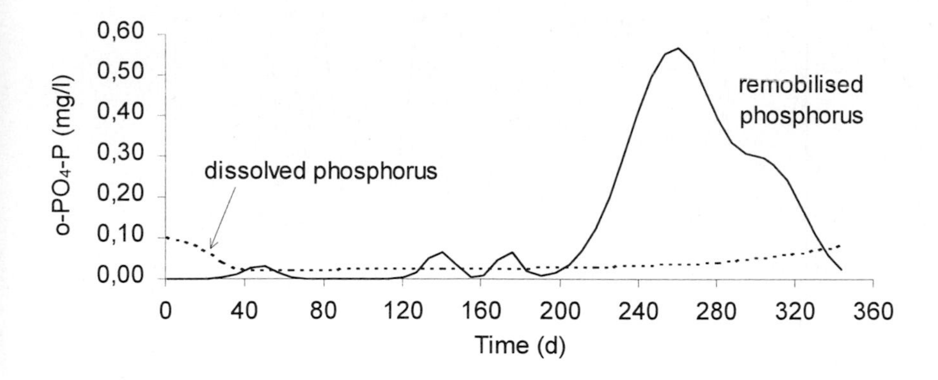

Fig. 9: Gaussian decomposition of phosphate-phosphorus for a riverine lake.

4 Results and Discussion

Investigations of sediments of different hypertrophic riverine lakes of the catchment by Schettler (Schettler, 1993) showed the interrelations between soluble reactive phosphorus (SRP) of pore water and pelagic waters under aerobic and anaerobic conditions. Under consideration of well-known chemical and microbial reactions the temperature-influenced turnover of nutrients was primarily supposed. For shallow riverine lakes of the River Havel the process of phosphate remobilisation is mostly described phenomenologically Mothes , Rohde , Schettler and Kalbe (Mothes, 1982; Rohde, 1983; Schettler, 1993; Kalbe, 1997) reported on temperature dependence of phosphate remobilisation in shallow lakes in North-East Germany. Schettler (Schettler, 1995) found analogous correlations during studies of shallow lakes in the watershed of the River Havel. With decreasing water temperature the phosphate remobilisation from sediment increases. No other input of phosphate could be observed by monitoring programs. Schettler (Schettler, 1995) stated that each time if the temperature of the upper sediment layer is higher than the temperature of the pelagic water the phosphate content of the overlying water increases and vice versa. Therefore, the correlation of the phosphate-sediment subsystem was studied by wavelet decomposition. Fig. 10 shows the comparison of both components at level 4 and 5. Between the warming up and cooling effects of the water body (dashed line) opposite events occur (solid line).

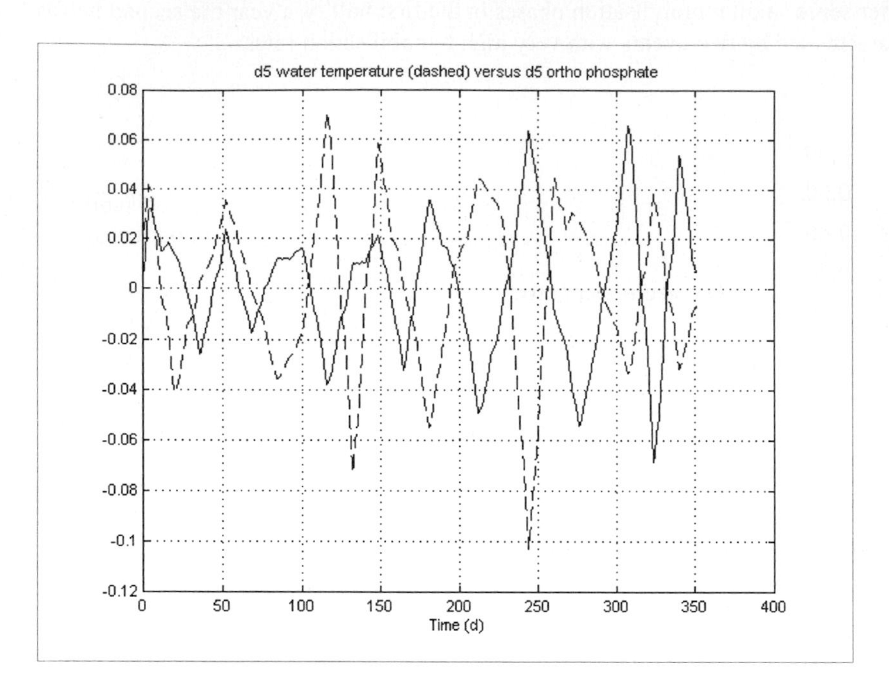

Fig. 10: Comparison of coefficients d4 and d5

In Fig. 11 this behaviour can be proven by the Residual Cross-Correlation (RCC). The highest absolute value could be found at lag 0 (negative).

From wavelet decomposition results the control of phosphate storage into sediment and phosphate remobilisation depends on the value and direction of the water temperature gradient. Analysing the gradients of the water temperature as a control variable of the phosphate remobilisation, four phases may be distinguished. In Fig. 12 an explanation of the different phases is given.

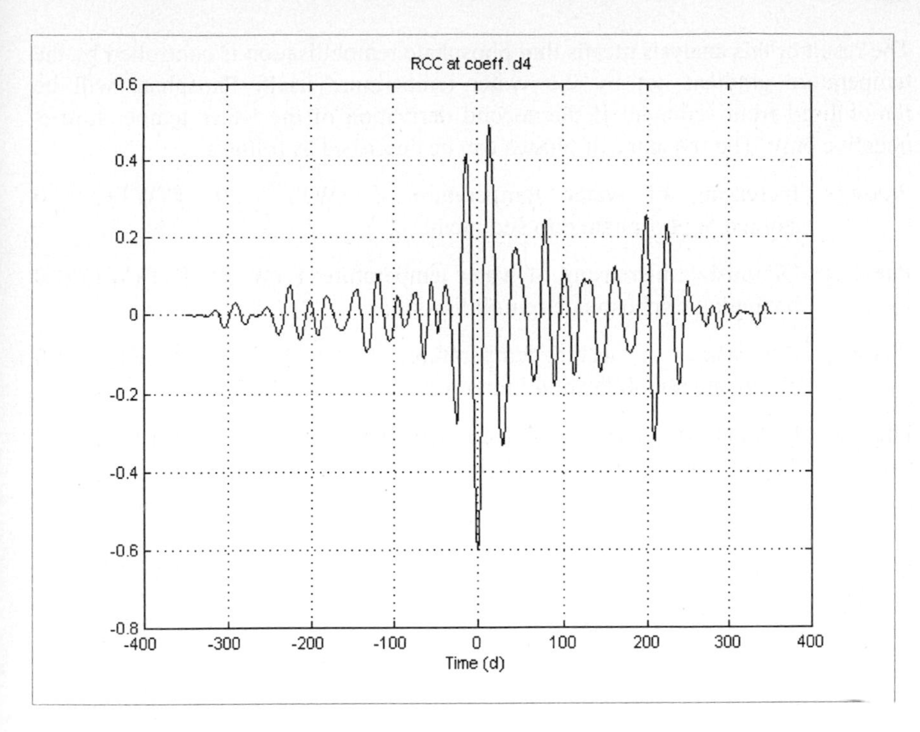

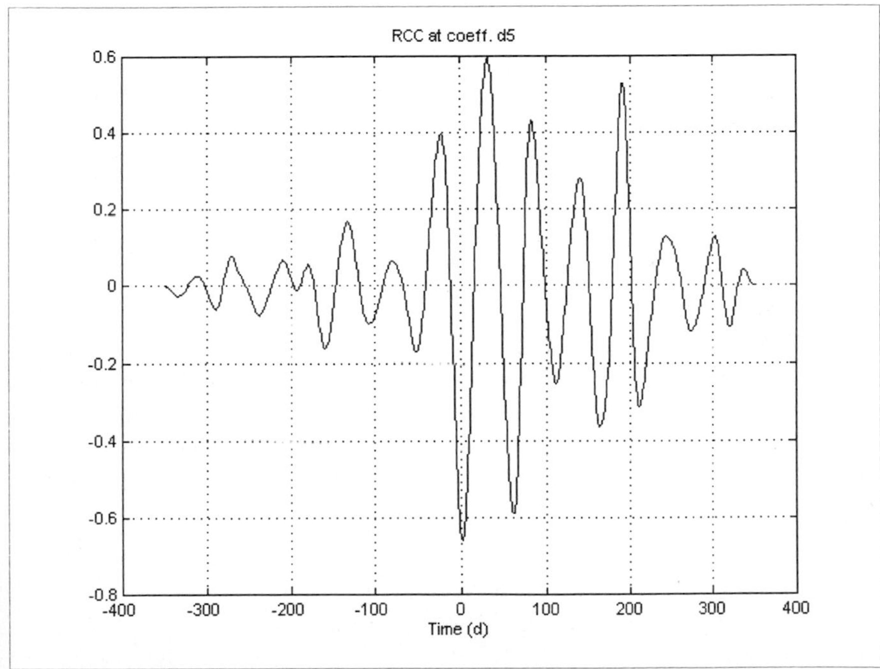

Fig. 11: Residual Cross-Correlation function for d4 and d5

The result of this analysis means that phosphate remobilisation is controlled by the temperature gradient not by the water temperature itself. Phosphate will be remobilised from sediment if the second derivation of the water temperature is negative only. The sequence of phases can be described as follows:

Phase 0: Increasing of water temperature: f' (WT) > 0, f''(WT) > 0 storage of phosphate into sediment;

Phase 1: Diminished increasing of water temperature: f' (WT) > 0, f''(WT) < 0 beginning of phosphate remobilisation;

Phase 2: Decreasing of water temperature: f' (WT) < 0, f''(WT) < 0 phosphate remobilisation increases;

Phase 3: Diminished decreasing of water temperature: f' (WT) < 0, f''(WT) > 0 stop of phosphate remobilisation.

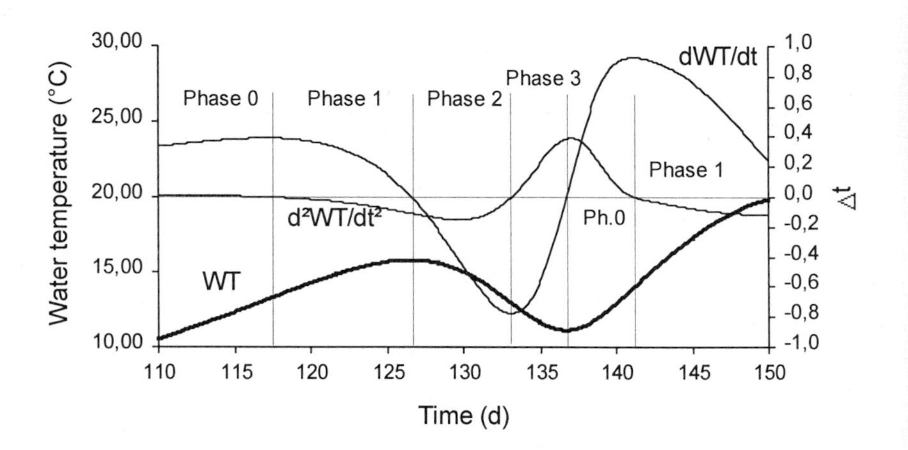

Fig. 12: Phases of phosphate remobilisation

For increasing water temperature the control process starts with phase 0. In the case of riverine lakes of the River Havel five events of phophate-phosphorus remobilisation from sediment could be find out (Fig. 13). Three events are too small for a measurable increase of phosphate in the pelagic water body. But two events show the strong influence of phosphate remobilisation from sediment onto the phosphorus content of the pelagic water zone.

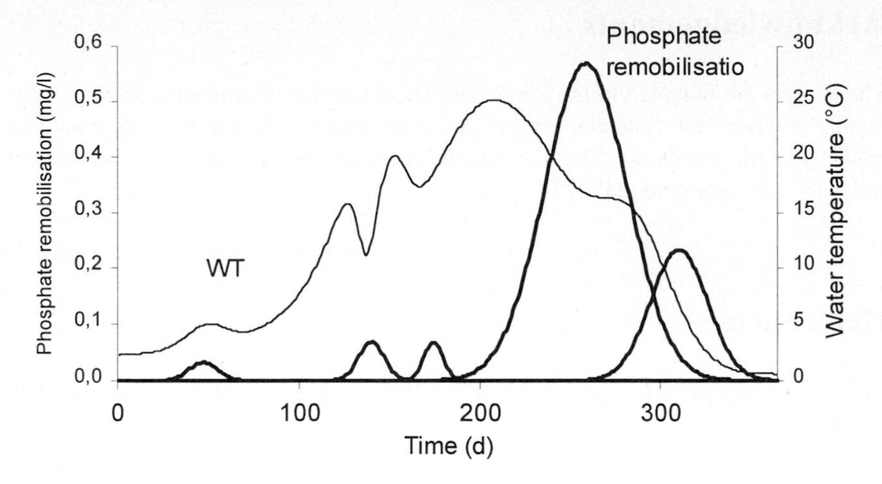

Fig. 13: Temperature control of phosphate remobilisation

5 Conclusions

Wavelet decomposition of water quality time series of the River Havel was used as a first attempt in the modelling framework of the River Havel. Time series models show the importance of diffuse pollution for riverine lakes. The amount of remobilised phosphate from sediment gives a strong nutrient basis for the eutrophication process. As a result of the modelling framework of the water quality subsystem the following statements can be made:

1. Phenomenological descriptions of the phosphate remobilisation process which causes diffuse pollution of riverine lakes are confirmed by wavelet decomposition of water quality time series.
2. The phosphate remobilisation process has to be included into water quality models for shallow riverine lakes and for river stretches.
3. Because of the dependence of temperature changes from other environmental changes, the process of diffuse pollution must be analysed intensively and has to be considered for sanitation programmes and regional developments in the catchment area.

Acknowledgements

The authors are indepted to Dr. Kalbe and Dr. Klose for helpful discussions. Also, thanks is given for financial support to carry out the research work and data analysis to the Fonds der Chemischen Industrie of the Verband der Chemischen Industrie e.V., grant no. 0400777.

References

Box, G.E.P.; Jenkins, G.M. and Reinsel, G.C. (1994): Time Series Analysis. Prentice Hall, Englewood Cliffs

Canale, R. P. (ed.) (1976): Modeling Biochemical Processes in Aquatic Ecosystems. Ann Arbor Science, Michigan

Chui, C.K. (ed.) (1992): Wavelet Analysing and its Applications. Academic Press, Boston

Daubechies, Ingrid (1988): Orthonormal Bases of Compactly Supported Wavelets. Comm.Pure Appl. Math., Vol 41, pp. 906-966

Dobesch, H.; Sulanke, H. (1970): Zeitfunktionen - Theorie und Anwendung. Verlag Technik, Berlin

Foster, I.; Gurnell, Angela; Webb, B. (1995): Sediment and Water Quality in River Catchments. Wiley, Chichester

Furrer, G.; Wehrli, B. (1993): Biogeochemical processes at the sediment-water interface: measurements and modelling. Applied Geochemistry: Environmental Geochemistry. Selected Papers from the 2nd International Symposium, Uppsala, Sweden, September, 16-19, 1991, pp. 117-119

Gnauck, A. (1998): Mathematical Modelling of Water Quality of Shallow Riverine Lakes. Ecol. Modelling (in prep.)

Hayes, F.R. (1955): The effect of bacteria on the exchange of radio-phosphorus at the mud-water interface. Verh. Int. Ver. Limnol. **12**: 111-116

Holden, A.V. (1961): The removal of dissolved phosphate from lake water by bottom sediments. Verh. Int. Ver. Limnol. **14**: 247-251

Hubbard, B.B. (1997): Wavelets, Birkhäuser Verlag, Basel

Kaiser, G. (1994): A Friendly Guide to Wavelets. Birkhauser, Boston, pp. 44-45

Klose, H. (1995): Die Eutrophierung der Havel und ihr bestimmender Einfluß auf Ökosystem und Nutzungen. In: Landesumweltamt Brandenburg (Hrsg.): Die Havel. Studien und Tagungsberichte, Bd. 8, Potsdam pp. 16-32

Lijklema, L. (1980): Interaction of orthophopshate with iron[III] and aluminium hydoxides. Environm. Sci. Technol. **14**: 537-541

Lijklema, L. (1983): Internal loading. Water Supply. **1**:35-42

Mallat, S. (1998): A Wavelet Tour of Signal Processing, Academic Press, London

Misiti, M.; Misiti Y.; Oppenheim, G.; Poggi, J.M. (1997): Wavelet Toolbox for Matlab®, The MathWorks, Inc.

Mothes, G. (1982): Sauerstoff- und Phosphatgehalt im Müggelsee als Indikation für Austauschprozesse. Acta hydrophys. **27**. 218-224

Parzen, E. (1970): Empirical Time Series Analysis. Holden Day, San Francisco

Patten, B.C. (Ed.) (1972): System Analysis and Simulation in Ecology. Vol. II, Academic Press. New York

Patil, R.V.; Pielou, E.C.; Waters, W.E. (1971): Statistical Ecology. Vol. I, II, II. Penn. State Univ. Press. London

Powell, M.T.; Steele, J.H. (1995): Ecological Time Series. Chapman&Hall, New York

Rohde, E. (1983): Veränderungen der Beschaffenheitsparameter von Wasser und Sediment eines hypertrophen Flachsees unter Einfluß einer Entschlammungsmaßnahme. Diss. Fak. Bau-, Wasser- u. Forstwesen, TU Dresden

Ryding, S.-O.; Forsberg, C. (1977): Sediments as nutrient source in shallow polluted lakes. SIL-UNESCO Symp. Amsterdam

Schettler, G. (1993): Dynamik und Bilanz des Schadstoffaustauschs zwischen Sediment und Wasserkörper in umweltbelasteten Havelseen. DFG-Bericht, Bonn

Schettler, G. (1995): Die Sedimente der Havelseen und deren jahreszeitliche Dynamik. In: Landesumweltamt Brandenburg (Hrsg.): Die Havel. Studien und Tagungsberichte. Bd. 8. Potsdam, pp. 46-57

Straškraba, M.; Gnauck, A. (1985): Freshwater Ecosystems - Modelling and Simulation. Elsevier, Amsterdam

Thomann, R.V. (1972): System Analysis and Water Quality Managment. McGraw-Hill, New York

In Situ Determination of Unsaturated Hydraulic Conductivity of Subsurface Soil in the Vadose Zone by the Technique of Modified Pressure Infiltrometer

Ali Mohammad Jafar[1], Yuji Takeshita[2] and Ichiro Kohno[2]

[1]Doctoral Student, Department of Environmental and Civil Engineering, Okayama University, Japan.
[2]Department of Environmental and Civil Engineering, Okayama University, Japan.

Abstract. *A simple and portable model which is based on real field situation and is comparable compatibility to the laboratory core sample model, the modified pressure infiltrometer technique, is introduced for the determination of field-saturated hydraulic conductivity, an essential parameter of the vadose zone. The field -saturated hydraulic conductivity,* K_{FS}*, is obtained from measurements of the steady flow rates. The field-saturated hydraulic conductivity is also obtained by this device using the falling head principle at the site of granite soil. The device is also used for field core sampling on which laboratory constant head, as well as falling head tests, are carried out. The field device is validated through comparisons to laboratory core sample experiments and other existing methods. This paper describes for the first time a versatile field device, which can be used for laboratory modelling and representing good performance for in situ determination of hydraulic parameters in a short time.*

Keywords. *In-situ testing, Laboratory testing, Field-saturated, Saturated Hydraulic conductivity, Matrix flux potential, Wetting front, 3D-flow*

1 Introduction

In situ determination of variably saturated sub-soil hydraulic parameter in real situations was, till now, a challenge for geo-technical engineers. Avoiding laboratory sampling disturbance and small scale laboratory specimens, field experimental procedure for finding hydraulic parameters in the soil of thevadose zone is one of the prerequisites for controlling and monitoring of applications such as the design of irrigation water systems for a particular area, pumping wells in water plants, drainage systems in building pits etc. Also, the gradual increasing global demands to protect the vadose zone water resource from present environmental problems due to pesticides and waste disposal. The most important hydraulic parameter of soil water movement is field-saturated hydraulic conductivity (K_{FS}). Any logical basis for characterising movement in the vadose zone must be based upon fundamental concepts of field hydraulic conductivity. Variable downward saturated flow through the porous medium is driven primarily by gravity, hydrostatic head and capillary force. Such flow is sometimes diverted in the vadose zone by barriers causing lateral transport, or accentuated by preferred pathways promoting rapid downward transport. Accurately predicting the attenuation and eventual location of solutes or constituents in the vadose zone are directly related to the wetting front movement of unsaturated flow of water. For actual measurement of water flow, behaviour in unsaturated soil, the measurement of hydraulic conductivity in the field is an essential task.

There are a lot of methods for the determination of hydraulic conductivity in laboratory and in the field. Nonhomogeneity and anisotropy of soils, fissures, tension cracks, and root holes commonly encountered in unsaturated soils, cannot be represented in small scale laboratory specimens. In reality, laboratory results of hydraulic conductivity are not representative values of the actual field situation. It is very difficult to simulate natural field conditions in the laboratory. Considering this point, field test results are more reliable for analysis of flow in the vadose zone. Among the direct and indirect field-methods for determination unsaturated hydraulic conductivity, ASTM standard guide D 5126-90 reviews alternative field methods for available techniques for measuring K in the vadose zone. In these techniques there are lots of limitations and assumptions for measuring hydraulic and transport properties at the unsaturated zone. Soil is assumed uniform, homogeneous and non-swelling in most of the experiments; but macropores, gradient in water content, soil bulk density, soil layering and changes in soil texture all occur near the soil surface, which can result in negative calculations of hydraulic conductivity. To minimise these limitations to some extent, our Modified Pressure Infiltrometer may be the most appropriate technique in the actual field situation for the determination of hydraulic conductivity. The main aim of our research is two-fold; firstly, to modify the existing device to make it more stable in the field situation and, secondly, to measure field hydraulic parameters in the real environment for predicting wetting front movement in the vadose zone by numerical simulation.

2 Methodology

2.1 Experimental Model

Our experimental model is a device designed to represent a simplified version of reality. This device can be used for the field experiment as well as in the laboratory experiment (see Fig.1 and Fig. 2). The previous device was in a different form. The Mariotte reservoir was fixed at the top of ring. Now the Mariotte reservior is separated from the ring. Consequently, the device becomes more portable and stable for the in-situ experiment. There is a possibility of disturbing the soil while the ring inserted into the soil surface; but by separating the Mariotte reservior, the degree of disturbance has been reduced to some extent. During field experiments, the soil is assumed incompressible by any force applyed from the surface. The possibilities of compressing the surface layer of the soil have been reduced in our modified arrangement.

The field model having the ring dimensions, 95mm inside diameter and 65mm depth of insertion, is attached with Mariotte reservoir which controls the positive constant head during in situ testing as well as laboratory core sample testing. The laboratory model is the modified assembling of field device. A bottom cap, a top cap and a spacer porous disk are assembled with the laboratory model. Here the constant head, as well as falling head principle, are applied. The Soil core sample having dimensions 95mm by 65mm, is be taken by inserting the ring into the soil surface of the vadose zone and laboratory testing is to be carried out by saturating soil sample and keeping the same field conditions. The clear stand pipe is very useful for recording the falling head field and laboratory data.

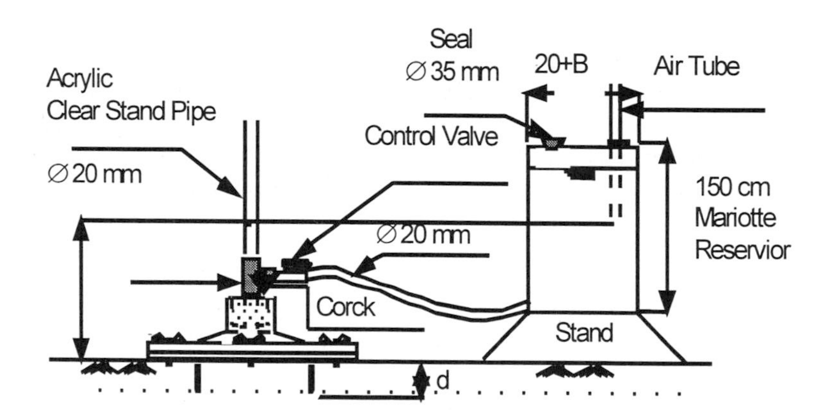

Fig. 1: Field Model of Modified Pressure Infiltrometer technique (Constant Head)

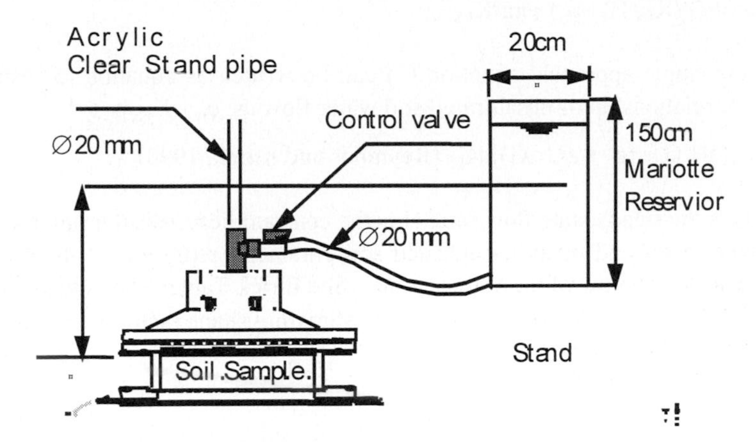

Fig. 2: Laboratory Model of Modified Pressure Infiltrometer technique (Constant Head)

For both cases, the field and the laboratory model, an air tube is used to control the constant head. The control valve is closed while testing the falling head in the field and laboratory.

2.2 Basic Equation

Considering a point source outside of a ring, an analytical expression for steady flow out of ring into rigid, homogeneous, isotropic, uniformly unsaturated soil is described as following:

$$Q_S=2\pi a(K_{FS}H+\phi_m)+\pi a^2K_{FS} \quad (1)$$

The above equation was developed considering a point source in a semi-infinite flow domain; but this was an approximation. Now, if the flow is considered within a ring, the flow geometry differs significantly from that of a point source.

The steady flow equation within the a ring is written as following:

$$Q_S= a/G\ (K_{FS}\ H+\phi_m)+\pi a^2K_{FS} \quad (2)$$

An implicit requirement of above equation is that no surface ponding (flooding) occurs outside the ring. The presence of a wetting front on the surface is, however, permissible and expected.

Now the field-saturated hydraulic conductivity, (K_{FS}), can be calculated by two approaches such as multiple head and single head. For multiple head approach, equation (2) can be written as:

$$Q_{s1}= a/G\,(K_{FS}\,H_1+\phi_m) +\pi a^2 K_{FS} \tag{3}$$

$$Q_{S2}= a/G\,(K_{FS}\,H_2+\phi_m) +\pi a^2 K_{FS} \tag{4}$$

For the single approach, equation (2) can be written as equation (5) using an important relationship of unsaturated soil water flow as $\phi_m= K_{FS}/\alpha^*$

$$Q_s =[(aH/G +\pi a^2 + a/ (\alpha G)]\,K_{FS} \text{ (Reynolds and Elrick, 1990)}, \tag{5}$$

where Q is the steady state flow rate, H is the constant head, a is the ring radius, G is a shape factor and α, an unsaturated soil parameter representing the capillary component of 3D-flow of the vadose zone. (See Elrick Table. 3 for site estimation of Alpha). Equation (2) can be written, for estimating shape factor, G as:

$$G= a(\,H\,K_{FS}+\phi_m)/(Q_s-\pi a^2 K_{Fs}) \qquad \text{Where} \quad H\geq 0 \tag{6}$$

The values of G were determined numerically substituting the proper values of K_{FS}, ϕ_m, H, d, and Q_S values from a numerical simulation. It was found that G is independent of H greater than 0.05m. (Positive head on soil surface). W.D. Reynolds and D.E. Elrick developed an empirical relationship among the shape factor, G , ring radius, a, and depth of ring insertion, d, into the soil surface as following.

$$G = 0.316\,d/a + 0.184 \tag{7}$$

This relation is also used for our calculation of hydraulic conductivity in the field.

3 Theory

In the vadose zone, soil hydraulic properties that control the flow into unsaturated soil from the Modified Pressure Infiltrometer are the field saturated hydraulic conductivity, K_{FS} and the unsaturated hydraulic conductivity, pressure head relationship, $K(\psi)$. The K_{FS} value applies to the field-saturated bulb and $K(\psi)$ relationship applies to the wetting zone. In unsaturated soil, hydraulic conductivity is significantly affected by combined changes in the void ratio and the degree of saturation of the soil. Water can be visualised as flowing only through the pore space filled with water. The air filled pores are non-conductive channels to the flow of water. Due to entrapped air in the field-saturated bulb, K_{FS} is less than the soil's truly saturated hydraulic conductivity, K_s, by a factor of two or more (Stephens et al., 1987, Constanz et al., 1988).

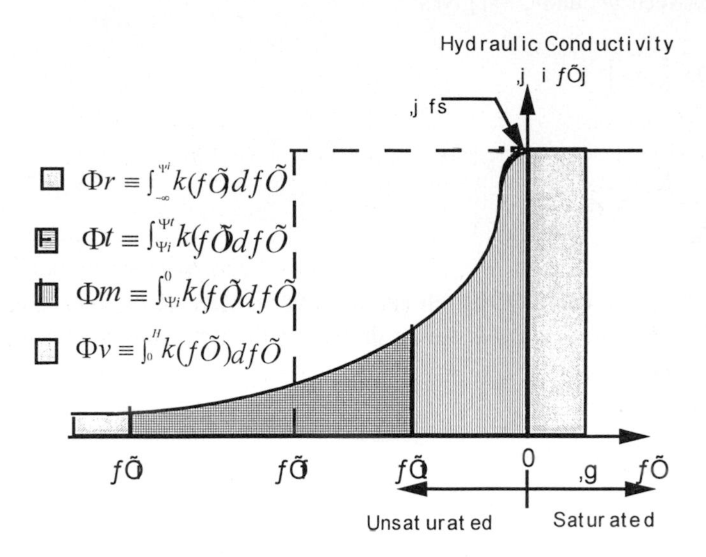

Fig. 3: Hydraulic conductivity and it's relation with various hydraulic parameters

Three of the most important factors governing liquid transmission in unsaturated soils are field-saturated hydraulic conductivity, K_{FS}, metric flux potential, ϕ_m, and sorptivity, S. K_{FS} or field-saturated hydraulic conductivity refers to the saturated hydraulic conductivity of soil containing entrapped air. K_{FS} is more appropriate than the truly saturated hydraulic conductivity for the vadose zone investigations.

The $K(\psi)$ relationship, known mathematically as the Kirchoff transform, has been shown by Gardner (1958) and others to be particularly useful for describing soil water flow (Fig. 3). Matrix flux potential, $\phi_{m,}$ is a measure of the relative importance of gravity and capillary for soil-water movement in a particular soil. Fine-textured soils, where capillary tends to predominate, have small Alpha values; and coarse-textured soils, where gravity effects manifest themselves more readily, have large Alpha values (Fig. 4). Theoretically Alpha (α^*) parameters can be explained by the Gardner (1958) exponential function:

$$K(\psi)=K_{FS}\exp(\alpha\psi) \text{ with } 0<\alpha<\infty \text{ and } \psi\leq 0 \quad (8)$$

Where α (m^{-1}) is an unsaturated soil parameter simply called the Alpha parameter. The general flux potential ϕ ($m^2.s^{-1}$) is defined as:

$$\phi(\psi)\equiv \int_{\Psi_i}^{\psi} K(\psi)d\psi \quad \text{Where } \psi_i \leq\psi \qquad (9)$$

Integrating between ψ_i and $\psi = 0$ gives:

$$\phi_m = \frac{K_{FS}}{\alpha}\left[e^{\alpha\psi}\right]_{\psi_i}^{0} \tag{10}$$

$$\alpha = \frac{K_{FS} - K_i}{\phi_m} = \lambda_c^{-1} \tag{11}$$

λ_c is the macroscopic capillary length (Philip, 1985). Here K_I is very small in comparison with K_{FS}, neglecting K_I for air dry soil in the field situation we get the following relation.

$$\alpha \approx \frac{K_{fs}}{\phi_m} \equiv \alpha^* = \left|\psi_f\right|^{-1} \tag{12}$$

ψ_f is the effective wetting front potential for Green and Ampt (1911) infiltration. α^*, α, λ_c, and ψ_f all of which are equivalent and represent single parameter estimates of the unsaturated hydraulic conductivity.

The values of Alpha depend on the soil type and the flow behavior is shown by an ideal line.

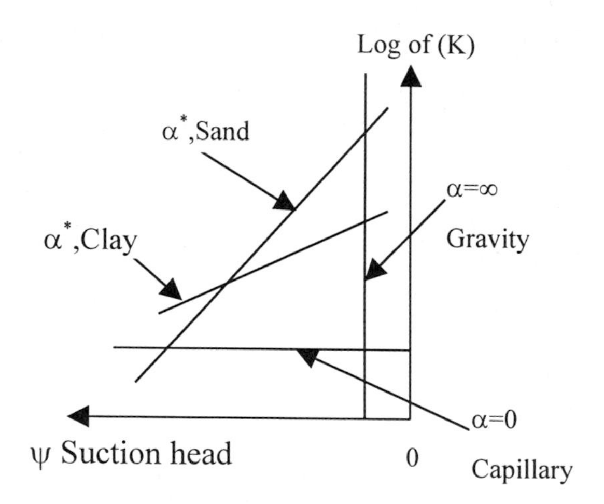

Fig.4: An Ideal relation of Suction head ~ lnK. Slopes indicate values of Alpha.

4 Site for In Situ Experiment and Soil Index Properties

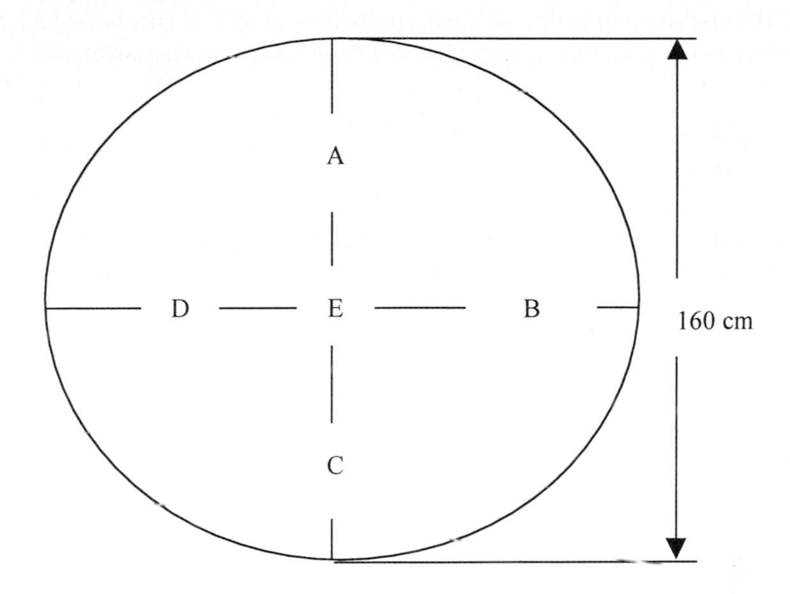

Fig. 5: Layout plan for an in situ experiment on granite soil at Okayama, Japan.

The site was selected at Okayama University campus. A hole with a 160cm diameter and 150cm deep was dug. By bringing granite soil from another place, it was deposited near the hole and to remove all gravel and stones having a diameter greater than 2mm, all the soil was sieved using a 2mm sieve. The hole was filled using this granite soil. During the filling of soil layer by layer, manual compaction was carried out. In this way, an artificial land having a cylindrical shape of 160cm diameter and 150cm deep was made by manual compaction at Okayama University campus. The site for field experiment was kept open under normal weather conditions for about six months. By natural rain fall and compaction this artificial land became a stable situation for field experiment. Five stations-A, B, C, D and E, having distance of 40cm from each other were chosen for the conducting in situ experiments. Soil index properties (i.e., soil grain properties and soil aggregate properties) had been checked and found as in Table-1. After checking the soil grain size, the Modified Pressure Infiltrometer device was set on the soil surface and the in-field temperature correction was carried out to adjust the field result. From the index properties of soil, it may be concluded that our experimental site is a medium loose and sandy soil. We also selected another place at dune sands of Tottori, Japan. The Soil at that location was a very dry and coarse grain sand. Soil index properties of dune sand were already checked.

Grain size analysis was determined by taking an air dry sample from the site of the granite soil from the artificial land made at Okayama University campus. There are a lot of methods for checking hydraulic conductivity from grain size curve. The prediction of hydraulic conductivity from grain size curves face a lot of limitations; but to get a preliminary idea, this grain size curve was developed.

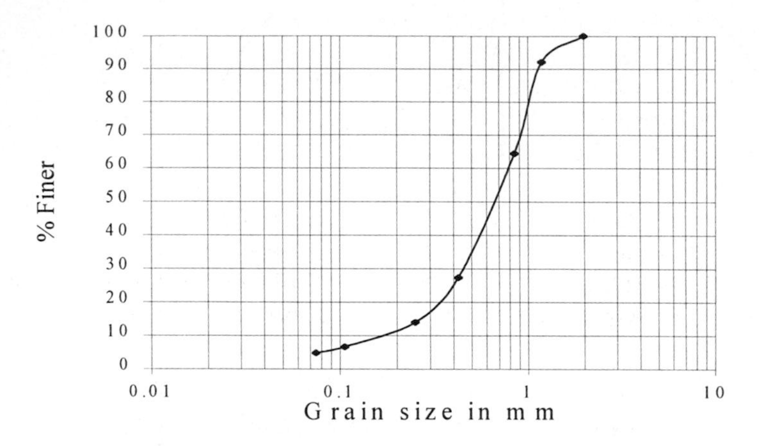

Fig. 6: Grain size curve of Granite soil of the site of in-situ experiment.

Description	Estimated value
Uniformity coefficient, C_u	4.8
Coefficient of Gradation, C_c	1.52
Bulk density (wet) ρ_b	1.83
Dry density, γ_d	1.66
Degree of saturation, S_r	84%
Specific Gravity, G_s	2.67
Void Ratio, e	0.608
Porosity, n	0.375

Table 1: Index properties of Granite soil.

5 In Situ Experiments

In situ experiments and laboratory experiments were carried out by Modified Pressure Infiltrometer technique by applying both the principles of constant head and falling head. Thc in situ experiments were conducted at two places. Firstly, a site was selected at Tottori Sand dunes and an experiment was carried out on natural ground of Grid land at the research site. Secondly, the experiment was conducted on a granite soil of artificial land at Okayama University campus. In both the cases, the initially soil conditions were dry but within 25 minutes a steady state flow was intriduced. In the case of Tottori dune sand, an instantaneous profile technique in the field was carried out to check the results of Modified Pressure Infiltrometer technique. In the case of granite soil, laboratory constant head method was carried out to compare the field results but it was difficult to control bulk density at the laboratory. Considering bulk density of field soil, a modified device of pressure infiltrometer is being developed to take intact soil cores (95mm in diameter and 65mm long) sample, then laboratory experiments on core sample werc carried out.

6 Calculation Approach

Considering the simultaneous equations (Multiple Head), the values of K_{FS} matrix flux potential, ϕ_m are calculated by solving Equations (3) and (4). Multiple head technique is on that equires independent measurements of hydraulic conductivity and flux potential are obtained; but limitations due to soil heterogeneity in the form of layering can give us unexpected values of field-saturated hydraulic conductivity (K_{FS}). But in our casc soil was uniform and homogeneous, so our results are acceptable.

Considering a single head approach and applying one positive head and site estimating the Alpha parameter, field-saturatcd hydraulic conductivity is calculated by using Equation (6). In the single head technique, only one water potential need to be applied to the infiltration surface of unsaturated soil. Although, this tactic avoids the occurrence of negative hydraulic conductivity and flux potential value. It also requires the independent measurement or estimation of Alpha (α^*) parameters. The previous study suggests that the single head pressure infiltrometer method yielded a field-saturated hydraulic conductivity (K_{FS}) which is usually accurate to within a factor of 2 when Alpha (α^*) is site-estimated and selected from the categories in Table 4 (Elrick et. al., 1989).

7 Experimental Results and Discussion

At first av disturbed sample was taken from experimental site and several tests were carried out by applying conventional laboratory methods of applying constant head. From these results it is clearly evident that the values of laboratory results are near about 10 times greater than actual field results. In the case of the conventional laboratory test the soil sample was fully saturated and situation was completely different from the real field; but in the field the soil was only partially saturated due to entrapped air. Later on, the Modified Pressure Infiltrometer was installed on the granite soil and field data was recorded. A good relationship between steady flow rate (Q_S) and applied head (H) was found (Fig.7). The consistency in steady flow rate to applied head results in better performance than any conventional field experiment.

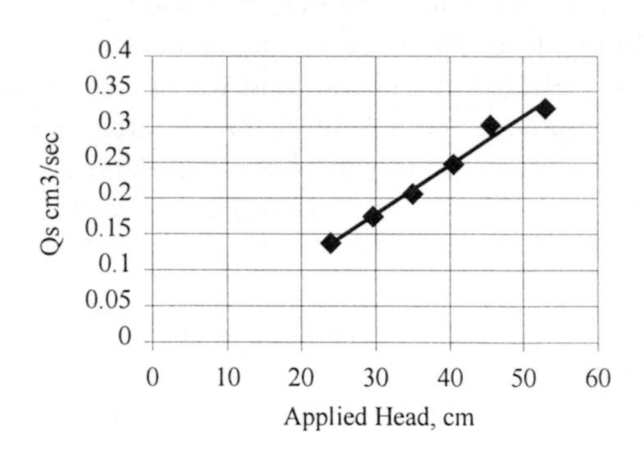

Fig. 7: Steady flow ~ Applied head at the field o Granite soil.

While the field experiment was being conducted there was no instability problem due to high head as the Mariotte reservoir is separated from the top portion of the ring which needs careful insertion into the unsaturated soil surface. In the previous technique of the Pressure Infiltrometer, the Mariotte reservoir was fixed on to the ring. As result it was very difficult to keep undisturbed conditions in the field during the ring insertion into the unsaturated soil surface of the vadose zone. Three forces are acting on unsaturated soil surface while the experiment is running in the field such as gravity force, hydraulic force and capillary force. When high head is applied, the hydraulic push has a great contribution on hydraulic conductivity. Constant head as well as falling head principles were applied in the field soil at the stations A, B, C, D, and E. Core samples were taken from each site and constant head as well as falling head principle were applied in the laboratory model.

Site	Constant Head, K_{FS} cm/s	Falling Head K_{FS} cm/s
A	4.5×10^{-4}	4.52×10^{-4}
B	4.9×10^{-4}	4.28×10^{-4}
C	5.68×10^{-4}	4.4×10^{-4}
D	5.58×10^{-4}	4.6×10^{-4}
E	6.12×10^{-4}	3.98×10^{-4}

Table 2: Field-saturated hydraulic conductivity.

Site	Constant Head, K_{LS} cm/s	Falling. Head, K_{LS} cm/s
A	6.846×10^{-4}	7.13×10^{-4}
B	6.646×10^{-4}	7.38×10^{-4}
C	6.480×10^{-4}	6.80×10^{-4}
D	7.25 x 10-4	7.50×10^{-4}
C	6.32×10^{-4}	9.68×10^{-4}

Table 3: Saturated hydraulic conductivity of core sample

To estimate the magnitude of the variation of the field and laboratory results at different stations, the mean value of K_{FS} and SDF (Standard Deviation Factor) was calculated. For the field experiment, a mean of K_{FS}=4.9x10^{-4}cm/s and SDF=0.0007, then for the laboratory experiment, a mean of K_{LS}= 7.2x10^{-4} and SDF= 0.0001. the laboratory results are 1.5 times greater than field results; this is because during the field experiment entrapped air created a disturbance in the flow of water within the pore space. The standard deviation is negligible because our experimental site was an artificial land of same type of granite soil. The standard deviation calculation of K_{FS}, based on multiple measurements at the site using the single-height approach is representative of the variability of the site.

Validity of the field results was checked by a laboratory core sample test; but the core sample test was conducted using a new laboratory model of Modified Pressure Infiltrometer technique. Therefore, we had to repeat the checking of the results of the core sample. It was done using conventional laboratory method. The mean value of thne field-saturated hydraulic conductivity (K_{FS}) of each station are compared to mean value of five core sample results. The trend line was drawn (Fig. 8). The trend line differs with the line of agreement by only about 5.5% which indicates a better performance of the Modified Pressure Infiltrometer for in situ experiments of unsaturated soil in the vadose zone.

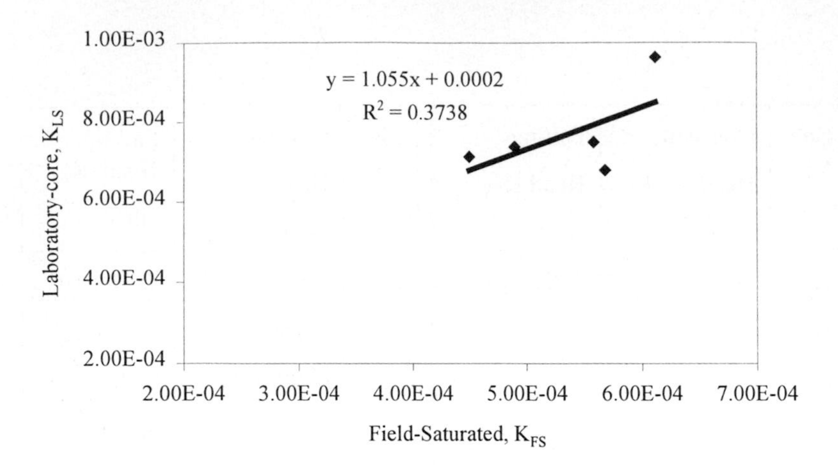

Fig. 8: Comparison of results of field-saturated hydraulic conductivity (K_{FS}) with the results of laboratory saturated hydraulic conductivity (K_{LS}) of core sample.

Calculation of field-saturated hydraulic conductivity for granite soil at our site was done by multiple head and single head approach.

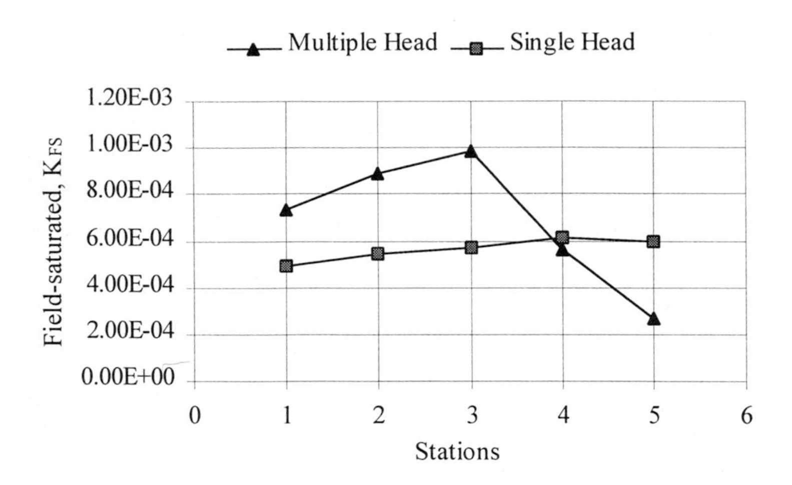

Fig. 9: Comparison of results of field-saturated hydraulic conductivity (K_{FS}) of 5 stations using multiple head and single head approach at the site of Granite soil.

It is observed from the results of the five stations at our site that single head approach of estimation of hydraulic conductivity is more consistence and less sensitive to soil variability (Fig. 9). For the granite soil of our site was homogeneous.

Soil texture/structure category	α^* cm^{-1}
Compacted structureless clayey soil	0.01
Fine texture clayey and unstructured	0.04
Most structure soils from clays through loam including fine sand	0.12
Coarse and gravelly sands	0.36

Table 4: (Elrick et.al., 1990)

Therefore, in both of the cases the results were acceptable; but single head results are more stable indicating good performance.

– Alpha selection for – In-situ test.	Results of K_{FS}
Elrick suggested value, α^*=0.36cm^{-1}	2.4x10^{-2}cm/sec
Estimated value, α^*=0.39cm^{-1}	2.7x10^{-2}cm/sec

Table 5: Alpha consideration

The Single head approach of the calculation of hydraulic conductivity needs independent measurement of the Alpha parameter or site estimation from Elrick, Table 4. The Alpha parameter was estimated (Fig 10) by fitting field data of lnK and suction head (ψ) of Instantaneous profile technique at the field of Tottri dune sand. The estimated value of Alpha parameter was found to be 0.39 cm^{-1}. According to Elrick`s (1990) suggestion(Table 4), the site estimation of Alpha value was 0.36 cm^{-1}.

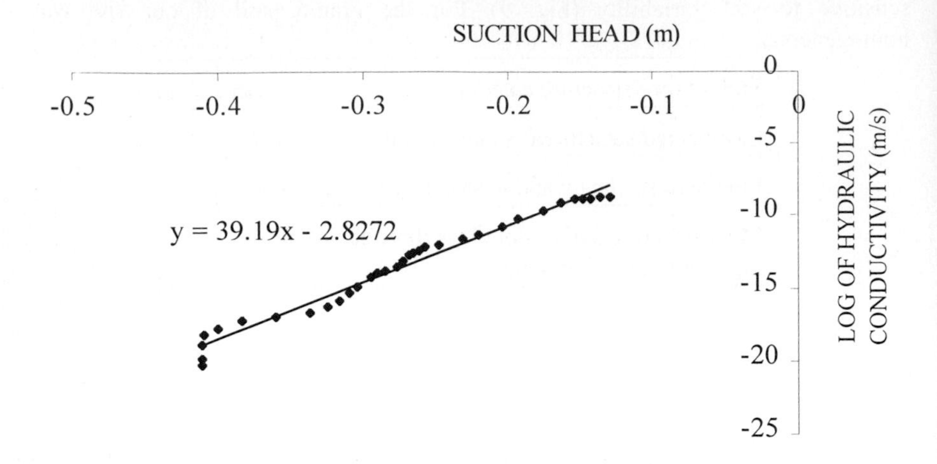

Fig. 10: Estimation of Alpha (α^*) for dune sand , Tottori, Japan.

The field-saturated hydraulic conductivity (K_{FS}) calculated by considering 0.36 cm^{-1} from Elrick table and by estimating from field data is shown in Table 5. The variation of the results of K_{Fs} is 12%, which represents acceptable values of Alpha in Table 4. Therefore, our estimation agrees with the suggested Table 4 of Elrick et. al., 1990.

The Influence of the Alpha parameter on K_{FS} was checked by using the field data of steady flow rate on granite soil at our site of artificial land. Our field parameters were as follows:

Steady flow rate (Q_s) = 0.22 cm^3/sec, Radius of ring = 4.75 cm, Positive head on soil surface = 50.6 cm, Shape factor (G) = 0.631 and the range of values of Alpha were from 0.1 to 1, assuming a rate of increment of 0.05. A relationship between field-saturated hydraulic conductivity and Alpha parameter was found (Fig. 11). From this relationship it is observed that the value of site estimation of Alpha parameter can assumed to be within a range from 0.25 to 0.30 cm^{-1}; because the calculated value of field-saturated hydraulic conductivity (K_{FS}) remains the same beyond the Alpha value of 0.3.

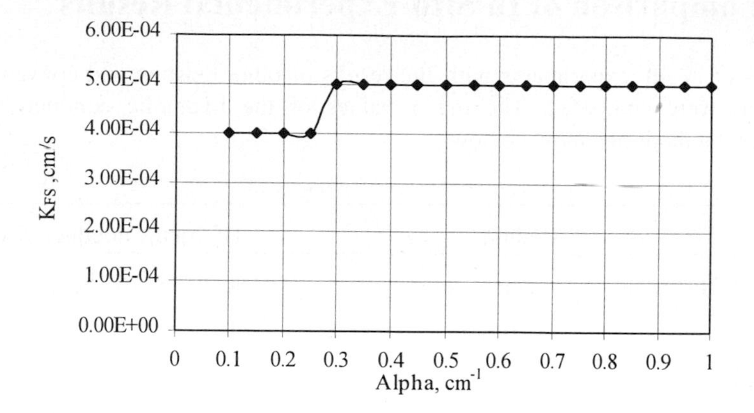

Fig. 11: Influence of Alpha values on the field-saturated hydraulic conductivity (K_{FS}) for granite soil, Okayama, Japan.

The influence of the Alpha parameter was also checked for the dune sand at Tottori, Japan.

The dune sand of our site has a high hydraulic conductivity and rate of steady flow was also very high. Steady flow rate, Q_s=10.72 cm^3/sec, Radius of ring = 4.75 cm, Positive head on soil surface = 23.50 cm, Shape factor (G) = 0.384 and the range of values of Alpha were from 0.01 to 1 assuming a rate of increment of 0.01. A relationship between field-saturated hydraulic conductivity (K_{FS}) and Alpha parameter is found (Fig. 12). From this relationship it is observed that site estimation of Alpha parameter for Tottori dune sand can be assumed to be 0.36 cm^{-1}.

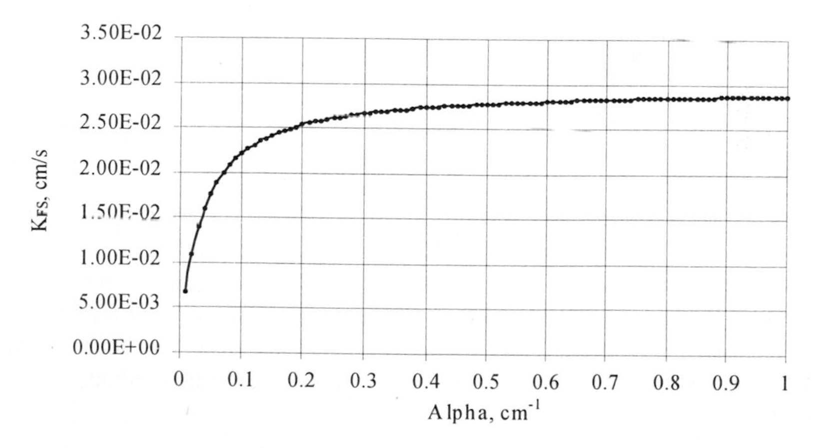

Fig. 12: Influence of Alpha values on the field-saturated hydraulic conductivity (K_{FS}) for dune sand, Tottori, Japan.

8 Comparison of In Situ Experimental Results

Results of in situ experiments with the results of other existing and conventional methods were compared. The mean values of the hydraulic conductivity of different methods are shown below:

Site	Methods	Hydraulic conductivity
Tottori dune sand	In situ testing by Pressure Infiltrometer	K_{FS}=2.8x10^{-2} cm/s
Tottori dune sand	Instantaneous technique	K_{FS}=2.0x10^{-2} cm/s
Okayama, granite soil	In situ testing by Pressure Infiltrometer	K_{FS}=5.3x10^{-4} cm/s
Okayama, granite soil	Laboratory P.I core-sample	K_{LS}=7.6x10^{-4} cm/s
Okayama, granite soil	Conventional Laboratory method	K_{LS}=3.6x10^{-3} cm/s

Table 6: Comparison of results of different methods.

The conventional laboratory method gives us a value 10 times higher than the field value. The soil was disturbed and conditions were different than core sample. Also the laboratory core sample result is 1.5 times greater than field result.

9 Advantages

1) The Modified Pressure Infiltrometer device can be used for determination of the field-saturated hydraulic conductivity (K_{Fs}) in the field as well as the saturated hydraulic conductivity (K_{LS}) in the laboratory. 2) The method is handy, less expensive and easy to install. 3) By using only one positive head (H), we can estimate field-saturated hydraulic conductivity (K_{FS}) within one hour. 4) The device is less susceptible to soil heterogeneity than other existing methods.

10 Conclusion

If we look at last two decades, it has become clear that the in situ measurement of soil hydraulic properties is essential for dealing with the complexities of water and solute movement in the field. It is therefore essential that accurate field methods be developed for measuring these properties. The main objective of this research work is to determine the field saturated hydraulic conductivity by developing a field model, which is compatible to laboratory methods. The field-saturated hydraulic conductivity (K_{FS}) of the sandy soil was determined and field results were verified by the Instantaneous Profile technique' and the 'Laboratory conventional constant head method'. The performance of the Modified Pressure Infiltrometer technique is reasonable for sandy soil. For granite soil field, results agreed with the results of core sample in laboratory, which gives us a good performance of the modified model. For core sampling, the ring thickness should be thin enough to minimize the disturbance of soil caused by displacement when ring is pushed into the soil. In addition, the simultaneous equation approach (multiple head solution) did not give us negative values of K_{FS}, demonstrating the reduced susceptibility of this device to soil heterogeneity. The new modified device is used for laboratory core sampling from which field bulk density (ρ_b), a very influencing factor for hydraulic conductivity, porosity(n), volumetric water content(θ) and degree of saturation(S_r), is estimated. Our new model may be useful in the field as well as in the laboratory applying the principles of constant head and falling head. The modified pressure infiltrometer technique is more versatile than any other conventional methods. As the method is handy and portable, we can create a large database for subsurface soil and this data base for hydraulic properties may be useful for computer simulations for monitoring the water movement and solute transportation in the vadose zone.

References

Soil Science Society of America, Special Publication Number-30, 1992

Soil science society of American Journal volume 53, no 2, March-April 1989, Estimation of unsaturated Hydraulic conductivity from field sorptivity measurement by White and K.M.Perroux

Water Resource Research Volume 30, 1994

Vadose zone Hydrology by Daniel B Stephen

Hands book of Vadose zone characterization and monitoring, Edited by L.G. Wilson, J .Glluen

Elrick, D.E.; Reynolds, W.D.; Parkin, G.W. and Fallow, D.J. (1992): Ponded Infiltration from rings and Auger holes: an historical perspective; Switzerland

Fallow, D.J.; and Elrick (1996): Field measurements of pressure heads with Guelph Infiltrometer. Published in soil sci. soc. Am J 60 ,1996

Reynolds, W.D. and Elrick, D.E. (1987): A laboratory and numerical assessment of the Guelph permeameter method. Soil Science, October 1987

Reynolds, W.D. and Elrick, D.E. (1990): Ponded infiltration from a single ring I. Analysis of steady flow, Soil Sci. Soc. Of American Journal 54 1233-1241

Reynolds, W.D. and Elrick, D.E. (1991): Determination of Hydraulic Conductivity using a tension infiltrometer. Soil Sci. Soc.of American Journal 55 : 633-639

Reynolds, W.D.; Elrick, D.E. and Clothier, B.E. (1985): The constant head well permeater: effect of unsaturated flow. Soil Sci 139 : 172-180

Reynolds, W.D.; Zebchuk and Topp, G.C. (1991): Field measurement of K_{sat} in a structured silty clay soil: Comparison of the Guelph permeameter and auger hole methods. Can J.Soil Sci. 71: 277

Stephens, D.B. (1992): Application of the bore hole permeameter p.43-68. In G.C.Topp et al (ed) Advances in measurement of soil physical properties: Bringing theory into practice. SSSA Spec. Publ. 30. SSSA, Madison, WI.

White, I. and Sully, M. (1987): A rapid in situ method for measuring soil hydraulic conductivity and field test of the bore hole permeameter. Water resource research 23 : 2207-2214

White, M. and Perroux, K.M.: Measurement of surface-soil hydraulic properties: Disc permeameter, Tension Infiltrometer and other techniques

Rule-based Spatial Query and Production as Part of the WHPA Simulation

Stefan Pühl and Peter-Wolfgang Gräber

Institute for Waste Management and Contaminated Sites (IAA)
Dresden University of Technology, D - 01062 Dresden, Germany
puehl@rcs.urz.tu-dresden.de; graeber@rcs.urz.tu-dresden.de

Abstract. *To allow spatial query and spatial production i.e. analysis in an Geographic Information System (GIS) environment a rule-based approach is suggested. It enables analysis based on spatial conditions of maps and additionally non-spatial facts. The rule base documents the spatial modeling process and can be reused for similar routine decisions. Examples form the delineation of well head protection areas (WHPA) illustrate this approach. The language for the rules is Lisp extended by spatial analysis functionality using a spatial vector structure (Open GIS Standard or ESRI shape files). It is implemented in Java.*

Keywords. *spatial analysis, rules, knowledge systems, GIS, Java, well head protection areas, groundwater simulation, decision support*

1 Introduction

1.1 Drinking-Water

Sensitive conflicts get a rise out of human beings when sustainable protection of water resources and its environment is taken into account. Often industry, commerce, farming, urban areas, increasing infrastructure, and unregulated tourism impair effective drinking-water protection.

Groundwater contributes the main part (64%) of the drinking-water production in Germany therefore groundwater protection is essential. The overall aim of protection is prevention –the same in groundwater protection. It prevents potential drinking-water from the contact with contaminating materials. Groundwater prevention is even more crucial because of very limited redevelopment chances; either extremely expensive or simply impossible.

1.2 Protection Areas

In the German state Saxony 68% of the drinking-water is groundwater. To protect the drinking-water there are 2055 protection areas, 13% of the whole state area; 2029 of them are for groundwater protection. Only 65% of these so called well head protection areas (WHPA) used for groundwater are fully delineated.

The delineation of WHPAs separated into three zones has direct social, economical, and ecological consequences in that area because of usability restrictions. This can lead to conflicts of interest involving water supply companies, land owners, and relevant authorities. In particular cases the scientific basis for these three protection zones is a hydrogeological report. This delineation is then enhanced by legal constraints, e.g. property rights, water laws, and groundwater regulations.

The relevant authorities in Saxony and in the other new states in East Germany have to check the protection areas delineated by the former GDR agencies. During this process, the need for further water supply, usability, and land development aspects are also considered. Hence, WHPAshave to be newly delineated if they cannot be enhanced.

Often, the relevant authorities do not have enough available capacity to do the delineation alone so that despite existing numerical models no simulations can be calculated. If this and other related problems are delegated, authorities must have tools to inspect efficiently the external solutions provided by engineering companies.

1.3 Decision Support System

The delineation of well head protection areas is based on data from biology, physics, chemistry, geology, hydrogeology, and engineering. Additionally, information from economics and administration is needed. Besides constant values (e.g. groundwater velocity) and matrix data (e.g. pump tests) there is a lot of spatial data (e.g. geology or simulation results like capture zones). This huge amount of data has to be collected, stored, and visualised for the decision maker. Normally, these decision makers are not computer experts so that a „friendly“ user interface and convenient automation is the key for acceptance.

The knowledge system component described here is embedded in a prototype of such a decision support system (Pühl, 1996) providing the WHPA groundwater simulation results, the HTML report structure, and the important interface to database and GIS functionality. It allows to draw conclusions based on these various types of data needed, especially to cope with spatial data.

2 Background and Overview

2.1 Rules

Rules are if-then statements or more formally actions A based on conditions C:

(C => A)

A set of these rules is called a rule base. The expert system using the rule base checks that every action A is performed if the condition C holds. In contrast to a sequential way of coding an operation, the order of the rules is irrelevant.

2.2 Query

There are two types of queries in a GIS: (1) database queries for attributes of a special spatial entity and (2) spatial queries to find relevant properties of the map: topology, relationships between spatial objects. The query represents the left hand side of the rule (LHS). It is the condition, a property of the map or additional parameters (facts). The performed action degrades to a message, a property report of the system describing the query results.

2.3 Production

Spatial production is creating new maps or new attributes. Often, these are results of combining different spatial and non-spatial data. Here obviously, production does not mean digitizing or scanning. Production is the right hand side of the rule (RHS). The action can create new map features, new maps or change them. It can take non-spatial facts into account and it can alter them, as well, using selected (LHS based) features.

2.4 WHPAs

The delineation of well head protection areas (WHPA) requires besides groundwater simulations (the spatial data describing different groundwater capture zones), additional non-spatial parameters and regulations, in a way standardized decisions. Therefore WHPA's are good candidates for a rule-based analysis approach combining query and production.

3 Questions

A naturally easy way to use rules is the implementation of regulations. These routine decisions are mostly well defined and consistent. With a rule-based approach questions and consequences like

1. Is the contaminant source in the WHPA?
2. For the sources the measurement accuracy is at least 1 m, if the groundwater velocity is greater than 5 m/d (higher risk).
3. How far away is the contaminant source located in the WHPA from the well?
4. If the groundwater velocity is smaller than 5 m/d, then combine all WHPAs and reduce them upstream to 600 m, calculated from one given well.

can be stated. After an outline of the rule-based concept, these questions are used to illustrate the approach providing „flying“ Lisp code based on the developed Java API.

4 Concept

The rules link the output to the input in a procedural and problem-oriented way: one rule base for one analysis procedure. Regulations or cartographic modelling operations are easily incorporated in a rule base. In short, the three computational steps are (as always)

- **Input** Maps, attributes associated with spatial entities (database), and additional knowledge (facts). These facts can be constants or derived parameters that do not necessarily depend on spatial attributes in the database.
- **Analysis** The checking or inference of the rule base (the procedural knowledge, in contrast to the non-procedural knowledge like maps or parameters) means applying the actions based on the queries, i.e. the productions to the given input. It is assumed that all rules are true.
- **Output** New maps, new facts, and additionally verbal descriptions are eventually produced by the rules.

Rules are a sophisticated way to ask the map, such as SQL is a way to ask for attribute and their values. For the database query the language SQL is practised, because normally a GIS operates with a relational database. For the spatial query a rule-based language (Lisp) is used. In SQL the results are tables. For spatial analysis the format is free, so a computer accessible format like XML or even HTML can be used besides new maps. This is convenient for CHI (= computer human interaction) as well as for further computer-based operations.

Yet the RHS of the rule can also be used for production, and not only for documenting the query results. SQL has a data definition language (DDL) and a data manipulation language (DML).The combination of both (e.g. select and create) makes this language very powerful. The rules have the same facility: conditions and actions. So a shortcut definition of analysis can be the very generic:

```
(query => production)
```

Lisp is used, because it is an established language for expert systems. Therefore only the new spatial functionality has to be documented, and the whole power of Lisp can be used with it. Normally, in SQL the database is created first (DDL) in order to retrieve the needle form the hay stack (DML). The rule-based analysis emphasizes the production (creation) based on query (selection) of given data. One does not use rules to create maps from scratch.

5 Query

The first example question was „Is the contaminant source is the WHPA?“. Using a pseudo code notation to separate the LHS from the RHS one may write:

```
(source in WHPA => message)
```

The message action qualifies this rule as a query. It is like looking on the map. The rule serves for computer recognition and further manipulation of the results. A rule that can really fly looks like this:

```
(defrule question-1 ""
    (point-shape (record ?ps-r) (OBJECT ?ps))
    (polygon-shape (record ?pys-r) (OBJECT ?pys))
    (test (point-in-polygon ?ps ?pys))
=>
    (printout t  "point #" ?ps-r " is in polygon #" ?pys-r crlf))
```

The condition contains the two needed objects: a *point* shape and a *polygon* shape. The slot (= variable or field of the shape) for the record holding the record number of the tested objects. The keyword test indicates a Boolean operation. The rule will be true, if every condition is sacrificed: there must be the particular shapes and the *point-in-polygon* function must be true. If so, a message is dropped.

6 Production

In the second example one wants to express a risk with „For the sources the measurement accuracy is at least 1 m, if the groundwater velocity is greater than 5 m/d.“. Again, a pseudo code notation might help to conceptualise the description as a rule:

```
(velocity > 5.0 => buffer with 1.0)
```

It is quite clear that the buffer is not up in the air. It has to be created in a certain *polygon* shape for further use, even if it is only for display purposes. A typical production needs to be performed.

The groundwater velocity is not a spatial entity. It is not coming from a spatial query like the example before. It is additional knowledge. This becomes possible because the rules are not limited to work on the map alone. Otherwise, we need to make at least two distinguished maps in advance, one for greater than 5.0 m/d, one for the rest. In a way the LHS is a query too, even if it is not a spatial one. To make it very plain, one can leave the condition open, i.e. the condition is always true:

```
( => buffer with 1.0)
```

This assures that the *buffer* is created. Then, the rule is nothing more than a statement (command) in a sequential way. The actual Lisp code for the second question as a rule looks like this:

```
(defrule question-2 ""
    (point-shape (OBJECT ?ps))
    (polygon-shape (name "test") (OBJECT ?pys))
    (test (> ?*gw-velocity* 5.0))
    (test (eq ?*done* false))
=>
    (bind ?*done* true)
    (set ?pys polygon (buffer point ?ps 1.0))
    (printout t  "Buffer, because v > 5.0 m/d" crlf))
```

There are the two shape files needed one for the *buffer* and one for the *points* to be buffered. In order to do it once (there is no additional condition to stop the buffering) a global done slot is tested. To facilitate the example the groundwater velocity is a global constant, like the done slot. The function used on the RHS of the rule

```
(set ?pys polygon (buffer point ?ps 1.0))
```

assigns the *buffer* directly for all *points* to the *polygon* shape called *"test"*. The slot in the condition of the rule tests for the name *"test"*. It shows the concept for accessing the shape: one can set all *polygons* at once (the whole shape) such as here or one can set individual *polygons* defined by the record number. E.g.

```
(set ?pys polygon ?py 0)
```

assigns a polygon *py* at the first record of the shape *pys*. Of course, the same is possible for the *points* and *arcs* as well.

7 The Combination of Query and Production: Analysis

The third question is „How far away is the contaminant source located in the WHPA from the well?". Obviously, the questions combines a query with an action (here a measurement). One might say measurement is a query, too. But if one applies it only for selected points, it becomes an action (which is a good performance gain). The pseudo code shows this

```
(source in WHPA => distance between well and source)
```

Again the real Lisp code is a little more clumsy:

```
(defrule question-3 ""
    (point-shape (record ?ps-r) (OBJECT ?ps))
    (polygon-shape (record ?pys-r) (OBJECT ?pys))
    (test (point-in-polygon ?ps ?pys))
=>
    (bind ?p (get ?ps point ?ps-r))
    (bind ?d (distance ?p ?*well*))
    (printout t  "point #"?ps-r " is in polygon #" ?pys-r crlf)
    (printout t  "with distance to the well: " ?d " m" crlf))
```

The LHS tests if the *point* is in the *polygon*, and the RHS performs the measurement action between the well (a point feature) and the *point* identified by its record number and of being in the *polygon*. Here, there is an example to access a particular point

```
(bind ?p (get ?ps point ?ps-r))
```

The new variable *p* gets the *point* from the shape *ps* with the record number of *ps-r*. If the record number is missing, the whole *point* shape is assigned, like in the *polygon* example in the last chapter.

8 Sequences and Functions

One might think at this point „Give me a break! I only need a few buffers, and that's it.“ or a selected measurement for the use in another application. After one has declared e.g. a *point* shape and a *polygon* shape

```
(defglobal ?*buffer* = (polygon-shape buffer))

(defglobal ?*points* = (point-shape test ascii))
```

one can use the spatial extended Lisp in this way, too, calculating buffers, areas, and distances

```
(set ?*buffer* polygon (buffer ?*points* 1.0))

(bind ?*py* (get ?*buffer* polygon 0))

(bind ?*a* (area ?*py*))

(printout t "Buffer area of #0: " ?*a* " m^2" crlf crlf)

(bind ?*p1* (get ?*points* point 1))

(bind ?*p2* (get ?*points* point 1))

(bind ?*d* (distance ?*p1* ?*p2*))

(printout t "  distance: " ?*d* " m" crlf)
```

There are no => anymore, indicating a rule-based execution. In short, the program calculates a *buffer* around a *point* and its area and prints the results. Then, it gets two points *p1* and *p2* from a *point* shape called *points*, calculates the distance between them, and delivers the results. Here, we are receiving the results on the screen. More sophisticated, this shall be ordered and put in a file for further use in a different application. Because the spatial results can be written in ESRI shapes they are immediately available e.g. in ArcView or Arc/Info.

In order to approach the last question „If the groundwater velocity is smaller than 5 m/d, then combine all WHPAs and reduce them upstream to 600 m, calculated from one given well.“, a function reduced-zone is written. This function is built with the analysis primitives in the following way:

```
(deffunction reduced-zone (?zone ?wells ?size)
    (bind ?uz (polygon-shape unified-zone))
    (bind ?wb (polygon-shape well-buffer))
    (bind ?rz (polygon-shape result-zone))
    (set ?uz polygon (union ?zone))
    (set ?wb polygon (buffer point ?wells ?size))
    (set ?rz polygon (intersect ?uz ?wb))
    (return ?rz))
```

The reduced zone *rz* is returned, so that the RHS of the rule can use the constructions

```
(defglobal ?*whpa* = (reduced-zone ?*zone* ?*wells* 600.0))
(?*whpa* writeASCII out)
```

to get the shape *whpa* and write it as an ASCII file. It is worth noting that the intermediate shapes like *uz*, *rz*, and *wb* are not saved. Thus, the function is a lot faster than the usual map creation for every step. The LHS of the rule looks much like the rule for question 2. The shapes zone and wells have to be declared there.

In the same way, the difference (minus)operation can be applied to create a function excluding certain well protected areas according to geological data. Functions like these are designed to form a library for the WHPA delineation regulations. It makes the reading and writing of rules more descriptive.

9 Application Programmer Interface (API)

One might directly access the functionality in another Java program. This is provided by a Java API. Basically, there are three parts: (1) the spatial data structure, (2) the spatial analysis functionality, and (3) the inference engine for the rules. There are two packages *grumio.format* for the geographical data structure and *grumio.spatial* for the analysis part. The third package is JESS (= Java Expert System Shell), the Lisp interpreter (F.-Hill).

9.1 Data Format

The spatial data structure consists of *points*, *multi-points*, *arcs*, and *polygons* and their shapes, basically a collection of the primitives with additional IO, e.g. ESRI shape file import and export.

```
GeographicPoint P = new GeographicPoint(3.0, 4.0);
PointShape ps = new PointShape();
```

Then one can read for instance into the empty *point* shape file with

```
ps.readShape("test");
```

a binary ESRI shape named *test.shp*. The database file (e.g. .dbf) for the attribute names is read separately.

9.2 Analysis Functionality

The spatial analysis part collects the functions such as area, *point-in-polygon*, and the set operations (union, intersection, and difference). It provides Java methods as user functions for the Lisp inference engine. And it can also be used directly as a Java API, e.g.

```
import grumio.format.*;
import grumio.spatial.*;
import java.io.*;

class AnalysisTester03 {
   public static void main(String args[])
      throws IOException
   {
      GeographicPoint[] P = new GeographicPoint[1];
      P[0] = new GeographicPoint(3.0, 3.0);
      System.out.println(P[0]);
      GeographicPolygon py = new GeographicPolygon();
      py = Buffer.calculate(P[0], 1.0);
      System.out.println(py);

      PointShape ps = new PointShape(P);
      GeographicPolygon[] pys = new GeographicPolygon[1];
      pys = Buffer.calculate(ps, 1.0);
      for (int i = 0; i < pys.length; i++)
                              System.out.println(pys[i]);
   }
}
```

This class can be complied as it is. The API allows the hard coding of specific calculations which are often needed. Of course, the Lisp engine as an interpreter makes this job a lot easier. One needs no compilation for every change that is made. This Java API is also used for such simple tasks as a ESRI shape ASCII converter (for WHPA simulation results).

9.3 Inference Engine

The Lisp engine is the third part. It runs the rules based on JESS (= Java Expert System Shell). With its interpreter, it loses some performance, but gains a lot of flexibility. The execution of the whole package can be embedded into other applications like ESRI's ArcView (via Avenue that allows an external process invocation) and because it is Java in the Web pages.

To sum there up, this new approach can be utilised in five ways:

1. **Rules** For completeness, this is the knowledge based decision support. Search algorithms and a group of other AI solutions can be used with spatial data.
2. **Sequences** Like a normal program, command after command, sequences produce a new map or additional parameters. Because the code is interpreted, the development cycle is faster. And the code can be better maintained for a slightly different problem. The code is also a perfect documentation of what is done.
3. **Functions in Lisp** The program in Lisp (a rule base or a sequential program) can be enhanced with user functions written in Lisp, the *defunction* construct. The spatial analysis primitives like *buffers* and set operations can be combined to form more powerful spatial analysis functions in order to build special purpose libraries. And, of course, they are parameterised for reuse.
4. **Mixed mode** One can use rules and sequential parts in one Lisp program, e.g. to assign the shape file to global variables and to do the analysis on this by facilitating the data access. User functions are accessible in both.
5. **Java API** The spatial analysis core API can be used in other Java programs or as a hard coded stand solo analysis (like the example). The second option the API offers is to write user functions in Java for the Lisp interpreter. Obviously, the compiled Java code is a performance gain compared with user functions in Lisp. The effort is worth it for heavily used operations.

10 Conclusion

10.1 Limits

The major limitations are found in the way of programming and their implementations: a Java based Lisp interpreter extended by spatial analysis functionality.

- **Rules** Not every decision is rule-based, even if it might be the most powerful way of describing knowledge. At least every arithmetically expressed decision (a number represents a certain non-valued feature such as risk, and the numbers are added up to draw conclusions) can be better written as rules. For example, adaptation is not possible without additional effort. In contrast, AI search algorithms are possible.
- **Performance** Two facts degrade the performance: Java and the Lisp interpreter. This makes it even worse, as they work normally on very big spatial data files.
- **Functionality** The package is not a fully fledged GIS. It implements spatial analysis functions on a vector data structure. This is its focus. There is always a GIS needed to georeference the input properly. Because of the shape file structure ESRI's ArcView can be a good companion.

The approach is not directly embedded into one of the major GIS products having the disadvantages mentioned above. The solution is based on the Lisp language and the connection with the GIS world is made by data means. That does not permit an integration e.g. via Avenue to get a button to reach the Lisp, or –if one so wishes– the Gisp (= geographical information search and production) interpreter to run a rule-based decision support analysis.

10.2 Advantages

The advantages (and limits) are based on the independent implementation of the spatial functionality.

1. **Defined derivation** Rules and commands describe the procedure. There is one open file with the rule base and its execution produces the new data. The input is variable. The procedure notation allows the reproduction of output data as a kind of self-documentation. This makes sensitivity analysis („What's happening if the area is separated in a slightly different way?") easier. A lot of manual work e.g. measuring different features, entering into the database, and format conversions can be programmed.

2. **Interactive Lisp** The interpreter enables faster development cycles because no compilation is needed. The written programs are easier to maintain and to adapt to other problems (additional to the data- procedure separation).

3. **Lisp and Java API extensions** The analysis primitives such as buffers and set operations can be combined to more sophisticated functions. These functions can be written in Lisp (interpreter based) or compiled in Java.

4. **Non-GIS input/output (IO)** The language frees the analysis from the dual GIS paradigm that data is either spatial (stored in shapes) or their attributes (stored in the map related database). The rule base or sequential program can take external data into account. It can also document the analysis in additional output data, such as a HTML file documenting which contaminant source is in a WHPA (spatial attributes from the database and the map) and in which velocity (external fact). The file can also have a special format for an other software package.

5. **Reduced spatially related IO** As not every intermediate map needs to be stored, they remain in memory, read-write activities are reduced to a minimum saving computational time (IO is the most costly part of the analysis) and hard drive space. If one wants to see and store temporary results, one can, but one dos not have to.

6. **System integration** The Lisp engine with its spatial extension is able to embedded into other application via Java. This allows platform independent Web-based interaction as well. In introducing a problem-oriented Java API for Lisp new functionality can be added, even GUI elements. SQL is also incorporated. The data transfer major GIS applications and special purpose applications such as simulators can talk to this analysis component.

7. **Procedure reusage** The separation of data and analysis enables reusage. The user can be asked for changing parameters –like new maps or external facts. The analysis functionality is scale independent therefore it can be used even if the scale of the maps are changing (of course, for map overlay they must be georeferenced).

Roughly, the idea is to complete an analysis –like data automation in the main GIS packages– based on a language. (The database query already has had the language SQL.) To allow for sophisticated analysis a rule-based language is used. To ease the development pain this is an interpreter. The vector data structure opens a wide range of application domains. With a common terminology for the spatial analysis functionality, this shall be easy to learn.

Acknowledgment

We thank Prof. Dr. Daniel Sui from Texas A&M University for his valuable discussions and advice.

References

Arnold, K. and Gosling, J. (1997): The Java Programming Language Second Edition, Addision Wesly, 1997

Chrisman, N. (1997): Exploring Geographic Information Systems, Wiley 1997

F.-Hill, E.: The Java Expert System Shell, http://herzberg.ca.sandia.gov/jess/

Pühl, S. and Gräber, P.-W. (1996): Ausweisung von Grundwasserschutzgebieten unter Verwendung von numerischen und wissensbasierten Methoden, in Fortschritte der Simulationstechnik, Vieweg 1996

Stefik, M. (1997): Knowledge Systems, Morgan Kaufman, 1997

Ullman, J. D. and Widom, J. (1997): A First Course in Database Systems, Prentice Hall, 1997